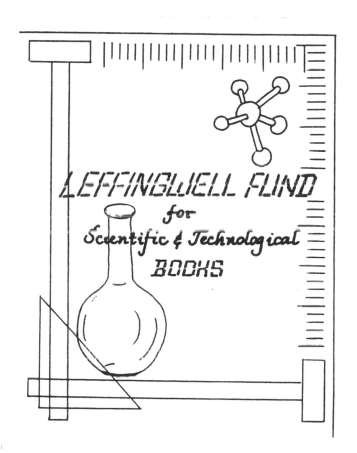

LEFFINGWELL FUND
for
Scientific & Technological
BOOKS

SUSTAINING LIFE

HOW HUMAN HEALTH DEPENDS ON BIODIVERSITY

SUSTAINING LIFE

HOW HUMAN HEALTH DEPENDS ON BIODIVERSITY

Edited by

ERIC CHIVIAN, M.D., AND AARON BERNSTEIN, M.D.

CENTER FOR HEALTH AND THE GLOBAL ENVIRONMENT
HARVARD MEDICAL SCHOOL

Foreword by Edward O. Wilson
Prologue by Kofi Annan

SECRETARIAT OF THE
CONVENTION ON BIOLOGICAL
DIVERSITY

UNITED NATIONS
DEVELOPMENT PROGRAMME

UNITED NATIONS
ENVIRONMENT PROGRAMME

THE WORLD
CONSERVATION UNION

OXFORD
UNIVERSITY PRESS
2008

OXFORD
UNIVERSITY PRESS

Oxford University Press, Inc., publishes works that further
Oxford University's objective of excellence
in research, scholarship, and education.

Oxford New York
Auckland Cape Town Dar es Salaam Hong Kong Karachi
Kuala Lumpur Madrid Melbourne Mexico City Nairobi
New Delhi Shanghai Taipei Toronto

With offices in
Argentina Austria Brazil Chile Czech Republic France Greece
Guatemala Hungary Italy Japan Poland Portugal Singapore
South Korea Switzerland Thailand Turkey Ukraine Vietnam

Published by Oxford University Press, Inc.
198 Madison Avenue, New York, NY 10016

www.oup.com

Oxford is a registered trademark of Oxford University Press

Library of Congress Cataloging-in-Publication Data
Sustaining life : how human health depends on biodiversity /
edited by Eric Chivian and Aaron Bernstein.
      p. cm.
Includes bibliographical references.
ISBN 978-0-19-517509-7
1. Biodiversity. 2. Environmental health.
I. Chivian, Eric. II. Bernstein, Aaron, 1976–
QH541.15.B56S96 2008
333.95′16—dc22 2007020609

The views expressed in this book are those of the authors and do not necessarily reflect those of the United Nations Environment Programme, the United Nations Development Programme, the Secretariat of the Convention on Biological Diversity, and the World Conservation Union or other cooperating institutions or agencies. The mention of a commercial company or product does not imply endorsement by any of the institutions or agencies involved in this book.

9 8 7 6 5 4 3 2 1

Printed in China
on acid-free paper

We dedicate this book to the millions of plant, animal, and microbial species we share this small planet with, and to our own species, Homo sapiens, *who first walked on Earth some 195,000 years ago and struggled to survive over the millennia to become the magnificent and extraordinarily powerful beings we are today.*

May we have the wisdom, and the love for our children and all children to come, to use that power to save the indescribably beautiful and precious gift we have been given.

FOREWORD

Ecologists have long used the metaphor of the canary in the mine to caution humanity. Like the delicate little birds once carried into coal mines following explosions or fires in order to detect poisonous gases, some sensitive plants and animals around us, by virtue of their sickness and dying, give early warning of dangerous changes in our common environment. The masterful presentations in *Sustaining Life: How Human Health Depends on Biodiversity* document beyond reasonable doubt that we ourselves are at risk of becoming a canary in today's world. In myriad ways humanity is linked to the millions of other species on this planet. What concerns them equally concerns us. The more we ignore our common health and welfare, the greater are the many threats to our own species. The better we understand and the more we rationally manage our relationship to the rest of life, the greater the guarantee of our own safety and quality of life.

Sustaining Life helps to fill a large and relatively unoccupied niche in environmental literature. Most people understand very well the dire effects of toxic pollution on their health. They also know that the ozone hole in the upper atmosphere is not a good thing, and that global warming, destruction of forests, and depletion of fresh water reserves are serious global threats. What has been harder to grasp, not only by the general public but also by most scientists, is the profound influence biodiversity has on human well-being. The reason is the prevailing world view that health is largely an internal matter for our species, and, with the exception of domesticated species and pathogenic microorganisms, the rest of life is something else.

The theme of *Sustaining Life*, in contrast, is that biodiversity matters profoundly to human health, and in almost every conceivable way. The mismanagement and destruction of species and ecosystems ongoing around the world mindlessly, and needlessly, lower the quality of the planet's natural resources, destabilize the physical environment, and can hasten the spread of human infectious diseases and the invasive enemies of the crops and forests on which our lives depend. There has been only a minimal effort to reverse this trend. In addition, bioprospecting, the exploration of biodiversity in order to open its mother lode of new pharmaceuticals, is still largely neglected and rudimentary. Little attempt has been made to utilize natural biodiversity to enhance public health. These various shortcomings produce the greatest burden for the developing countries, where 80 percent of humanity lives and most health crises erupt.

The shift in world view recommended by the authors of *Sustaining Life* is predicated on the increasingly obvious principle that humanity, having evolved as part of the web of life, remains enmeshed within it. We do not float above the biosphere in some higher spiritual or technoscientific plane. Life swarms around us, and even in us: Most of the cells in our bodies are not human but bacterial; more than 700 species live within our mouths alone, a specialized community that helps prevent the invasion of pathogen species. An estimated four million bacterial species occur in a ton of fertile soil, comprising ten billion or so organisms to each gram of weight. Although invisible, the collectivity of these organisms in soil and elsewhere is vital to our continued existence. Similarly, while a few thousands of the millions of insect species in the world afflict us as pests and disease carriers, we depend on the rest for our very lives. If beneficial insects did not flourish, most of the land ecosystems of the world would collapse and a good part of humanity would perish with them.

For many reasons, not least our own well-being, we need to take better care of the rest of life. Biodiversity, the authors of *Sustaining Life* argue with compelling urgency, will pay off in every sphere of human life, from medical to economic, from our collective security to our spiritual fulfillment.

Edward O. Wilson

PROLOGUE

One of the main reasons the world faces a global environmental crisis is the belief that we human beings are somehow separate from the natural world in which we live, and that we can therefore alter its physical, chemical, and biological systems without these alterations having any effect on humanity. *Sustaining Life* challenges this widely held misconception by demonstrating definitively, with the best and most current scientific information available, that human health depends, to a larger extent than we might imagine, on the health of other species and on the healthy functioning of natural ecosystems.

Biological diversity—the variety of life on Earth—is at the heart of our efforts to relieve suffering, raise standards of living, and achieve the U.N. Millennium Development Goals (a set of eight goals adopted by U.N. member nations for the year 2015 to promote human health and well-being around the world—they range from the goal of halving extreme poverty and hunger, to ensuring environmental sustainability, to combating HIV/AIDS, malaria, and other diseases). We cannot do without the countless services provided by biodiversity: pollinating our crops; fertilizing our soils with nitrogen, phosphorus, and other nutrients; providing millions of people with livelihoods, medicine, and much else. Advances in medicine, including treatments for currently untreatable diseases, would not be possible without the powerful pharmaceuticals derived from plants, animals, and microbes or without the knowledge gained from other species in biomedical research. We must conserve and sustainably use this pillar of human life. Yet biodiversity is declining at an unprecedented rate and is woefully underappreciated as a resource and as an issue meriting high-level attention.

This publication is an attempt to help the world change course. I applaud the Center for Health and the Global Environment at Harvard Medical School for assembling the international scientific team, from developed and developing countries alike, that produced this seminal work. I am delighted that this educational effort is also the product of close cooperation with a number of U.N. agencies, including the Secretariat of the Convention on Biological Diversity, the U.N. Environment Programme, and the U.N. Development Programme.

Written in straightforward, nontechnical language that any reader can understand, *Sustaining Life* is meant to educate and inform. But it also aims to convey a sense of urgency around the issue and, ultimately, to convince policy makers and the public that our future health and prospects—indeed, our very lives—depend on addressing this challenge with all our creativity and will, so that we do not deprive future generations of the opportunity to benefit from Nature's wealth.

Kofi Annan

PREFACE

Edward O. Wilson once said about ants, "We need them to survive, but they don't need us at all." The same, in fact, could be said about countless other insects, bacteria, fungi, plankton, plants, and other organisms. This fundamental truth, however, is largely lost to many of us. Rather, we humans generally act as if we were totally independent of Nature, as if we could do without most of its creatures and the life-giving services they provide, as if the natural world were designed to be an infinite source of products and services for our use alone and an infinite sink for our wastes.

During the past fifty years or so, for example, our actions have resulted in the loss of roughly one-fifth of Earth's topsoil, one-fifth of its land suitable for agriculture, almost 90 percent of its large commercial marine fisheries, and one-third of its forests, while we now need these resources more than ever, as our population has almost tripled during this period of time, increasing from 2.5 to more than 6.5 billion. We have dumped millions of tons of chemicals onto soils and into fresh water, the oceans, and the air, while knowing very little about the effects these chemicals have on other species or, in fact, on ourselves. We have changed the composition of the atmosphere, thinning the ozone layer that filters out harmful ultraviolet radiation, toxic to all living things on land and in surface waters, and increasing the concentration of atmospheric carbon dioxide to levels not present on Earth for more than 600,000 years. These carbon dioxide emissions, caused mainly by our burning fossil fuels, are unleashing a warming of Earth's surface and of the oceans and a change in the climate that will increasingly threaten our health and the survival of other species worldwide. And we are now consuming or wasting or diverting almost half of all the net biological production on land, which ultimately derives from photosynthesis, and more than half of the planet's renewable fresh water.

We are so damaging the habitats in which other species live that we are driving them to extinction, the only truly irreversible consequence of our environmental assaults, at a rate that is hundreds to even thousands of times greater than natural background levels. As a result, some biologists have concluded that we have entered what they are calling "the sixth great extinction event," the fifth having occurred sixty-five million years ago when dinosaurs and many other organisms were wiped out. That event was most likely the result of a giant asteroid striking Earth; this one we are causing.

Most disturbing of all, as a result of all of these actions taken together, we are disrupting what are called "ecosystem services," that is, the various ways that organisms, and the sum total of their interactions with each other and with the environments in which they live, function to keep all life on this planet, including human life, alive.

We have done all these things—our species, *Homo sapiens*, one species out of perhaps ten million on Earth, and maybe even many times more than that, behaving as if these alterations were happening someplace other than where we live, as if they had no effect on us whatsoever.

This heedless degradation of the planet is driven by many factors, not the least of which is our inability to take seriously the implications of our rapidly growing populations and of our unsustainable consumption of its resources, largely by people in industrialized countries, but increasingly by those in the developing world. Ultimately, our behavior is the result of a basic failure to recognize that human beings are an inseparable part of Nature and that we cannot damage it severely without severely damaging ourselves.

This book was first conceived in 1992 at the Earth Summit in Rio de Janeiro, when the largest collection of world leaders ever assembled until that time, along with tens of thousands of concerned policy makers, scientists, environmentalists, and others, gathered to set ambitious goals for controlling global climate change and for conserving the world's biological diversity. What we recognized then, and what is even more widely apparent now, is that, in contrast to the issue of global climate change, which has seen significant attention paid to the potential consequences for human health, with chapters devoted to this topic in all the major international reports, the same has not been true for the issues of species loss and ecosystem disruption.

This general neglect of the relationship between biodiversity and human health, we believe, is a very serious problem, for not only are the full human dimensions of biodiversity loss failing to inform policy decisions, but the general public, lacking an understanding of the health risks involved, is not grasping the magnitude of the biodiversity crisis and not developing a sense of urgency to address it. Tragically, aesthetic, ethical, religious, even economic arguments have not been enough to convince them.

To address this need, the Center for Health and the Global Environment at Harvard Medical School proposed that it coordinate an international scientific effort to compile what was known about how other species contribute to human health, under the auspices of the United Nations, and to produce a comprehensive report on the subject. Happily, the U.N. Environment Programme, the U.N. Development Programme, and the Secretariat of the Convention on Biological Diversity agreed to co-sponsor this project, and at a later time, the International Union for the Conservation of Nature and Natural Resources joined them. The result is this book, *Sustaining Life: How Human Health Depends on Biodiversity.*

We have focused much attention in *Sustaining Life* on seven groups of organisms in order to illustrate what their loss and, by extension, the loss of countless other organisms mean for human health. We have focused particular attention on amphibians, which are among the most threatened of any group of organisms on the planet, with almost one-third of some 6,000 known species in danger of extinction, and more than 120 believed to have already gone extinct in the past few decades. There is no evidence in the fossil record that such a high rate of extinction among amphibians, which have been on Earth for more than 350 million years, has occurred in the past, so it is believed that this loss is a new, and human-caused, phenomenon.

We have given many examples in the book of how amphibians contribute to human medicine—from the vitally important chemicals they contain that may lead to new pain killers and drugs to treat high blood pressure, to the central roles

Southern Gastric Brooding Frog (*Rheobatrachus silus*). Tadpole being delivered from mother's stomach. (© Michael J. Tyler.)

they have played, and continue to play, in biomedical research. Amphibians may, for example, help us figure out ways to prevent bacteria from developing resistance to our antibiotic treatments, a rapidly escalating phenomenon that is causing great alarm among physicians as they struggle to keep one step ahead of their patients' infections. We provide here yet another example to help the reader understand the magnitude of our loss with a loss of amphibians:

Gastric brooding frogs (*Rheobatrachus vitellinus* and *R. silus*), the only amphibians known to raise their young in their stomachs, were discovered in the 1980s in undisturbed rainforests in Australia. The female swallows her fertilized eggs, which then hatch in her stomach. When the hatchlings become fully developed tadpoles, they are "delivered" to the outside world, propelled by their mother's vomiting, where they continue their development into adult frogs.

The stomachs of all vertebrate species, including frogs, contain cells that secrete acid and enzymes such as pepsin to begin the process of digesting food. There are also compounds that stimulate emptying of the stomach so that its contents can be moved along into the small intestine where further digestion takes place. The ingestion of food triggers the release of these compounds. Preliminary studies with gastric brooding frog tadpoles demonstrated that they secrete a substance, or substances, that both inhibits acid and pepsin secretions and prevents stomach emptying so that they do not end up being digested by their mother. But these studies, which might have led to important new insights for treating human peptic ulcers, a disease that affects more than twenty-five million people in the United States, could not be continued because both species of *Rheobactrachus* became extinct.

Scientists with expertise in a wide range of disciplines, from industrialized and developing countries alike, have been involved in putting this book together. We have done so because we are convinced that it can help people understand that human beings are an integral part of Nature, and that our health depends ultimately on the health of its species and on the natural functioning of its ecosystems. We have done so because all of us hope that our efforts will help guide policy makers in developing innovative and equitable policies based on sound science that will effectively preserve biodiversity and promote human health for generations to come. And we have done so, finally, because we all believe that life on Earth is sacred and that we must never give up in trying to preserve it, and because we all share the conviction that once people recognize how much is at stake with their health and lives, and with the health and lives of their children, they will do everything in their power to protect the global environment.

Eric Chivian, M.D.
Aaron Bernstein, M.D.

NOTES FOR READING

Common names for species are capitalized throughout the book, so that it is clear, for example, that we are talking about the species called the Green Frog, rather than just any green frog.

The term "bacteria" is capitalized when it refers to the domain Bacteria, one of the three major categories of life in the three-domain model (the other two are the Archaea and the Eukarya). Any general member or members belonging to the domain Bacteria are referred to as lowercase "bacterium" or "bacteria," respectively.

Latin names are given in parentheses after the common species names and are in italics, as in Green Frog (*Rana clamitans*). Common names can differ markedly by region and by country.

Weights and measures are generally given in both U.S. and metric amounts. Which one comes first is determined by which units were used for the original statistic.

Suggested readings are given at the end of each chapter, along with relevant websites, for those wishing to explore subjects in greater depth. We chose readings that are easily available online or in bookstores or libraries.

Scientific references from the peer review literature, for those wishing to consult primary sources, are given in a separate reference section at the end of the book, chapter by chapter, with those cited in the text listed first by number, in the order they are cited, followed by other references that have been consulted, in alphabetical order.

Because we needed to cover an enormous range of subject areas in order to provide a comprehensive examination of the dependency of human health on biodiversity, we have asked a number of experts to write sections for some of the chapters to complement and enrich the work of the chapter authors. In addition, large portions of the volume have been peer reviewed by leading figures in their respective fields. Contributing authors and peer reviewers are identified at the end of the book, by chapter.

The reader may notice that there is some repetition in the chapters of key concepts, for example, about the impacts of global climate change on biological systems. This has been deliberate, because we expect that most people will read chapters individually rather reading the book from cover to cover. As a result, we wanted each chapter to be able to stand on its own.

Finally, we have tried to stay away from technical terms and concepts, because our goal has been to make the information contained in this book understandable to all. When such terms are used, we have provided definitions immediately after them, rather than in a separate glossary.

Sustaining Life was certainly written for scientists, physicians, and public health professionals who need to understand the fundamental connections between human health and Nature. We hope they will use it in their research, their teaching, and in their practice of medicine. But it was primarily written for the general reader and for policy makers, so that they could appreciate what we are in danger of losing with a loss of biodiversity. Ultimately, it is they who will determine whether or not humans are successful in protecting the natural world.

Acknowledgments

Close to 300 people have had a hand in the making of this book—as authors, editors, reviewers, advisors, research assistants, research librarians, staff assistants, illustrators, photographers, and financial benefactors. It will not be possible to thank everyone as they should be thanked, for that would require another book. But we must recognize the following people:

We first acknowledge our remarkable fellow chapter authors, who have used their great expertise to help translate the science into compelling terms that an interested, but not necessarily a scientifically trained, reader could comprehend. We thank them for their extraordinary skill and hard work and for their commitment to the project and generosity to us.

Then, our wonderful research and staff assistants—Margaret Thomsen, Emily Huhn, Chris Golden, Joe Orzall, and Charlotte Hadley—who worked with the authors and with us to find just the right figures and tables for our needs, who researched the countless questions we raised and helped us keep track of the endless details involved in writing a book.

We must here also thank Google, the ISI Web of Knowledge, PubMed, and Wikipedia, whose parallel universes we have inhabited for significant portions of the past few years. And we have to mention those geniuses who conceived of and developed the Internet and the World Wide Web, changing all of our lives forever, without whom this book would not have been possible.

Our superb illustrators and designers, photo researchers, and Photoshop magicians, working at the Public Broadcasting Station WGBH and as freelancers—Tong Mei Chan, Elles Gianocostas, Lisa Abitbol, Doug Scott, and Deborah Paddock—deserve special mention and our deepest gratitude, as do our enormously skilled research librarians at Harvard—Jack Eckert, Mary Sears, and Dana Fisher.

There are a large number of other people all over the world whom we should recognize and thank—scientists, photographers, illustrators, and others who went out of their way so that this project would succeed. We cannot even list all their names. But several stand out, those we could always count on: Alejandro Alvarado, Karl Ammann, Adam Amsterdam, Joshua Arnow, Michael Balick, Robert Barlow, Julia Baum, Dami Buchori, Virginia Burkett, Marc Cattet, Gordon Cragg, John Daly, Robert Diaz, Giovanni Di Guardo, Andrew Durocher, Elaine Elisabetsky, Frank Epstein, Jonathan Epstein, Paul Epstein, Bill Fenical, Toshitaka Fujisawa, Beatrice Hahn, Brian Halweil, Jim Hanken, Ray Hayes, Hans Herren, Nancy Hopkins, Clinton Jenkins, Fred Kirschenmann, Donald Klein, Thomas Kristensen, Mike Lannoo, Richard Levins, Tom Lovejoy, John Marchalonis, Michelle Marvier, Roz

Naylor, Ralph Nelson, Fernando Nottebohm, Judy Oglethorpe, Toto Olivera, Norman Pace, Andrew Price, Anne Pringle, John Reganold, Callum Roberts, Noel Rowe, Gary Ruvkun, Carl Safina, Bill Sargent, Anja Saura, Scott Schliebe, Chris Shaw, David Sherman, Ann Shinnar, Louis Sibal, Sigmund Sokransky, Melanie Stiassny, Amos Tandler, Else Vellinga, Burt Vaughan, David Wake, Diana Wall, LaReesa Wolfenbarger, Richard Wrangham, Junko Yasuoka, and Michael Zasloff. Thank you all. We hope we shall be able to return the favor someday.

We are honored by the co-sponsorship of this project by the U.N. Environment Programme (UNEP), the U.N. Development Programme (UNDP), the Secretariat of the Convention on Biological Diversity (CBD), and the International Union for the Conservation of Nature and Natural Resources (IUCN), and could not have had better collaborators, colleagues, and friends who represented these agencies and organizations and who helped us with the content of the book and with shepherding it through their vetting processes—Hiremagular Gopalan and Maiike Jansen at UNEP, Charles McNeill at UNDP, Jo Mulongoy at the CBD, and Jeff McNeely at the IUCN. And we are deeply indebted to Kofi Annan for his Prologue and to Ed Wilson for his Foreword, and to both of them for their wisdom and outstanding leadership over many years in working to protect the global environment.

We also thank our editors at Oxford University Press, first Kirk Jensen, and in the last few years Peter Prescott, for being as excited about this book's potential as we were from its beginnings and for their unquestioning faith in us during its long gestation. Greater patience hath no men.

To our loving, supportive families, we can never repay you for your insightful critical commentaries; for enduring too many nights of take-out, and our frequent absences, both mentally and physically; and for always trusting that in the end it would be worth it.

And finally, this book has been made possible through the enormously generous financial support, for which we are deeply grateful, of many individuals, foundations, and corporations, including

3M
Arkin Foundation
Arnow Family Fund
Baker Foundation
Bristol-Myers Squibb
The Chazen Foundation
Nathan Cummings Foundation
Carolyn Fine Friedman
Richard & Rhoda Goldman Fund
Clarence E. Heller Charitable Foundation
Johnson & Johnson Family of Companies
Henry A. Jordan, M.D.
J. M. Kaplan Fund
Leon Lowenstein Foundation
John D. and Catherine T. MacArthur Foundation
New York Community Trust
Newman's Own Foundation
Josephine Bay Paul and C. Michael Paul Foundation
The Pocantico Conference Center of the Rockefeller Brothers Fund
Rockefeller Financial Services—Anonymous Donor
Roger and Vicki Sant
Silver Mountain Foundation
Jennifer Small
Threshold Foundation
V. Kann Rasmussen Foundation
Lucy A. Waletzky, M.D.
Wallace Genetic Foundation
Wallace Global Fund
Shelby White
Winslow Foundation

Thank you everyone, Merci, Gracias, 謝謝, شکریہ.

The Editors

CONTENTS

This is the token of the covenant which I make between Me and you and every living creature that is with you, for perpetual generations . . . and the bow [rainbow] shall be in the cloud; and I will look upon it, that I may remember the everlasting covenant between God and every living creature of all flesh that is upon the earth.

—GENESIS 9:16

There is not an animal on the earth nor a being that flies on its wings, but [forms part of] communities like you.

—KORAN (QUR'AN) 6:38

All this world is strung on me like jewels on a string. I am the taste in the waters, the radiance in the sun and moon, the sacred syllable Om that reverberates in space, the manliness in men. I am the pleasant fragrance in earth, the glowing brightness in fire, the life in all beings.

—BHAGAVAD GITA VII:7–9

My love to the footless, my love to the two-footed, my love to the four-footed, my love to the many-footed . . . All sentient beings, all breathing things, creatures without exception, let them all see good things, may no evil befall them.

—"GRADUAL SAYINGS" OF THE BUDDHA

Sustaining Life

CHAPTER 1

WHAT IS BIODIVERSITY?

Stuart L. Pimm, Maria Alice S. Alves, Eric Chivian, and Aaron Bernstein

Although there is substantial controversy about the circumstances in which it was said and about the exact wording of the original remark, J.B.S. Haldane, one of the most prominent and brilliant evolutionary biologists of his time, when asked what one could conclude about the Creator from studying His work, is reputed to have said, "He had an inordinate fondness for beetles."

Biological diversity, or biodiversity for short, is the variety of life on Earth—its genes, species, populations, and ecosystems. Human actions that have degraded land, bodies of fresh water, and the oceans have already caused biodiversity to decline sharply, and even greater losses are expected if humanity continues its current unsustainable use of natural resources. Although such activities as the release of greenhouse gases have exacted heavy, and in some cases potentially catastrophic, tolls on the global environment, the loss of biodiversity is the only truly irreversible consequence of environmental degradation. When a gene, a species, a population, or an ecosystem is lost, it is gone forever.

When considering the loss of biodiversity, species loss has become the most widely used measure. The subject of biodiversity loss is, however, broader and more complex than this, because there is diversity at other levels of organization as well. For example, genetic diversity exists among members of an individual species, and a species can lose some of this diversity when local populations are lost even though the species itself has not gone extinct. There is also diversity at higher levels, above the species level, in the genera (the plural of genus), families, orders, classes, phyla, and kingdoms to which species belong, and in the types of ecological communities, or ecosystems, they are a part of. A loss of diversity, or in function, at any one of these levels may be independent of such losses at another level. For example, an ecosystem may shrink dramatically in area and lose many of its functions, even though all of its constituent species may manage to survive.

This chapter provides an overview of the current status of the world's biodiversity. It is intended to provide a baseline for the chapters that follow that will, in turn, examine some of the threats to biodiversity, the ways we depend on it, and how our health and our lives may be endangered by its loss.

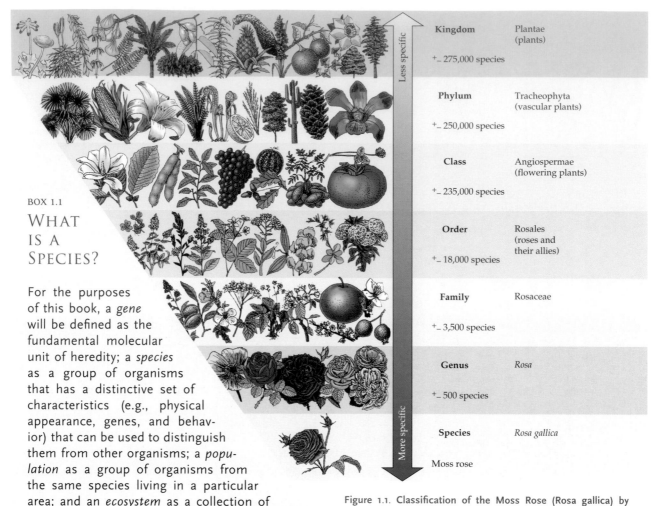

Kingdom	Plantae (plants)	
+_ 275,000 species		
Phylum	Tracheophyta (vascular plants)	
+_ 250,000 species		
Class	Angiospermae (flowering plants)	
+_ 235,000 species		
Order	Rosales (roses and their allies)	
+_ 18,000 species		
Family	Rosaceae	
+_ 3,500 species		
Genus	*Rosa*	
+_ 500 species		
Species	*Rosa gallica*	
Moss rose		

Less specific → More specific

BOX 1.1
WHAT IS A SPECIES?

For the purposes of this book, a *gene* will be defined as the fundamental molecular unit of heredity; a *species* as a group of organisms that has a distinctive set of characteristics (e.g., physical appearance, genes, and behavior) that can be used to distinguish them from other organisms; a *population* as a group of organisms from the same species living in a particular area; and an *ecosystem* as a collection of different species, the physical environment in which they live, and the sum total of their interactions.

The standard definition of a species is based on the ability of members of the same species, under natural conditions, to mate and produce fertile offspring, whereas members of different species cannot. So even though horses and donkeys can mate and produce mules, they are considered to be different species, because the mules are sterile. But this definition is meaningless for species that reproduce asexually. How can we tell them apart? And how can we determine, even for those species that reproduce sexually, which is which, if we know little about their reproductive habits, which is usually the case? Most species are known from only one or, at most, a few individuals, or from only one location.

Different species generally have different appearances and behave in different ways. But such differences may be too subtle to decipher, as is seen in the case of Neotropical Skipper Butterflies

Figure 1.1. Classification of the Moss Rose (Rosa gallica) by Species, Genus, Family, Order, Class, Phylum, and Kingdom. Starting with kingdom, one can remember these categories in descending order by the mnemonic "Kings Play Chess On Fancy Glass Stools." (From Purves et al., *Life: The Science of Biology*, 4th ed. © 1995 Sinauer Associates, Inc.)

(see figure 1.3). And in these cases, we have come to rely on molecular methods, which have made it possible to analyze the unique genetic compositions of different species and, by this means, to distinguish between them (see "The Microbial World" in this chapter for a description of these methods). The reader should understand, however, that the science of identifying species is not at all clear-cut, that even within the same species, there is a diversity of genotypes (the genetic composition of an individual) and phenotypes (the observable traits and behavior of an organism), and that this diversity is in a state of constant flux, as organisms evolve in response to their environments.

Carl Linnaeus Dressed as a Laplander After Returning from Lapland in Northern Sweden. ~1735–1740. (Mezzotint engraved by H. Kingsbury after Martin Hoffman. In the Public Domain http://www.ucmp.berkeley.edu/history/linnaeus.html.)

BOX 1.2

CARL LINNAEUS

We owe a great debt to the Swedish botanist and physician Carl Linnaeus, also known as Carolus Linnaeus or Carl von Linné, who in the mid-eighteenth century devised an ingenious way of using Latin names to identify individual species, which is the method that scientists still use today. Linnaeus first named the genus with an initial capital letter, for example *Acer*, and then the species, all in small letters, for example, *saccharum*, as in *Acer saccharum* (Sugar Maple) or as in our species, *Homo sapiens*. Geographical variants of species, also called subspecies, are often described by a third name, as well, for example, *Pan troglodytes troglodytes*, a subspecies of the Common Chimpanzee from West–Central Africa that is believed to be the original source of the HIV/AIDS virus (as discussed in chapters 6 and 7). When the genus has already been mentioned or when it is clear which genus is being considered, it is often abbreviated, as in *A. saccharum*. If one is talking about an organism or organisms but knows only the genus, the convention is to name the genus, followed by an abbreviation for species, as in *Acer* sp. or *Acer* spp. Some genera, such as *Anopheles* mosquitoes, have special two-letter abbreviations (in this case *An.*) so that they are not confused with their close cousins, the *Aedes* mosquitoes (abbreviated *A.* or *Ae.*). *An. gambiae* (*Anopheles gambiae*) is the most deadly carrier of malaria in Africa today, whereas *Ae. aegypti* (*Aedes aegypti*) is the main carrier of dengue fever and yellow fever viruses.

In formal usage, the Latin name is followed by the name of the taxonomist who described it. When it is followed by L., it indicates that the name was given by Linnaeus. Taxonomists are scientists who classify and group organisms into hierarchical categories called taxa (singular taxon), such as their genus and species, or the family, order, class, or phylum to which they belong.

This system has allowed for worldwide recognition of particular species, which may have many different common names in different countries and in different languages. A species' Latin name will be written in italics in this book.

The Relatedness of Life

One way to assess the breadth of biodiversity is to make a family tree for all life on Earth. Such a tree, as is illustrated in Ernst Haeckel's extraordinary figure published in 1866 (figure 1.2), gives a sense both for the total variety of life known at that time and for how organisms were thought to be related to one another. Published only seven years after Darwin's *Origin of Species* (Haeckel dedicated his book to Darwin, as well as to Wolfgang Goethe and Jean Lamarck), Haeckel's book, *Generelle Morphologie der Organismen*, drew upon Darwin's theory of evolution and related Haeckel's own ideas about how various life forms evolved from shared common ancestors.

Our newfound ability to compare genes from different organisms has revolutionized the construction of such a tree and, as a result, our appreciation of biodiversity. This revolution has come about primarily on two fronts. The first involves the organization of the broadest groupings of life, which was what Haeckel did. The second deals with how best to define a species, especially species of microbial organisms. In most basic biology courses, life is still grouped into one of five kingdoms: Animals, Plants, Fungi, Protists (the simplest of the eukaryotes, the name for those organisms whose cells have nuclei), and Monera (which are prokaryotes, or one-celled organisms without nuclei or other membrane-enclosed organelles such as mitochondria or chloroplasts). Haeckel's classification scheme was essentially the same—Animalia, Plantae, and Protista were the three main branches, with Monera defined as protists, and Fungi as plants. But in contrast to the five-kingdom model first devised by Robert H. Whittaker in 1969,[1] Haeckel's did not distinguish between prokaryotic and eukaryotic organisms. The ability to recognize the fundamental differences between these two groups was an outgrowth of increasingly sophisticated microscopes, unavailable to Haeckel.

Both the five-kingdom model and Haeckel's categorize organisms mainly by their observable form and component parts, or their phenotype, a way of classifying living things that is now being complemented, by many scientists, with methods that put greater emphasis on an organism's genes. This conceptual change has come about from a recognition that different organisms may look the same but, in fact, have strikingly different genetic compositions that serve to identify them as separate species (figure 1.3). It has also resulted from a realization that the traditionally defined kingdoms do not represent equal divisions of life—some kingdoms may be more closely related than others. For instance, the genetic evidence shows that animals, plants, fungi, and protists are all more closely related to one another than any one of them is to the prokaryotic Monera.

Another configuration of the Tree of Life with greater fidelity to these genetic distinctions was first proposed by Carl Woese in 1990. Woese's tree has three major branches: Eukarya, Archaea, and Bacteria (figure 1.4). This model, which has found growing acceptance among scientists, represents a major shift from the traditional division of living things into five kingdoms, grouping all life instead into these three branches, known as domains. The three-domain map illustrates that most genetic diversity is likely to be found among the microbes; that, in evolutionary terms, vascular plants (e.g., corn from the genus *Zea*), fungi (e.g., the genus *Coprinus* belonging to the mushroom family, the *Coprinaceae*, known as the

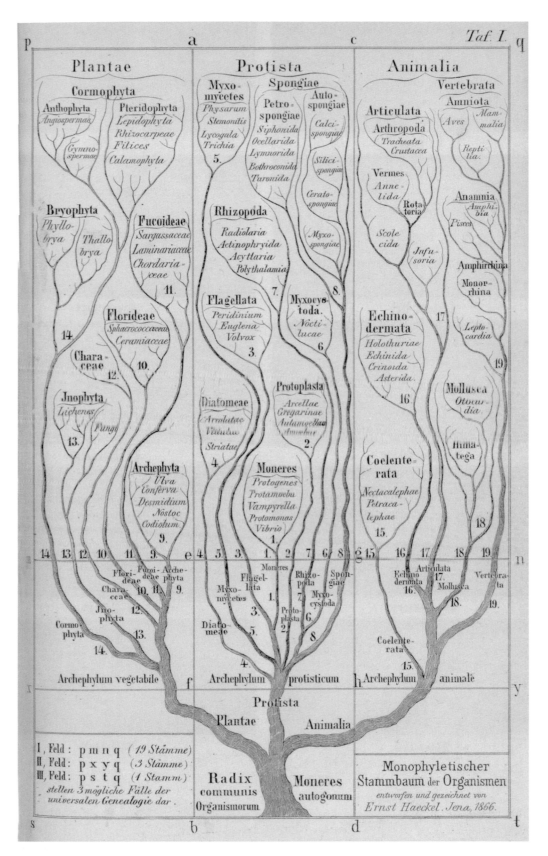

Figure 1.2. Ernst Heinrich Philipp August Haeckel's Tree of Life. (From his *Generelle Morphologie der Organismen*. G. Reimer, Berlin, 1866. Boston Medical Library in the Francis A. Countway Library of Medicine.)

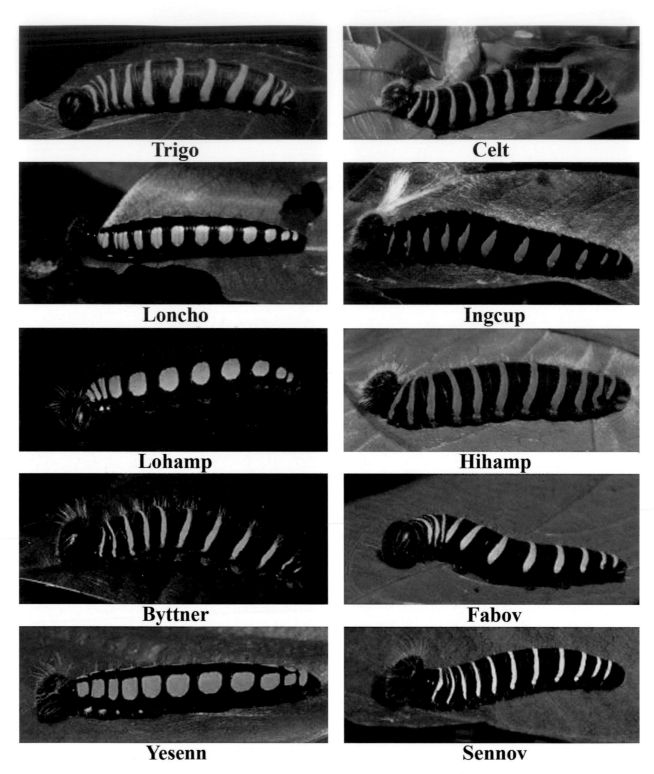

Trigo

Celt

Loncho

Ingcup

Lohamp

Hihamp

Byttner

Fabov

Yesenn

Sennov

Figure 1.3. The Caterpillars of Ten Different Butterfly Species Comprising What Was Originally Thought to Be One Species—the Neotropical Skipper Butterfly (*Astraptes fulgerator*). Years of morphological study, combined with DNA bar-coding of museum specimens (a method of using a short DNA sequence from a particular portion of an organism's genome, assigning specific colors to the nucleotides, and then arranging them in order as stripes, like the universal product code, as a means of identifying different species), identified ten different species among butterflies that were all thought to belong to one species, *A. fulgerator*. The adults of these species differ very subtly and could not be distinguished one from the other, but their caterpillars have distinctive appearances and somewhat differing ecological preferences, and they feed on different plants. The caterpillars have been given interim names based on their primary food plants or color characteristics. Hidden diversity was revealed by combining DNA bar-coding with traditional methods of species classification. Such cryptic species may be prevalent in tropical regions. (From Hebert et al. *Proceedings of the National Academy of Sciences of the USA*, 2004;101(41):14812–14817. © 2004 National Academy of Sciences.)

Figure 1.4. Three-Domain Map. This map, made up of the prokaryotes—Archaea and Bacteria—and the eukaryotes (organisms that possess a nucleus), groups organisms according to their genetic differences. Species are classified along a continuum that connects them back to a single common ancestor. (Reprinted with permission from Norman Pace, "A Molecular View of Microbial Diversity and the Biosphere." *Science*, 1997;276:734–740. © 1997 American Association for the Advancement of Science.)

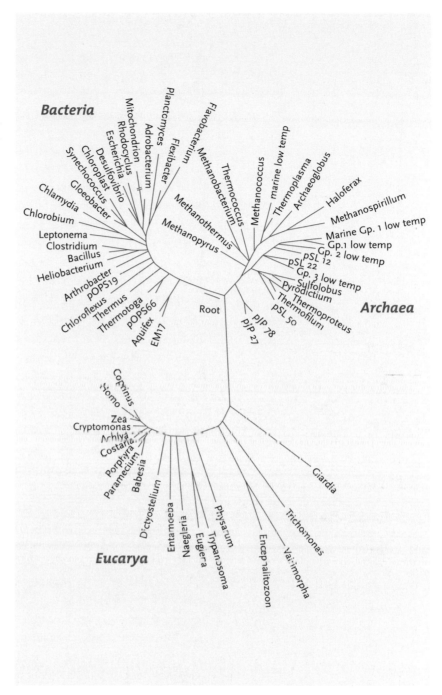

ink caps), and higher animals (e.g., our genus *Homo*) have all evolved relatively recently; and that all organisms can ultimately be traced back to a single common ancestor. In a sense, the three-domain model stands the classical Haeckel "Tree of Life" on its head, with humans—in fact, with all plants, animals, and fungi—at the very "bottom" of the tree, at least in terms of biodiversity and the length of our existence on Earth, instead of, as we like to see ourselves, at the very "top." The three-domain model has stimulated much-needed research into the evolutionary relationships among organisms in general and into the world of microorganisms in particular.

The Microbial World

Determining just how species-rich the microbial world is remains difficult for several reasons. For one, most microbes cannot be grown in culture, which has made research directed at establishing their identities a challenge. By making rough estimates with a microscope of the numbers and types of bacteria in samples of seawater, lakes, sediments, and soils, for example, it was found that less than 1 percent of the total were able to be cultured. (New RNA analyses of bacterial communities have come to the same conclusion.) Nor can they generally be reliably distinguished one from the other based on their appearance—one bacillus (a type of bacterium shaped like a rod), for example, looks pretty much the same as any other.

Scientists have turned to genetics in the belief that comparing the genes of various bacteria and archaea may be a more rigorous way of telling them apart. One of the most reliable genes to examine encodes a kind of RNA in ribosomes called 16S rRNA (ribosomes are the factories within cells where protein synthesis occurs). The composition of the 16S rRNA gene has remained relatively constant in most organisms for millions of years. Over time, however, slight changes in the gene have accrued as species have evolved, with the result that each species ends up possessing a distinct 16S rRNA gene. By comparing the subtle differences in this gene between organisms, scientists can distinguish one species from another. They can also gauge how closely related they are to each other and when they last shared a common ancestor. This was the method used to put together the three-domain map shown in figure 1.4. The 16s rRNA gene also happens to be short, which makes it easy and inexpensive to produce many copies for sequencing studies via the polymerase chain reaction or PCR (see section on the bacterium *Thermus aquaticus* on page 179 in chapter 5).

Using these new technologies, scientists have scoured Earth, taking microbial samples from every conceivable locale: from the surface waters of the Sargasso Sea near Bermuda to volcanic vents in the deep oceans, from Lake Vostok which is buried under Antarctic ice, to hot water geysers in Wyoming's Yellowstone National Park, from rainforest soils in Brazil to desert soils in the U.S. Southwest, and they have begun to understand how vast and how diverse the microbial world is.

The number of individual microbes on Earth is thought to be a high as 4 to 6 times 10 to the 30th power (4–6×10^{30}), a count that some have said may be one billion times more than the total number of stars in the universe! Most of them are thought to live in subsurface layers of the land and the oceans.[2] But no one really knows how many microbial species there are, even to the nearest order of magnitude (estimates run from ten million to as high as one billion distinct species).

In a single cubic meter of soil alone, there may be millions of different microbial species, and yet only 6,000 species of bacteria and archaea have been formally named. What is not clear, however, is whether in another cubic meter of soil of the same type, or of a different type, or one found in some other part of the world, the same bacterial species would be found or whether they would be different ones. That is, it is not clear whether bacterial species are in general so widely distributed that the same ones end up being found everywhere. What is clear is that most of the biodiversity on

Earth, on land and in the oceans, is most likely microbial, and that scientists know almost nothing about microbial diversity. What they do know is startling. For example, research done by Sigmund Sokransky, Bruce Paster, and their colleagues at the Forsyth Institute in Boston has identified more than 700 distinct bacterial species, as well as an assortment of archaea, fungi, and amoebas, that reside in the human mouth (with the total number of organisms in the mouth estimated to surpass six billion). And some 800 distinct microbial species, almost all of which are bacteria (including thousands of strains), have been found living in the human intestine (see box 3.1 on microbial ecosystems in chapter 3).

The Archaea

The Archaea were the first prokaryotes. While they resemble the Bacteria, their prokaryotic cousins, under a microscope, the Archaea can be distinguished by their unique biochemistry. In some ways, they are more like we eukaryotes than they are like bacteria. For example, they wrap their DNA in proteins called histones, as do we, while bacteria use other types of proteins. In other ways, they are very different. They make their membranes out of ether-linked lipids, while bacteria (and we humans) make ours out of ester-linked lipids. We know very little about archaeal species, but they are believed, like the Bacteria, to be extremely diverse.

Originally, archaea were thought to be confined to extreme environments, but that is not the case. Archaea inhabit many other places as well, such as temperate soils, the roots of plants, and even the human mouth and intestine, as mentioned above. Many, though, are indeed capable of living under extreme conditions and are called "extremophiles," perhaps because they evolved when their environments were extreme by today's standards and have maintained the biochemical adaptations that had allowed them to live in these environments. Take, for instance, one of the archaea known only as "strain 121," because it can live at a scorching 121 degrees Celsius, well above the boiling point of water. This remarkable microbe was found in a volcanic deep sea vent along the Juan de Fuca Ridge in the northeast Pacific Ocean, at a depth of 7,447 feet (2,270 meters, or about 1.4 miles).[3]

Archaea have also been found in saturated salt brines, such as the intensely salty waters of the Dead Sea and the Great Salt Lake; at the very deepest points in the oceans, such as the 6.8 mile (11 kilometer) deep Mariana Trench off the coast of the Philippines (for comparison, the depth of the Mariana Trench is more than 200 times the height of the world's tallest building, Taipei 101, in Taiwan), where they are subjected to enormous pressures and an absence of oxygen; thousands of feet underground; and in harshly acidic or alkaline conditions. Two archaeal species (of the genus *Picrophilus*) were discovered in another volcanic vent, this one off of northern Japan, living in extremely acid conditions, at a pH of less than 1 (the same pH as 0.1 molar sulfuric acid). Another archaeon, *Deinococcus radiodurans*, can survive a radiation dose of 1,750,000 rads. In comparison, an exposure of 500 to 1,000 rads is enough to kill a human. Some archaea can even live on methane or sulfur instead of oxygen.

In spite of rapidly growing knowledge about the microbial world in recent decades, and that most of the biological diversity on Earth is clearly microbial, this chapter, and others in this book, focuses primarily on the macroscopic, multicellular world of plants and animals, for that is still what we know best.

DETERMINING RATES OF SPECIES EXTINCTION

Startling figures have been published about how many species are going extinct each year or even each hour. All of these estimates require a determination of how many species exist, a number about which there is considerable

BOX 1.3

ARE VIRUSES ALIVE?

Scientific debate continues about whether or not viruses should be classified as life forms. Because this book is about biodiversity, for completeness sake, we briefly discuss this question here. For one thing, viruses employ DNA or RNA, the basic genetic material for all life on Earth, and like all other living things, they contain proteins. They also reproduce, although as far as is known, they cannot do so by themselves, but instead require the genetic machinery of a host cell in order to multiply. And they evolve in response to their environments, as can be widely appreciated from studying, for example, the evolution of influenza viruses (see "Diversity of Pathogens" in chapter 7, page 308). Whether or not to call viruses alive, however, depends on how one defines what it means to be alive, for example, whether living things have to be composed of cells (viruses are not) or have to reproduce on their own (viruses cannot). But, as is true with the issue of whether or not bacteria can be classified as species, or whether or not microbes can form ecosystems, it all depends on where one draws the line. While the details of this debate are far beyond the scope of this chapter, viruses are being treated by some biologists as if they belong to the Tree of Life. They are classified, for example, in much the same way as other organisms, even up to their Latin names.

Some have grouped them by their geometric structures, or by the identity of the host organisms they infect. But the most commonly accepted classification scheme for viruses, one developed by Nobel Prize–winning biologist David Baltimore, is based on their type of DNA or RNA. For example, viruses can be classified by whether they carry their genetic material as a single-stranded or double-stranded piece of DNA or RNA. Some viruses, known as retroviruses, such as the HIV virus, store their genes as RNA only for it to be made into DNA when the virus infects a cell. Various other infectious agents share some characteristics of viruses, including viroids, satellites, and prions (prions are infectious proteins discussed in box 7.1 in chapter 7), but these agents have even less of the generally accepted characteristics of life forms than viruses do.

The taxonomy of viruses is similar to that of cellular organisms, with the exception that the highest level of classification for viruses is the order to which they belong. If we look at the Ebola Virus from Kikwit in the Democratic Republic of Congo (formerly called Zaire), for example, it is classified as follows: Order Mononegavirales, Family Filoviridae, genus *Ebolavirus*, species *Zaire ebolavirus*.

A recent study demonstrated that a virus called *Acidianus* Two-Tailed Virus, or ATV, that grows inside an extremophile archaeon in a volcanic hot spring in southern Italy produces two protein-containing tails on its own, without the presence of the host cell. (These tails may assist the virus's locomotion in these environments where host-cell density is low.)[a] This finding suggests that other viruses may likewise be more independent of host cells than we realize and have characteristics that we would define as being alive.

uncertainty. Moreover, because extinctions have always been part of Earth's history, before and after humans arrived on the scene, one must ask whether the number of present-day extinctions is unusual. We tackle each of these topics in turn. In doing so in this and in subsequent chapters, we will demonstrate that the current rate of species extinction is unprecedented in human history, that it is the result of human actions, that it rivals some of the great extinction events of the geologic past, and that, unchecked, it constitutes a grave and growing threat to human health.[4,5]

How Many Species Are There?

About 1.5 million species have been identified and given scientific names (other more recent estimates have put this number at 1.75 million),[6] but only about 100,000 of these—including some terrestrial vertebrates, flowering plants, and invertebrates with pretty shells or wings—are popular enough for taxonomists to know them well. Birds and mammals are particularly well known, with roughly 10,000 avian and 4,300 mammalian species described. Many new species are found each year, though a taxonomist cannot always be certain whether the specimen in hand has not already been given a name by someone else in a different country or, sometimes, by someone from a previous century. Some of these newly discovered species make news, such as a new baleen whale found off Japan in 2003, new types of deep sea squids identified in 2001 at various ocean sites, 361 new species (mostly insects) identified in the inland rainforests of Borneo from 1999 to 2004, a new giant deer species discovered in a remote nature reserve in Vietnam in 1992 (belonging to a group called Muntjac deer, and now almost extinct due to loss of habitat), and a new monkey found in South America in 2002, *Callibebus bernhardi*. Numerous other species discoveries in both accessible and inaccessible habitats may not be as widely publicized. These include, among others, some new frogs and insects found in the

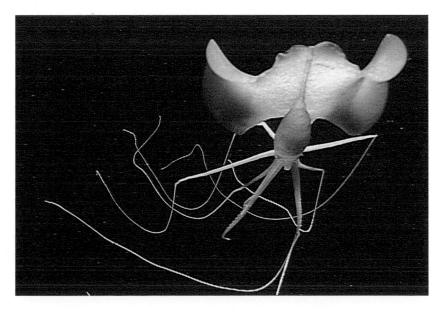

Figure 1.5. Newly Discovered Squid. This large squid (an unnamed species), estimated to be 4 to 5 meters (roughly 13 to 16.5 feet) in length, was encountered by the ROV (remotely operated vehicle) *Tiburon* at a depth of 3,380 meters (more than 2 miles) off the coast of Oahu, Hawaii. (© 2001 Monterey Bay Aquarium Research Institute.)

High Andes in Bolivia; hundreds of newly described species of fish, plants, and other organisms from the deep oceans; new species of fish from the depths of the Amazon River; and newly identified species of fish and crustaceans in tropical peat swamps in Southeast Asia.

Estimates of how many species are presently living on Earth range from six to fifteen million or more, but tend to cluster around ten million. These figures not only exclude microbes, but are also likely to significantly underrepresent species that are small in size, such as mites and nematodes (round worms), and those found in inaccessible or difficult to study places such as the oceans. In these cases, enormous species biodiversity may be missed.

MARINE SPECIES

Although the oceans cover 71 percent of the planet's surface and make up more than 95 percent of the volume of the biosphere, only 250,000 to 300,000 marine species have been described, a fraction of the numbers found on land.[7,8] Less than 5 percent of the oceans have been explored, a result of their vast expanse and inaccessible depths, and much less effort has been spent, compared to that on land, examining life in the seas. When concentrated efforts have been made, we begin to get a glimpse of the enormous extent of marine biodiversity. In 2004, for example, as a part of the ten-year Census of Marine Life project (www.coml.org/), researchers from seventy countries have added approximately 13,000 new species to total counts, including 106 newly described species of marine fish. And there is growing evidence that when the smallest marine organisms—the minute arthropods, worms, phytoplankton, zooplankton, bacteria, and archaea—are included in these tallies, marine biodiversity may approach, and even exceed, levels found on land.[9] A sampling from the Sargasso Sea near Bermuda tends to support this hypothesis: Hundreds of new microbial species were found in surface water samples taken from only four sites.[10]

For coral reefs, species counts are prone to similar underestimates, particularly for smaller organisms that are likely to have more limited geographic distributions. Believed to contain the greatest species diversity of all shallow-water marine ecosystems and often called "the rainforests of the seas," tropical coral reefs could harbor as many as 950,000 species, more than ten times the number (some 93,000) that have been described to date.[11]

Uncertainties about the magnitude of marine biodiversity are even greater for the deep oceans. The average depth of the world's oceans is 12,467 feet (3,800 meters, or more than 2.3 miles), and the soft sediments that blanket the deep ocean's bottom comprise 65 percent of Earth's surface, representing the largest area of habitat of any kind on Earth.[12] These sediments may contain even more species than are found in shallow waters, a prediction that has been supported by the remarkable diversity of organisms found in even small samples of deep sea mud. More than 800 species were found, for example, in just 215 square feet (21 square meters) of ocean bottom sampled off the U.S. East Coast, almost 60 percent of which were previously unknown.[13] In some Australian sediment samples, more than 90 percent of the species found had never been seen before. As a result of these and other sampling expeditions, some

Figure 1.6. Ammonite Fossils. The oceans have been home to an enormous diversity of life for hundreds of millions of years. These ammonites, belonging to the Class Cephalopoda, which includes present day species of the Chambered Nautilus (*Nautilus pompilius*), as well as octopus, squid, and cuttlefish, were extremely numerous, widespread, and diverse in ancient seas. This was particularly true during the Mesozoic Period, which lasted from 245 million years ago to the end of the Cretaceous Period, 65 million years ago, at which time ammonites, along with the dinosaurs, went extinct. (From Jean Charles Chenu, *Illustrations conchyliologiques, ou descriptions et figures de toutes les coquilles connues vivantes et fossiles, classées suivant le système de Lamarck*, Vol. 4, Plate 12, 1842–1853. From the collections of the Ernst Mayr Library, Museum of Comparative Zoology, Harvard University.)

TABLE 1.1. NUMBER OF NAMED DISTINCT LIVING SPECIES BY GROUP
(EXCLUDING PROKARYOTES)

GROUP	NUMBER OF NAMED SPECIES (IN THOUSANDS)
Protozoa	40
Algae	40
Plants	270
Fungi	70
Animals	
Vertebrates	45
Nematodes	15
Mollusks	70
Arthropods (total)	855
Crustaceans	40
Arachnids	75
Insects	720
Other animals	95
Total	1,500

Adapted with permission from Robert M. May, The dimensions of life on Earth, in *Nature and Human Society: The Quest for a Sustainable World*, Peter H. Raven (editor), National Academy Press, Washington, DC, 1997.

have predicted that the number of deep sea organisms, including worms (e.g., polychaetes [segmented marine bristleworms] and nematodes), crustaceans, and mollusks, may range anywhere from 500,000 species to more than 10 million.[13]

While there may be uncertainty, as there is on land, about the number of species that live in the oceans, there is none about the richness of marine life at higher taxonomic levels. The oceans are home to forty-four phyla (see figure 1.1 for an example of a phylum) compared to only twenty-eight on land. Of all thirty-three known animal phyla, thirty-two inhabit the sea, while only twelve live on land. This richness most likely reflects the fact that life first appeared in the oceans and that it existed there for almost three billion years before moving onto land. It also suggests that the variety of body plans and of genetic, biochemical, physiological, and metabolic patterns and pathways is likely to be greater in the oceans than it is on land, and that the oceans offer an extremely rich, and largely unexplored, resource for biomedical research.

Extinction Rates Before Humans

Extinctions are natural events and occurred long before humans walked on Earth. Thus, any claims of human-caused biodiversity loss must take into account what we know from past "background" or "natural" extinction rates. Studies of once abundant and widespread marine species that dominate the fossil record show that these species generally lasted from one to ten million years before going extinct naturally. Mammalian species are thought to last on average about 2.5 million years, and at least for some rodent species, there is evidence about what determines their natural cycles of emergence and extinction. These cycles seem to be associated with long-term, periodic changes in climate that result from variations in the path of Earth around the Sun and in the tilt of its axis of rotation (called the Milankovitch cycles, after the Serbian astrophysicist Milutin Milankovitch, who developed the mathematical explanation for Earth's natural, long-term, periodic, climatic cycles of heating and cooling).[14]

Suppose that extinctions were spread out over time and did not occur simultaneously. (Simultaneous extinctions occur as the result of, e.g., a single catastrophic event like the asteroid that landed in the Gulf of Mexico sixty-five million years ago and that is believed to have been the cause of the extinctions at the end of the Cretaceous Period, when dinosaurs and ammonites largely disappeared.) Then, based upon findings from the marine fossil record, we would expect one species out of a sample of a million to go extinct about every one to ten years. The background extinction rate can then be estimated at approximately one extinction for every million species each year, which, given the marine fossil data, is a number at the higher end of the range (the lower end would be one extinction per million species every ten years). This number can be transformed into a ratio that is stated as "one extinction per million species-years." We think of this as the background rate of extinction.

For birds, recent extinctions (i.e., those that have occurred in the last 2,000 years) run about one or a few per year.[5] The total number of bird species is about 10,000, so this translates roughly into an estimated 100 extinctions per million species per year, or about 100 times the background rate. Extinction rates for other groups of animals and plants, such as amphibians, primates, and some gymnosperms, tend to be even higher.

It is reasonable to ask how one can apply an extinction rate derived from the marine fossil record to modern-day birds or, for that matter, to any other groups of organisms alive today. To address this question, one must bear in mind that the species that are most prone to current extinction are both localized and rare. So, data from ancient marine organisms are likely to underestimate, if anything, current extinction rates among nonmarine organisms, because these ancient marine populations were once very widespread and very abundant and therefore less vulnerable to extinction. Speciation rates (i.e., the rate at which new species develop) and the creation of what are called "molecular phylogenies," which are maps of evolutionary relationships among organisms based on comparative analyses of their genes and proteins (using methods similar to 16S rRNA analyses—see page 10 and figure 1.4, the three-domain map, which attempts to create a molecular phylogeny for all life on Earth), can be used to supplement estimates for background extinction rates. These data broadly support an average origin time for modern species of around one million years ago.

First Contact with Humans

There is good evidence that early humans were responsible for species losses on a large scale. The prevailing explanation for the disappearance of dozens of large mammalian genera between 10,000 and 50,000 years ago is that human hunting was the main cause. Important differences separate these early human-caused extinctions from those of the present time. Past extinctions typically were among large, predator-naive, land animals living in limited geographic areas, and they seem to have been mainly the consequence of overexploitation. The threats to species in the latter half of the twentieth century and today, such as habitat degradation and destruction, overharvesting, pollution, and global climate change, occur on a global scale, and as a result, they tend to cause widespread extinctions that occur in all types of organisms in habitats around the world. These differences render suspect arguments that attempt to downplay concern over current species loss by pointing to the role that humans may once have played in the distant past.

Recent Extinction Rates

In the past 500 years, a total of 844 species have been listed by the International Union for the Conservation of Nature and Natural Resources (IUCN; see box 1.4 for more information on the IUCN) as having gone extinct (if we also count the 122 species of amphibians that the Global Amphibian Assessment [www.globalamphibians.org] lists as probably extinct, this number approaches 1,000), but it is clear that numerous others have not been counted. The extinction status for most plants, animals, and microbes is poorly known, even for those species that have been identified, and most biologists believe that only a small percentage (10 to 15 percent or less) of the total number of species alive today (see the above section "How Many Species Are There?") have been discovered to date. Historically, most extinctions that have been identified have occurred on oceanic islands, but over the past twenty years, continental extinctions have become as common as those on islands. The consensus of scientists is that the current rate of species extinctions is on average somewhere between 100 and 1,000 times greater than prehuman levels, and that we are moving toward an extinction rate that is on average 10,000 times greater.

The three case histories presented below, which typify recent extinctions (i.e., during the past few hundred to few thousand years), support the claim that present extinction rates are unusually high.

PACIFIC ISLAND BIRDS

Polynesians colonized Pacific islands between 4,000 and 1,000 years ago. Their imprint on these islands is fresh and provides unambiguous evidence of massive human-caused extinctions. For example, the bones of many bird species persist into, but not through, layers of archaeological digs that show the earliest presence of human beings. Along with other archaeological evidence, this indicates that these bird species went extinct after humans appeared, but not before. There is little doubt that early colonists ate the large, probably tame, and often flightless birds in great numbers. They also introduced rats onto the islands, and the rats, too, would have found the tame birds easy pickings. With only Stone Age technology, the Polynesians may have exterminated as many as 1,000 bird species, representing about 10 percent of the world's total at that time. In some places, they may have exterminated all the bird species they encountered.

In Hawai'i, we know of forty-three bird species that once lived on these islands only from their bones. But because bird bones are fragile and easily pulverized, we may never find the bones of all the species that went extinct. How many might be missing? One way to answer this question is to look at the number of bird species known to have been alive in the past 200 years whose bones have not been found in archeological digs. Using this number, one can then arrive at an estimate of

Figure 1.7. The O'o (*Moho nobilis*) of Hawai'i, Which Is Now Extinct. (From Baron Lionel Walter Rothschild, *The Avifauna of Laysan and the Neighbouring Islands*. R.H. Porter, London, 1893–1900.)

how many bird extinctions may have occurred without leaving a trace, and that number is 40.

James Cook landed on the Hawaiian Islands in 1778, and European settlement, with introductions of cattle and goats, began shortly thereafter. The livestock destroyed native plants that were as unprepared for large mammalian herbivores as the birds were for the rats introduced by the Polynesians. More bird extinctions occurred following European settlement. Today, our only records of some eighteen other species of birds, not included in the above counts, are the specimens collected and preserved by nineteenth-century naturalists. The total number of Hawaiian bird species that are believed to have gone extinct since humans first arrived on the islands in the fourth and fifth centuries A.D. is therefore 101 (40 + 43 + 18). What remains today? A dozen bird species are so rare that there is little hope of saving them. If we cannot find these species, then they probably cannot easily find each other either, in order to reproduce. Another dozen we can find, but their numbers are so small that their future survival is uncertain. Of an estimated 136 species that once existed before human contact, only eleven now survive in populations that suggest they have a future.[15]

Birds have not been the only casualties. Of 980 native Hawaiian plants, 84 are extinct and 133 have wild populations of fewer than 100 individuals. Hundreds of land snails and several reptiles have also been documented as being among the victims of human settlement.

As the Polynesians colonized the Pacific, from New Zealand north to Hawai'i and east to Easter Island, they exterminated not only 1,000 bird species but a large number of other species as well. All Pacific islands, with the exception of those most remote, are thought to have had several species of pigeons and parrots each, many of which have completely disappeared.

Pacific islands are not unusual. In the last 300 years, the islands of Mauritius, Rodrigues, and Réunion in the Indian Ocean have lost thirty-three species of birds, including the dodo, thirty species of land snails, and eleven species of reptiles, and St. Helena and Madeira in the Atlantic Ocean have lost thirty-six species of land snails. Island species are particularly vulnerable to extinction, because islands often lack large predators of, as well as significant numbers of native species competitors for, or diseases of, alien species. Thus, when alien species are introduced, especially alien predators, the results can be devastating. The ranges of most island species are also generally small, so there is a greater chance that habitat destruction will destroy their entire habitat. This raises the obvious question: Do we find evidence of massive extinctions only on islands?

SOUTH AFRICAN FLOWERING PLANTS

To begin to answer this question, let us look at the Cape Floristic region, a small area of the southern tip of Africa that possesses enormous plant diversity. On a per area basis, this region has more species of plants than any other in the world, including the Amazon rainforest. It is home to a distinct and unusual flora composed of several vegetation types, the most common and species-rich of which is known as the fynbos (which gets its name from the Afrikaans word for "fine bush," describing the appearance of many fynbos plant species). Of the approximately 8,500 plant species that have been identified in this region, thirty-six have become extinct in the last century, and some 618 species are threatened, that is, are likely to become extinct within at most a few decades (see box 1.4 for a description of the IUCN and categories of endangerment). Invading alien plants—particularly Australian wattle trees (from the genus *Acacia*)—and the conversion of natural areas to agriculture are the two major identified causes that are endangering species and causing their extinction in the region.

AUSTRALIAN MAMMALS

Of sixty recent mammalian extinctions, nineteen are from Caribbean islands, yet another instance of high extinction rates for islands. But eighteen more were on the continent of Australia, and these represented around 6 percent of all its known nonmarine mammalian species. The extinctions have been equally divided between two areas. The first is the southern arid zone—a sparsely inhabited area of mostly spinifex desert (spinifex is a coarse grass with sharp-pointed leaves) where there is also extensive livestock grazing. The second is the wheat belt of the southern tip of western Australia, where 95 percent of the natural woodland has been cleared. Another forty-three mammalian species are no longer found in one-half or more of their former ranges. Some survive only because they live on protected offshore islands. Medium-sized ground dwellers weighing between 35 grams (a little more than an ounce), such as mice and rats, and 5.5 kilograms (about 12 pounds), such as wallabies (animals that belong to the same family as kangaroos but are smaller in size), have been hardest hit. Those that have had greater survival success include bats, some arboreal (living in trees) species such as opossums and gliders (gliders are marsupials [mammals that have a pouch where newly born offspring complete their development] that are able to glide on air currents, sometimes for great distances), and those that use rock piles for shelter.

Three causes are thought to be behind the extinctions: the destruction and fragmentation of natural habitats, the introduction of livestock and other species, and recent changes in the frequency, severity, and duration of fires. Domestic farm animals may have also destroyed vegetation cover and caused extensive soil erosion and compaction. Introduced rabbits are competitors with some native species for some food resources already in decline. The predatory Red Fox (*Vulpes vulpes*) introduced in the 1860s, perhaps for fox hunting, may well have destroyed populations of some small mammals, even in remote areas. Supporting this hypothesis is the fact that foxes are absent from areas of the continent that have the fewest extinctions, and that fox control programs have been successful in halting the decline of some small mammal populations.

These three examples of recent species extinctions are for both plants and animals and have taken place on both islands and continents. They are typical of other extinctions in groups of terrestrial animals and plants that have been well studied, all of which bear a distinct human fingerprint.

BOX 1.4

THE IUCN AND THE RED LIST

The 2006 Red List of Threatened Species includes 16,118 species from a broad range of taxonomic groups (including vertebrates, invertebrates, plants, and fungi) that are, or may become, threatened with extinction. But this number is clearly only a small fraction of the total of threatened species, because only 40,168, or roughly 2.5 percent of the approximately 1.5 million known species, have been fully assessed, and as stated above, there may be 10 million species or more in all. Within the major groups of vertebrates for which we have good data, roughly 12 percent of birds, 20 percent of mammals, almost one in three amphibians, almost one in three

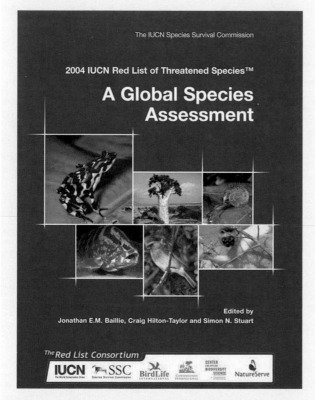

IUCN Red List Cover. In 2004, the IUCN Red List of Threatened Species was released in conjunction with the published book *A Global Species Assessment*. The next assessment is due in 2008. Each year the Red List of Threatened Species is published as an online searchable database that can be found at http://www.iucnredlist.org. (© 2006 International Union for the Conservation of Nature and Natural Resources.)

primates (see section on primates in chapter 6), and 40 percent of turtles and tortoises are threatened. Approximately one-fifth of all known amphibians are Critically Endangered or Endangered (the two highest levels of threat under the Red List system). Only two plant phyla—the cycads (Cycadophyta) and the conifers (Coniferophyta), both belonging to the group Gymnosperms—have been fully assessed: 52 percent and 25 percent, respectively, are threatened (see section on gymnosperms in chapter 6).

The term "threatened" has a specific scientific meaning when referring to a species' risk of extinction. A species is considered "threatened" if it is Critically Endangered, Endangered, or Vulnerable. Placement into one of these groups depends upon an assessment of the adequacy of the species' habitat and of the size and growth (or decline) of its population. For example, a species is considered Critically Endangered when the best available evidence indicates that more than 80 percent of its population has been lost, either during the past decade or over the course of the past three generations (whichever is longer); when the total number of mature, living individuals is typically less than 250; or when its habitat extends less than 10 square kilometers (almost 3.9 square miles) and is shrinking. Critically Endangered species have about a 50 percent risk of going extinct in the wild within ten years, or three generations. The criteria for listing a species as Endangered or as Vulnerable follow similar parameters of population and habitat, with Endangered species carrying a 20 percent risk of extinction in 20 years or five generations (up to a maximum of 100 years), whichever is longer, and Vulnerable species having a 10 percent extinction risk in 100 years.

Our knowledge of the status of threatened species has significant gaps. Vertebrates have been relatively well studied, with about 40 percent of known species assessed. But we still know very little about the degree of threat for a majority of invertebrates, plants, and fungi (and almost nothing about threats to bacteria or archaea) or about the status of most species living in freshwater and marine habitats. For many groups of organisms, there are few or no data about their threatened status. Only about 1,200 insects, for example, out of an estimated 950,000 described species have been assessed by the IUCN. The 2006 Red List tells us, however, that the number of species threatened with extinction is increasing across almost all the major taxonomic groups that have been assessed.

Surveys are conducted on a regular basis to determine the number of the world's species threatened with extinction. The most authoritative of these surveys is put out every few years by the International Union for the Conservation of Nature and Natural Resources (IUCN), now known as the World Conservation Union, based in Gland, Switzerland. For more than thirty years, members of the IUCN's Species Survival Commission (SSC), a network of more than 7,000 scientific experts, have been evaluating the conservation status of species and subspecies on a global scale. The SSC publishes its findings regularly as the Red List of Threatened Species.

Suppose that all these "threatened" species will become extinct in the next 100 years (many would go sooner, of course). If so, then future rates of extinction for birds, with 12 percent of the roughly 10,000 or so known bird species considered "threatened," would be 1,200 extinctions per million species per year, or more than 1,000 times background rates. Similar estimates obtain for other groups. The conclusion is that extinctions have been high in the recent past and that their rate is rapidly accelerating.

This conclusion leads to some obvious questions. First, will speciation (the development of new species) create new species to offset these losses? In the geological past, there have been five episodes, the five major extinction events, when, as the consequence of various factors, large fractions of Earth's biodiversity were eliminated. For example in one of these events known at "The Great Dying," which occurred about 250 million years ago during the Permian period, an estimated 90 to 95 percent of all marine species were eradicated over a time scale of one million years or more. We know from this extinction event, and from others in the geologic record, that biodiversity did eventually recover, but that it did so over millions to tens of millions of years.

Second, can we predict where species will become extinct (and perhaps improve our estimates of how many will do so) by looking at the detailed causes of their extinction? Sometimes, but most often this is not possible. For example, accidentally or deliberately introduced species can cause extinctions, but whether they will or will not is generally hard to know. Predicting future extinctions from these introductions, or from other factors such as global climate change, may not now be possible, because the cascade of events that follow them often involves so many variables, and because our understanding of the physical, chemical, and biological systems that are affected is generally so inadequate.

Just because we have difficulty with these predictions, however, does not mean the threats are insignificant. In fact, we have seen from the above examples of species extinctions, both on islands and on continents, the devastating effects from introduced species. And it is becoming increasingly clear that global climate change will threaten the survival of large numbers of species in coming decades, with some estimates saying its direct effects, and those that are indirect, for example, by changing habitat, will result in the extinction of some 15 to 37 percent of species by the year 2050.[16]

Predicting the magnitude of extinctions from habitat destruction, the factor usually cited as being the most important cause of current and expected extinctions (in birds, e.g., it is implicated in about 75 percent of the approximately 1,200 species that are threatened), may be somewhat easier. Habitat destruction is continuing and, in some cases, accelerating, such that some common species may lose their habitats within decades.

SECONDARY EXTINCTIONS

Once one species goes extinct, it is likely that many others will go extinct as a result. Some are easy to understand. For every bird or mammal or insect that goes extinct, those species of parasites or bacteria that can live on and/or in no other host will also disappear (see box 3.1 on microbial ecosystems in chapter 3). An example may be seen with some termite species, which have within them flagellated protozoa that are, in turn, associated with different types of bacteria. Presumably, these species of termites, protozoa, and bacteria, having co-evolved, are highly specific to one another, so if the termite went extinct, so would the protozoa and the bacteria. Other changes can be quite complicated. Species are bound together in ecological communities to form a food web of interactions. Once a species is lost, other species that fed upon that species or that benefited from it, competed with it, or were food for it would also be affected. These species, in turn, may affect yet other species. Ecological theory suggests that the patterns of secondary extinctions may be highly complex and thus difficult both to demonstrate and predict.

Not surprisingly, then, few clear-cut examples of secondary extinctions have been documented. Perhaps the best example involves the butterflies of Singapore. About 95 percent of the island's forests have been cut, and about half of its roughly 400 species of butterflies have been lost. The main cause is obviously habitat destruction, but the various butterfly species differed in how vulnerable they were. Those that had a wide variety of food plants generally fared better than those that were more specialized, feeding on just a few species. Specialized species, it seems clear, will go extinct when they lose the one (or few) species on which they feed.

THE LOSS OF POPULATIONS AND GENES

While much of the concern over the loss of biodiversity centers on the global loss of species, most of the benefits that biodiversity confers depend on local species populations.[17] An obvious example is a forest that provides protection to a city's watershed. While no species might go extinct globally if the forest were to be cut down, there would be a loss of the ecosystem services the forest provided, for example, in preventing soil erosion and in filtering out pollutants in air and groundwater. Simply put, it is the local loss of diversity that is important in this case. In addition, populations also supply genetic diversity, since different populations across a species' range will differ to varying degrees in their genetic composition. Such genetic diversity has great value for agriculture, for example, when plant breeders rely on the diversity of genes in the wild relatives of crops to provide these crops with resistance to various diseases. When populations of species are eliminated locally, some of their genes may become extinct globally.

The average number of different populations per species has been estimated for well-known species to be roughly 220.[18] If this average applied to all species, it would suggest that there may be more than two billion species populations globally, of which, it is estimated, 160 million (8 percent) are lost each decade. If present trends continue, while many species may be saved in protected areas (e.g., in national parks and zoos), those species will be just remnants of their once geographically extensive and genetically diverse selves.

CONCLUSION

Conservationists justifiably place tropical forests on land and coral reefs in the ocean at the top of their priority lists because they are thought to hold such large fractions of the world's known species. These ecosystems are also likely to hold a significant proportion of the world's species populations. Nonetheless, a comprehensive strategy for saving biodiversity needs also to save some ecosystems that may contain fewer species, not only because of the various services they may provide, but also because of the distinctive ecological composition and evolutionary information they contain. Tundra, temperate grasslands, lakes, polar seas, estuaries, and mangroves are all good examples. Some of these major habitat types, such as tropical dry forests and Mediterranean-climate shrublands are, on average, even more threatened than are tropical moist forests.[19] The Florida Everglades or Brazil's Pantanal, for example, do not rank as places with a high concentration of species, but they achieve prominence because flooded grasslands are scarce globally and are uniformly vulnerable. Other regions attain prominence because of the biological phenomena they house, such as the Arctic tundra and its migratory shorebirds, polar bears, and caribou.

Species diversity provides a kind of insurance policy for ecosystems, buffering them against such stresses as temperature changes, diseases, and pests that can result in species loss and ecosystem disruption. This is known as "ecosystem resilience" or "ecosystem reliability" and is a function of there being a diversity of responses to stressors among diverse organisms. For example, some coral species are more resistant to bleaching than others, so a reef with greater coral species diversity may suffer less bleaching and be more likely to survive the stress of sea surface temperature warming.[20] Population and genetic diversity may also confer stability.

Genetically and spatially diverse forests of willow species (*Salix* spp.) have greater resistance to infestation by the Willow Beetle (*Phratora vulgatissima*), while mixed populations of Sockeye Salmon (*Oncorhynchus nerka*) are better able to maintain productivity despite changes in climatic conditions that affect their freshwater and marine environments.[21,22] Honeybee colonies provide still another example of diversity conferring a survival advantage. Beehive temperatures must be regulated to ensure the well-being of those inhabiting it, particularly the brood. As temperature rises outside the hive, worker bees inside will start to fan hot air out. The temperature at which any given bee starts to fan is genetically programmed (the male bees that mate with the queen to produce the colony are the source of its diversity). Hives that have a greater diversity of fanning thresholds among worker bees because of greater genetic diversity have core temperatures that are more gradually adjusted and more stable.[23]

There is also a greater likelihood, with higher species diversity, of redundancy at the level of functional groups and, as a consequence, of greater ecosystem resilience.[24] That is, if one or more species in an ecosystem is lost or is no longer able to perform its functional role, for example, as a decomposer of leaf litter or as a pollinator of certain plants, other species present that perform these same functions may be able to take their place.

The functioning of an ecosystem is thus critically dependent on the biodiversity of its constituent species and populations, and it is this functioning that determines the ability of ecosystems to provide the essential goods and services that keep humans and all other species on the planet alive. This is the subject of chapter 3.

But first we will look at how human activity threatens biodiversity.

◆ ◆ ◆

SUGGESTED READINGS

ARKive, www.arkive.org (electronic archive of photographs, moving images, and sounds of endangered species and habitats).

The Beak of the Finch, Jonathan Weiner. Vintage Books, New York, 1995.

Biodiversity II, M.L. Reaka-Kudla, D.L. Wilson, and E.O. Wilson (editors). John Henry Press, Washington, DC, 1996.

From So Simple a Beginning: The Four Great Books of Charles Darwin, Edward O. Wilson (editor). W.W. Norton & Company, New York, 2006.

The Future of Life, Edward O. Wilson. Alfred A. Knopf, New York, 2002.

Precious Heritage: The Status of Biodiversity in the United States, Bruce A. Stein, Lynn S. Kutner, and Jonathan S. Adams (editors). Oxford University Press, New York, 2000.

The Sea Around Us—An Illustrated Commemorative Edition, Rachel Carson. Oxford University Press, New York, 2003.

Swift as a Shadow: Extinct and Endangered Animals, Rosamond Purcell. Houghton Mifflin, Boston, 1999.

Tree of Life Project, tolweb.org/tre/phylogeny.html (interactive website where one can explore relationships between organisms).

The World According to Pimm: A Scientist Audits the Earth, Stuart L. Pimm. McGraw Hill, New York, 2001.

See also IUCN Red List of Threatened Species (redlist.org), U.N. Environment Programme World Conservation Monitoring Center (www.unep-wcmc.org), U.N. Convention on Biological Diversity (www.biodiv.org), European Union Nature and Biodiversity homepage (europa.eu.int/comm/environment/nature/), and Convention on Trade in Endangered Species of Wild Fauna and Flora (CITES) (www.cites.org).

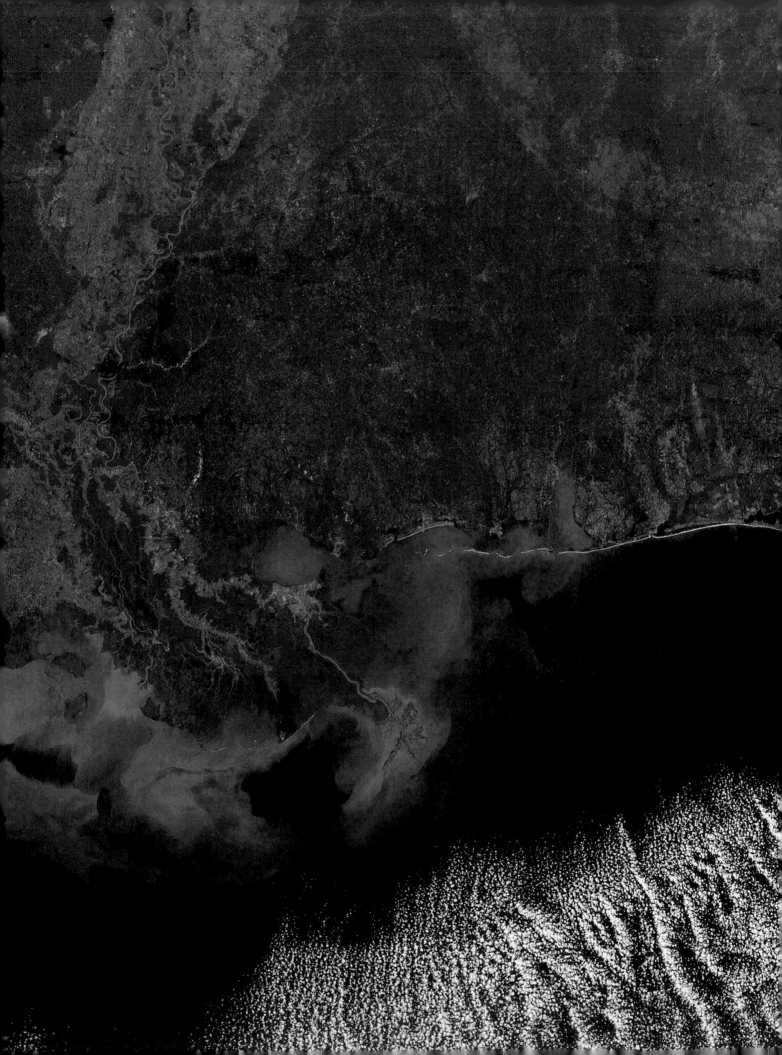

CHAPTER 2

HOW IS BIODIVERSITY THREATENED BY HUMAN ACTIVITY?

Eric Chivian and Aaron Bernstein

> *He [modern man] commonly thinks of himself as having been here since the beginning—older than the crab—and he also likes to think he's destined to stay to the bitter end. Actually, he's a latecomer, and there are moments when he shows every sign of being an early leaver, a patron who bows out after a few gaudy and memorable scenes.*
>
> —E.B. WHITE, *Second Tree from the Corner*

Although species have gone extinct since life began, what distinguishes present-day extinctions from those that have occurred in the past is a distinctive human fingerprint. This chapter considers how human activity has resulted in environmental changes that are known to threaten species. Although each of these changes is discussed as if it were acting in isolation, the reality is that threatened species most often come under pressure from several environmental assaults at once. In some cases, these changes may even work synergistically, such that their combined impact is greater than the sum of their individual effects. They can also act in such a way that one insult sets the stage for another, as may have occurred with the demise of some species of harlequin frogs in Costa Rica, where climatic changes are thought to have predisposed them to chytrid fungal infections[1] (see chapter 6, page 212).

Such a one-two punch, or one involving several human-caused factors acting together, may threaten many species on Earth. For example, research performed in northwestern Ontario lakes has shown that climate change and acid rain may act together to make water clearer and thus more easily penetrated by harmful ultraviolet (UV) radiation. Dissolved organic carbon, which consists of a variety of natural compounds that come from soils and plants, serves as an important UV radiation shield for aquatic life. More than twenty years of observation by David Schindler and his colleagues at the University of Alberta has demonstrated that the total amount of dissolved carbon in the lakes is lowered both by droughts associated with climate change (which result in a reduction of organic carbon flowing into the lakes from the surrounding land) and by acidification of the water by acid rain.[2,3] As a result, some aquatic species, such as those whose young develop in shallow waters where dissolved organic carbon is most reduced, are put at risk from exposure to increased UV radiation, already at higher levels from stratospheric ozone depletion. UV radiation has also been shown to damage aquatic food webs, decreasing photosynthesis and growth in some aquatic algae,[4] and harming some aquatic invertebrates.[5]

It may be difficult for some to read this chapter without having a sense of anger or despair, born out of recognizing how we humans are driving to extinction the very organisms that allow us to thrive on this planet. But before we can act effectively as individuals and as groups to reverse this self-destructive trend, guidelines for which are provided in chapter 10, we must first understand, as fully as we are able, the ways we threaten the survival of other species. That is the goal of this chapter, which, to our knowledge, is one of the most comprehensive reviews available on this subject. We start with habitat loss, which is currently the most serious threat to biodiversity.

HABITAT LOSS: ON LAND

Humans have already altered to varying degrees nearly half of Earth's land surface, and in the next thirty years, this number will likely rise to about 70 percent.[6] Given that the IUCN currently lists habitat loss as a contributor to the endangerment of nearly 50 percent of all threatened species, it is clear that habitat loss will remain a leading driver of species endangerment and extinction in coming decades. Some of the major factors resulting in habitat loss are considered below.

Deforestation

Forests range from hot, dripping rainforests to dry woodlands that merge into savannahs, from conifer forests in temperate regions to those that grade into tundras. They also include temperate deciduous forests (deciduous trees are those that lose their foliage, e.g., maples, oaks, and birches, for some part of the year). What qualifies as deforestation is equally diverse, ranging from absolute clear-cutting to selective logging to sustainable harvesting, and to the deforestation that accompanies damage by fires. Estimates of annual global deforestation tend to converge for tropical humid forests around the figure of 120,000 square kilometers (km²), or about 46,300 square miles. Only about half the original 14 to 18 million km² (5.4 to 7 million square miles) of tropical humid forests remain, with much of the clearing having been done in the last fifty years. For tropical dry forests, annual rates of deforestation are around 40,000 km² (around 15,400 square miles), or about 1 percent of what remains today.[7] These estimates are for forests that have been cleared. Estimates of tropical forests burned, selectively logged, or harmed by being near new forest edges are even larger than those for clear-cut areas. Some of these areas grow back, but most are left as almost useless land, capable of supporting only a fraction of their original diversity.

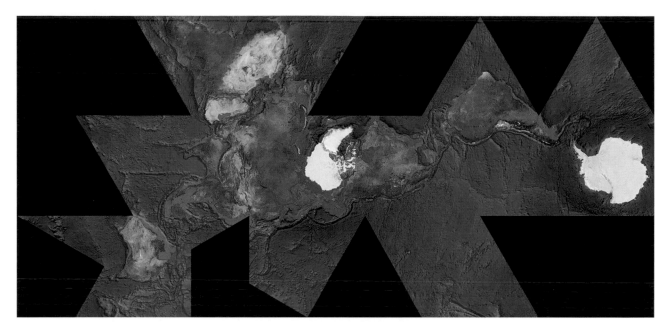

Figure 2.1. Dymaxion Map of Earth. The Dymaxion map, created by Buckminster Fuller, is a global projection onto the surface of an icosahedron, a three-dimensional shape made up of twenty planar faces, each of which is an identical equilateral triangle. The icosahedron has then been cut and unfolded into two dimensions. Fuller claimed that the Dymaxion projection, which we call the Fuller Projection in this chapter, had advantages over all others: There is less distortion of the relative sizes and shapes of land masses; the map demonstrates that there is no "right way up" in the universe, no "up" or "down," no "north" or "south"; and it shows the continents as an almost continuous land mass (which had been true when they were joined together as a supercontinent called Pangea before 250 million years ago), rather than as groups of continents separated by oceans. (© 1938 Buckminster Fuller. Spaceship Earth Satellite Map © 2002 Jim Knighton and Buckminster Fuller Institute. The word "Dymaxion" and the Dymaxion Map are trademarks of the Buckminster Fuller Institute. See www.bfi.org.)

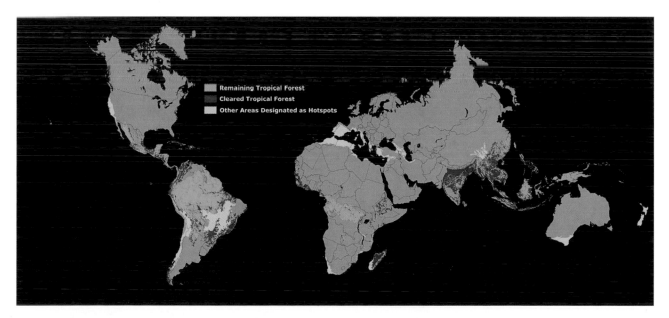

Figure 2.2. Global Fuller Projection Map Showing Tropical Deforestation and Biodiversity "Hotspots." This map has been created by repositioning the Fuller Projection continents, and by using the Environmental Systems Research Institute's (ESRI) ArcGIS software, which allows GIS data to be plotted onto a Fuller Projection. It shows both cleared and remaining tropical forests, along with other areas that are designated as biodiversity "hotspots." Scientists have been showing increasing interest in Fuller Projections, because of their ability to preserve the relative sizes and shapes of land masses. (Map created by Clinton Jenkins, using Fuller world projection; originally appeared in S.L. Pimm and C. Jenkins, Sustaining the variety of life. *Scientific American*, 2005;293(3):66–73.)

Figure 2.3. Satellite Images of Rondonia, Brazil, in 1975 and in 2001. Note the massive clearing of tropical rainforest, over a period of less than thirty years, in this region on both sides of parallel, regularly spaced, newly constructed roads, resembling what is often described as a "fish-bone pattern." (Courtesy of U.S. Geological Survey.)

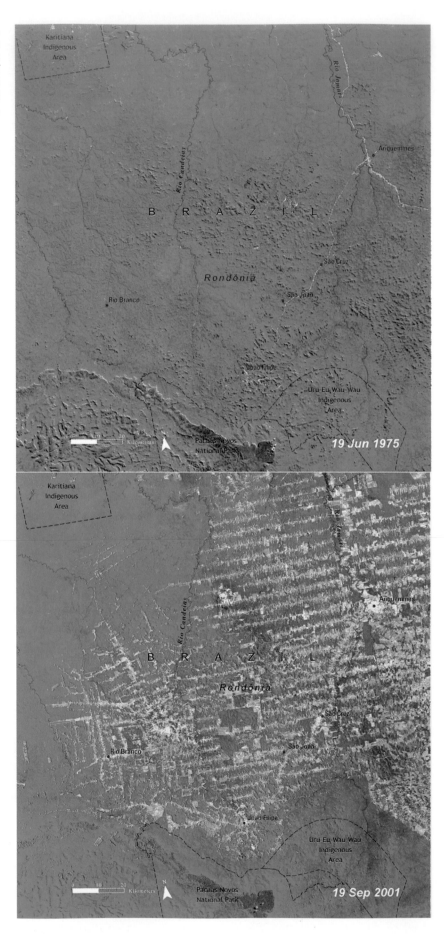

In other regions, such as in Asian Russia, up to 5,000 km² of forests are cut each year, most of which is clear-cutting. Although these logged areas are generally replanted or left to reforest naturally, such practices have led to widespread forest degradation in the region. Temperate forests are recovering in some places, such as the eastern United States, and some are encouraged by the appearance of tree plantations around the world. Globally, some 2 million km² (about 770,000 square miles) of these plantations have been planted, most of which are made up of pine or *Eucalyptus* trees. While these plantations have the outward appearance of forests, they cannot be considered natural forest ecosystems any more than single-crop farm fields can be considered natural substitutes for the ecosystems that they have replaced, such as the prairies in the U.S. Great Plains or the pampas of Argentina. Both planted systems have markedly reduced levels of biodiversity compared to their natural forbears. The growing trend toward large tree plantations instead of sustainably managed natural forests, as has been occurring in such countries as Finland, where a great proportion of forests are now even-aged stands of Scots Pine and Norway Spruce, will endanger large numbers of forest species.[8]

Population growth and migration into forested areas, government subsidies paid for forest clearance, corruption, improved harvesting technology, and, in some areas, human indifference to their destruction drive the overharvesting of tropical forests.

Most clearing of tropical rainforests (perhaps as much as 70 percent according to United Nations figures from the 1990s) occurs to make way for agriculture, including livestock grazing. While the newly established farms can provide habitat to support some species that had once lived in the rainforests, most cannot survive in their new homes. In addition, because many rainforest organisms can live only in limited areas of the forest that meet their peculiar needs for temperature, water, and food, and nowhere else (see discussion below on endemic species), when those areas are cut or burned down, those organisms are lost. Indeed, according to the IUCN's 2006 Red List, forest clearing for crops and livestock is a threat to more than 20 percent of terrestrial species.

Other Threats to Species on Land

Cities may impinge upon and pollute habitats, making them less suitable for their flora and fauna. Presently home to about half the world's population, cities are growing by 2 percent each year, so that urban populations, according to the U.N. population division, will grow to 60 percent of the world's total by the year 2030, with even greater proportions in the developing world.[9] Also, the building of dams, irrigation projects, and other water development activities can disrupt the integrity of habitat and threaten species. While each of these activities brings great benefits to humanity, they all come with significant costs to species and ecosystems (see chapter 3 for a more detailed discussion of threats to ecosystem services).

Discerning the role of habitat loss among other drivers of species extinctions, such as introduced species and hunting, can sometimes be difficult. For example, some bird species such as the Passenger Pigeon (*Ectopistes migratorius*), the Carolina Parakeet (*Conuropsis carolinensis*), and Bachman's Warbler (*Vermivora bachmanii*) became extinct in the forests of eastern North America following massive deforestation in the nineteenth century. But for the Passenger Pigeon, it was over-hunting that ultimately led to its demise, with forest losses contributing both directly and indirectly, by serving to concentrate the birds into smaller and smaller areas, thereby making them more vulnerable to hunters. (See section on overexploitation on page 42 for further discussion of the Passenger Pigeon.)[10]

Other parts of the world, notably Europe, have not become centers for species extinctions despite extensive land transformations. Habitat destruction undeniably causes different numbers of extinctions in different places. Why should this be so?

Endemic Species and "Hotspots"

Another way to ask this question is: What are the features common to centers of human-caused extinctions? Each of the areas in the three case studies of extinction presented in chapter 1—Hawai'i, the Cape Floristic region, and Australia—and several others not mentioned, holds a high proportion of species found nowhere else. Scientists call such species *endemics*. Remote islands are rich in endemics. For example, endemics constitute 90 percent of Hawaiian plants and 100 percent of Hawaiian land birds. But continental areas can also be rich in endemics. About 70 percent of the plants in the Cape Floristic region, 74 percent of Australian mammals, more than 90 percent of North American fish, and the great majority of North American freshwater mollusks are endemic to those regions. In contrast, only about 1 percent of Britain's birds and plants are endemics, and all of eastern North America in recent times has had only about thirty-five endemic birds (including the three mentioned above that are now extinct).

Past extinctions are so concentrated in small, endemic-rich areas that an analysis of global extinction is effectively a study of extinctions in only a few extinction centers. Why is this the case? Consider some simple models of extinction. The simplest one supposes only that some species groups are more vulnerable than others. This model does a poor job of predicting global patterns for the following reasons. First, the model predicts that the more species that are present, the more there will be to lose. Yet the number of species an area contains is not a good predictor of the number of extinctions. Relative to continents, islands have few species, yet they can suffer many extinctions. Second, if island birds were intrinsically vulnerable to extinction, then Hawai'i and Britain, with roughly the same number of breeding land birds and both with widespread habitat modification, would have suffered equally. Hawai'i had more than 100 extinctions; Britain, only three.[11]

All the Hawaiian species were found only on the islands; none of the British species was. This suggests another model of extinction, the so-called "cookie-cutter" model, where something destroys, or "cuts out," a randomly selected area. Species that were in this cut-out area but that were also found elsewhere survive, for they can recolonize. Only some of the endemics go extinct, the proportion depending on the extent of the destruction. In this model, where habitat destruction "cuts out" areas, the number of extinctions correlates weakly with the area's total number of species but strongly with the number of its endemics. And this seems to be the case. Small endemic-rich areas, called centers of endemism, contribute disproportionately to the total number of extinctions, with endemic species that have small, geographically concentrated ranges being the most at risk. The localization of endemics, then, is the key variable in understanding global patterns of recent—and future—extinctions.

Hotspots are centers of endemism that have unusually high levels of habitat destruction. Currently hotspots make up only about 1.4 percent of Earth's total land surface (their original area had been almost ten times greater than it is now), yet they contain more than a third of all known mammals, birds, reptiles, and amphibians. Only slightly more than a third of hotspot habitat is presently protected in any way. Sixteen of the twenty-five areas are forests, with most of these being tropical forests. Even for the three that are relatively undisturbed—in the Amazon, the Congo, and New Guinea—only about half the original tropical forest remains. As a consequence of high levels of habitat loss, these twenty-five hotspots are where the majority of threatened and recently extinct species are to be found.[7]

HABITAT LOSS: IN THE OCEANS

Although scientists are uncertain about the extent of marine biodiversity, they have no doubts about the growing impact that humanity is having on the oceans. More than 50 percent of the world's population lives within 100 miles (60 kilometers) of the coast, and the figure could rise to 75 percent by the year 2020.[12] It is hardly surprising, then, that coastal waters are becoming increasingly polluted and are suffering large scale losses of wetland habitat. Moreover, some 95 percent of marine fish catches come from continental shelf regions, and these fisheries end up consuming a quarter to a third of all the primary production in these areas. (Primary production, the total amount of organic compounds produced by photosynthetic organisms harvesting energy from the Sun, constitutes the base of the food web, with herbivores consuming these organisms, called primary producers or autotrophs, and being consumed, in turn, by carnivores.)[13] It is in these same coastal waters that the majority of known marine biodiversity resides.[14]

As on land, the peak of marine biodiversity lies in the tropics, particularly in coral reefs. Coral reefs are home to almost 100,000 marine species (although, as stated in chapter 1, the total of coral reef species may be more than nine times that number), including an estimated 4,000 to 5,000 species of fish, which comprise almost 40 percent of the world's known marine fishes.[15] Though their combined area is just 0.2 percent of the ocean surface, coral reefs fringe approximately one-sixth of the world's shorelines.[16]

The global center of marine biodiversity lies in the Southeast Asian archipelago, encompassing the Philippine and Indonesian islands. This region, sometimes referred to as the Coral Triangle, supports the greatest concentrations of known marine species found anywhere on the planet, both in the coral reefs and in the vast expanses of its mangroves and seagrass beds. In the Atlantic Ocean, the Caribbean holds the greatest biodiversity. As on land, those reef areas that are the most threatened, including those in Southeast Asia and the Caribbean, are the same ones that hold the greatest number of endemic species. And, as on land, these marine hotspots are the most in danger.[15]

An estimated 20 percent of the planet's reefs have already been destroyed by human activities, and an additional 50 percent are threatened and at risk of collapse.[17] Threats

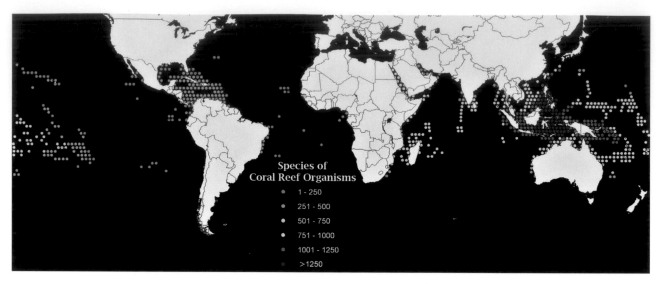

Figure 2.4. Species of Coral Reef Organisms in Mercator Projection. The map shows the richness of species that live on coral reefs, including fish, corals, snails, and crustaceans. Each dot represents an equal area grid cell that covers approximately 50,000 square kilometers (slightly more than 19,000 square miles). Note how coral reef species are concentrated in the southwestern Pacific Ocean, surrounding the islands of Indonesia and the Philippines. (Map created by Clinton Jenkins with permission, using data from Callum Roberts and colleagues.)

to reefs come from overfishing and damaging fishing practices, such as by using cyanide or dynamite; coastal development and pollution by sewage, agricultural runoff, and toxic substances; land erosion and silting; direct physical damage; the harvesting of coral for limestone, jewelry production, and for other industries; and the acidification of seawater by higher atmospheric carbon dioxide dissolving in water (impeding formation of the coral skeleton). But, most of all, reefs are threatened by warming sea surface temperatures from global warming that can cause coral bleaching and lead to various lethal infectious diseases (see section on cone snails in chapter 6, page 257). It is believed that such widespread impacts on coral reefs will translate into large numbers of extinctions, especially for coral reef species that have limited ranges.

Marine Habitat Loss from Fishing

In addition to coral reefs, several other marine ecosystems have been imperiled by human activity. Overfishing is widely regarded as the single greatest threat to marine biodiversity (this topic is also covered in the section on overexploitation below). Some fishing practices are particularly destructive. Bottom trawling, for example, in which weighted nets are dragged across the sea floor, destroys critical ocean floor habitat for developing organisms, undermining the marine food web. Dragging heavy trawling gear across the seabed has been likened to the clear-cutting of forests on land, but the scale of destruction in the case of bottom trawling is significantly greater. Trawls are estimated to scour nearly 1.5 billion hectares (~5,800,000 square miles) of continental shelf habitat each year, an area roughly one-tenth the size of the entire land surface of Earth, and 1,000 times the area of forests that are lost each year.[18] Much of that area has been hit by trawls before, sometimes many times, resulting in less diverse and less structurally complex habitats than those

Figure 2.5. Coral Reef: Hard Corals Just Beneath the Water's Surface Off the Island of Sulawesi in Indonesia. (© Fred Bavendam/Minden Pictures.)

that existed before the onset of trawling. Trawling has had such a negative impact on shallow-water seafood stocks that in recent years trawlers have had to move to the deep oceans. Some 40 percent of trawling now occurs at depths beyond the continental shelves. Today's trawlers can penetrate to depths as great as 2 kilometers (1.24 miles), and their deep-water catch can be found in supermarkets worldwide.[19] In a story that parallels what occurred in shallow waters, the trawlers that first fished these deep and virgin grounds brought up as much coral as fish. Yet, after only a few years, trawling these same areas yielded relatively little coral bycatch because the seafloor habitats had already been stripped bare.

Deep-water trawling has especially devastating effects on slow-growing marine organisms and their habitats, in particular, those found in deep-sea or "cold-water" coral reefs. These reefs, essentially unknown and unexplored until the 1990s, are found off the coast of all the world's continents, in waters as deep as 1,000 meters (3280 feet). Nowhere is the analogy of clear-cutting forests more appropriate than it is for the tops of deep-water seamounts and steep continental slopes where deep-sea corals can be found. Dense and diverse communities of invertebrates that have taken thousands of years to develop are being cleared to bare rock in these habitats in the space of only a few years. Some of the largest sea fans brought up in trawls are hundreds to thousands of years old.[19] Not only are these deep and unseen habitats being destroyed, but along with them, there are almost certainly widespread extinctions, with many species highly vulnerable to habitat loss likely to be disappearing far more quickly than we can identify them.[20]

It has often been assumed that marine species are resilient to extinction, because they are believed to produce abundant offspring that disperse widely over large geographic ranges. But this is not so. Many marine species produce relatively few young and have limited dispersal, and a significant fraction have highly restricted geographic ranges.

HABITAT LOSS: FRESH WATER

Despite the fact that rivers, lakes, and wetlands cover less than 1 percent of Earth's surface and hold only about 0.01 percent of its water, they harbor extraordinary concentrations of biodiversity. Freshwater fishes alone comprise almost one-quarter of all vertebrate species, and when amphibians, aquatic reptiles (e.g., crocodiles and turtles), and aquatic mammals (e.g., otters, river dolphins, and water shrews) are added to this freshwater fish total, current data indicate that as much as one-third of all vertebrate species are confined to freshwater habitats. Unfortunately, knowledge of the total diversity of fresh waters, particularly for invertebrates and microbes, is incomplete, and to date there has been no comprehensive global analysis of freshwater biodiversity comparable to those for terrestrial systems. Even with this knowledge gap, it is quite evident that freshwater systems are exceedingly rich in species.

For most taxonomic groups, freshwater species richness tends to be greatest in tropical regions. However, this is not universally the case. For many invertebrate groups, the temperate United States is home to a significant portion of the world's freshwater species diversity. For example, the fresh waters of the United States harbor some 60 percent of the world's known crayfish species, 30 percent of its freshwater mussels, 40 percent of its stoneflies, and 30 percent of its mayflies, though these percentages may be a bit overstated given that the United States also has relatively more freshwater species researchers than other parts of the world.[21]

Freshwater habitats are among the most endangered in the world, and decline of freshwater biodiversity outpaces that in both terrestrial and marine systems. The particular fragility of freshwater biodiversity is illustrated by the Living Planet Index, a measure of global trends in vertebrate populations from 1970 to 2000. While the overall index fell 40 percent during that period, the terrestrial and marine indices each fell by around 30 percent—it was the freshwater index, which fell by 50 percent, that brought the total index percentage down. Among North America's rich fish fauna, for example, some 364 species are considered to be either Endangered or Critically Endangered. This figure represents a 45 percent increase in endangerment over the previous decade and translates into more than 30 percent of all native fishes being under threat.[22] Freshwater mussels have fared even more poorly, and of the 300 species known to live in U.S. waters, some 67 percent are vulnerable to extinction or are already extinct.[21]

Figures such as these are far from unusual, and even in temperate regions once far from centers of human impact, rates of attrition are surprisingly high. In the snow-fed waters of the Nepalese Himalayas, for example, 42 percent of fish species are considered in danger of extinction.

The situation in the tropics, where much freshwater biodiversity is concentrated, is less well studied but is certainly every bit as severe. For example, the rich fish fauna

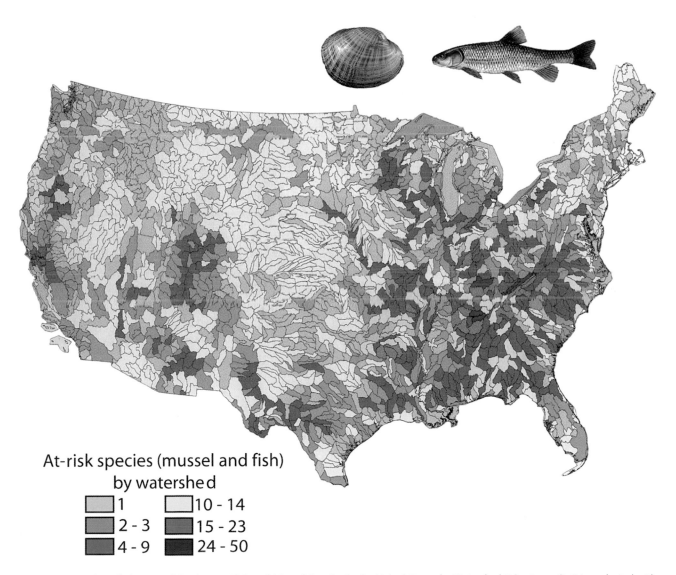

At-risk species (mussel and fish)
by watershed
- 1
- 2 - 3
- 4 - 9
- 10 - 14
- 15 - 23
- 24 - 50

Figure 2.6. **Number of Threatened Freshwater Fish and Mussel Species in the United States by Watershed Prior to 1998**. (Map adapted with permission from L.L. Masters et al., eds., *Rivers of Life: Critical Watershed for Protecting Freshwater Biodiversity*. © 1998 NatureServe and the Nature Conservancy, Arlington, Virginia.)

of peninsular Malaysia has undergone mass attrition, mainly as a result of habitat degradation resulting from deforestation for timber. After a four-year intensive collecting effort, only 45 percent of species historically recorded from the peninsula could be found there. And in Mexico, while thirty-six fish species were considered at risk in the 1960s, within ten years that figure had risen to 123, and it continues to rise.[23]

What makes these freshwater ecosystems so vulnerable to human activities and environmental change is in part a reflection of the disproportionate richness of inland waters, but it is also a result of fresh water being a pivotal resource underpinning the welfare of our own species. Fresh water is essential for human health, food production, hydropower generation, transportation, and economic growth and development. It is also an important focal point of many cultures and religions. During the twentieth century, global human population increased fourfold; during that same period, water withdrawn from freshwater ecosystems increased eightfold. Humans currently appropriate half of the estimated annual global runoff (which broadly equates with net precipitation on land), and the growing water requirements of increasing human populations have transformed rivers and lakes around the world. As a result of large-scale water extraction, mostly for irrigated agriculture, the natural flow of one or more of the planet's great rivers on almost every continent, such as the Colorado, Ganges, Nile, Indus, and Yellow rivers, have been so reduced that they no longer flow to the sea during the dry season.

In addition to draining rivers, we also impound and redirect them. In some places such as China, people have been doing this for millennia. The manipulation and control of the water supply have been, and continue to be, an enduring preoccupation of our species. Construction of massive engineering systems for water storage and flow regulation has transformed the planet's waterways. Our total impact has been extraordinary. Today, some 41,000 large dams more than 15 meters (about 49 feet) high, and countless smaller ones, regulate the flow of more than 60 percent of the world's rivers and retain some 10,000 cubic kilometers (km^3) (2,399 cubic miles) of water on land, a figure that represents more than five times the volume of all of the world's rivers. As of 1998, an additional 349 dams more than 60 meters (about 197 feet) high were planned or under construction around the world. Impacts are, of course, not limited to freshwater biodiversity. Worldwide some eighty million people have been forcibly relocated as a result of these mammoth water engineering projects. With the completion, projected for 2009, of the Three Gorges Dams in China, a further 1.9 million displaced peoples will be added to that tally. Dam construction can also cause outbreaks of some infectious diseases, such as schistosomiasis (see chapter 7, page 300). Globally, water impounded in artificial reservoirs since the 1950s represents a 700 percent increase in the standing stock of river water held on land. This massive redistribution of weight, according to some published reports, has contributed to measurable changes in Earth's rotation and in its gravitational field.[24]

Such enormous changes in hydrology, coupled with harmful changes in water chemistry that result from water storage in reservoirs (e.g., by trapping nutrients and preventing them from flowing downstream), riparian deforestation (i.e., alongside rivers, streams, lakes, and ponds), widespread introductions and/or escapes of exotic species, and the overexploitation of species intensify the threat to freshwater biodiversity.

Further damage comes from pollution. The productivity of freshwater systems is largely driven by the capacity of water to act as a solvent, but this capacity also makes them vulnerable, as their positions in landscapes turn lakes, rivers, and wetlands into prime recipients of nutrient runoff from sewage and fertilizers, and of toxic substances, such as heavy metals, pesticides, and human medicines. Unlike marine waters, fresh waters usually lack the volume necessary to sufficiently dilute contaminants or to mitigate their impacts. Eutrophication (the overenrichment of aquatic systems with nutrients) and algal blooms are now widespread, and the massive loss of global wetlands, due to draining and "reclamation," has undermined the capacity of many freshwater systems to absorb these organic inputs (see section on pollution, below).

Yet another threat to freshwater systems comes from the mining of coal, the world's greatest substrate for electricity generation. Before being sold, coal is washed to remove impurities, including soil and rocks. The wastewater from this process contains a variety of toxins, including heavy metals such as mercury, lead, and arsenic. In Kentucky and West Virginia, billions of gallons of this wastewater have been stored in about 500 impoundments known as sludge lagoons or slurry ponds. With heavy rains (which are predicted to increase in both number and intensity with global warming), these containment structures periodically collapse. The largest of such events in recent times occurred in 2000, when at least 300 million gallons of the Big Branch slurry impoundment breached its confinement and ran into the Big Sandy River in Kentucky and eventually into the Ohio River. (In comparison, the *Exxon Valdez* spill released eleven million gallons of oil into Prince William Sound.) The toxic slurry created an aquatic dead zone that stretched for 20 miles. It also contaminated the water supply of 27,000 people.[25]

Mountaintop coal mining literally decapitates mountaintops with explosives to get at coal seams below, dumping the rubble into neighboring valleys, where it buries headwater streams and contaminates stream and river systems. More than 400,000 acres in Appalachia have been affected by surface coal mining, including mountaintop mining, along with some 1,200 miles of streambeds.[26] Still further consequences for freshwater ecosystems come from the burning of coal to generate electricity. Burning coal is a major source of mercury and of air pollutants, such as sulfur dioxide, a major cause of acid rain, and produces more global warming CO_2 than any other fossil fuel. In fact, coal-burning power plants in the United States produce as much CO_2 as all of America's cars, trucks, buses, and planes combined.[26]

Extending through great distances and across landscapes, freshwater ecosystems have extensive interfaces with their surroundings—from headwaters to estuaries, from their main channels to surrounding floodplains, and from their surface waters to the groundwater beneath. Each of these interfaces also varies over time, both with season and with longer term cycles. Because of these complex relationships, "fencing off" isolated stretches of freshwater systems to safeguard them is futile, and protection of one or a few water bodies is unlikely to maintain ecosystem integrity or its harbored biodiversity. In practice, relatively large areas of land need to be managed in order to fully protect even relatively small water bodies. To be successful, protection of freshwater biodiversity requires control of the upstream network, the surrounding land, the riparian zone, and downstream reaches. As such, a healthy and stable freshwater ecosystem that sustains its biodiversity may be the best indicator of whether modifying that ecosystem and consuming its fresh water are sustainable in the long term.

OVEREXPLOITATION

On Land

Overexploitation—whether by hunting, fishing, or collecting—refers to harvesting that occurs at a rate exceeding the ability of an organism to maintain its population numbers. In some cases, overexploitation has led to extinctions—the Great Auk (*Pinguinus impennis*) and the Passenger Pigeon (*Ectopistes migratorius*) are prime examples of this. In the 1830s, John James Audubon wrote that "the light of noonday was obscured as by an eclipse" by huge flocks of Passenger Pigeons in the United States. Breeding colonies blanketed all the treetops in some areas in swaths that were up to forty miles in length and that sometimes weighed so heavily upon the trees they uprooted them. But the pigeons were hunted in such extreme numbers for meat and for sport (in one competition, the winner killed 30,000 birds!) that by the 1890s they were so rare that they could no longer reproduce fast enough to avoid extinction in the wild. On September 1, 1914, Martha, the world's last Passenger Pigeon, died in the Cincinnati Zoo.[27]

Overharvesting has also nearly wiped out some plant species that produce important human medicines. This has been the case, for example, for the Rosy Periwinkle (*Vinca rosea*), a plant found in Madagascar that is the source of two alkaloids, vinblastine and vincristine, that have revolutionized the treatment of Hodgkin's lymphoma and acute leukemia, respectively, helping, in combination with other chemotherapeutic agents, to turn them from diseases that were generally fatal into ones for which complete cures are possible in many patients. Before plantations were set up to grow the Rosy Periwinkle, its wild populations were in danger of being lost. Overharvesting also threatens the African Cherry Tree, *Prunus africana*, found in mountainous regions of Africa, where it is known locally as *Omugoote* or *Entasesa*. Believed to be effective for treating many conditions, including fevers, malaria, and chest pain, *P. africana* has been most widely harvested for its use in reducing enlargement of the prostate gland (a condition called benign prostatic hypertrophy that is very common worldwide in men older than 50). The bark of *Prunus africana* has been so overcollected, especially in Cameroon and Madagascar, and exported in such great quantities (3,225 tons in 1997) to supply medicinal extracts (sold as "pygeum"), that scientists have warned that the tree will be extinct in the wild in five to ten years unless sustainable cultivation and harvesting practices are implemented[28] (chapter 4 provides more examples of medicines that were nearly lost due to species extinction).

Another form of overexploitation that endangers many species is the trade in live animals, both legal and illegal, for food, pets, zoos, and biomedical research. In the United States, in 2002 alone, according to the Centers for Diseases Control and Prevention, the number of live animals that were imported (there may be equally large numbers that were never identified) included 47,000 mammals, 379,000 birds, 2 million reptiles, 49 million amphibians, and 223 million fish. The Beijing office of the Convention on International Trade in Endangered Species of Wild Fauna and Flora (or CITES—see appendix B) estimated that the income from global trade in endangered animals and plants exceeded 10 billion U.S. dollars each year, making it the third largest source of black-market income after drugs and guns.[29]

The practice of hunting and eating bushmeat (which can refer to any species of terrestrial wildlife, usually from the tropics, that are consumed for food) also threatens some species, particularly nonhuman primates. People have been hunting bushmeat for millennia. The difference between what has been standard practice in many parts of the world for eons and now is one of quantity. Major increases in the demand for bushmeat have been driven by expanding populations and the need for food, and by new access to parts of the forest that had previously been inaccessible. According to the Bushmeat Crisis Task Force (www.bushmeat.org), "80% of all animal-based protein consumed in Central Africa, and as much as 50% of the daily protein intake for rural and urban families," comes from bushmeat. In the Congo Basin alone—which includes the Republic of Congo, the Democratic Republic of Congo, Cameroon, the Central African Republic, Gabon, and the Republic of Equatorial Guinea—an estimated 60 percent of mammalian species are hunted unsustainably. The kill includes duikers (small antelope species that live in forests or in dense bushland) and other antelopes, various rodent species such as Cane Rats (*Thryonomys swinderianus*) and Porcupines (*Hystrix cristata*), and birds, snakes, lizards, monkeys, chimpanzees, and gorillas. The quantities consumed, as well as those shipped abroad to an ever-growing, bushmeat-eating, international market, are staggering. For Central Africa, it is estimated that more than one million tons of forest animals are killed for food each year. That's enough to supply more than thirty million people with a quarter-pound "bushmeat burger" each day for a year.[30] Because of bushmeat hunting, in tandem with deforestation and diseases such as Ebola, some species of nonhuman primates are Endangered or Critically Endangered (see chapter 6, page 231, and chapter 7, page 315, for further discussion of bushmeat hunting and its implications for human health).

The SARS (severe acute respiratory syndrome) epidemic of 2003, caused by a virus that eventually infected 8,098 people worldwide and resulted in 774 deaths, is thought to have begun in China following a human exposure to an infected wild

Figure 2.7. Himalayan Palm Civets *(Paguma larvata)*. (Photograph courtesy of the Yokohama City Bureau of the Environment, Kanagawa Prefecture, Japan, www.city.yokohama.jp/me/ kankyou/dousyoku/nogeyama/tenji/hakubishin. html.)

Himalayan Palm Civet (*Paguma larvata*), also known as the Masked Palm Civet, in a live-animal meat market in Guangdong Province.[31] Four species of Chinese horse-shoe bats of the genus *Rhinolophus* were also found to carry the SARS virus and may, in fact, be the virus's natural reservoir in the wild, passing the virus on to the civets (bats are also sold in these markets as food, and their feces are consumed as traditional medicines).[32] Nevertheless, more than 10,000 of the Palm Civets were destroyed in an attempt to halt the disease. The emergence of this new zoonosis (a zoonosis is an infectious disease that can be transmitted from animals, both domestic and wild, to humans) serves to underscore the extensive practice in China and in some other Asian countries of the killing and eating of large numbers of wild animals, including many species of mammals, reptiles, amphibians, birds, and marine organisms, some of which are endangered, both for food and for their presumed benefits for health, longevity, wisdom, and potency. The proliferation of wildlife restaurants, a sign of new prosperity in China, may further endanger many species in the region.

Overharvesting in the Oceans

Coral reef habitats are also subject to intensive exploitation. Most coral reefs fringe the coasts of developing countries, where population pressures and poverty are driving more people to get their protein from the sea. In their efforts to sustain dwindling catches, people are supplanting traditional fishing methods with extremely damaging ones, for example, using dynamite to kill fish, as well as poisons such as cyanide, both of which threaten reefs. Use of such destructive fishing techniques is most widespread in Southeast Asia, deep within the heart of the world's major hotspot of marine biodiversity.[33]

Species with restricted geographic ranges are most likely to be threatened with extinction. But those that are more widespread have also undergone steep declines, as a consequence of overexploitation. Many species of tropical grouper, for example, while having broad ranges, spawn in large seasonal aggregations and are thus vulnerable to being wiped out by targeted fishing.[34] Other marine species have become so highly valued that it now pays to pursue them to the limits of their ranges, even in locations where they are rare.

Rarity also drives demand for other marine products, such as ornamental shells (see section on cone snails in chapter 6, page 257), aquarium fish, and corals. Because they are not traded for food, statistics are lacking about the numbers of traded organisms involved. It is clear, however, from even a cursory examination of how many shops specialize in marine curios in the United States and Europe that the volumes traded are very large and growing. These outlets are supplied from a relatively small number of countries, with the Philippines and Indonesia among the largest exporters.

OVERFISHING

Overexploitation is the single greatest threat to marine species, particularly over the last fifty years or so when industrial fishing practices, which make use of vessels capable of harvesting and storing more than 1,000 tons of seafood with each outing, have markedly expanded the capacity and reach of fish catching. However, overfishing is not a new phenomenon. It was also a feature of previous centuries, when some of the largest and most abundant vertebrates were stripped from the oceans.[35]

It took only thirty years in New England, for example, to eliminate what were originally abundant whale populations, and by the early nineteenth century, because their local stocks were exhausted, New Englanders had to make voyages lasting three to four years to the Hawaiian Islands, and to other distant regions, in order to find and hunt whales. Similar examples of such early overexploitation come from many other parts of the world. Steller's Sea Cows (*Hydrodamalis gigas*) in the Bering Sea were driven to extinction by the year 1768, just twenty-seven years after they were discovered by shipwrecked Russian voyagers. In the Galapagos, fur seals were exploited to the verge of extinction by the nineteenth century.[36]

Nineteenth-century accounts from tropical Australia describe bays and seagrass meadows that once thronged with Dugongs (*Dugong dugon*), marine mammals related to manatees that are about 9 feet long and weigh approximately 800 pounds. Migrating herds containing perhaps hundreds of thousands of animals stretching over distances of three to four miles were reported.[35] The total population of Dugongs in the late nineteenth century was, by current estimates, between 1 and 3.6 million individuals. Today, Dugong populations are greatly reduced as a result of decades of overhunting. They remain highly vulnerable to habitat loss, because they frequent coastal waters, depend on seagrass beds for food (which are increasingly threatened in many parts of the world), and have low reproductive rates. In the Torres Strait, which separates Papua New Guinea from the Cape York Peninsula in Northeast Australia—the most important Dugong habitat in the world—population estimates in 1996 were fewer than 28,000.[37]

When the first settlers reached the New World, the Chesapeake Bay was a cradle of abundance. It was visited by Gray Whales (*Eschrichtius robustus*), dolphins, otters, manatees, Atlantic Sturgeon (*Acipenser oxyrinchus*), sharks, rays, turtles, and alligators, and it supported vast oyster populations that kept its waters clean. Today, as a result of overharvesting and overdevelopment of the watershed and coastline, there are few reminders of this bountiful diversity, and the Chesapeake is plagued by harmful algal blooms and oxygen-depleted water.[35]

Farther north, in the Gulf of Maine and the Grand Banks off Newfoundland, the continental shelf once abounded with Atlantic Cod (*Gadus morhua*). At the time John Cabot planted the flag for England in 1497, these giant predatory fish dominated these regions. They hunted in enormous schools, and for four centuries we have exploited them with hooks and nets and traps. The cod in this region comprised one of the richest fisheries the world had ever seen. The slide in their numbers began with the introduction of the trawl to New England in the early twentieth century. The collapse of Canadian cod stocks in the early 1990s is widely considered to be one of the most catastrophic in the history of fishing. In 1988, 479,141 tons of Atlantic Cod were caught in Canada; in 1995, a mere 12,490 tons were harvested.[38] The harvesting that led to this decline was so intense that it decreased the age at which the cod reproduced. In the mid-1980s, for example, cod living off Newfoundland and Labrador began to breed by six years of age; by the mid-1990s, they began at five. Cod not only were breeding earlier but were smaller in size. These changes have come about as the result of genetic changes in cod, presumably from the selection pressure of overfishing.[39] A moratorium on cod fishing has been imposed over much of eastern Canada since 1992. Despite this moratorium being in place for more than a decade, Atlantic Cod have shown no signs of a comeback, and the fish are now so scarce in Canada that an expert panel of scientists has recommended that they be listed under that nation's federal Species at Risk Act.

The ecosystem transformations that we began centuries ago with harpoon and spear, we continue today with factory trawlers, purse seines (a net that encircles its catch and then gets pinched off at the bottom), and long-lines (a kind of fishing line with thousands of hooks that can extend dozens of miles—every year more than a billion hooks are set on long-lines).[40] Globally, the world's fisheries are in serious trouble. Studies performed in the world's major ocean regions have concluded that the total mass of large predatory marine fish is only about 10 percent of what it was about forty to fifty years ago, at the advent of industrialized fishing practices. In addition, the fish that we are catching are younger and are also from lower on the marine

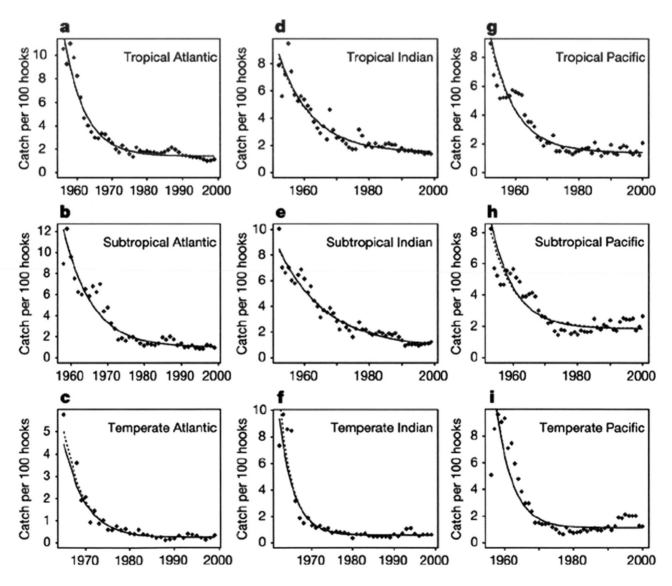

Figure 2.8. Collapse of Large Commercial Fisheries in the Atlantic, Pacific, and Indian Oceans over the Last Forty Years. The graphs are drawn showing the "catch per 100 hooks" of large commercial fish. (Reprinted with permission from R.A. Myers and B. Worm, Rapid worldwide depletion of predatory fish communities. *Nature*, 2003;423:280–283. © 2003 Macmillan Publishers Ltd.)

food chain, because the predators at the top of the food chain have largely been fished out of the oceans.[41,42] If present trends continue, there will be still further collapse of global fisheries, serious consequences for marine ecosystems, and widespread extinctions. The U.N. Food and Agriculture Organization (FAO) has estimated that about 50 percent of commercial fisheries are being harvested at their maximum potential catch, and a further 24 percent are already overexploited.[43]

INTRODUCED SPECIES

Humans have been moving species around the world and introducing them into new environments for millennia, carrying with them their plants, seeds, and domesticated animals, as well as pests and microbes that cause epidemic disease. But the numbers and geographical extent of these introductions have increased exponentially in modern times, along with greatly expanded travel and trade. Most of these transported organisms, which either arrive accidentally as "hitchhikers"—for example, on clothing, in used tires, in wooden pallets, or in the ballast water of ships—or are deliberately introduced, do not create problems in their new environments. But some introduced species become invasive, disrupting ecosystems and threatening the survival of other species, sometimes driving them to extinction. Some invasive species, including some insects, seem to thrive particularly well in already degraded environments.[44–46] After habitat destruction, the issue of invasive alien species is thought by many to be the greatest current threat to biodiversity.

Some species may be introduced as a result of human activity but without any direct human involvement. For example, they can be carried by wind over long distances, as may have occurred during the severe hurricanes in the fall of 2004, when a fungus causing Soybean Rust (*Phakopsora pachyrhizi*) is thought to have traveled from Brazil to the continental United States, posing a risk to soybean crops and perhaps to some other plant species as well (*P. pachyrhizi* can infect some ninety-five species), in Arkansas, Mississippi, Alabama, and Florida.[47] Other infectious microbes can also be carried by the wind. The soil fungus *Aspergillus sydowii*, which caused a fatal infectious disease in some Caribbean sea fans, was found to have originated in the Sahara Desert and the Sahel in Africa, from where it may have been carried in giant dust clouds by trade winds to the Caribbean.[48] In both of these cases, where winds may have transported microbes to invade new environments, human-caused global warming, with its propensity to exacerbate extreme weather events, such as larger and more severe hurricanes from warming sea-surface temperatures,[49] and longer, more intense droughts (e.g., enlarging the clouds of dust traveling from Africa to the Americas or from China to the western United States), may have played a part in their spread. Global warming may also result in threats to native species when they are forced to compete with invaders that have migrated into their ranges in order to find more compatible climatic conditions. These species movements into new regions as a result of global warming have major implications for agriculture, and for diseases of domestic animals, wildlife, and humans (see section on global warming below).

Possible invasion from genetically modified (GM) organisms are a further example. Pollen from GM plants may fertilize their wild cousins, creating transgenic species that have the capacity to outcompete their wild parents. Although there are as yet no confirmed reports of such invasions, the evidence that GM canola genes and GM maize genes have found their way into commercial non-GM canola plants and native maize plants, respectively, is not reassuring. GM fish such as salmon could also pose enormous risks to wild species, driving them to extinction by successfully interbreeding with them and producing less fit offspring. (See chapter 9 for further discussion of GM organisms.)

Our deliberate introductions have a long and generally embarrassing history. In the 1890s, some New Yorkers introduced 100 European Starlings (*Sturnnus vulgaris*) into Central Park, because starlings were mentioned in the Shakespeare play *King Henry the Fourth*, and there was a desire to bring all the birds that showed up in the Bard's work to the United States. These New Yorkers were simply following a long-established tradition, much like that which led, for example, to hunters introducing exotic birds and mammals to shoot, or fishermen bringing to their waters challenging new fish to catch.

While English garden birds in New York, originally intended as quaint curiosities, have caused some problems (e.g., clogging of jet engines and damaging crops), some intentional introductions have been devastating. Invading organisms threaten other species in many ways—they may compete with them for space or for food, displace them, consume them, act as parasites, or transmit diseases to them, all of which may result in a decline or extinction of local populations or of an entire species.

Consider what happened to the genus *Partula*, which contains fifty-eight species of tree snails that live on the Society Islands, northwest of Tahiti. In the late 1970s, a predatory snail, *Euglandina rosea*, was imported from Florida to control the African Land Snail (*Achatina fulica*) that was itself introduced onto the islands a decade earlier. Unfortunately, *E. rosea* preferred *Partula* species to the alien African snails it was intended to control, and it ate to extinction in the wild fifty-four of the fifty-eight *Partula* species. Many of these species were salvaged in captive breeding programs, but these programs did not guarantee survival: A parasite, for example, wiped out the last few remaining *Partula turgida* in 1996 in the London Zoo.[50]

Kudzu (*Pueraria lobata*), a vine brought from Japan to the southeastern United States to help prevent erosion along the banks of newly constructed roads, provides another example of disastrous consequences from the introduction of a new, invasive species and demonstrates how transportation corridors can facilitate the spread of invasives. Kudzu has blanketed large areas with its rapid, unimpeded growth, completely smothering native trees and other plants and damaging power lines. It now covers some seven million acres in the United States, a figure that is expected to double in the next ten years.[51]

But most introductions are not deliberate. Rats of several species were stowaways on oceanic voyages, including probably those of Columbus, and certainly those of the Polynesians as they colonized Pacific islands. The Pacific island of Guam has lost almost all of its native forest birds as well as most of its lizards to a tree snake, *Boiga irregularis*, thought to have arrived there as a cargo stowaway on U.S. transport ships during World War II (the snake is aggressive and poisonous enough to hospitalize about fifty people a year on Guam).[52] Infected humans took HIV—the virus that causes AIDS—from Africa to the rest of the world. In much the same way, early European colonists of the New World and the Pacific introduced other infectious organisms, such as the spirochete bacterium that causes syphilis and the virus that causes smallpox, to populations that had no resistance to them. Smallpox alone may have killed half or more of the Native American population in the few decades following the arrival of Spanish conquistadores to the Americas.

The health of ecosystems has also suffered from invasive species. The Zebra Mussel (*Driessana polymorpha*), a coin-sized, black-and-white striped mollusk that was brought inadvertently to the U.S. Great Lakes aboard ships from Russia in 1986, has proliferated so widely in the lakes and other waterways in nineteen U.S. states that it has threatened these bodies of fresh water by lowering oxygen levels and outcompeting many other organisms, such as freshwater clams, for food.[53,54] The American Comb Jellyfish (*Mnemiopsis leidyi*), also known as the Comb Jelly, or in Britain as the Sea Gooseberry or Sea Walnut, has traveled in the opposite direction. It is believed that it was transported in the bilge water of ships going from the Chesapeake Bay in the United States, first to the Black Sea, and finally to the Caspian Sea, where it caused a collapse of the anchovy fishery.[55] The International Maritime Organization has estimated that some 7,000 different species are transported in the roughly three to five billion gallons of cargo ship ballast water that is moved around the world each year.[56]

Two alien species are major threats to Lake Victoria in Africa. The Nile Perch (*Lates niloticus*), introduced to create a major fishery in the lake, has driven several native fish species to extinction and has devastated the economies of fishing villages around the lake that had depended on them.[57] And the Water Hyacinth (*Eichhomia crassipes*) has spread so extensively in the lake, as it has in other lakes and rivers all over the world (though there is evidence that it may be less widespread in recent

Figure 2.9. Water Hyacinth (*Eichhomia crassipes*) Invading a Lake. (Courtesy of David Sanger Photography.)

years), that it has threatened many species by choking them out of surface waters and by reducing levels of dissolved oxygen.[58]

Invasive alien species have received greater attention in recent years by many agencies and organizations, including the U.N. Environment Programme, the International Maritime Organization, and the IUCN, as the problem accelerates. But even greater attention must be paid if we are to avoid the enormous costs involved from such consequences as clogged water pipes and fishery collapse, and the great harm that invasive species can cause to biological systems, threatening other species, disrupting ecosystems, and posing risks to human health.

INFECTIOUS DISEASES

A multitude of infectious diseases, caused by viruses, bacteria, fungi, protozoa, and other organisms such as mites, threaten the survival of many species. In some cases, disease results from the introduction of new pathogens directly transported by humans, such as the protozoan that causes avian malaria (*Plasmodium relictum*) or the virus that causes avian pox (*Poxvirus avium*), both of which were inadvertently introduced into Hawai'i (about 100 years after the mosquito vector that transmits them was introduced) and resulted in major population declines in Hawaiian endemic birds. Pathogens may also be introduced into new areas indirectly by human activity, for example, as in the case of Soybean Rust mentioned above, or that of Eastern Oyster Disease (*Perkinsus marinus*), which extended its range from Long Island to Maine in the mid-1980s when sea temperatures warmed (from strong El Niños adding to the effects of global warming). In other situations, warming sea temperatures may lead to diseases caused by pathogens that may already be present but that become infectious when the resistance of the host organism is compromised, which may be the case in some corals that succumb to a variety of infections following bleaching events (see section on cone snails in chapter 6, page 257).

To get a sense of how infections can devastate land-based species, consider the following two examples. First, the European Honeybee (*Apis mellifera*), which was originally brought to the United States in the eighteenth century, has been decimated in the wild due to a variety of factors, including an assault from the blood-sucking mites of the genus *Varroa*. Although it is not certain how and when these mites first came to the United States (they were originally discovered in Maryland in 1979), it is thought that they arrived with honeybees that were imported to crossbreed with native strains. Mites have taken a tremendous toll on honeybees in the United States, both in colonies and in the wild, with populations crashing by 50 to 90 percent in the mid-1990s in the span of just two years. Honeybees are also threatened by a variety of other human actions, including pesticide applications, competition from Africanized honeybees (a hybrid between European bees and an introduced African honeybee), and habitat loss.[59] The loss of pollinators and the implications of this loss for agriculture are explored further in chapter 8.

The Asian Chestnut Blight Fungus (*Cryphonectria parasitica*), introduced into the United States around the turn of the twentieth century, probably via an imported Chinese chestnut tree (which has a natural resistance to the fungus), was first confirmed at the Bronx Zoo in 1904. The fungus spread rapidly and effectively wiped out all adult American Chestnut trees (*Castanea dentata*) from eastern United States forests.[60] The American Chestnut was once one of the most valuable trees in the forest, because of its production of food for both humans and wildlife and its light, strong, and straight timber, which made it an unparalleled construction material. Many colonial houses and barns in New England in the eighteenth and nineteenth centuries were framed with chestnut beams.

An increase in diseases has also been noted among marine organisms, such as turtles, corals, marine mammals, sea urchins, and mollusks, a trend that evidence suggests will continue in coming years. Several factors may be involved, including the expansion of marine aquaculture, an increase in ship ballast water dumping into coastal ecosystems, the collapse of some marine fisheries, warming ocean temperatures, the exposure of marine organisms to pathogens from domestic animals, and a possible weakening of marine organisms' defenses due a buildup of human-released chemicals in their tissues.

In addition to Eastern Oyster Disease mentioned above, several further instances of infections contributing to marine species decline have been documented. Since 1987, there have been at least eight known epidemics in marine mammals around the world caused by viruses from the genus *Morbillivirus*, the same family of viruses that causes measles, canine distemper, and rinderpest. The outbreaks have caused massive die-offs in several species, including 18,000 Harbor Seals (*Phoca vitulina*) and a few hundred Grey Seals (*Halichoerus grypus*) along northern European coasts in 1998, and another 21,000, mostly Harbor Seals, in essentially the same area in 2002; thousands of Baikal Seals (*Phoca sibirica*) in Lake Baikal in 1988; hundreds of Striped Dolphins (*Stenella coeruleoalba*) along the coasts of Spain, France, Italy, Greece, and Turkey in 1991 and 1992; and more than 50 percent of Bottlenose Dolphin populations (*Tursiops truncates*) living off the Atlantic coast of the United States from 1987 to 1988 (with a much smaller epidemic to follow along the Gulf of Mexico in 1994).[61]

Outbreaks of several lethal infectious diseases of corals have also occurred in recent years, generally after corals have been bleached by warming ocean temperatures. In some areas, such as the Caribbean, corals seem to be particularly at risk. The once dominant coral in many Caribbean reefs, *Acropora cervicornis*, has virtually disappeared due to disease. Diseases that infect multiple coral species, such as black band disease (which affects forty-two known species), white plague types I and II (22 species), and *Porites* pox (10 species), are of the greatest concern.[62]

Still other infections threaten species, ranging from Serengeti lions to the American Elm. New research is just beginning to understand the complex chain of ecological events that lead to outbreaks of infectious diseases in wild and domesticated plants and animals. This topic is treated in greater depth in chapter 7.

POLLUTION

Humanity has in the past century brought about unprecedented alterations to the chemistry of the natural world. Ecosystems that have developed intricate relationships among their resident organisms over millions of years are now being exposed to changes in the acidity of their surroundings, higher concentrations of heavy metals, and a host of completely new synthetic compounds, many with potent biological activity, such as pesticides and herbicides. The natural cycles of carbon, nitrogen, phosphorus, and water that had existed in relatively stable equilibria are now all profoundly altered as a result of human actions. All these changes have major ramifications for life on Earth and may threaten the survival of some vulnerable species.

Nutrients

Few elements have as much influence on life as nitrogen. It is a key element in controlling the function of many terrestrial, freshwater, and marine ecosystems, as well as the composition and diversity of their resident species. For example, many plant species that are adapted to low levels of nitrogen in water and in soils can be lost, along with the species that depend on them, including microorganisms and herbivores (and their predators), when these levels are significantly altered. Seventy-eight percent of the atmosphere is composed of nitrogen, but in its gaseous form (N_2), it is not biologically available for most living things. Soil bacteria are required to convert or "fix" this nitrogen from the atmosphere into different chemical forms (collectively known as reactive nitrogen) that organisms are then able to use. Today, the use of fertilizers in agriculture and the burning of fossil fuels have substantially increased the availability of nitrogen to biological systems.

Modern conventional agriculture has come to rely on nitrogen fertilizers to bolster food production, and these fertilizers have been an essential ingredient in keeping up with the food demands of a rapidly growing human population. However, not all fertilizer applied to crops gets used by them. By some estimates, as much as 50 percent of the nitrogen applied washes off into groundwater or into waterways and has the unintended effect of "fertilizing" everything that lives in its downstream path. Because many of these waterways eventually end up in the oceans, this nitrogen has had major effects on estuaries and on coastal ocean ecosystems, altering their biodiversity and their functioning.

One of the most devastating impacts from excessive nitrogen release has been the creation of marine "dead zones." Nitrogen-rich agricultural runoff and sewage discharge have played a large role in creating the nearly 150 marine dead zones around the world. The nitrogen fertilizes phytoplankton (microscopic plants and algae that live near the water's surface), leading at first to a boom, and then to a bust, in their populations. As these microorganisms die, bacteria feed on them and consume oxygen in the process, thereby depleting the water's oxygen content and making it uninhabitable for fish, plants, and other organisms. The number of dead zones has doubled every decade from 1960 to the present. Most are found near the coasts of wealthier countries; nearly one-third are off of the United States. The world's largest is in the Baltic Sea, reaching an area of up to 84,000 km^2 ($\sim$32,400 square miles, more than the size of Lake Victoria, the second largest lake in the world). Other large "dead zones" are found in the northwestern Black Sea and in the northern Gulf of Mexico.[63]

Some algal species can threaten wildlife in other ways, producing what are referred to as harmful algal blooms (HABs), which can either physically damage other organisms (e.g., by clogging the gills of fish) or poison them by exposing them

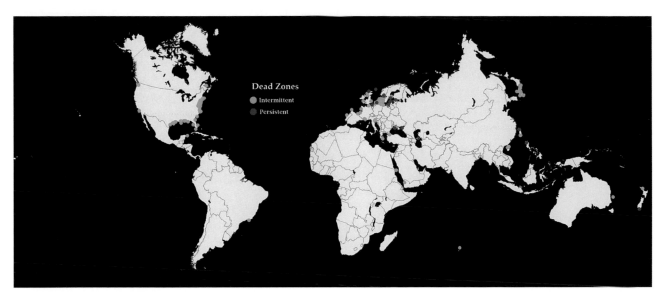

Figure 2.10. Global Map of Marine Dead Zones in Fuller Projection. Intermittent dead zones include those that are annual in appearance, as well as those that reappear at other regular or irregular intervals. This map has been created using the Environmental Systems Research Institute's (ESRI) ArcGIS software, which allows GIS data to be plotted onto a Fuller Projection. (Map and land mass arrangement created by Clinton Jenkins. Adapted with permission from Diaz et al., A global perspective on the effects of eutrophication and hypoxia on aquatic biota, in *Proceedings of the 7th International Symposium on Fish Physiology, Toxicology, and Water Quality, Tallinn, Estonia, May 12–15, 2003*, G.L. Rupp and M.D. White (editors) EPA 600/R-04/049. U.S. Environmental Protection Agency, Ecosystems Research Division, 2004, 1–33.)

to various toxins. The precise causes of HABs are not fully understood, but scientists believe that warming ocean temperatures combined with excess nutrient inputs from agricultural runoff and from sewage discharge may increase their frequency and global spread. The introduction of new algal species, through the discharge of ship ballast water or through changes in water circulation patterns from human barriers such as dikes, or from natural processes such as hurricanes, may also contribute to HABs.

HABs may kill very large numbers of fish—a single bloom, for example, is capable of wiping out an entire fish farm containing hundreds of tons of fish. Furthermore, HAB toxins may work their way up the marine food web, putting species higher on the food chain at greater risk. Brown Pelicans (*Pelecanus occidentalis*) and West Indian Manatees (*Trichechus manatus*), both endangered species, have fallen victim to HABs. In humans, the toxins are typically ingested with contaminated seafood and can cause severe neurological symptoms or even death.[64]

Phosphorus, also a component of agricultural fertilizers and sewage, can produce algal blooms and dead zones when released into fresh water, as nitrogen does in salt water. (Nitrogen is the main growth-limiting element in marine ecosystems; in freshwater ecosystems, it is phosphorus.) Algal blooms can form in fresh water at very low phosphorus concentrations, even at levels that are about one-tenth those generally found in agricultural soils. These blooms may also produce toxins that are directly harmful to other aquatic organisms, but more commonly, they produce an opaque film that blocks sunlight, threatening the survival of plants that live beneath the water's surface and of the various organisms that depend on them. The loss of these plants, combined with the oxygen consumed when the algal bloom ultimately dies, depletes the water's oxygen content, leading to freshwater dead zones and to fish kills.[65]

Persistent Organic Pollutants

A vast, uncontrolled experiment has been taking place with life on Earth over the past sixty years. Tens of thousands of human-made chemicals have been introduced and distributed around the globe. Many of these clearly have been of enormous benefit to humanity. However, some, including a group of chemicals known collectively as persistent organic pollutants, or POPs, have proven themselves to be significant hazards. POPs last for years or decades in the environment before they are broken down. They can travel long distances in air or water, and they can accumulate in fatty tissue, allowing them to become progressively more concentrated as they move up a food chain. POPs can mimic hormones and other biologically active molecules and, as a result, may affect reproductive capabilities, cause cancer, suppress the immune system, and interfere with the development and function of the nervous system in many animal species, including our own.

Perhaps no organisms are more at risk from POPs than predators at the top of the marine or freshwater food chains that have a high percentage of their body weight present as fat. For example, Polar Bears that live in POP-contaminated areas, such as the Svalbard archipelago off the northern coast of western Russia, have many times the concentrations of POPs in their tissues than do Polar Bears living elsewhere[66] (see section on Polar Bears in chapter 6, page 224). Similarly, Beluga Whales (*Delphinapterus leucas*) that live in the mouth of the St. Lawrence River in Canada have been plagued by high rates of infections, cancer, and reproductive failure, thought to be due, at least in part, to their high POP burdens (mainly PCBs, DDT, and an insecticide called mirex), in addition to significant concentrations of lead and mercury. (Belugas are at the top of this aquatic food chain, consuming fish that have concentrated pollutants originating in the river and the Great Lakes.) In the 1980s and early 1990s belugas that washed up on the shores of the St. Lawrence River contained such high levels of pollutants that they qualified as toxic waste according to Canadian government standards.[67]

Many POPs, even in minute concentrations, can affect reproduction through alterations in hormone levels that can affect fertility, reduce offspring viability, and cause malformations in reproductive organs. Some male American Alligators (*Alligator mississippiensis*) in Florida's Lake Apopka, for example, were found in 1994–1995 to have very low levels of the male hormone testosterone and a 25 percent reduction, on average, in the size of their penises. In 1980, Lake Apopka was the site of a major chemical spill, with large amounts of the pesticides DDT and dicofol ending up in the lake. In addition, the lake suffered from runoff from surrounding agricultural areas and from a nearby sewage treatment plant. Alligators in the lake, at the top of the food chain, were found to contain high levels of DDT metabolites, and subsequent lab experiments demonstrated that these chemicals were the likely cause of the lowered testosterone levels and of the alligators' sexual deformities.[68,69]

Other examples of the threats that POPs may pose to wildlife include the local extinction of trout from the U.S. Great Lakes in the 1960s, presumably from the toxic effects of dioxins on their eggs, and the near collapse of some raptor populations such as those of the American Bald Eagle (*Haliaeetus leucocephalus*) and the Peregrine Falcon (*Falco peregrinus*), possibly from the disruptive effects of DDT on eggshell formation. In addition, populations of minks, otters, seals, and several marine bird species have all been adversely affected by POPs.

The Stockholm Convention, which went into force during 2004, aims to eliminate the threat posed by twelve of the most dangerous POPs. While the convention is a key step forward, these chemicals are in many cases very difficult to remove from the environment (some pesticides, last applied in the 1970s and banned since then, still reside in wildlife and in ourselves). Moreover, the convention, while having a mechanism to add new chemicals, does not presently address several new POPs, such as flame retardants known as PBDEs that are very widespread in the environment and are beginning to show up in the tissues of a wide variety of organisms.

Pharmaceuticals

Since the mid-1990s, the populations of three species of vultures—the Oriental White-Backed Vulture (*Gyps bengalensis*), the Long-Billed Vulture (*Gyps indicus*), and the Slender-Billed Vulture (*Gyps tenuirostris*)—in the Indian subcontinent have plummeted by more than 90 percent. The cause of the vulture deaths remained a mystery until early 2004, when an international research team identified diclofenac, a painkiller and anti-inflammatory veterinary medication (also used in humans to treat arthritis), as the culprit.[70] Vultures ingested the medication by eating dead livestock that had been treated, and although diclofenac was therapeutic for the livestock, it caused kidney failure and death in the vultures, bringing all three species to the verge of extinction. In 2005, when India banned veterinary diclofenac use, the population size of all three species had dwindled to less than 3 percent of their levels a decade earlier.[71]

Figure 2.11. White-Rumped, or Oriental White-Backed, Vulture *(Gyps bengalensis)*. This is one of three vulture species whose population numbers have collapsed, likely due to the veterinary medicine diclofenac. (Courtesy of Ronald M. Saldino.)

Vultures in southern Asia and in many other parts of the world serve as major scavengers of wildlife and livestock carrion. In India, they also scavenge human corpses left for them by members of the Parsi faith. Some are concerned that as the vultures are lost, the ecological niche they occupy will be filled by feral dogs that can carry rabies, possibly leading to a wider epidemic of the disease in humans than already exists. More people die of rabies in India, some 30,000 a year according to the World Health Organization, than in any other country in the world. A new and effective substitute for diclofenac, meloxicam, has been developed that does not appear to harm vultures, but it is more expensive and is not yet widely used.[72]

Stories such as these have raised questions about the wisdom of administering some other medicines to livestock, such as hormones and antibiotics, because of their potentially serious ecological and human health impacts. Some livestock feeding operations, for example, rely on hormones to promote rapid growth, and these hormones can end up in groundwater, soils, and aquatic ecosystems, where they can affect the reproductive ability of some exposed fish species. The human health effects from long-term exposures to such growth hormones, especially in children, are not well understood.

Antibiotics are also very widely given to livestock and used in aquaculture. According to a 1998 study by the U.S. National Academy of Sciences, 19 million pounds (9,500 tons or about 8,618 metric tons) of antimicrobials are used every year in the United States alone for nontherapeutic purposes in cattle, swine, and poultry—more than six times the amount, some three million pounds, used annually in the United States for treating people.[73] While some antibiotics are given to treat specific infections or are used on a chronic basis to prevent the development and spread of infections in overly crowded conditions, most are given, especially to pigs and poultry, to promote more rapid growth using less feed. There are very great potential costs to this practice. For one, antibiotic resistance can develop in the infectious organism, both to the antibiotic that has been used and to others in its class. Given that antibiotic resistance is already a crisis in human medicine, continuing the nontherapeutic use of antibiotics in livestock production, especially antibiotics that have chemical structures and actions similar to those used to treat human infections, is extremely ill-advised, because it may generate even more strains of antibiotic-resistant bacteria that can infect humans.

For these reasons, and others, the World Health Organization has recommended that such nontherapeutic antibiotic use in livestock be banned. Potential ecological problems exist, as well. It has been shown, for example, that antibiotics can cause adverse effects in some aquatic plants and can alter the diversity, growth, and activity of some soil bacteria. In addition, the release of antibiotics into coastal marine ecosystems can harm some crustaceans and other organisms.

The same is true for human medicines, which are released in vast amounts into the environment around the world via wastewater discharge. This is a serious and escalating problem that is just beginning to receive scientific scrutiny. In 2002, for example, the U.S. Geologic Survey found that 80 percent of U.S. streams sampled contained a host of drugs, including antidepressants, hormones, and steroids. Similar contamination has been found in many other countries.[74] It should come as no surprise that these human medications have major effects on other species, often at low concentrations, given how close we are biochemically to other organisms. If the synthetic estrogens in birth control pills such as ethynylestradiol, for example, end up in waterways, as shown in lab experiments, they can kill trout at chronic exposures of concentrations of one part per billion and can reduce the fertility of male trout by half at levels 100 times smaller.[75] In some rivers in the state of Colorado, male White Sucker fish (*Catostomus commersonii*) living downstream from sewage treatment plants have been found to be developing female sexual organs, presumably from estrogens in the wastewater.[76] And in the south branch of the Potomac River in the state of West Virginia, male bass are bearing eggs.[77]

Antidepressants such as fluoxetine (Prozac), widely used and highly persistent in the environment, have been shown to cause developmental abnormalities in aquatic organisms, although these effects occur at concentrations that are ten times or more greater than those found in municipal sewage effluents.[78] Even the most modern sewage treatment plants in the most highly industrialized countries are not equipped to filter out or to breakdown the enormous amounts of human medicines and their metabolites that end up being released into the environment, for which the impacts on wildlife (and on people) are still largely unknown.

Every chemical reaction that occurs in Nature depends upon the acidity of the environment in which it occurs. Since the Industrial Revolution, humanity has altered the acid–base balance of the biosphere, principally by the combustion of fossil fuels. This combustion produces sulfur and nitrogen compounds that fall back to Earth as acid rain and snow, acidifying soils and bodies of fresh water. It also releases sufficient quantities of CO_2 to acidify the world's oceans, because CO_2 becomes carbonic acid (H_2CO_3) when it dissolves in water[79] (see section on the acidification of seawater, page 69).

The problems of acid deposition for terrestrial and freshwater organisms and ecosystems are well documented. Some forests in the northeastern United States and Canada and, in particular, some stands of Red Spruce (*Picea rubens*) and Sugar Maple (*Acer saccharum*), have suffered severe diebacks (reduced growth leading to mortality) because of acidic soils produced in the latter half of the twentieth century.[80] These forests lie downwind from coal-fired power plants, hundreds and sometimes even a thousand or more miles away, whose acidic gases ultimately come to rest on the trees' needles, leaves, and soils. The acidification results in shifts in the balance of nutrients available to the trees, most specifically in a depletion of available calcium in the soils, thus interfering with calcium dependent cellular processes and making the trees more vulnerable to infections, pest infestations, temperature stresses, and death.

Freshwater species likewise have had to cope with more acid environments, because much of the byproducts of fossil fuel combustion wind up in fresh water. Such acidic water can be directly toxic to fish, but it may also harm organisms by bringing about changes in the water's chemical composition. One of the most important of these changes involves increased concentrations of aluminum in the water (acidic water leaches out aluminum from soils), which is toxic to some freshwater organisms. These combined exposures —to increased acidity and increased levels of aluminum—may explain an observation that has been made many times over: Lakes and streams with higher acidity have reduced fish biodiversity. Acidic waters may also increase the bioaccumulation of mercury in freshwater fish and can result in greater human lead exposures because more lead is leached from lead pipes and solder into drinking water by the higher acidity.

The rapid accumulation of atmospheric CO_2 predicted for coming decades will also accelerate acidification of the world's oceans, which are thought to have already absorbed roughly one-third of the CO_2 produced by humans since the Industrial Revolution began, making the oceans more acidic. Such acidification could lead to widespread species loss and major effects on marine ecosystems.[79] (See "Acidification of Seawater" under "Global Climate Change," below.)

Limits on nitrogen and sulfur emissions from power plants have significantly lowered acid deposition in the industrialized world and have improved water and soil quality. In parts of the developing world, however, particularly in China and India, both heavily reliant on fossil fuels, the reverse is true, with levels of acid deposition growing rapidly.

Heavy Metals

Many naturally occurring metals, such as lead, mercury, cadmium, and arsenic, are known to be toxic to plants and animals. Lead and mercury, in particular, pose a unique danger to biodiversity, because human activities have released them into the environment at concentrations well above those that would be present by natural processes alone. The combustion of gasoline in motor vehicles and the burning of coal in power plants are now the main sources worldwide for environmental lead and mercury, respectively. Lead has also entered wildlife through the use of lead ammunition in hunting and lead sinkers in fishing, and mercury has been introduced into agricultural soils via various fungicides used to treat seeds.

Lead has been demonstrated to be toxic to a wide variety of animals—birds, frogs, fish, turtles, and many mammals, including cattle, deer, bats, and fur seals. It is also highly toxic to humans. Lead kills nerve cells and can cause severe damage to nervous systems, especially those of developing organisms. In vertebrates, it gets stored in bone and therefore may be particularly dangerous during fetal development, because of increased bone turnover during pregnancy and, as a result, fetal exposures to higher blood lead levels. Such exposures can cause behavioral abnormalities and decrease the survival of offspring. Bird species, especially waterfowl and birds of prey, can build up lethal lead concentrations in their bodies through ingesting lead buckshot.

Mercury is also a potent nerve toxin that can result in impaired coordination, responsiveness, hearing, vision, smell, and cognition. It can also damage sperm production. Mercury poses a risk to organisms when it is converted by aquatic bacteria from elemental mercury, which is essentially not biologically available, to an "organic" form known as methylmercury that is, and that can remain in tissues for months. In fish and other aquatic organisms, methylmercury can alter behavior; impair growth, development, and productivity; and result in death. Methylmercury is also highly persistent in the environment and can bioaccumulate, becoming progressively more concentrated as it moves up the food chain. In some fish at the top of the food chain, mercury concentrations can reach levels that are more than one million times those found in their surroundings. Organisms everywhere on Earth, from Arctic Ringed Seals (*Phoca hispida*) and Hong Kong's Humpback Dolphins (*Souse chinesis*) to Canadian birds of prey and Lake Victoria fish, have been found to have harmful mercury levels in their tissues. Indeed, any fish-eating animal, including humans, runs the risk of experiencing mercury toxicity.

In addition, mercury can affect organisms that form the base of the food chain, such as microbes whose respiration can be impaired. Such findings prompted a European Commission working group on mercury to state, "There are strong indications that the present concentrations of mercury over large areas in Europe are increased to levels that may affect the decomposition of organic matter and have an adverse effect on the recycling of important nutrients."[81]

Mercury may have other worrisome effects on ecosystems as well. It can induce chromosomal abnormalities, and it may be detrimental to plants, not only by inhibiting the ecosystem services provided by some plant-associated bacteria, but also by altering the composition of chlorophyll molecules, replacing magnesium at their centers, thereby disrupting photosynthesis.

Herbicides and Pesticides

Each year, according to estimates from the U.S. Environmental Protection Agency, 5.7 billion pounds (about 2.6 billion kilograms) of pesticides are released into the global environment.[82] Clearly, these chemicals have been extremely important to agriculture and public health. However, pesticides are, by their very nature, toxic substances and thus can cause harm to many species, including so-called "nontarget organisms," that come into contact with them.

The most recognized scenario in which pesticides and herbicides sicken or kill non-target organisms involves the local release of a chemical in sufficient quantities to be acutely toxic to inadvertently exposed organisms. In the 1990s, for instance, the pesticide monocrotophos, intended to control grasshoppers and other pests on alfalfa crops, was responsible for the death of more than 6,000 Swainson's Hawks in Argentina.[83]

However, more subtle effects of these chemicals on wildlife have become apparent in recent years. The herbicide atrazine, widely used in the United States (~75 million pounds are applied each year) but banned in seven European Union countries, has been shown to change the sex of Leopard Frogs (*Rana pipiens*) and to slow their gonadal development at levels of only 0.1 parts per billion, a concentration that is found in rainwater essentially everywhere in the United States.[84] In lakes, ponds, rivers, and streams that receive runoff from land where atrazine has been applied, concentrations can reach significantly higher levels. There are great concerns about the impacts of this herbicide on exposed amphibian populations (see the section on amphibians in chapter 6).

Plastics

Because they are cheap, durable, waterproof, and light, plastics have become the major material for producing consumer goods in the last twenty years. Worldwide in 2005, more than 100 million tons (~250 billion pounds) of plastic were manufactured.[85] An estimated one million tons are dumped into the oceans every year. Surveys of debris that washes up on coasts around the world demonstrate that plastics make up most of the marine litter worldwide, usually representing 60 to 80 percent of all garbage that washes ashore.[86]

Plastics are particularly lethal to marine animals. Fishing nets, ropes, and packaging bands such as six-pack holders can get entangled in gills and other body parts, causing strangulation. Plastic bags and balloons can be ingested by sea turtles and pilot whales, which mistake them for jellyfish or squid. Styrofoam, which crumbles into small particles in the ocean that resemble fish eggs, are often fed to the chicks of marine birds. Plastics can also concentrate toxic substances such as PCBs and pesticides and release them when ingested.[87] Kofi Annan, former Secretary General of the United Nations, stated that marine litter, the majority of which is composed of plastics, "is killing up to a million seabirds and 100,000 sea mammals and turtles each year."[88]

Virtually no marine animal is safe from plastics in the world's oceans, and some species, such as the Laysan Albatross (*Diomedea immutabilis*), have been particularly hard hit. In a survey of 251 Laysan Albatross chicks both dead and alive, only six did not contain plastics. Among those that had died, the average burden was 24 grams, or approximately 1 percent of their body weight.[89]

Figure 2.12. Plastics Ingested by a Laysan Albatross (*Diomedea immutabilis*). Laysan's beaches (Laysan is an island at the tail of the Hawaiian archipelago) are covered with plastics, including chemical lightsticks that fishermen use to attract swordfish and tuna, beads, fishing line, buttons, checkers, disposable cigarette lighters, toys, PVC pipe and other PVC fragments, golf tees, dishwashing gloves, and felt-tipped markers. The vast majority come from Asian countries more than 2,000 kilometers (about 1,243 miles) away. (Courtesy of Steven Siegel, Marine Photobank.)

Microscopic plastic pieces and fibers, concentrations of which have tripled in the environment since the 1960s, are a newly recognized and growing concern. They are showing up in beach and seabed sediments and are being consumed by zooplankton and other organisms at the base of the marine food web.[90]

ULTRAVIOLET RADIATION

Before around 500 million years ago, life was confined to the oceans, as ultraviolet (UV) radiation from the Sun, which primarily damages DNA but can also affect the stability of proteins and cell membranes, passed through Earth's atmosphere in doses sufficient to prevent life from colonizing the land. Seawater acts as a partial UV filter, protecting marine life. When blue-green algae in the oceans evolved photosynthesis, a process that produces sugars, energy, and oxygen from water, carbon dioxide, and sunlight, the oxygen they produced began to form the ozone layer in the stratosphere with the help of UV radiation (UV catalyzes the conversion of atmospheric oxygen [O_2] to ozone [O_3]). Sometime after 500 million years ago, when the amount of oxygen in the atmosphere was still only a fraction of what it is today, the ozone layer had grown sufficiently dense to block enough UV from striking Earth's surface that life could exist on land. It is thought that plants called psilophytes were the first settlers on the land around 450 million years ago. Animals, such as springtails (Collembola), minute invertebrates that mostly live in the soil, did so at a later time, with the earliest known fossils dated to about 400 million years ago (for more about Collembola, see chapter 8, page 346).

This bit of history captures the importance of solar UV radiation, and of the ozone layer that filters it, to terrestrial life on Earth. Ozone depletion that began in the mid-twentieth century led to the formation of a markedly thinned ozone layer over Antarctica (metaphorically called an "ozone hole"), with lesser thinning over the Arctic, and still less over other parts of Earth. The Montreal Accords, perhaps the most successful international environmental treaty ever put into effect, has strictly limited or banned outright most chemicals that destroy the ozone layer, such as chlorofluorocarbons (CFCs), putting it on a path to recovery. The original prediction was that full recovery would be achieved by around 2050, but emerging evidence suggests that global warming may delay the recovery.[91] Furthermore, some chemicals, such as the potent, broad-spectrum agricultural pesticide methyl bromide, known to damage the ozone layer, are still heavily used in such countries as the United States, and there are parts of the developing world, such as China and India, where ozone-damaging CFCs are still being widely employed in air conditioning.

UV radiation can have a host of detrimental effects on organisms. It can impair the ability of phytoplankton and other marine plants to reproduce and photosynthesize (some produce increased amounts of pigments to protect themselves from higher UV levels), and because these organisms lie at the base of the food web, high levels of UV may reduce biomass in the oceans, though this has yet to be documented. UV has direct effects upon organisms that are higher up on the food chain, as well, such as amphibians (see chapter 6, page 207) and fish. Fish eggs and larvae are susceptible to UV damage, but what, if any, impacts this is having in the wild remains unclear. Plants, in general, tend to be smaller and have less developed root systems with increased UV exposure. Studies with maize and *Arabidopsis thaliana*, a member of the mustard plant family that includes cabbages and radishes, have demonstrated that UV-induced DNA damage can be passed on, and even amplified, in offspring.[92] UV radiation is also well known to cause skin cancer and cataracts in some domestic animals, and presumably in some wildlife, and to impair immune system function, all of which can also be seen in humans.[93] In sum, UV radiation is harmful to most life, and its levels that reach Earth's surface have risen because of stratospheric ozone depletion, but we have only begun to see and understand how UV will affect biodiversity on a global scale.

WAR AND CONFLICT

In reviewing the causes of species endangerment and ecosystem disruption, scientists rarely mention war and conflict. Yet their effects can be devastating to populations of wildlife and to already threatened environments such as tropical forests. An estimated 160 wars have been fought in the past sixty years, the majority of which have been regional conflicts among various political, religious, tribal, or ethnic factions, rather than wars between nations. Large displacements of human populations have resulted, with refugees, sometimes numbering in the hundreds of thousands, moving into areas where they attempt to "live off the land," practicing slash-and-burn agriculture and engaging in extensive deforestation and large-scale hunting of wildlife.[94] Wars also result in the proliferation of land mines and unexploded munitions (some dating from World War II) that endanger wildlife and domestic animals, as well as people, and may result in widespread damage to ecosystems when they explode. As many as 100 million land mines may still be active (and may remain active for decades), deployed around the globe, with such countries as Cambodia carrying an unusually heavy burden, having an estimated six to ten million scattered across its territory.[95]

Chemical pollution is another outgrowth of war, either from spills or from deliberate release, such as occurred during the Vietnam War. From 1961 to 1969, the U.S. military sprayed an estimated 100,000 tons of highly concentrated defoliant chemicals containing dioxins on the forests and croplands of Vietnam, Laos, and Kampuchea (Cambodia) to strip away vegetation and expose enemy troops, and to destroy crops.[96] Dioxins, potent carcinogenic and endocrine-disrupting toxins that may last for decades in the environment, have been associated with marked declines of wild carnivores, ungulates (hooved animals, including oxen), and elephants in such countries as Vietnam.[97]

We mention here two recent examples of the heavy costs of wars and conflict on species and ecosystems. The first involves the conflict in Iraq over the past seventeen years. In this case, the usual environmental impacts of war—from the original 1991 Gulf War and from the current war that began in 2003—as a result of fires and explosions; the disturbance of fragile desert habitat by trucks, tanks, and other heavy equipment; and the various chemicals associated with modern-day warfare that are released into the environment were greatly added to by Saddam Hussein's destructive actions. In the 1991 war, Iraqi troops set fire to more than 600 Kuwaiti oil wells and to trenches filled with oil, releasing great plumes of black smoke[98] that were thought to have disoriented some of the estimated one billion migrating ducks, geese, gulls, and cranes on their way back to Eastern Europe.[99] In addition, the oil that was also deliberately spilled by Hussein into the waters and marshlands of southern Iraq killed an untold number of birds, as well as other aquatic wildlife. But the single greatest environmental disaster wrought by Saddam Hussein was his draining and destruction of more than 90 percent (an estimated 5,200 square miles or roughly 13,500 km^2) of the ancient marshlands that lie between the Tigris and Euphrates Rivers, the largest wetland ecosystem in the Middle East and western Eurasia (nearly twice the size of the original Everglades in Florida) and home to the Marsh Arabs, descendents of the 5,000-year-old civilization of the Babylonians and Sumerians. The marshes were also the permanent home of millions of birds, including rare species such as the Marbled Teal (*Marmaronetta angustirostris*) and the Basrah Reed Warbler (*Acrocephalus griseldis*),[100] as well as a stopover for millions more migrating from Siberia to Africa[101] and a major spawning ground for fisheries. Although recovery efforts have begun, it is believed that populations of many species were severely affected, if not wiped out, with the loss of the wetlands. The full magnitude of this environmental disaster, however, has yet to be determined.

The second example has to do with civil wars in Africa over the past twenty-five years, which have had disastrous effects on wildlife populations in some areas. In these cases, large numbers of heavily armed military and guerrilla forces have occupied national parks and wildlife preserves, where they poached wildlife for food or used them as items of trade for arms and ammunition or for other goods or services. In the 1979 civil war in Uganda, for example, a great many African elephants and other large mammals were killed.[94] The same was true during the Rwandan civil war from 1990 to 1994 and during its aftermath, when there was widespread hunting of buffalo and antelopes.[102] The recent wars and unrest in the Democratic Republic of Congo (DRC) have contributed to markedly reduced wildlife populations in several protected areas. Of particular concern has been the poaching there of Northern White Rhino (*Ceratotherium simium cottoni*) in Garamba National Park (by poachers from Sudan), elephant in many areas including Salonga National Park, and Grauer's Gorilla (*Gorilla beringei graueri*) in Kahuzi-Biega National Park, where uncontrolled mining of coltan (a rare mineral used in cell phones and other electronic devices) has led, in addition, to the opportunistic poaching of elephants for meat and ivory.[103] Bonobos have also been extensively poached in the DRC.[104] And over the past two decades, secondary to various wars, populations of African elephants in Angola, Mozambique, the Sudan, Somalia, and Uganda have all declined significantly.[94]

Armed conflict often disrupts the livelihoods of rural communities. If they can no longer farm, they become much more dependent on natural resources, including wildlife, during and after conflicts, placing a further burden on the environment. Besides direct impacts on wildlife and habitats, wars can have serious impacts on infrastructure important for conservation, and often can drastically reduce conservation capacity. It sometimes takes years after a conflict is over to rebuild human capacity and reestablish donor confidence, and yet it is during these transition times immediately following a conflict, when access is reopened to areas rich in natural resources, that environmental damage is most often done. It is also at these times that governments and local communities are often weakened and unable to control unsustainable exploitation.[105]

Military preparation by itself is also a factor in the loss of biodiversity. In the United States, for example, as well as in other countries, military bases and other installations have a history of polluting the air, soil, and groundwater with toxic chemicals, including radioactive waste, and of damaging natural ecosystems with bombing and construction. All are likely to threaten native species. It is not at all reassuring that the U.S. Department of Defense, on the grounds of national security, has been exempted by the U.S. Congress from provisions of the Endangered Species Act, the Marine Mammal Protection Act, and the Migratory Bird Protection Act.[106,107] Some have argued that the definition of national security must include broad protections of the environment and its ecosystem functions.

GLOBAL CLIMATE CHANGE

Earth's climate is determined by complex interactions that involve the Sun, oceans, atmosphere, land, and living things. The composition of the atmosphere is particularly important because certain gases (including water vapor, carbon dioxide, methane, halocarbons [carbon compounds containing the halogen elements chlorine, fluorine, and/or bromine], ozone, and nitrous oxide) absorb heat originating from the Sun that has radiated back from Earth's surface. As the atmosphere warms, it in turn radiates heat back to the surface, much like the enclosing glass does in a greenhouse, to create what is commonly called the "greenhouse effect." This is a natural phenomenon. If it did not occur, Earth's temperature would be much like that on Mars, which has an average surface temperature of approximately −63 degrees Celsius (−81 degrees Fahrenheit), rather than what it is, around 14.6 degrees Celsius (about 58 degrees Fahrenheit). Changes in the composition of the atmosphere alter the intensity of the greenhouse effect. Such changes, which have occurred many times in the planet's history and are the result of natural cycles related to Earth's rotation about its axis and its orbit around the Sun, have helped determine past climates and will affect future climates, as well.

Through changing the composition of the atmosphere and extensively modifying the land surface, humans are exerting a major and growing influence on the planet's climate. No meaningful debate remains in the scientific community about our role in causing changes to Earth's climate. For more than 600,000 years (and perhaps for millions of years) before the latter half of the nineteenth century, when the Industrial Revolution began in earnest, the concentration of carbon dioxide (CO_2) in the atmosphere did not exceed 280 parts per million by volume (abbreviated as ppmbv). During the last 150 years or so, however, CO_2 levels have risen by more than 35 percent, such that in 2008, they were more than 380 ppmbv. This increase has resulted from the burning of coal, oil, and natural gas, which releases CO_2 into the atmosphere, and from the destruction of forests around the world, which serve to take it up. Rising concentrations of CO_2 and other greenhouse gases are intensifying Earth's natural greenhouse effect. Global projections of population growth and assumptions about future energy use indicate that the CO_2 concentration will continue to rise, likely reaching between two and three times its mid-nineteenth-century level by 2100, unless a dramatic change in the trajectory of the world's energy consumption occurs. This doubling or tripling will occur in the space of somewhat more than 200 years, a brief moment in Earth's geological history.[108]

As we add more CO_2 and other heat-trapping gases to the atmosphere, the world is becoming warmer. This changes other aspects of climate as well. Historical records of temperature and precipitation have been extensively analyzed in many scientific studies. These studies demonstrate that the average global surface temperature has increased by more than 1 degree Fahrenheit (0.6 degree Celsius) since the latter part of the nineteenth century. About half of this rise has occurred since the late 1970s. Nineteen of the twenty warmest years since global temperatures were first accurately recorded in 1856 have occurred since 1980. In 1998, the average global surface temperature set a new record by a wide margin, exceeding that of the previous record year, 1997, by about 0.3 degree Fahrenheit (0.2 degree Celsius). This 1998 record was surpassed in 2005. Higher latitudes have warmed more than equatorial regions, higher altitudes have warmed more than those at sea level, and nighttime temperatures have risen more than daytime temperatures.[108]

With the warming of Earth, more water evaporates from the oceans and lakes, eventually to fall as rain and snow. During the twentieth century, annual precipitation increased about 10 percent in the middle and high latitudes. The increased evaporation and precipitation has caused more torrential rains and flooding in some areas, and droughts in others. The warming is also causing permafrost to thaw and is melting sea ice, snow cover, and mountain glaciers. Global sea level rose 4 to 8 inches (10 to 20 centimeters) during the twentieth century, because ocean water expands as it warms and because melting glaciers are adding water to the oceans. There is also growing evidence, although not conclusive, that global warming has contributed to the increased strength and duration of El Niño events over the past decade.[109,110]

According to the Intergovernmental Panel on Climate Change, or IPCC, co-recipient of the 2007 Nobel Peace Prize (a group of more than 2,000 of the world's leading scientists that surveys the scientific literature to assess Earth's present and future climates—their reports are considered the benchmark for predictions of climate change), scientific evidence confirms that human activities are the cause of a substantial part of the warming experienced over the twentieth century. New studies indicate that temperatures in recent decades are higher than at any time in at least the past 1,000 years.[111] It is very unlikely—the IPCC Fourth Assessment Report says the chances are between 1 and 10 percent[108]—that these unusually high temperatures can be explained solely by natural climate variations. In this chapter, and in the book as a whole, we refer to these human-caused changes in Earth's climate both as "global warming" and "global climate change," or at times, just "climate change." A more accurate phrase would be "global warming with associated changes in global climate."

With warming of the global climate over the last century, animals and plants have responded in many ways, some of which have already been alluded to in this chapter and are presented in others. Plants leaf out or flower earlier, migratory birds arrive earlier in the spring, and species ranges move toward the poles or to higher altitudes. Some ecosystems, such as alpine meadows, cloud forests, arctic tundra, and coral reefs, are especially sensitive to warming, and species in these regions may be particularly at risk.

Average global surface temperatures are expected to increase further from 1.1 to 6.4 degrees Celsius (around 2 to 11.5 degrees Fahrenheit) by the year 2100. The magnitude and the rate of this increase, unprecedented for the last 10,000 years, will threaten the survival of many species, especially those unable to migrate to new ranges or otherwise adapt. Global climate change, by itself or acting synergistically with other environmental changes secondary to human activity, could well become the factor most responsible for species extinctions over the next 100 years. A recent study, for example, looking at a sample of some 1,000 species from terrestrial regions from Mexico to Australia, representing a total of 20 percent of the planet's land area, has predicted that given the most probable climate-warming scenarios, about one-quarter of these (the range is 15 to 37 percent) will be at risk of extinction by the year 2050.[112]

In this section we look at some models and provide a number of examples of how global warming and the associated changes in global climate are now affecting, and how in the future they might affect, species extinctions around the world. There is no topic more important for us to consider, because these human-caused global changes are accelerating rapidly and may become the main driver of species loss and ecosystem disruption in coming decades. The reader should understand that the examples given below have occurred as the result of an average warming of global surface temperatures of only around 1 degree Fahrenheit during the past 150 years or so, and that the upper limit of warming predicted by the IPCC for the year 2100 is more than ten times this amount. To appreciate how enormous this change would be, consider that this degree of average warming for the surface of Earth, more than 10 degrees Fahrenheit, matches that which has occurred since the end of the last ice age, about 18,000 thousand years ago, when significant areas of North America, Europe, and Asia were covered by glaciers of ice one mile thick.

A more comprehensive review of climate change and species loss than what is given below is not possible here but can be found in such sources as *Climate Change and Biodiversity* listed in the Suggested Readings at the end of this chapter.

Range Change on Land

As the climate warms, some species will need to migrate to higher latitudes or higher altitudes to find habitats in which they can survive. Some mobile species, such as various types of birds and butterflies, will be able to change their ranges fast enough to keep up with temperature and other climatic changes. An analysis of some species of birds and butterflies, for example, has shown that, on average, they have moved their ranges toward the poles by around 6 kilometers (3.75 miles) per decade since 1960.[113,114] Other less mobile species will not be able to move fast enough, will find significant barriers in their way—such as roads, cities, and farms—or will have no place to go to escape the changing climate if they, for example, already live near the poles or at the tops of mountains. Many species will be lost as a result.

Two examples illustrate the impacts of climate change on species' ranges. The first involves the distribution of a beautiful butterfly that lives in southwestern parts of the United States, Edith's Checkerspot Butterfly (*Euphydryas editha*), a species that has been studied by many researchers for long periods of time. What has been found is that the butterfly has shifted its range northward and to higher altitudes over several decades. Despite their mobility, the butterflies are still at risk, because at the northern edge of their range they are being threatened by habitat loss from human activity, and at the southern edge they are being wiped out by warming temperatures and drier conditions that have significantly reduced populations of the plants they feed upon, causing them to starve. They are caught in an extinction vise, squeezed from the north by habitat loss and from the south by climate change.[115]

The second involves vascular plants in the Alps (vascular plants are those that have specialized water-carrying tissues and include, among other groups, ferns, flowering plants, and gymnosperms such as conifers). For more than ninety years, European botanists have been studying these plants, so there was a record for comparison when scientists began to survey their present-day ranges. In 1994, researchers from the University of Vienna demonstrated that at more than two-thirds of the sites they studied, vascular plant species had moved their ranges up the mountains by an average of 4 meters (around 13 feet) per decade over the past 70 to 90 years, in response to an average regional warming on these mountains of 0.7 degree Celsius (or about 1.3 degrees Fahrenheit). The clear implication was that those species that had reached the summit would have no place to go if there were further warming, and if they could not adapt and were found nowhere else, they would become extinct.[116]

However, things are far more complicated than the ability of an individual species to migrate. The impacts of climate change can involve cascades of events that, given our often very basic level of understanding of biological systems, are enormously difficult to anticipate. Species do not exist in isolation. They are parts of ecosystems in which they depend on a wide range of other species for their survival, for example, on those species that make up their food supply. They are also highly affected by the presence of predators and of those species with which they compete. And, of course, these food species, predators, and competitors are themselves dependent on an enormous array of other species. Because different species will have differing migration responses to climate change, there will be a consequent breakdown in relationships among species and in the functions of the ecosystems they comprise, in both the new and old environments, threatening many species with extinction.

Similar range shifts accompany warming ocean temperatures for those marine species able to migrate. Seeking colder waters, they have moved toward the poles or to greater depths. In one study, the distributions of nearly two-thirds of the dozen North Sea fish species that were examined, including both commercially important fish such as Atlantic Cod (*Gadus morhua*), Haddock (*Melanogrammus aeglefinus*), and the Common Sole (*Solea solea*) and those not targeted by fisheries such as the Snakeblenny (*Lumpenus lampretaeformis*), have shifted in latitude or in depth over the past twenty-five years. Cold-water fish have moved farther north, while some warmer water fish have moved into the North Sea, whose waters have warmed by an average 0.6 degree Celsius (about 1.1 degrees Fahrenheit) from 1962 to 2001. What was of concern to the researchers was that the fish that changed their ranges were generally those with shorter life cycles, and that some of these were already threatened by being overharvested commercially. There was also concern that as temperatures in the North Sea continue to warm, resident species will respond differently in changing the timing of their biological cycles and in migrating northward, altering in unpredictable ways the composition and species interactions in ecosystems that are already highly stressed by human activity.[117]

Warming sea surface temperatures cause coral reefs worldwide to lose their symbiotic algae and appear "bleached," making them vulnerable to various fatal infections (see also discussion of infectious diseases, above). When ocean temperatures exceed the mean summer maximums by as little as about 1 degree Celsius (1.8 degrees Fahrenheit), bleaching invariably occurs. The acidification of ocean waters from CO_2 also threatens coral reefs. Because these ecosystems are among the most diverse of all marine habitats, their loss would threaten the survival of countless organisms.

Warming oceans may also threaten species in a variety of other ways, such as by contributing to the formation of harmful algal blooms (in association with excessive nutrient discharge from sewage or agricultural runoff), for example, red tides that expose some species, such as the West Indian Manatee (*Trichechus manatus*) or the Red Snapper (*Lutjanus campechanus*), to their lethal toxins.

But perhaps the issue of greatest concern is that warming oceans may interfere with the upwelling of deep, nutrient-rich waters, containing nitrogen, phosphorus, and iron, in many parts of the world, which are essential for the growth of phytoplankton.[118–120] Phytoplankton (photosynthetic, mostly microscopic organisms) are at the base of the marine food chain, and losses in their populations could have dire consequences for all marine life. For example, krill, tiny shrimplike organisms that feed on phytoplankton and other zooplankton like themselves, would be affected. Occupying all the world's oceans, krill are able to convert the photosynthetic energy from phytoplankton into a form that many marine organisms up the food chain depend on for their survival, including numerous species of fish, squid, penguins, seals, and baleen whales. Krill populations have been falling in many parts of the world as oceans have warmed, for example, by 80 percent since the 1970s over the entire southwest Atlantic sector, and so have a variety of other marine organisms that eat krill.[121,122] The western coast of the United States has also been severely affected, presumably from warming oceans, with significant declines in krill and other zooplankton, in schooling baitfish such as Pacific Herring (*Clupea pallasii*) and Surf

Smelt (*Hypomesus pretiosus*), and in some seabirds that feed on these fish, such as Brandt's Cormorant (*Phalacrocorax penicillatus*) and the Common Murre (*Uria aalge*). These birds washed up by the tens of thousands in the states of Oregon and Washington in the spring of 2005, with evidence that they had starved to death.[123]

Phytoplankton and krill are also major carbon sinks, taking up carbon dioxide from the atmosphere and depositing it in long-term stores on the ocean floor, where it can remain undisturbed for thousands of years.[124] The warming of oceans from global climate change, by reducing phytoplankton and krill populations, can therefore become a major positive feedback for greenhouse warming.

MELTING SEA ICE

As described in chapter 6 (see page 225), the melting of sea ice from warming sea surface temperatures threatens the survival of Polar Bears in the Arctic. This occurs by starvation, because seals, their major food, are more easily able to avoid capture by the bears when they surface for air if there are larger areas of open water. With a loss of nutrition, polar bears have lower reproductive rates, and their cubs are less able to survive to adulthood. Melting sea ice will also affect Polar Bear populations, because the number of Ringed Seals (*Phoca hispida*), their main food, is likely to drop as temperatures continue to warm. Ringed Seals are born and suckled in the early spring on the ice, and prior to weaning, which usually occurs when the ice begins to break up, they depend on intact ice sheets as platforms to rest on between dives and to hide from land predators in dens under the snow. An added threat to the seals is that with warming temperatures, the roofs of their dens are more likely to collapse, exposing seal pups to predators such as Arctic Foxes (*Alopex lagopus*) and Polar Bears. Early melting of sea ice also forces the pups to swim before they are able to protect themselves from marine predators or to withstand the stress of prolonged exposure to cold waters, and warming ocean temperatures bring in new subpolar species that the seals are not adapted to feed on.[125] Because Ringed Seals, the most abundant of all Arctic seals, are considered a keystone species, losses in their populations are likely to threaten many other organisms in Arctic ecosystems.

Melting ice in the Antarctic will also exact a heavy toll. Some Antarctic penguins, such as Adélie Penguins (*Pygoscelis adeliae*), are highly threatened by retreating sea ice, because krill abundance depends on the amount of winter sea ice cover (as well as on the amount of phytoplankton available for the krill to eat), and krill are the main source of food for Adélies.[121] The largest increases in temperatures on the planet, a warming of some 6 degrees Celsius (almost 11 degrees Fahrenheit) over the past fifty years has been recorded in the western Antarctic Peninsula, where at some study sites Adélie populations have crashed by nearly 70 percent over thirty years as krill have declined. Adélies are also threatened by increased snowfall in parts of the Antarctic, a consequence of climbing ocean temperatures, that can waterlog their eggs when the snow melts, killing the chicks inside.[122]

Emperor Penguins (*Aptenodytes forsteri*), the beloved stars of the movie "March of the Penguins," have also declined in population over the past fifty years, with the declines being linked to an unusually warm period in the Antarctic during the 1970s. Emperors mainly feed on Antarctic Krill (*Euphausia superba*), fish (mainly the Antarctic Silverfish [*Pleuragramma antarcticum*]), and squid, all of which are reduced in population size when there is less sea ice. Although reduced sea ice tends to increase egg-hatching success in Emperor Penguins, because the distance they have to march to get to feeding grounds is also reduced, the availability of food is thought to be the critical factor in Emperor Penguin survival.[126] Further warming in the Antarctic is likely to result in still larger Emperor Penguin population losses.

Ten of the world's seventeen penguin species are listed as threatened in the IUCN's 2006 Red List of Endangered Animals; all of these ten live in the Southern Hemisphere. Global climate change is endangering this group of seabirds, which have been around since at least fifty to sixty million years ago.

Acidification of Seawater

The release of CO_2 into the atmosphere by human activity affects marine ecosystems not only by greenhouse warming as described above but also by changing the acidity of ocean waters. This has become a source of growing concern.

One measures acidity by what is called a pH scale, where pure water has a neutral pH of 7. Recordings lower than 7 mean that what is being measured is acidic, with each drop by one unit referring to a tenfold increase in the concentration of hydrogen ions, while those higher than 7 mean that it is alkaline, with each unit increase standing for a tenfold decrease in hydrogen ions. The pH of seawater measures from around 8 to 8.3, so the oceans are naturally somewhat alkaline. In the past few decades, only about 50 percent of the CO_2 released by human beings has stayed in the atmosphere. Of the remainder, about 20 percent has been taken up by terrestrial plants, and about 30 percent by the oceans.[127] When CO_2 is absorbed by seawater, it forms carbonic acid, as it does in carbonated soda water, by the chemical reaction $CO_2 + H_2O \rightarrow H_2CO_3$.

After a period of pH stability that has lasted for hundreds of thousands of years, human activity has now made the oceans more acidic (compared to preindustrial times) by a pH change of about 0.1 units. This may not seem like much, but if human CO_2 emissions, mainly from the burning of fossil fuels, continue at the rate predicted by the IPCC, the pH will become even more acidic, by an additional 0.3 pH units by the year 2100, a change in the acidity of the oceans that has not occurred for more than twenty million years, and one that would take many thousands of years for natural processes to reverse.[128]

The implications for marine life and for humans are potentially catastrophic. Greater acidity will reduce the concentration of carbonate ions in seawater, interfering with the ability of some marine organisms to form skeletons, shells, and other hard body parts out of calcium carbonate ($CaCO_3$), threatening their survival. These include some of the most abundant, diverse, and important life forms in the oceans, such as reef-forming corals, crustaceans, mollusks, and certain plankton such as the foraminifera and the photosynthetic coccolithophores that are at the base of the marine food chain. Calcareous green algae, echinoderms, and bryozoans may also be affected. Coccolithophores form giant blooms in the spring and summer before sinking to the ocean bottom, carrying with them absorbed carbon, both as sugars from photosynthesis and as calcium carbonate in their microskeletons. Here is another potential positive feedback loop: Higher atmospheric carbon dioxide levels, by making oceans more acidic, could reduce coccolithophore populations (by interfering with their skeletal formation), thereby reducing a major marine CO_2 sink and leading to still higher atmospheric CO_2 concentrations.

In addition, much as a piece of chalk, which is made from calcium carbonate, disintegrates in a glass of vinegar (a mild acid), so too will the calcium carbonate shells and skeletons of many marine organisms when certain levels of acidity are reached. Corals, pteropods (small marine snails that are key elements of the food chain), corraline algae (commonly found in reef communities), and organisms in the cold, deep ocean waters of high latitudes may be the most at risk.

Disruption of Biological Cycles

The potential impact of global climate change on the timing of such biological events as the arrival and departure times for migratory birds, the breeding of amphibians, or the flowering of plants is a subject of great research interest. It is also a topic that has generated very serious concern in the scientific community because of the potential risks in disrupting natural cycles and of decoupling relationships among different species that have co-evolved over many thousands of years. The problem is that one species may alter its timing more or less than another on which it depends, each one perhaps responding to different environmental cues that serve to set its biological clocks. For example, if a migratory bird arrives at its spring habitat earlier than its usual time because of an unusually warm winter, but its primary food species is not yet present, it may starve. If a flower opens too early in response to warming temperatures, but the main species that pollinates it has not yet developed, it may not be pollinated, and its seeds may not form.

The Great Tit (*Parus major*), a bird that is widely distributed in woodlands across Europe and Asia, has been extensively studied at the De Hoge Veluwe National Park in the Netherlands over the past twenty-five years. During this time, mid-spring temperatures (between April 16 and May 15) in the park have warmed by about 2 degrees Celsius (3.6 degrees Fahrenheit), altering the timing of some biological cycles more than others. While there has been no change on average from 1980 until 2004 in the time of year that Great Tits lay their eggs, in when their eggs hatch, or in when chicks are first able to fly and leave their nests, the cycles of Winter Moth caterpillars (*Operophtera brumata*), a major source of food for Great Tit chicks, have changed. The caterpillars now reach their peak of growth on average two weeks earlier than they did in 1980, so now it is only the earliest Great Tit chicks that are adequately fed. Winter Moth caterpillars, in turn, feed on young, tender oak leaves, and "bud burst" (when the oak leaves first open) in De Hoge Veluwe National Park has occurred on average ten days earlier in 2004 than in 1980. One would expect Great Tit and Winter Moth populations to decline given these changes, and indeed, measurements seem to indicate that this is happening, but it is not yet clear whether these declines are part of natural cycles or whether they are caused by climate-change triggered decouplings.[129]

The effects of biological cycle disruptions have gone so far as to select for organisms that may be genetically better suited to the new temperature conditions. For example, Pitcher-Plant Mosquitoes (*Wyeomyia smithii*) in eastern North America over the past thirty years may have undergone a genetic shift in their response to day lengths (so that they enter winter dormancy at the right time) as a consequence of global warming.[130] Likewise, North American Red Squirrels (*Tamiasciurus hudsonicus*) in the Yukon have advanced the time of their spring breeding during the last decade, the result, it is believed, of genetic selection in response to warming temperatures.[131] Other species may also be able to evolve rapidly and adapt to global warming, and this possibility has been encouraging to some conservation biologists, because these species, and perhaps many others, might thereby dodge extinction. But no one knows whether Pitcher-Plant Mosquitoes and North American Red Squirrels are highly unique in their ability to adapt, or whether this may be a more widespread phenomenon, and no one knows whether these genetic adaptations will actually decrease the risk of extinction, either for these species or for others with which they interact.[132]

Complex Interactions

The examples given in this section involve interactions among several factors, acting in combination with climate change or triggered by its effects, that together serve to threaten species. As may be obvious from the discussion above, the same is most likely also true for other, and perhaps for most, examples in which global climate change contributes to species declines and extinctions. We have included the examples that follow in a separate section only because the various factors involved may be somewhat better understood.

As discussed in more detail in the section on amphibian losses in chapter 6, some amphibians in the Cascade Mountains of the U.S. Pacific Northwest, such as the Western Toad (*Bufo boreas*), may be threatened by climate change because of how it leads to shallower lakes and ponds, thereby resulting in B. *boreas* tadpoles being exposed to more ultraviolet B radiation (UVB) and ultimately succumbing to infections caused by a fungus called *Saprolegnia ferax*. In this case, several factors seem to be affecting the toads: reduced precipitation secondary to climate change, increased UVB, and a lethal fungal infection. A similar story in thought to be the case for Harlequin Frogs in the Monteverde Cloud Forest Preserve in Costa Rica, also discussed in chapter 6. Here, the frogs were subjected to changes in temperature and humidity caused by climate change, creating optimal conditions for, and seemingly making the frogs more vulnerable to, infections caused by another fungus, *Batrachochytrium dendrobatidis*, which led to their demise.[133]

Climate change can affect pathogens and species survival in other ways, as well. For example, in the Arctic and sub-Arctic, warming temperatures have affected nematode parasites in Musk Oxen (*Ovibos moschatus*), such that now they complete their life cycles in one year instead of two. Increased infestations by nematodes are having a significant impact on Musk Oxen survival and on their ability to reproduce.[133]

Three examples of the effects on climate change on ecosystems and on the survival of large numbers of organisms that live in them are given below. All illustrate the difficulty of predicting the effects of climate change on biological systems.

- *Marine Dead Zones* (see chapter opening figure): In 1993, above-average rainfall during the first half of the year and torrential rains during the summer, events that will increase in number and intensity as the atmosphere continues to warm, caused the Mississippi River to overflow its banks, flooding some 9.3 million hectares (around 23 million acres) of farms, towns, and cities in nine U.S. Midwest states and washing into the river tons of industrial and agricultural chemicals as well as human and animal wastes. The size of the dead zone in the Gulf of Mexico, beginning at the mouth of the river, increased dramatically following these events, as a result of excessive nutrient and toxic chemical exposures (see page 52 for a more complete discussion of the formation of dead zones in coastal marine ecosystems). Many fish, mollusks, and countless other organisms perished.

- *Forest Pest Infestations*: Warming of the Kenai Peninsula in the state of Alaska over the past thirty years has affected the life cycle of the Spruce Bark Beetle (*Dendroctonus rufipennis*), allowing more to overwinter and speeding up their reproductive cycle to one year instead of two or three. Exploding populations of beetles have attacked White Spruce (*Picea glauca*) and Lutz Spruce (a White–Sitka hybrid) trees, resulting in the destruction of more than three million acres of spruce forests in the Kenai. The trees were also more vulnerable, because they were less able to mount their usual defense against the beetles—an outpouring of sap, which clogs the beetles' larval channels in the trees—secondary to drought conditions from the warming. (See chapter 6, page 251, for a map of these forest losses and for a more detailed discussion.) Large numbers of species are clearly threatened by such massive destruction of these forests and by the subsequent fires that ignited in the stands of dead timber.[134]

    There has been a similar effect due to warming on the life cycle of the Mountain Pine Beetle (*Dendroctonus ponderosae*) in Rocky Mountain pine forests in the western United States, with the beetle reproducing in one year instead of two. In this case, the beetle wreaks its destruction on the pines by transmitting a fungal disease called Pine Blister Rust (*Cronartium ribicola*).

- *Drought and Coastal Wetlands*: Prolonged droughts secondary to climate change may set the stage for stressors, which are relatively harmless when they occur in isolation, to act in concert and become deadly. For example, a severe three- to four-year drought in some southern parts of the United States beginning in 1999 resulted in an unprecedented die-off of salt marshes along more than 1,500 kilometers (about 940 miles), totaling more than 250,000 acres, of the Southeast and Gulf coasts. Experimental evidence suggests that the following sequence of events most likely transpired. The drought stressed the dominant plant in the marsh, Cordgrass (*Spartina alteriflora*), by causing marsh soils to be too dry, too salty, too acid, and perhaps also too concentrated in toxic chemicals (coastal wetlands are highly effective sponges for toxic compounds, including heavy metals, discharged by urban centers). At the same time, Periwinkle Snail (*Littoraria irrorata*) populations had increased, perhaps because of declines in populations of their major predator, Blue Crabs (*Callinectes sapidus*), and these snails feed on a fungus (genus *Fusarium*) that lives on Cordgrass. As they eat the fungus, the snails pierce holes into the Cordgrass, already weakened by drought, facilitating the fungus's infection of the plant, ultimately converting the salt marshes into mudflats.[135] Because salt marshes are among the most diverse of all coastal ecosystems, this massive die-off clearly affected large numbers of resident species.

In making the argument for the importance of preserving biodiversity, it has been necessary to begin with a discussion of the impacts of human activity on individual species. But as we discuss in chapter 3, this importance derives largely from the contributions that species make to the structure and functioning of ecosystems.

◆ ◆ ◆

SUGGESTED READINGS

Climate Change and Biodiversity, Thomas E. Lovejoy and Lee Hannah (editors). Yale University Press, New Haven, Connecticut, 2005.

Environmental Endocrine-Disrupting Chemicals: Neural, Endocrine, and Behavior Effects, Theo Colburn, Frederick Van Saal, and Polly Short (editors). Princeton Scientific Publications. Princeton, New Jersey, 1998.

Global Warming and Biological Diversity, Robert L. Peters and Thomas E. Lovejoy (editors). Yale University Press, New Haven, Connecticut, 1992.

The Great Reshuffling: Human Dimensions of Invasive Alien Species, Jeffrey A. McNeely (editor). International Union for the Conservation of Nature and Natural Resources, Gland, Switzerland, 2001.

A Green History of the World: The Environment and the Collapse of Great Civilizations, Clive Ponting. Penguin Books, New York, 1991.

How Many People Can the Earth Support? Joel E. Cohen. W.W. Norton & Company, New York, 1995.

Impacts of a Warming Arctic. Arctic Climate Impact Assessment, Susan Joy Hassol. Cambridge University Press, Cambridge, U.K., 2004.

Marine Conservation Biology: The Science of Maintaining the Sea's Biodiversity, Elliott A. Norse and Larry B. Crowder (editors). Island Press, Washington, D.C., 2005.

A Plague of Rats and Rubbervines: The Growing Threat of Species Invasions, Yvonne Baskin. Island Press, Washington, D.C., 2002.

Silent Spring, Rachel Carson. Houghton Mifflin Company, Boston, 1962.

Song for the Blue Ocean, Carl Safina. Henry Holt & Company, New York, 1998.

The Trampled Grass: Mitigating the Impacts of Armed Conflict on the Environment, J. Shambaugh, J. Oglethorpe, and R. Ham. Biodiversity Support Program, Washington, D.C., 2001; see www.bsponline.org.

The Unnatural History of the Sea, C. M. Roberts. Island Press, Washington, D.C., 2006.

CHAPTER 3

ECOSYSTEM SERVICES

Jerry Melillo and Osvaldo Sala

A human being is part of the whole, which we call the "Universe": a part limited in time and space. He experiences himself, his thoughts and feelings as something separated from the rest, a kind of optical delusion of his consciousness. This delusion is a kind of prison for us, restricting us to our personal desires and affection for a few persons nearest us. Our task must be to free ourselves from this prison by widening our circle of compassion to embrace all living creatures and the whole of nature in its astonishing beauty.

—ALBERT EINSTEIN, *Ideas and Opinions*

Earth's mosaic of ecosystems—forests, grasslands, wetlands, streams, estuaries, and oceans—when functioning naturally, provides materials, conditions, and processes that sustain all life on this planet, including human life. The benefits that all living things obtain from ecosystems are called "ecosystem services." Some are very familiar to us, such as food and timber that are essential for our lives and important parts of the global economy. What are equally important, but certainly less well recognized, are the array of services delivered by ecosystems that do not have easily assigned monetary values but that make our lives possible. These include the purification of air and water, the decomposition of wastes, the recycling of nutrients on land and in the oceans, the pollination of crops, and the regulation of climate.

Ecosystem services are generated by a complex of natural cycles, ranging from the short life cycles of microbes that break down toxic chemicals to the long-term and planetwide cycles of water and of elements such as carbon and nitrogen that have sustained life for hundreds of millions of years. Disruption of these natural cycles can result in disastrous problems for human beings. If, for example, the natural services that result in the control of pest populations ceased—that is, if the life cycle of some natural pest enemies were altered, or if they were eliminated in some areas—there could be devastating crop failures. If populations of bees and other pollinators crashed, society could face similar dire consequences. If the carbon cycle were badly disrupted, rapid climate change could threaten whole societies. We tend to take these services for granted and do not generally recognize that we cannot live without them, nor can other life on this planet.

The ecosystems of the world deliver their life-sustaining services for free, and in many cases, they involve such complexity and are on a scale so vast that humanity would find it impossible to substitute for them. In addition, we often do not know what

(left)
Hand Pollination of Apple Blossoms in Nepal. Bees in Maoxian County, at the border between China and Nepal, have gone extinct, forcing people to pollinate apple trees by hand. It takes twenty to twenty-five people to pollinate 100 trees, a task that can be performed by two bee colonies. (From Farooq Ahmad and Uma Patrap, International Center for Integrated Mountain Development, Nepal.)

species are necessary for these services to work, or in what numbers and proportions they must be present.

As the world's human population grows and nonsustainable, per-capita consumption of all kinds of materials increases, ecosystems are being degraded and their capacity to deliver their services is being compromised. The degradation of the world's ecosystems is a "quiet crisis," largely hidden from view, but the consequences of this degradation are potentially catastrophic for human beings.[1] In this chapter, we review the character of ecosystem services, present examples of current work that attempts to provide an economic valuation of these services, and discuss how human activity is threatening them.

THE CHARACTER OF ECOSYSTEM SERVICES

Ecosystems services can be divided into four major categories: provisioning, regulating, cultural, and supporting. For the purposes of this chapter and this book, we look at these services predominantly from a human perspective. *Provisioning services* are the products obtained from ecosystems and include foods and medicines. *Regulating services* are the benefits people obtain from ecosystem controls of climate, plant pests and pathogens, animal diseases (including those that affect humans), water quality, soil erosion, and much more. *Cultural services* are the nonmaterial benefits that people obtain from ecosystems: recreational, aesthetic, spiritual, and intellectual. And *supporting services* are those necessary for the production of all other ecosystem services and include the production of new organic matter by plants through photosynthesis (called "primary production") and the cycling of life-essential nutrients such as carbon, nitrogen, phosphorus, and other elements required for the chemistry of life.

PROVISIONING SERVICES	REGULATING SERVICES	CULTURAL SERVICES
Products obtained from ecosystems	Benefits obtained from environmental regulation of ecosystem processes	Nonmaterial benefits obtained from ecosystems
• food • fuel wood • fiber • medicines	• cleaning air • purifying water • mitigating floods • controlling erosion • detoxifying soils • modifying climate	• aesthetics • intellectual stimulation • a sense of place

SUPPORTING SERVICES

Services necessary for the production of all other ecosystem services
- primary productivity
- nutrient cycling
- pollination

Figure 3.1. A Sampling of Ecosystem Services.

BOX 3.1

Microbial Ecosystems: Editors' Note

Although the accepted concept of what an ecosystem is, and the one we use in this chapter and throughout this book—that an ecosystem is the sum total of all the organisms in a specific environment and their interactions with each other and with the nonliving components of that environment—includes microbes, scientists generally define ecosystems in macroscopic terms, and primarily by the plants and animals they contain, for these are what we see and what we know best. But it is becoming increasingly clear that most biodiversity on Earth is microbial, that microbes mediate many ecosystem services that sustain life,[a] and that possibly no multicellular organism exists without one or more microbial species living symbiotically on it and/or in it, some of which are necessary for its survival. (Symbiosis is the interaction between two organisms that are living together in an intimate association—this can be mutualism, where both organisms benefit from this interaction; commensalism, where one organism benefits and the other is not affected; or parasitism, where one organism benefits at the expense of the other.)

There is also a wider appreciation that there is another kind of symbiosis at work as well, one that involves the relationship between whole cells and what are called organelles within them, some of which had originally been independent organisms, such as chloroplasts. In converting energy from the sun by photosynthesis and in storing it, all plants depend on chloroplasts. It has become clear that chloroplasts were originally cyanobacteria, which over time, and on several different occasions, were incorporated as integral parts of early algal and plant cells hundreds of millions of years ago. Carl Woese, who devised the three-domain model for classifying life on Earth (see figure 1.4 in chapter 1), is credited with some of the early molecular work showing both chloroplasts and mitochondria to be bacterial in origin.[b,c] Chloroplasts possess their own DNA and replicate independently of the cells they inhabit, although much of their genome now resides within their host cell's nuclei. Starting in the 1960s, Lynn Margulis championed into widespread acceptance the theory that early prokaryotic organisms became organelles in eukaryotic cells, greatly expanding upon an idea that had first been proposed in the late nineteenth century by the German scientist Andreas Schimper.[d]

A similar story can be told about mitochondria, the energy factories that fuel almost all modern plant and animal cells. These mini power generators were originally primitive bacteria that similarly joined with larger cells and became essential plant and animal cellular

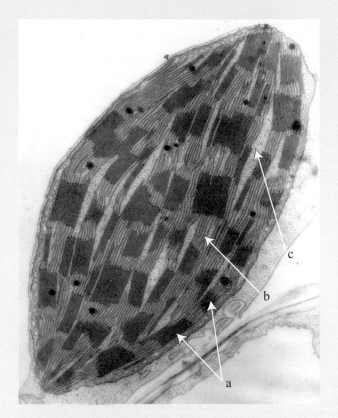

Chloroplast Inside a Higher Plant Cell: Electron Micrograph of a Chloroplast in Cross Section. (a) These flattened hollow disks, each of which is called a thylakoid, together form a stack called a granum. Chlorophyll molecules in the thylakoid membranes initiate the photosynthetic process when they absorb photons from sunlight. (b) Lamellae, the membrane structures that connect the thylakoids from granum to granum. (c) Stroma, the semifluid material that contains the chloroplast's DNA, as well as RNA and enzymes and is the site where carbon dioxide is transformed into glucose and where chloroplast proteins are made. (© Imperial College London; electron micrograph taken by A.D. Greenwood, early 1970s, Department of Botany, Imperial College; print given by J. Barber and A. Telfer; www.bio.ic.ac.uk/research/nield/expertise/chloroplast.html.)

organelles. Mitochondrial DNA reflects its bacterial origins: It is circular (whereas our eukaryotic DNA is linear), it uses a bacterial type of apparatus and code book to be translated into proteins, and its composition resembles bacterial DNA much more than it does that from multicellular organisms.[e] In spite of these differences, mitochondria have woven their way into the fabric of the cells they inhabit. As with chloroplasts, much of their DNA has been transferred to their cells' nuclei, a place that may be safer for genetic material to reside and where its replication can be completed with greater fidelity, and like chloroplasts, mitochondria replicate independently.[f]

One feature that is of great interest in animals that reproduce sexually is that because some of their mitochondrial DNA is confined to the cytoplasm, it is contained in the ovum but not in sperm. As a result, one can begin to trace the maternal lineage of an offspring by analyzing its mitochondrial DNA. This is being done to trace the origins of some human populations. Using mitochondrial DNA, for example, it has been possible to trace some 40 percent of all present day Ashkenazi Jews to four maternal lineages who lived in Europe 3000 years ago.[g]

The field that encompasses these insights about the interrelationships among microbial organisms and the multicellular organisms they inhabit, a field that owes much to the work of Thomas Brock (the discoverer of *Thermus aquaticus*—see chapter 5, page 179) in the 1960s and the Dutch microbiologist Martinus W. Beijerinck decades earlier, has been called "microbial ecology," and it is currently in a state of explosive growth as a discipline. From this field a new concept is emerging, where entire ecosystems can be individual organisms themselves, or even their organ systems or portions of them, and the inhabitants of these "ecosystems" are largely microbes (although other organisms such as mites occupy them as well)—bacteria, archaea, fungi, algae, protozoa, and viruses. The importance of this way of thinking about ecology cannot be overstated.

- It illustrates that multicellular life on Earth may exist only in association with communities of microbes and that there may be no such thing as a totally independent multicellular organism.

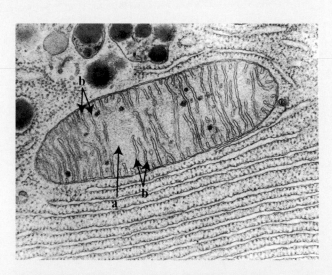

Transmission Electron Micrograph of a Mitochondrion in Cross Section from a Human Pancreatic Cell. (a) The mitochondrial matrix is the space within the inner membrane of the mitochondrion that contains mitochondrial DNA. The citric acid cycle, or Krebs cycle, in which a cell can produce energy from glucose and oxygen, takes place here. (b) Cristae, the folds in the mitochondrial inner membrane, are where electron transport takes place, a process that is a eukaryotic cell's most efficient means of production of adenosine triphosphate, or ATP. ATP is the principal energy storage molecule for all cells. (Photo by Keith R. Porter. © Photo Researchers, Inc.)

- It raises interesting questions about the definition of a species, which assumes that each organism contains only a single genome. Joshua Lederberg, who shared the 1958 Nobel Prize in Physiology or Medicine, has suggested, for example, that the human genome should perhaps include the collective genomes of all of our resident microbes, our so-called microbiome.[h]

- It leads to a deeper understanding about health and disease in animals and plants, including in ourselves, our livestock, and our crops.

- It provides a fuller picture of how immune systems work, where resident beneficial microbes, accepted as "self" as opposed to being rejected as "nonself," may help regulate the development of immune system components, such as Paneth cells in the human small intestine, and trigger their responses (see below).

- It suggests that in our attempts to assess the impacts of environmental changes on organisms, we need to take into account how those changes will affect not only the external environment of organisms, but also their internal environments.

- It makes clear the vital connections we humans, as well as all other multicellular organisms, have with the microbial world, with which we have co-evolved.

- It calls into question the wisdom of such practices as using antibiotic resistance genes in genetically modified foods or of giving antibiotics indiscriminately to aquacultured organisms and to livestock.

- And it challenges the long-held notion that microbes are mostly harmful and that we should attempt to rid ourselves and our immediate surroundings of them, such as by the use of antibacterial soaps and personal hygiene products, actions that are both futile and potentially unhealthy.

In addition to examples presented elsewhere in this book, such as the relationships of vascular plants with their mycorrhizal fungi and of legumes with their nitrogen-fixing bacteria (as detailed in chapter 8), models of "microbial ecosystems" in multicellular organisms are many. Perhaps the best known are the ruminants, such as cows and goats, which possess an organ called a rumen that is filled with billions of anaerobic bacteria, anaerobic fungi, and ciliated protozoa (which themselves have hydrogen-utilizing, methane-generating bacteria within them). These complex microbial communities carry out the process of digesting cellulose and other polysaccharides, breaking these compounds down into simpler sugars that ruminants can then absorb in their intestines.

Wood-eating termites are also dependent on microorganisms. They have flagellated protozoa in their intestines that, in turn, are living symbiotically with many different types of bacteria that surround them and live within them, all of which serve to break down the indigestible components of wood—lignin and cellulose—into digestible compounds for the termite. New studies of bacteria in the gut of the termite *Reticulitermes speratus* have identified more than 300 different species in each individual, and there are estimates that the number may be as high as 700![i]

Corals depend on their resident zooxanthellae, microscopic photosynthesizing organisms that provide them with oxygen and nutrients, to survive.[j] Without them, corals appear "bleached" and become vulnerable to fatal infections. The larvae of some oysters and barnacles will not settle and metamorphose into adults until they are colonized by specific bacteria. And cocoa trees are protected from certain fungal diseases by the presence of other fungi that live within their tissues.[k]

We, too, are colonized by a vast, dynamic, and complex world of microbes—on our skin and our eyes, and in all our organs that communicate with the outside world, such as our ears, mouth, nose, trachea, lungs, gastrointestinal tract, and vaginal canal. The number of bacteria in our intestines alone is on the order of 100 trillion, which is about ten times the total number of human cells in our bodies, and these bacteria together are thought to contain 100 times more genes than the entire human genome.[h] Let us look briefly at three of these "ecosystems."

SKIN

Our skin is heavily populated with a wide assortment of bacteria, fungi, and mites, the microscopic arthropods that live in our sebaceous (oil) glands and hair follicles. Different regions of the skin have different numbers of microbial flora. For example, the moist areas of our armpits and the spaces between our toes may harbor as many as 10 million bacteria per square centimeter (about 65 million per square inch), whereas dry areas such as our forearms may contain only one hundred thousandth that number. It is as if one were comparing a rainforest with a desert.[l] And the species themselves may also differ from one skin environment to another.

One recent study of skin microbes on the forearms of six healthy people identified a total of 182 bacterial species belonging to 91 genera. There was a great deal of variation from one person to another, with only four species found on all six subjects. Forearm skin microbial populations also changed over time, with many of the original species no longer present, being replaced by others, when the subjects were tested again 8 to 10 months later.[m] Some skin microbes have been found

Electron Micrograph of a Hair Follicle Mite. Hair Follicle Mites (*Demodex folliculorum*) are thought to be present in a large proportion of people. They live, generally unnoticed, mostly in short hair follicles, such as in those of eyelashes and eyebrows. They also live in the nose and ears, where they feed on secretions and cellular debris. Besides *D. folliculorum*, another skin mite species, *Demodex brevis*, inhabits sebaceous or oil glands in our skin. Whether these mite species play a beneficial role for us under normal conditions, for example, by ingesting dead cells and microbes, is not known. They are found in greater numbers in patients with various skin diseases, but it is not clear whether their increased populations are a consequence, rather than a cause, of these conditions. (Photo © Andrew Syred, Microscopix Photolibrary. Data source: B. Baima and M. Sticherling, Demodicidosis revisited. *Acta Dermato-Venereologica*, 2002,82(1).3.)

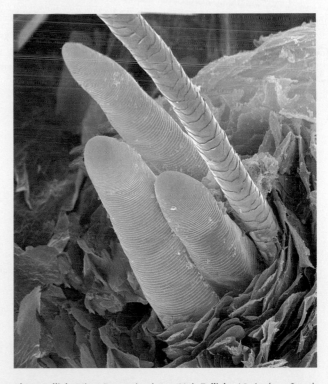

Three Follicle Mites Burrowing into a Hair Follicle. (© Andrew Syred, Microscopix Photolibrary.)

Scanning Electron Micrograph of a Subgingival Microbial Community in the Human Mouth. Here we see biodiversity of an oral biofilm demonstrating at least three different bacterial species—two types of bacilli, the rod-shaped organisms (one longer and of greater diameter in cross section than the other), and a type of coccus, or ball-shaped bacterium. (From Ziedonis Skobe, Forsyth Institute.)

to secrete antimicrobial compounds, such as the bacteriocins,[n] that may complement the action of others on the skin, such as psoriasin, a peptide produced by human epithelial cells that has the ability to control the growth of certain Gram-negative bacteria. But despite these and other new insights about microbes that live on our skin, relatively little is known about their diversity or about the role they play in maintaining skin health or in causing disease.

MOUTH

The "ecosystem" of the human mouth has been much better studied than that of the skin, chiefly by Sigmund Socransky, Bruce Paster, Anne Haffajee, and their colleagues at the Forsyth Institute, a dental research center in Boston. Following observations made by the Dutch microscopist Anton van Leeuwenhoek, who studied scrapings of plaque from his own mouth in 1683 and made what may be some of the first drawings of bacteria, these researchers have estimated (based on 16S rRNA studies—see chapter 1, page 10) that there are more than six billion microbes in the human mouth, comprising more than 700 species.[o] Almost all of these are bacteria, but there are also archaea, fungi, amoebas, and viruses. We focus here on the bacteria, for they have been the best characterized.

Each person is thought to have a characteristic set of oral microbes, and each part of the mouth—the tongue, soft palate, gums, and teeth, for example—shows a different composition. The subgingival space, that gap between the base of teeth and the inside of the

gums, has been the most extensively studied region, because of its role in periodontal disease, a condition of gum inflammation that can result in tooth loss. The loss of teeth is a significant public health problem worldwide, especially for older populations. Almost three in ten adults older than age 65 in the United States, for example, have lost all of their teeth, primarily because of periodontal disease and tooth decay.[p] Understanding the role that gum microbes play in periodontal disease takes on added importance because of growing evidence that periodontal infections may be associated with atherosclerotic vascular disease, including of the coronary arteries.[q]

A biofilm, consisting of layers of microbes held together by a mucus matrix, coats the mouth's tissues. In the subgingival space, one type of biofilm covers teeth and forms dental plaque; another lines the inside of the gums. One could consider then that not only is the mouth itself an "ecosystem," but so are its various regions, as well as the microenvironments within these regions, each of which contains a different array of microorganisms. These biofilms are extremely resistant to physical removal, such as by dental flossing and tooth brushing, rapidly reestablishing themselves. They can also be highly resistant to antibiotics. Subgingival biofilms are firmly attached to the gums and the teeth and serve to protect them from disease by preventing pathogenic bacteria and other organisms from gaining a foothold. Some commensal oral bacteria have also been found to secrete antimicrobial toxins that kill pathogens.[r,s] Others have been shown to stimulate human epithelial cells lining the gums to produce their own antimicrobial peptides, known as beta-defensins.[t] Studies are under way to determine whether specific species of bacteria and other microbes in these biofilms are associated with specific diseases such as periodontal disease (in which, e.g., some archaea have been shown to play important roles)[u,v] and oral cancers,[w] and whether regular screening of one's oral flora may serve as an early indicator for these diseases.

INTESTINE

There has also been intense interest in the microbial organisms of the human intestine, where the vast majority of the microbes in our bodies reside. Molecular studies based on the same rRNA techniques used with oral bacteria, along with what is called "fluorescent in situ hybridization" (which identifies DNA sequences by attaching fluorescent antibodies to them), have revealed that there are on the order of 800 distinct microbial species, most of which are bacteria, that live in our small and large bowels, comprising thousands of strains or subspecies.[x] Various archaea, viruses, yeasts, and protozoa also reside in our intestines, with estimates, for example, that there are some 1,200 different types of viruses alone in our feces.[y] The true extent of the diversity of these other organisms, however, remains unknown. (We should note here that debate surrounds whether rRNA studies accurately measure the number of different microbial species that actually reside in specific environments. Knowing whether a given piece of rRNA represents a normal resident of an individual's bacterial community, or whether it was, in the case of the human bowel, e.g., ingested on a piece of food, is difficult to decipher, as is determining whether the bacterium identified is actively functioning in the microenvironments in which it is discovered.) The composition of these intestinal communities has been found to differ not only between individuals, but between different regions of the intestine in the same individual, and between the luminal (the interior space) and mucosal (the surface lining) areas in the same regions.[z]

What they are all doing there is a question that is beginning to occupy a large number of researchers around the world. Some of the services provided by intestinal microbiota are clear. For one, it has long been known that some intestinal bacteria help us break down otherwise indigestible polysaccharides, complex carbohydrates found in plants, into easily absorbed sugars. They also produce vitamins for us, such as vitamin K (as well as very small amounts of the B vitamins—B_{12}, folate, and thiamin).[aa] While we obtain some vitamin K from a number of foods, including leafy greens and other vegetables, our main source comes from bacteria in our intestines. Vitamin K is a key co-factor in pathways that control blood clotting and in the formation of human bone through its action on a protein called osteocalcin.

Research on mice raised with sterile intestines has shed further light on some of the roles played by our intestinal flora. Germ-free rodents must consume around 30 percent more calories to maintain the same body weight compared with normal animals. They are also more susceptible to infection.[bb] Investigators who added Bacteroides thetaiotaomicron (a bacterium that is 1,000 times more abundant in our intestines than the much more widely studied E. coli [see section on E. coli in chapter 5] and that comprises some 25 percent of all our intestinal bacteria) to the intestines of germ-free mice have discovered several remarkable things. B. thetaiotaomicron has been found to monitor concentrations in our guts of a simple sugar called fucose that it uses for energy and to signal our intestinal cells to manufacture more of this sugar when supplies are low. In return, B. thetaiotaomicron performs an array of essential "ecosystem services." For one, the bacteria are major players in the breakdown of polysaccharides

(in fact, much of the bacterium's genome, sequenced in 2003, is devoted to this process).[cc] They also help form the protective layer of mucus coating intestinal epithelial cells, which provides both a physical barrier, preventing these cells from being injured, and which, along with the presence of tight cellular junctions, blocks bacteria from crossing the single-cell-thick epithelial layer to invade other tissues.[bb,dd] B. thetaiotaomicron may protect our bodies from infections in others ways, as well, by interacting with special cells in the small intestine called Paneth cells that are known to secrete a variety of antimicrobial compounds such as defensins (which are thought to help fight food-borne and water-borne bacterial infections);[ee] by competing directly with potential pathogens for space and nutrition, thus preventing their colonization; and by producing their own antimicrobial substances, including lactic acid, hydrogen peroxide, and potent antimicrobial peptides such as bacteriocins. Finally, B. thetaiotaomicron helps stimulate the growth of new blood vessels, a process called angiogenesis, crucial to the intestines' ability to absorb nutrients.[ff] Studying this angiogenic role of B. thetaiotaomicron may lead to new insights about how human intestinal cancers form and how to treat them.

Given that we and all other organisms on Earth live in a world composed primarily of microbes with which we have co-evolved complex and dynamic interdependent relationships, it is critically important that we enlarge our definition of ecosystems to include them. By this perspective, ecosystems exist at multiple levels of organization—from the microbial population level, where different genetic strains of a bacterial species, for example, fill different biological niches, say, within one layer of a human subgingival biofilm; to the microbial species level, where different bacterial species inhabit different layers of this biofilm; to the tissue level, where the makeup of a microbial community on the tongue is different from that lining the gums of the subgingival space; to the organ level, where the flora of the mouth is distinct from that on the skin; to the level of the organism as a whole. Individual organisms are, in turn, parts of communities that are arranged at progressively higher levels of organization, ultimately at the level of what we have traditionally referred to as an ecosystem, such as a temperate forest or a coastal marine wetland.

Until we begin to see ecosystems along such a continuum, we will fail to appreciate the vital and central role played by microbes in our lives and in the lives of all other species on Earth, both in health and in disease; we will pay insufficient attention to the enormous diversity and complexity of organization that exists at all the various sublevels of traditional ecosystems; and we will ultimately have a superficial and incomplete understanding of how ecosystems function to sustain the living world.

Provisioning Services

For millennia, people have harvested Nature's bounty for nourishment, shelter, and fuel. They have also used plant products to treat a range of illnesses, including malaria and other maladies. Many of the goods harvested from aquatic and land ecosystems are traded in economic markets. For example, at the beginning of the twenty-first century, the annual world fish catch from the oceans and from fresh water was about 130 million metric tons (about 143 million tons), valued in the range of 100 billion U.S. dollars.[2] Fish is a core component of peoples' diets in many parts of the world, such as in Africa and Asia, where some 20 percent of the population depends on fish as a primary source of protein.[3]

Land ecosystems, including grasslands and forests, are also important sources of marketable goods. Grasslands help supply people with a wide range of animal products, including meat, milk, wool, and leather. Forests supply people with many things, including food, timber, and wood for fuel. Fruits, nuts, mushrooms, honey, and many other foods are also extracted from forests. Wood, bamboo, grasses, and other plant materials are used to construct homes and other buildings. Organic material from trees and other plants supplies about 15 percent of the world's total energy consumption; in developing countries, it supplies almost 40 percent.[4,5] In addition, natural products extracted from many hundreds of forest and nonforest plants are used by industry. Examples include oils, resins, dyes, tannins, and insecticides.

Regulating Services

CLEANING AIR

Both plants and soil microbes are involved in cleaning the air we breathe. Plant canopies, especially forest canopies, function as filters of particulates in the air and as chemical reaction sites that help regulate the composition of the atmosphere.[6] The major sources of atmospheric particulates are (1) the combustion of coal, gasoline, and fuel oil, (2) cement production, (3) lime kiln operation, (4) incineration, and (5) the burning of crops. These human activities produce fine particles less than 100 μm (micrometers) in diameter, such as black carbon, and coarse particulates (greater than 100 μm), such as dust. (A micrometer is one millionth of a meter, or approximately 0.00004 inches, so 100 μm is approximately 0.004 inches.) Plant canopies capture a variety of particulates, ranging from harmless sea-salt aerosols near the oceans to dangerous lead particles alongside roads in countries, both industrialized and developing, where lead is still being used as a gasoline additive.

Plant surfaces, particularly moist leaf surfaces, are sites where a wide range of chemical reactions occur.

Figure 3.2. Marsh Arab Reed House. This *mudhif*, a typical floating house of the Marsh Arabs of southern Iraq, is made from reeds tied and woven together, as it has been for thousands of years in this area. Saddam Hussein largely destroyed these marshes, but major international efforts are under way to help them recover. (From Nik Wheeler Photography.)

Forest canopy purifies air by filtering particulates and providing chemical reaction sites where pollutants are detoxified

Forest canopy and leaf litter protect the soil surface from the erosive power of rain

Forest trees and plants store carbon and help slow human-caused global climate change

Forests help maintain the water cycle and stabilize local climates

Forests provide goods such as food, timber, and medicines

Forests provide critical habitat for plants, animals, and microbes

Forest tree roots bind soils and help prevent erosion

Deep forest soils store large volumes of water

Forest soils purify water, acting as a massive filter

Figure 3.3. Temperate Forest Ecosystem Services. (Original photo by Jin Young Lee, www.dreamstime.com. Text design by Tong Mei Chan.)

There, polluting compounds such as nitric oxide, the precursor of ground-level ozone, produced mainly by automobiles and power plants, can be transformed into harmless compounds.[7] Some soil microbes are also capable of many of these transformations. One group of microbes known as "methanotrophs," for example, that live in well-drained, well-aerated soils and belong to the Archaea (see chapter 1, page 11), break down methane,[8] a powerful greenhouse gas that is involved in global warming.

PURIFYING WATER

Many well-vegetated upland areas, freshwater wetlands, and estuaries function to purify water. The purification processes involved can be biological, physical/chemical, or a combination of the two.

Upland Areas

Forests, shrublands, and grasslands that occur in upland areas throughout the world are important sources of clean water for human use. The journey of water through these ecosystems is like slowly dripping water through a massive filter. The rain that falls on many of these ecosystems often contains substantial amounts of chemicals, such as inorganic nitrogen (in the form of ammonium or nitrate compounds), and other inorganic and organic compounds. As it percolates through the soil, the water is stripped of many of these chemicals, both by being taken up by plants and microbes and by coming into contact with chemically reactive sites on clay and on organic matter to which such compounds bind. For example, in healthy middle-aged forests in New England, rain enters with an average nitrogen load of about eight pounds per acre each year. Stream water leaving these forests often contains less than one-tenth the concentration of nitrogen that was present in the rainfall.[9]

Freshwater Wetlands

Since the dawn of civilization, freshwater wetlands have been absorbing and recycling nutrients from human settlements. This ecosystem service is performed by a variety of wetland ecosystem types, including those that occupy lowland areas along streams and rivers, and those that border lakes. As water flows through these wetlands, plants, microbes, and sediments filter out nutrients, such as nitrogen and phosphorus, from the water column. Plants take up these nutrients and incorporate them into root, stem, and leaf material. Microbes transform a water-soluble form of nitrogen into gaseous forms that are biologically inactive and harmless to the environment. And physical and chemical processes in sediments, such as those involving the adsorption (an accumulation of a substance on the surface of a solid, forming a molecular film) of phosphorus to particles, function to purify water.

Nutrient retention and processing, characteristic of natural wetlands, have been exploited in reconstructing former wetlands, such as is now occurring in the marshlands of southern Iraq,[10] or in the building of new ones, such as those being developed by some coastal cities and towns. They are constructed so that water flows slowly over sediments and through vegetation, giving them time to strip the water of

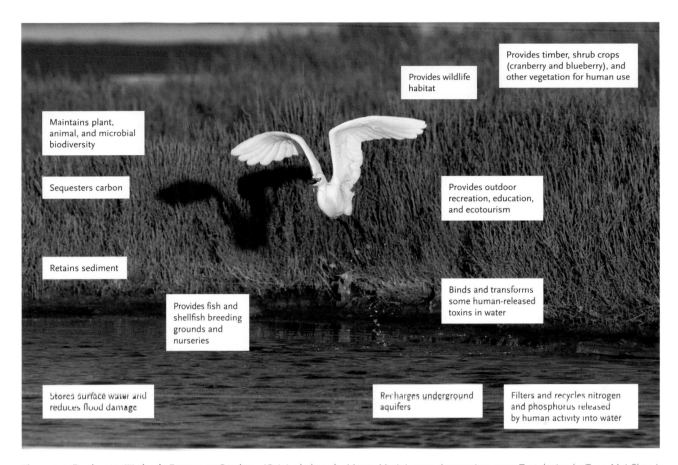

Figure 3.4. Freshwater Wetlands Ecosystem Services. (Original photo by Mauro Marini, www.dreamstime.com. Text design by Tong Mei Chan.)

The following labels appear in the figure:

- Provides wildlife habitat
- Provides timber, shrub crops (cranberry and blueberry), and other vegetation for human use
- Maintains plant, animal, and microbial biodiversity
- Sequesters carbon
- Provides outdoor recreation, education, and ecotourism
- Retains sediment
- Provides fish and shellfish breeding grounds and nurseries
- Binds and transforms some human-released toxins in water
- Stores surface water and reduces flood damage
- Recharges underground aquifers
- Filters and recycles nitrogen and phosphorus released by human activity into water

nutrients.[11] In addition to controlling the rate of water flow, managers of constructed wetlands often keep vegetation in a rapid growth phase through periodic harvesting in an effort to maximize the amount and the speed of nutrient uptake. They also regulate oxygen levels in the sediments to increase the loss of gaseous nitrogen, and manipulate the supply of soluble iron and aluminum to enhance the rate of phosphorus removal.

Constructed wetlands also have the ability to remove human-made compounds, including some that are toxic, from flowing water. At a U.S. Environmental Protection Agency research laboratory in Athens, Georgia, for example, studies have shown that an enzyme produced by the invasive Parrot Feather (*Myriophyllum brasiliense*), a freshwater plant that can spread rapidly to clog rivers, ponds, and irrigation channels, effectively breaks down trinitrotoluene, better known as TNT.[12] This has led to several successful pilot projects in which constructed wetlands were able to remove the chemical from water that had been contaminated by military firing ranges. (See also "Binding and Detoxifying Pollutants in Soils, Sediments, and Water," below.)

Estuaries

Bivalve mollusks in estuaries, including mussels, clams, and oysters, act as filtering systems that can remove suspended materials and that consume algae secondary to eutrophication (the overgrowth of algae in aquatic ecosystems resulting from excessive levels of human-released nutrients). An often-cited example is the filtering capacity of Eastern Oysters (*Crassostrea virginica*) in the Chesapeake Bay. For centuries, oysters in the bay were so numerous that they could filter its complete volume in approximately a three-day period. The result of this massive filtering activity was to maintain clear and oxygen-rich waters.

A combination of pollution, habitat destruction, overharvesting, and other pressures has dramatically reduced the oyster population of the Chesapeake Bay, and those of other major estuaries along the U.S. East Coast. For the Chesapeake, the decline has been so great that it now takes almost a year for the oysters to filter the bay, more than 100 times as long as it did as recently as about 100 years ago.[13] The result of this decline has been the loss of a critical ecosystem service, the filtering of water that has been essential to maintaining the quality of the bay. Its waters are now murkier and poorer in oxygen concentrations and in aquatic life.[14]

MITIGATING FLOODS

For millennia, many regions of the world have been subject to extreme weather events, including periods of excessively heavy rainfall and the short-term flooding of relatively flat areas, known as "flood plains," that border lakes, rivers, and streams. Flood plains include a variety of habitats such as forests and wetlands. Some flood plains bordering major rivers are vast, such as those of the Mississippi River, whose flood plain is up to 130 kilometers (80 miles) wide in some areas. Examples of other large river flood-plain ecosystems include the Sudd swamps on the White Nile in Sudan and the Okavango River wetlands in Botswana. Unaltered flood plains serve as habitat for many plant and animal species. For example, the Gran Pantanal of the Paraguay River in South America is home to an estimated 600 species of fish, 650 species if birds, and 80 species of mammals.[15]

Flood plains are one of Nature's "safety valves." Following excessive rains, floodwaters flow over riverbanks and into the forests, wetlands, and other habitats that constitute flood plain ecosystems. Some of the water is soaked up by the soil. In time, the floodwaters recede, leaving behind a new supply of nutrient-rich sediments that enhance the flood plain's fertility and make these ecosystems among the most productive in the world.

Many ancient civilizations—for example, in Mesopotamia, Egypt, China, and India—used flood plains as agricultural sites, taking advantage of the periodic enhancement of soil fertility by the flood-related deposition of nutrient-rich sediments. As human populations have grown, development pressures on flood plains in many parts of the world have increased, resulting in a compromising of the ability of flood plains to absorb floodwaters.

The Mississippi River and its tributaries, which drain about one-third of the lower forty-eight states, flooded during the summer of 1993 following above average rains in the first half of the year and an unusually high number of torrential rain events during the summer. Such extreme precipitation events, including both torrential rains and droughts, are predicted to increase in frequency and intensity as a result of global warming.[16] The loss of the flood plain alongside the river compounded

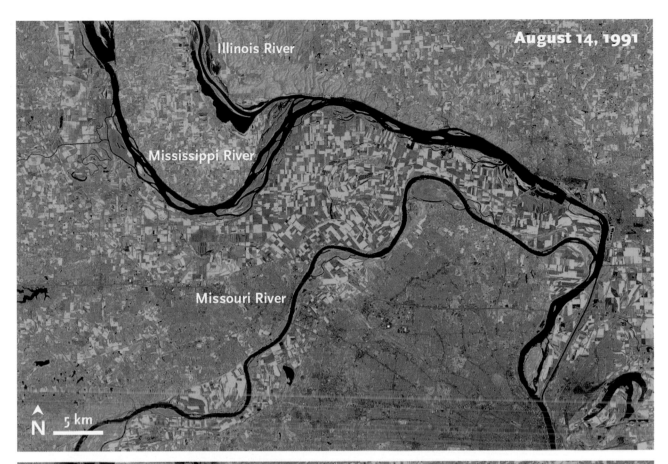

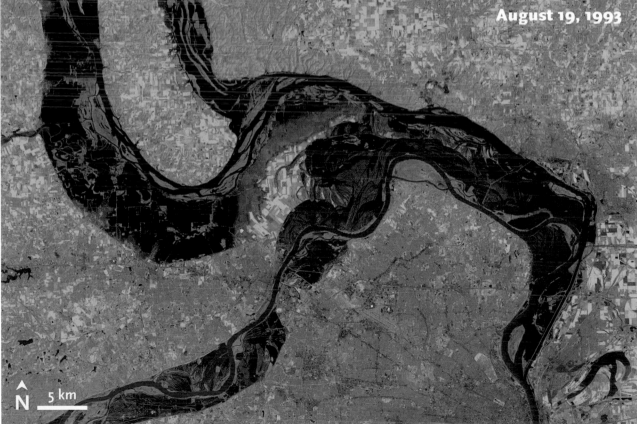

Figure 3.5. St. Louis, Missouri, Before and After the Summer Flooding of 1993. Satellite photos taken August 14, 1991, and August 19, 1993. The Illinois, Missouri, and Mississippi Rivers breached their banks secondary to torrential rains and to the compromising and development of the rivers' flood plains, flooding millions of acres of towns, cities, and farmland. (Courtesy of NASA Earth Observatory Photo.)

Figure 3.6. Flooded Farm near Hillview, Illinois, on the Illinois River, Summer 1993. (Courtesy of Jim Wark, Airphoto.)

the devastation. Floodwaters spread over 9.3 million hectares (23 million acres), inundating farms, towns, and cities in nine midwestern states—North Dakota, South Dakota, Nebraska, Kansas, Minnesota, Iowa, Missouri, Wisconsin, and Illinois. The toll was enormous—50 people were killed, more than 70,000 homes were lost, 8.7 million acres of farmland were damaged, and total property losses were estimated at $12 billion.[17]

The high cost of the flood damage has been attributed to three practices: the drainage of flood plain wetlands, the building of permanent structures on flood plains, and the construction of levees to keep floodwaters from spilling over. In the Midwest, one of the original, and most important, changes to the landscape was, in addition, the loss of beaver dams. Beavers had shaped the flood plain landscape for thousands of years prior to European settlement. The seventeenth- and eighteenth-century fur trade brought the beaver to the verge of extinction in Illinois by the mid-nineteenth century (the beaver is once again abundant there due to its reintroduction). With the loss of beaver dams, and the start of intensive farming that required the draining of wetlands, came unimpeded tributary flow into the Mississippi River and increased flooding.[18]

During the past century, the drainage of wetlands in the U.S. Midwest intensified to produce more farmland and home sites. The flood-moderating service of these wetlands was not recognized. Missouri, Illinois, and Iowa, the three states that suffered the most damage from the 1993 floods, have less than 15 percent of their original wetlands.[19]

Building permanent structures such as barns, houses, and factories on flood plains also increases damage and the accompanying financial losses when floods occur, for two reasons. First, they are valuable properties. Second, these structures and their associated roads, parking lots, and other paved surfaces reduce the area of soils and sediments that are able to absorb floodwaters. If forest or other natural vegetation covers a flood plain, the floodwaters spread over the land slowly, and the land absorbs much of the water. Because land in a developed flood plain is less able to absorb excess water, the water spreads more rapidly and extensively.

Finally, hundreds of levees were built along the Mississippi and its tributaries to hold floodwaters back from the flood plain. Although levees may save lives and property where they are built, they cause floodwaters upstream to surge, damaging farms and towns that are less protected. In addition, they prevent the periodic deposition of sediment in the flood plains that replenishes the soil, serving to maintain its levels and its ability to absorb floodwaters. The building of levees and the subsequent loss of marsh soils, along with the draining, destruction, and development of freshwater wetlands, is thought to be, in part, responsible for the massive flooding of areas of New Orleans following Hurricane Katrina.[20]

CONTROLLING EROSION

Inland Sites

Vegetation provides natural protection for soils against erosion in several ways. First, the plant canopies intercept rainfall and reduce the force with which rainwater hits the soil surface. Second, roots bind soil particles in place and prevent them from washing down slopes. Third, old root channels help to minimize the powerful force of surface runoff by routing water into the soil, like drain pipes. Animal burrows serve the same function.

The U.N. Food and Agriculture Organization (FAO) has estimated that during the closing decade of the twentieth century, erosion damaged or destroyed each year between ten and twenty million acres of the world's cropland. Erosion has affected some areas more than others. In China by 1978, erosion had forced the abandonment of about one-third of all arable land. Erosion rates in many parts of Africa are estimated to be nine times higher than erosion rates in Europe.[2]

Erosion can affect human health in both direct and indirect ways. By reducing the area of croplands, erosion may contribute to food shortages and compromised nutritional states among people in some developing countries. Erosion can also directly cause deaths through mudslides. For example, intense rains falling on steep slopes that were cleared of their forests in the Caribbean and throughout Central and South America have resulted in thousands of people dying in massive mudslides in recent decades, including those accompanying Hurricane Mitch in 1998.[21]

Large-scale mudslides became the signature of Hurricane Mitch, which grew to become the Atlantic basin's fourth strongest hurricane ever, with sustained winds of 180 mph for more than 24 hours. The hurricane stalled off the coast of Honduras from October 27, 1998, until the evening of October 29, dropping up to 25 inches of rain in

Figure 3.7. Mudslide in the Hamlet of Rincon Argentino, in Tecpan, Guatemala, October 5, 2005. This mudslide, a result of torrential rains from Hurricane Stan, killed four children and left fifteen people missing. Mudslides in places were a half a mile wide and 15 to 20 feet deep. (Courtesy of Reuters/Mario Linares.)

one six-hour period in some places. The heavy rains led to widespread flooding and mudslides that resulted in 33,000 homes being destroyed, at least 7,000 deaths and 5,000 missing, and thousands of cases of cholera, malaria, and dengue fever.[22]

Hurricane Stan, which dumped torrential rains on Guatemala and other parts of Central America from September 29 through October 5, 2005, killed more than 1,036 people in Guatemala alone; left 130,000 homeless and three million without power, water, and other basic services; destroyed crops and livestock; and damaged nearly 2,500 miles of roads, cutting off many regions from outside help.[23] A U.S. Geological Survey research team reported that many of the deadly mudslides occurred in areas where the forests had been cleared to make way for agriculture.

Ocean Edge

Mangrove forests and salt marshes are the most common ecosystems found in many coastal areas. They perform an important ecosystem service by buffering the land

BOX 3.2

The Tsunami of December 26, 2004

The great Southeast Asian Tsunami of December 26, 2004, which killed more than a quarter million people, left millions homeless, and caused widespread devastation in Indonesia, Thailand, India, Sri Lanka, and other countries, provides an important case study for the role that natural coastal ecosystems may play in the physical protection of people and land against storm surges. Preliminary studies in Sri Lanka suggested that in areas where coral reefs, vegetated coastal sand dunes, and healthy mangrove forests were intact, damage to the coastal zone was lessened. And investigations in Thailand, particularly in the most affected province of Phang Nga, demonstrated that mangrove forests and seagrass beds significantly mitigated the destructive force of the tsunami.[a-c] Model simulations have supported the role of coral reefs in buffering the impacts of tsunamis.[d] But some researchers have stated that while mangroves and coral reefs dampen the destructive action of normal storm-generated waves, their protective roles during tsunamis are less clear, and that distance from the epicenter, elevation and distance from the shore, shoreline profiles, wave characteristics, and other factors may be as or more important in determining levels of destruction on land.[e]

Figure 3.8. Photo of Mangroves in Southeast Florida. The strong, dense branches and roots of mangroves break up the force of waves and storm surges and stabilize coastlines. (Courtesy of U.S. National Oceanic and Atmospheric Administration.)

against ocean storm surges. Plants in these ecosystems stabilize submerged soil sediments, thereby preventing coastal erosion. Scientists at the Mangrove Ecosystems Research Centre in Hanoi, North Vietnam, for example, have compiled evidence that mangroves are more effective than concrete sea walls in controlling the raging floodwaters from tropical storms.[24] Forested areas can also help dampen the force of hurricanes and protect coastal communities, as was seen with Hurricane Felix and its 160-mile-per-hour winds that came ashore in northern Nicaragua and southern Honduras on September 4, 2007.

However, both salt marshes and mangrove forests are rapidly being destroyed. To uninformed people, salt marshes have often appeared to be worthless, empty stretches of land. As a result, many of them have been used as waste dumps or have been filled in with dredged sediments to form artificial land for building homes or industrial complexes. Mangroves are also under assault from other forms of coastal development, such as from shrimp aquaculture and unsustainable logging.[25] Some countries, such as the Philippines, Bangladesh, and Guinea-Bissau, have lost 50 percent or more of their mangrove swamps.[26] Losses of salt marshes and mangroves have consequences beyond the loss of ocean storm buffers. Most important, these land-margin ecosystems are among the most productive breeding grounds and nurseries for commercially important fish (see section on aquaculture in chapter 8, page 373) and are important habitats for many species of birds.

BINDING AND DETOXIFYING POLLUTANTS IN SOILS, SEDIMENTS, AND WATER

One of the consequences of our industrial and agricultural activities is that we have spread, both intentionally and unintentionally, heavy metals and radioactive elements worldwide, a result of having mined them for a variety of purposes. We have also released into the global environment, in varying concentrations, tens of thousands of man-made chemicals—pesticides, medicines, industrial chemicals, household products, and other compounds—some of which degrade very slowly, accumulate in the food chain, and eventually end up in our own tissues (see the section on pollution in chapter 2, page 51). In some places where they are deposited, they reach toxic levels that can render such areas unusable by humans and a danger to many other forms of life.

Scientists are using a variety of vascular plants that have the capacity to concentrate potentially toxic elements without doing themselves harm, to clean up and restore these contaminated areas. The India Mustard Plant (*Brassica juncea*), for example, can accumulate lead, chromium, cadmium, nickel, zinc, copper, and selenium; the Alpine Pennycress (*Thlaspi caerulescens*) binds zinc and cadmium; and the Common Sunflower (*Helianthus annuus*) can capture some radioactive substances.[27–29] These "abilities" are being put to use in what is called "bioremediation" or "phytoremediation." Such plants are particularly abundant in tropical or subtropical regions, perhaps because high metal concentrations in their tissues may confer some degree of protection against plant-eating insects and microbial pathogens that are common in these regions. Other species of plants are also being

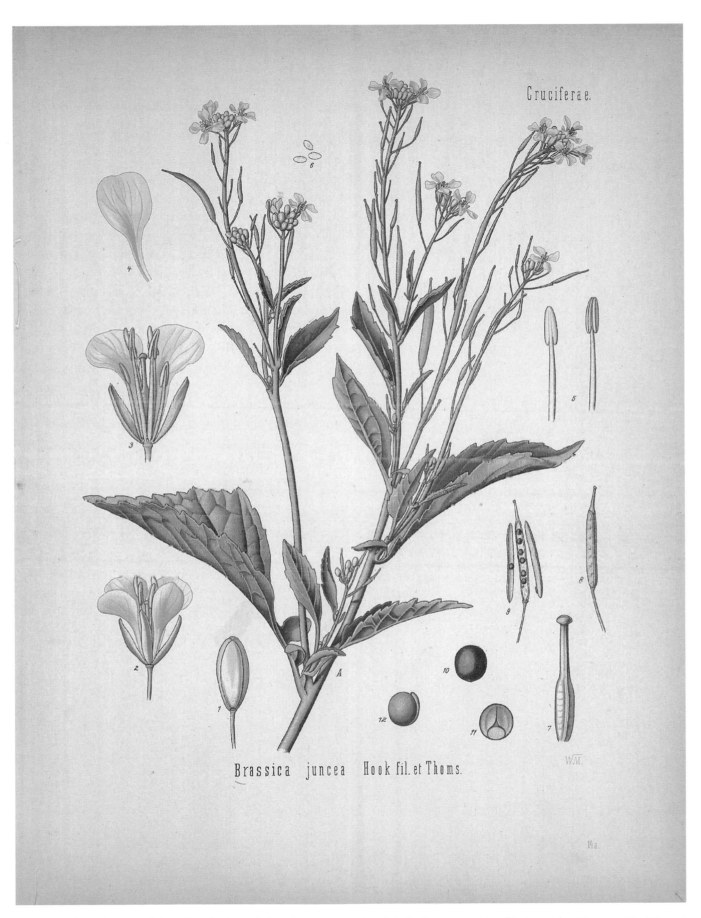

Figure 3.9. The India Mustard Plant (*Brassica juncea*). Brassica juncea can absorb in its tissues a variety of toxic metals. (© 1995–2004 Missouri Botanical Garden. From F.E. Kohler's *Medizinal-Pflanzen*, Gera-Untermhaus, 1887, www.illustratedgarden.org/mobot/rarebooks/.)

investigated for their ability to bind toxic substances, including alfalfa, tomato and pumpkin vines, bamboo, Cordgrass, willow and poplar trees, and even the invasive Kudzu.[30] And some nonplant species, such as the lichen *Trapelia involuta*, which can concentrate uranium in its tissues,[31] and some fungi, such as "white rot fungi" (particularly *Phanerochaete chrysosporium*) and "brown rot fungi" (notably, species of *Gloeophyllum*), which are able to accumulate heavy metals,[32,33] are also being studied for their bioremediation potential.

Sometimes the presence of plants growing on contaminated sites signals their ability to accumulate toxic substances. Recently, Brake Ferns (*Pteris ensiformis*), common in southeastern parts of the United States and in some other parts of the world, were found growing at a central Florida lumberyard where soils were heavily polluted with arsenic from wood preservatives (e.g., those used in "pressure-treated wood"). The Brake Ferns were taking up the arsenic in their tissues. Another arsenic-binding plant, the invasive, aquatic Water Hyacinth (*Eichhornia crassipes*), has been used to remove arsenic from drinking water.[34] Arsenic-contaminated drinking water is a problem in many parts of the United States, especially in the West and in Alaska. It is also a significant problem in other parts of the world, such as Bangladesh, where more than 60 percent of the groundwater contains high concentrations. In Bangladesh, millions of people have been exposed to arsenic levels that increase their risk for acute toxic effects, such as vomiting, esophageal and abdominal pain, and bloody "rice water" diarrhea; and chronic effects, such as keratosis (a thickening of the skin), changes in skin pigmentation, and cancers of the skin, lungs, bladder, and kidney.[35]

Figure 3.10. Brake Fern (*Pteris ensiformis*). (Courtesy of National Parks Board, Singapore.)

Two phytoremediation examples stand out as models for environmental cleanup and public health protection. The first involves an experiment conducted in a small pond near the ill-fated Chernobyl nuclear power plant in the Ukraine. The pond, like other areas surrounding Chernobyl, was heavily contaminated with strontium-90, cesium-137, and other toxic radioactive substances that had been released during the reactor fire in 1986. Scientists grew Sunflowers (*Helianthus annuus*) on Styrofoam rafts floating in the pond, with their roots dangling in the water like those of lettuce plants growing in hydroponic tanks. The Sunflowers were found to rapidly accumulate radioactive strontium and cesium in their tissues to levels that were several thousand times higher than concentrations in the water.[36]

Another noteworthy success story was the cleanup of a lead-laced tract of land at the DaimlerChrysler company complex in Detroit. The cleanup process was straightforward. First, the top four feet of soil were moved to a nearby site and planted with India Mustard and Sunflowers, both of which can accumulate lead. The lead concentration in the soil was reduced by 43 percent as a result of these plantings, which brought the site into compliance with both federal and state regulations. The project cost about half of what it would have cost to cart the 5,700 cubic yards of soil to a hazardous waste landfill. Instead, the cleanup crew had to dispose of only a few cubic yards of lead-rich plant material.[37]

Some microorganisms in naturally functioning estuarine and marine ecosystems are also able to perform the ecosystem service of detoxifying wastes generated by humans, such as petroleum and petroleum byproducts, such as gasoline, that are spilled into these environments on a regular basis. Many of the component compounds present in these spills carry health risks for humans and for many other organisms. When these compounds adhere to sinking particles, they settle to bottom sediments, where in some settings, naturally occurring microbes, such as the marine bacterium *Alcanivorax borkumensis* SK2, are able to detoxify them, ultimately turning them into carbon dioxide and water.[38]

Microorganisms are also being investigated to turn other man-made chemicals into harmless substances. One, for example, an anaerobic bacterium named BAV1, has been found to break down vinyl chloride, a hazardous industrial chemical present in about one-third of all toxic waste Superfund sites in the United States.[39] Vinyl chloride can causes neurological symptoms such as dizziness and headaches with acute exposures, and a rare form of liver cancer with longer term exposures. Other microbes are able to degrade some pesticides, such as malathion, atrazine, and DDT, as well as such herbicides as 2,4,5-trichlorophenoxyacetic acid (commonly known as 2,4,5-T),[40–43] and to reduce the harmful effects of some radioactive elements.[44]

CONTROLLING PESTS AND DISEASE-CAUSING PATHOGENS

A pest is any organism that interferes in some way with human welfare. A variety of weeds, insects, rodents, bacteria, fungi, and other organisms compete with humans for food, affect fiber production, or spread disease. Croplands and pests go together. One estimate is that croplands support more than 50,000 species of plant pathogens, 9,000 of insects and mites, and 8,000 of weeds. The loss of productivity in these

Figure 3.11. Eucalyptus Tree Hedgerows Separating Wheat Fields in Western Australia. (© Oil Mallee Company of Australia.)

managed ecosystems can be very high. Worldwide crop yields are reduced by about one-third by pests and disease.[45] And the losses could be much higher if it were not for ecosystem services that result in keeping populations of pests and disease-carrying organisms under control.

Sometimes we have to learn the value of natural controls the hard way. A story from China illustrates this point. In the 1950s, at the time of Mao Tse-Tung's Great Leap Forward, Chinese officials became concerned that flocks of birds were devouring large amounts of grain. To stop this attack on an already imperiled food supply, government officials declared that sparrows (which ultimately ended up meaning "any small perching bird") were "enemies" and therefore candidates for eradication. Millions of Chinese set about killing birds. Their success was frightening. Over several days in 1958, an estimated 800,000 birds were killed in Beijing alone. Major pest outbreaks resulted from this bird eradication program and led to significant crop losses. The mistake was ultimately realized and the bird killing was halted.[46]

Maintaining natural pest control sometimes requires understanding how landscape diversity relates to this critically important ecosystem service. Let us look at one example, that of hedgerows, which are linear stands of small trees or shrubs, natural or planted, that separate fields or pastures. In the southern German state of Bavaria, mosaics of such hedgerows and forest plantations border agricultural lands. They are modern Germany's most diverse woody habitats, with up to thirty species of woody plants and many species of herbivorous insects. The insects are for the most part specialists, and they feast on the woody plants in the hedgerow, while largely

Figure 3.12. Adult Vidalia Beetles (*Rodolia cardinalis*) Feeding on Cottony-Cushion Scale Insects. Note the small reddish beetle larva on the back of the scale insect. (© Photo by Jack Kelly Clark, University of California Statewide IPM Project.)

ignoring the crops. The presence of the insects in the hedgerows attracts generalist predators and parasites, which feed not only on them, but also on aphids in the nearby grain fields. It is because of the hedgerows and their unique food webs that northeast Bavaria is one of the few places in Germany where farmers do not need to spray for wheat aphids. (See chapter 8 for more discussion of insects and other organisms that benefit crops.)

Using natural pest control as a model, scientists have tried to develop biological control mechanisms to replace pesticides. Biological controls involve the use of naturally occurring disease organisms, parasites, or predators to control pests. The use of a beetle to control Cottony-Cushion Scale (*Icerya purchasi*) is a good example. Cottony-Cushion Scale is a small insect that sucks the sap from the branches and bark of many fruit trees, including citrus trees. A native of Australia, it was accidentally introduced into the United States in the 1880s. An American entomologist figured out that another organism from Australia, the Vidalia Beetle (*Rodolia cardinalis*), was very effective in controlling scale. Because the beetle feeds exclusively and voraciously on the scale, its introduction almost eliminated Cottony-Cushion Scale in orchards within a few years. Today, both the scale and the beetle are present in very low numbers in U.S. orchards, and the scale is not considered an economically important pest.[47]

MODIFYING REGIONAL AND LOCAL CLIMATE

While climate plays a major role in the distribution of vegetation globally, vegetation also has a major influence on local and regional climates. For example, the rainfall in the Amazon Basin is, in part, a consequence of the existence of the region's forests. About half of the mean annual rainfall in the basin is recycled by the forests themselves via evapotranspiration—a process that accounts for the total amount of water transferred from plant-covered surfaces of Earth to the atmosphere, which combines evaporation from open bodies of water and from the soil, with "transpiration," the movement of water within plants and its eventual loss to the atmosphere as water vapor. Computer modeling studies suggest that extensive deforestation in the Amazon could dramatically reduce rainfall in the region so that the forests might not be able to reestablish themselves. At the local scale, trees create "microclimates" by providing shade and surface cooling associated with evapotranspiration. Deforestation can also result in climatic change in areas adjacent to the forest, with losses in rainfall that can affect agriculture and the availability of water in these areas.[48-50]

Storing Carbon and Stabilizing the Climate

Land ecosystems of the world are large storehouses of organic carbon. Estimates place these carbon stores in the range of 2,100 billion metric tons, or BMTs (about 2,300 billion or 2.3 trillion tons). About 600 BMTs of this are stored in plant tissue, and 1,500 BMTs are stored in the soil as organic matter.[51] Recent analyses of the global carbon cycle indicate that carbon stores, in at least some land ecosystems, are growing, albeit in small annual increments. Furthermore, it is argued that this growth in plant and soil carbon stocks is slowing the buildup of carbon dioxide in the atmosphere, thereby slowing the rate of climate change and providing the valuable ecosystem service of stabilizing the global climate system.[52]

Over the past decade, environmental policy makers have recognized the important role that terrestrial carbon sinks play. In fact, as part of the U.N. Framework Convention on Climate Change, policy makers have sought to increase the size of these sinks through direct management actions as a way of slowing climate change. It is important to note that while terrestrial carbon sinks are enormously important over the near term in taking up our carbon dioxide emissions, we should not rely on them to bail us out over the long term. By the middle of the twenty-first century, they may be so reduced in size that their contributions in taking up carbon, relative to the amounts released from the burning of fossil fuels, may be minimal.

Cultural Services

RECREATION

Outdoor recreation contributes to human well-being around the globe in many different ways. Recreational opportunities on land include activities such as hiking, photography, camping, backpacking, large and small game hunting, bird watching, wildlife viewing, bicycling, and off-road vehicle use. Water-based recreational activities include fishing, boating, water skiing, and swimming. In the United States in 1995, almost 95 percent of the population sixteen or more years of age participated in some form of outdoor recreation.[53] In a recent poll in the United States, more than 65 percent reported that they use the outdoors for health and exercise, relaxation, and stress reduction.[54] Tourism, or as it is now called, "eco-tourism," centered on wildlife and nature reserves, is one of Africa's fastest-growing industries, although it will be sustainable only when the needs of native communities are taken into account, and when a fair proportion of the profits received are used to benefit local populations.

PSYCHOLOGICAL, EMOTIONAL, SPIRITUAL, AND INTELLECTUAL VALUES

The value of leisure in natural settings to humans is multiple and includes (1) personal psychological benefits such as better mental health, personal development and growth, and personal appreciation; (2) psychophysiological benefits such as improved cardiovascular health; (3) social and cultural benefits, such as community satisfaction, reduced social alienation, tighter family bonding, a greater nurturance of others, increased cultural identity, and diminished social problems by at-risk youth; (4) economic benefits such as reduced health costs, increased productivity, less work absenteeism, and decreased job turnover; and (5) environmental benefits such as improved relationships with, and a greater understanding of, our dependency on the natural world. Edward O. Wilson's "biophilia" hypothesis suggests that many of these benefits may derive from our innate and hard-wired bond with other living organisms.[55]

Our natural world is a thing of beauty largely because of the diversity of living forms found in it. Artists have attempted to capture this beauty in drawings, paintings, sculpture, and photography, and it has inspired poets, writers, architects, and musicians to create works reflecting and celebrating the natural world. This work has led to fulfillment and rejuvenation for the artists and their audiences.

Nature also provides for many people great spiritual value. This is true not only for those who believe that all living things are God's creation and must be treated with everlasting reverence, but also for those who do not believe in God at all, but who nevertheless regard life, in all its beauty and variety and mystery, with a profound sense of awe and wonder. Life on Earth may be as sacred to many nonbelievers as it is to any deeply religious person who worships God.

Trying to understand the incredible complexity of biological systems and how organisms have evolved over more than 3.5 billion years is for some of us the most challenging, fascinating, and fulfilling way we could ever imagine to use our powers of observation and our intellects. It may also be the most important, because it will be through our own understanding of the natural world, and that by countless others, and through our collective efforts to help policy makers and the public understand its vulnerability to our nonsustainable behaviors, that we may have a chance to protect the global environment.

Supporting Services

PRIMARY PRODUCTION

Net primary production (NPP) is the amount of plant material generated during a year through the process of photosynthesis. It represents the energy base that powers all ecological processes and, as a result, underlies the capacity of ecosystems to provide all other ecosystem services.

For the world's land ecosystems, NPP is estimated to be about 120 billion metric tons (about 132 billion tons) of new organic matter each year in the form of plant leaves, stems, and roots. This material, in turn, functions as the material and energy base for all of the provisioning, regulating, and cultural services provided to humans by land ecosystems. The annual NPP of the oceans is similar in magnitude, and it supports services such as marine fisheries and the cycling of nutrients in the oceans.[56]

Scientists estimate that currently, human beings consume, degrade, or co-opt about 40 percent of all terrestrial NPP. This has major implications for other species and for ecosystems. For example, we are selectively harvesting plants to the point of causing local and regional extinctions, and we are clearing rainforests, altering the climate in those areas and reducing the viability of adjacent ecosystems.[57,58]

NUTRIENT CYCLING

The global cycles of carbon, hydrogen, oxygen, nitrogen, phosphorus, and sulfur and perhaps as many as twenty-five other elements sustain life on Earth. As these elements move through the environment, either in organic or inorganic forms, they affect other basic ecosystem processes, such as photosynthesis and the decay of organic materials by microbes, and thus affect the way the world works in fundamental ways.

Human activities such as agricultural intensification, urbanization, industrialization, and the introduction and removal of species alter the flow of elements through the environment. These alterations contribute to major environmental

TABLE 3.1. MACRONUTRIENTS AND MICRONUTRIENTS

MACRONUTRIENTS	MICRONUTRIENTS		
Carbon	Arsenic	Iodine	Tin
Hydrogen	Barium	Iron	Tungsten
Oxygen	Boron	Manganese	Vanadium
Nitrogen	Bromine	Molybdenum	Zinc
Phosphorus	Chlorine	Nickel	
Sulfur	Chromium	Selenium	
Calcium	Cobalt	Silicon	
Magnesium	Copper	Sodium	
Potassium	Fluorine	Strontium	

problems such as climate change, acid precipitation, photochemical smog, and "dead zones" in the oceans. We will need to better manage these element cycles if we are going to be successful in using the global environment in sustainable ways.

Table 3.1 lists the elements believed to be essential for animals, microbes, and plants. The elements required in large quantities are referred to as macronutrients. Six elements—carbon, hydrogen, oxygen, nitrogen, phosphorus, and sulfur—are the major constituents of living tissue and comprise 95 percent of the biosphere.

POLLINATION AND SEED DISPERSAL

Pollination

Flowering plants and their animal pollinators work together in Nature. Because plants are rooted in the ground, they lack the mobility that animals have when mating. Many flowering plants rely on animals to help them mate. Bees, beetles, butterflies, moths, hummingbirds, bats, and other animals transport the male reproductive structures, called pollen, from one plant to another, in effect giving plants mobility. One of the rewards for pollinators is food—nectar (a sugary solution) and pollen. Plants often produce food that is precisely correct for a specific pollinator.[59] The nectar of flowers pollinated by bees, for example, usually contains between 30 and 35 percent sugar, the concentration that bees need in order to make honey. Bees will not visit flowers with lower sugar concentrations in their nectar. Bees also use pollen to make "bee bread," a nutritious mixture of nectar and pollen that is eaten by their larvae.[60]

Seed Dispersal

For millions of years, animals have consumed fruits and scattered piles of seed-rich dung across wide expanses of the landscape. This critical ecosystem service helped the large-fruited trees to populate their habitats and migrate across the land in response to a variety of disturbances, including climate change.

As Yvonne Baskin notes, the distribution of plants in both temperate and tropical regions of the Americas may be quite different today from what it was during the Pleistocene epoch, tens of thousands of years ago, because of the demise of large fruit-eating animals that lived then. For example, the lowland forests of Central America lost mastodon-like gomphotheres, giant ground sloths, and other massive consumers of fruits, seeds, and foliage. Without these animals to disperse their seeds, fruit trees may have lost significant portions of their ranges over the millennia.[46] Two renowned botanists, Dan Janzen of the University of Pennsylvania and Paul Martin of the University of Arizona, have suggested that the same thing may have occurred when temperate North America lost its fruit-eating megafauna. Trees bearing large fruits such as the Osage Orange (*Maclura pomifera*), Pawpaw (*Asimina triloba*), and Persimmon (*Diospyros virginiana*) over the millennia grew progressively more sparse and limited in range.[61]

Today, toucans, agoutis, monkeys, fruit bats, and other frugivores (fruit eaters) and seed dispersers provide a critical ecosystem service that helps to maintain the biodiversity of terrestrial ecosystems and their essential life-giving services.

THE ECONOMIC VALUE OF ECOSYSTEM SERVICES

Numerous examples illustrate that ecosystem services have significant economic value. Here we consider three such examples, one that involves the delivery of clean water to New York City, and two others that involve the pollination of cash crops.

Clean Water for New York City

New York City has traditionally been known for its clean drinking water, which has been ranked as among the best in the United States. It originates in the watersheds of the Catskill Mountains. In recent years it has deteriorated in quality because of sewage and agricultural runoff that have overwhelmed the watersheds' natural purification systems. When the quality dropped below the standards of the U.S. Environmental Protection Agency, New York City's administration began to investigate the cost of replacing the natural system with an engineered filtration plant. Estimates for building the filtration plant ranged from six to eight billion U.S. dollars, with annual operating costs of around $300 million—a great deal of money to pay for what once cost nothing.[62]

These high capital and operating cost estimates prompted further thinking about the problem. The result of the reanalysis showed that it would be far cheaper (a one-time cost of $1 billion) to restore the integrity of the watershed's natural purification services, than to build the filtration plant. Faced with these alternatives, the city decided to restore the watershed. In 1997, it raised the necessary funds by issuing a

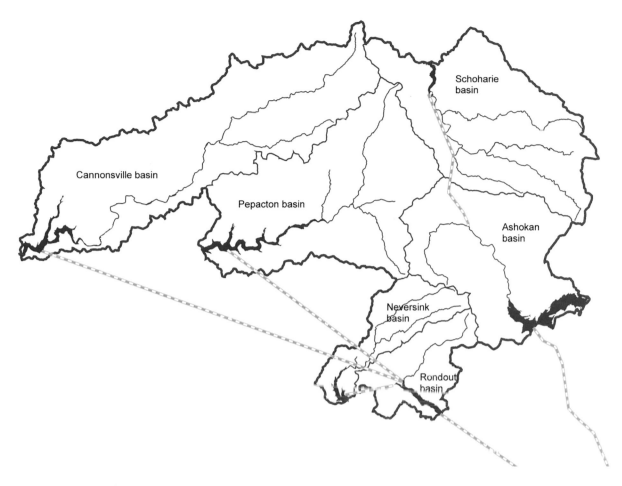

Figure 3.13. Map of the Catskill Watershed. The six major drainage basins are shown, along with their reservoirs and aqueducts en route to New York City. (Map created by the Catskill Center for Conservation and Development, December 2005. Data source: New York City Department of Environmental Protection, Bureau of Water Supply, Global Information Systems 2005.)

bond, both to purchase land in the watersheds of the Catskills and to halt its further development, compensating landowners for restrictions on private development, and subsidizing improvement of their septic systems.[63]

In this case, the citizens of New York City saved their clean water supply by preserving the natural watershed that created it, and saved billions of dollars in the process.[64] They also protected other valuable ecosystem services, such as the watershed's ability to provide flood control and to serve as a carbon sink to help mitigate global warming. Their success is serving as a model for other municipalities, such as Rio de Janeiro in Brazil.

Pollination of Cash Crops

COFFEE IN COSTA RICA

A good example of the economic value of the ecosystem service of pollination can be found on coffee farms of Central America. Working in Costa Rica, a team of World

Palmae
(Cocoineae)

Elaeis guineensis L.

77

Figure 3.14. Oil Palm Tree (*Elaeis guineensis*). (© 1995–2005 Missouri Botanical Garden. From F.E. Kohler's *Medizinal-Pflanzen*, Gera-Untermhaus, 1887, www.illustratedgarden.org/mobot/rarebooks/.)

BOX 3.4

PALM OIL, EDITORS' NOTE

Millions of acres of Oil Palms have been planted in the tropics—in Asia, Africa, Latin America, and Oceania, and many millions more are being planned for the industrial production of palm oil. Palm oil is used in a wide range of foods, such as bread, margarine, cookies, and crackers (where it may be labeled only generically as "vegetable oil"), and in such products as lipstick, toothpaste, and soap.[a] As many as one in ten products on supermarket shelves contain palm oil. Demand is also rapidly growing for palm oil as a source for biodiesel fuel for electric power plants and vehicles, as substitutes for fossil fuels become increasingly attractive. While Oil Palms can be grown and harvested sustainably for local populations, as has been demonstrated in many parts of Africa and in some countries in South America, problems arise when its industrial-scale commercial cultivation requires massive deforestation.

Countries such as Malaysia are developing large-scale palm oil biodiesel production for export, mainly to the European Union, where interest in such biofuels is very high. The usual scenario is that rainforests are cut down and burned to make way for palm oil plantations (the severe forest fires in Indonesia in 1997, it should be noted, were triggered by such burning), threatening countless species because of deforestation.[b] Biodiversity is further eroded as a result of the high levels of chemical inputs—herbicides and fertilizers—that these palm tree monocultures require. Most disturbing is the deforestation that has been the consequence of new palm oil plantations in Indonesia and Malaysia, where 80 percent of the world's palm oil comes from, that is wiping out the rainforest homes of Orangutans (and other species) in these countries, 90 percent of which have already been destroyed. A recent report by the Center for Science in the Public Interest has concluded that palm oil plantations are the main threat to Orangutan survival on the island of Sumatra.[c]

A further problem with palm oil production in Southeast Asia comes from the draining and burning of large areas of peatland to establish new plantations, resulting in enormous, and previously uncounted, carbon emissions into the atmosphere.[d] A study by scientists from two Dutch organizations, Wetlands International and Delft Hydraulics, has concluded that the draining and burning of peatland in Indonesia for palm oil production currently releases more than two billion tons of carbon into the atmosphere each year, catapulting Indonesia into the position of the world's third leading producer of greenhouse gases.[e]

Deforestation in Malaysia, in part secondary to the proliferation of palm oil plantations, may also have played some role in the outbreak in Malaysia of Nipah virus disease in 1998, causing larger numbers of fruit bats carrying the disease to search for food, outside of the forests, in fruit trees bordering pig farms. With their excreta containing the virus, the bats were able to infect the pigs, which then passed the virus onto people. Such widespread deforestation can also result in the emergence and spread of other vector-borne infectious diseases carried by mosquitoes and snails (see chapter 7, page 294).

The interest in palm oil, and in some other tropical oils such as coconut oil, has been fueled in part by concerns about the use of *trans* fatty acids, which can raise cholesterol, in food, and the belief that palm oil would not have such effects. But several studies have contradicted this belief, showing that palm oil raises cholesterol levels and promotes heart disease.[f-j] Palm oil production may therefore be as unhealthy for human beings as it seems to be in some parts of the world for the environment.

Wildlife Fund researchers found that preserving forest fragments around coffee farms boosted their crop yields and raised average incomes by about $62,000 per year, roughly 7 percent of the average farm's annual income. The preserved forest provided a reliable source of bees to help pollinate the plants. Coffee-plant flowers near the forested areas received twice the number of bee visits and double the amount of pollen transfer compared to flowers farther away. The increased pollination led to 20 percent greater yields and 27 percent fewer deformed coffee beans.[65,66]

PALM OIL IN MALAYSIA

Often a pollinator's worth to a crop is apparent only when it is missing or added. The story of Oil Palms in Malaysia illustrates this point. The African Oil Palm (*Elaeis guineensis*) was introduced to Malaysia from the forests of Cameroon in West Africa in 1917. At that time, the weevil that pollinated the palm was not brought along with the trees. For decades, the palm growers of Malaysia relied on expensive, labor-intensive hand pollination, much like the apple growers of Maoxian County in Nepal, as illustrated in the opening figure for this chapter. In 1980, the weevil was imported to Malaysia. The presence of this natural pollinator soon boosted fruit yield in the palms by 40 to 60 percent, and also generated substantial savings in labor, amounting to approximately $140 million per year.[67]

THREATS TO ECOSYSTEM SERVICES

A variety of factors affect ecosystem services. In this section, we review some of the major ones: climate change, deforestation, desertification, urbanization, wetland drainage, pollution, dams and diversions, and invasive species.

Climate Change

The Fourth Assessment Report of the Intergovernmental Panel on Climate Change (or IPCC) projects that human-induced climate change will lead to a warming of the surface of Earth on average of between 1.1 and 6.4 degrees Celsius (around 2 to 7 degrees Fahrenheit) by 2100.[68] The report also indicates that climate change will result in the disappearance or fragmentation of some natural ecosystems in particular areas. The ecosystem services that are expected to be lost are likely to be costly or impossible to replace.[16]

Climate change will affect terrestrial, freshwater, and marine ecosystems. When models that project future climates are coupled with those that project the future distribution of Earth's major terrestrial vegetation types, some dramatic changes show up in the simulations. For example, a recent simulation developed by the Hadley Centre of the U.K. Meteorological Office, using a highly regarded model that couples climate with terrestrial ecosystems, projects that climate change over the twenty-first century will result in the disappearance of much of the Amazon rainforest due to hotter and dryer conditions, and in its replacement by tropical savanna, a tree-grass mix similar to what exists today along the southern and southeastern edges of the Amazon Basin.[69] The loss of the rainforest will diminish the ability of the region to supply forest products, including timber, food (fruits and nuts), and medicines extracted from plants, animals, and microbes. In addition, the disappearance of rainforests will reduce the region's capacity to store carbon and will, in fact, result in the release of a large amount of carbon to the atmosphere that had been stored in organic forms in the forests' trees and soils. The newly released carbon will result in further warming and so is considered a positive feedback to the climate system, with further warming promoting more carbon release, leading to more warming, and so on. And the loss of the rainforests will affect local and regional climate, resulting in drier conditions.

Climate change will also have major effects on freshwater ecosystems and the services they deliver in some parts of the world, with probable changes in the amount, timing, and distribution of rain, snowfall, and runoff, leading to changes in water availability. An interesting example is related to the effects of climate change on snow-packs. Snowpacks serve as natural water storage in mountainous regions and pole-ward portions of the globe, releasing their water in spring and summer. Snowpacks will likely decrease as the climate warms, despite increasing precipitation, because scientists predict that more precipitation will fall as rain and that snowpacks will develop later and melt earlier. As a result, peak stream flows will very likely come earlier in the spring, and summer flows will be reduced, at times dramatically. Potential impacts of these changes include the increased possibility of flooding in winter and early spring, and more water shortages in summer.[16] In regions where summer flows are dramatically reduced, and where competition for water resources is high, the in-stream ecosystem services, such as providing habitat for fish, will be disrupted and, in extreme cases, lost.

Unique marine ecosystems such as coral reefs will also be adversely affected by climate change, because they do not do well outside of a relatively narrow temperature envelope. The last few years have seen unprecedented declines in the health of coral reefs. During the 1998 El Niño, there were record sea-surface temperatures and associated coral bleaching (resulting from loss of the algae that live within the corals and that are required by them for survival). Up to 70 percent of the coral may have died in a single season in some regions (see chapter 2, page 35).[70]

When we lose corals, we also lose important ecosystem services. Coral reefs provide important habitat for fishes, render protection for coastal areas against storm surges, and are important recreation sites for tourists. Reefs are also one of the largest global storehouses of marine biodiversity, with untapped genetic resources.

Deforestation

The most serious problem facing the world's forests is deforestation. When forests are destroyed, they no longer provide us with ecosystem goods and services, and in the tropics, their destruction threatens the cultural and physical survival of native peoples. Deforestation often results in decreased soil fertility and increased soil erosion, with critical plant nutrients, such as nitrogen, being flushed from soils into streams in deforested watersheds.[71] Uncontrolled soil erosion, particularly on steep slopes, can affect the production of hydroelectric power as silt builds up behind dams. Soil erosion can also result in increased sedimentation of waterways, harming downstream fisheries. In drier areas, deforestation contributes to the formation of deserts through the process of desertification (see below). When a forest is removed, the total amount of surface water that flows into rivers and streams actually increases. However, because this water flow is no longer regulated by the forest, the affected region experiences alternating periods of floods and droughts.

Deforestation is a major factor in species loss. Many tropical species, in particular, have limited ranges within forests, making them especially vulnerable to habitat modification and destruction. Migratory species, as well, including birds and butterflies, suffer significant losses.

Deforestation leads to changes in both regional and global climate. Trees pump out substantial amounts of water into the air, which falls back to Earth as precipitation. When a large forest is cleared, rainfall may decline and droughts may become more frequent in the region. Tropical deforestation may also contribute to global warming by causing the release of stored carbon into the atmosphere as carbon dioxide.

Calculating the current rate of deforestation is beset by a number of difficulties, including a lack of adequate satellite coverage, disagreements over definitions, and other problems (see also chapter 2, page 30). For the period 1980–1995, the FAO estimates that forested areas in the industrialized world increased by about 2.7 percent, while they decreased by 10 percent in developing countries. As discussed in chapter 2, estimates of deforestation for tropical humid forests are around 120,000 square kilometers per year (about 46,300 square miles), and for tropical dry forests, around 40,000 square kilometers per year (about 15,400 square miles).[72]

Desertification

Desertification, the degradation of once-fertile arid and semiarid land into nonproductive desert, involves the loss of biological or economic productivity and complexity in croplands, pastures, and woodlands. It is due mainly to climate variability and unsustainable human activities, the most common of which are overcultivation, overgrazing, deforestation, and poor irrigation practices.[73] Seventy percent of the world's drylands (excluding the hyperarid deserts), or about 3.6 billion hectares (about 8.9 billion acres), is degraded. While drought is often associated with land degradation, it is a natural phenomenon that occurs when rainfall is significantly below normal recorded levels for a long period of time.[74]

By definition, drylands have limited freshwater supplies, but precipitation can vary greatly during the year in these regions. In addition to this seasonal variability, wide fluctuations occur over years and decades, frequently leading to drought. Over the ages, dryland ecosystems have become attuned to this variability in moisture levels, with plants, animals, and microbes able to respond quickly to its presence or its absence. For example, satellite imagery has shown that the vegetation boundary south of the Sahara can move by up to 200 kilometers (about 124 miles) when a wet year is followed by a dry one, and vice versa.[75]

People have survived in dryland areas by adjusting to these natural fluctuations in climate. The biological and economic resources of drylands—notably soil quality, freshwater supplies, vegetation, and crops—are easily damaged. People have learned to protect these resources with age-old strategies, such as by adopting nomadic lifestyles in agricultural practices and in the raising of livestock. However, in recent decades these strategies have become less practical with changing economic and political circumstances, population growth, and a trend toward more settled communities. When land managers cannot, or do not, respond flexibly to climate variations, desertification is the result.

Desertification causes a reduction in a variety of ecosystem goods and services. Food production is undermined. If desertification is not stopped in an area, malnutrition, starvation, and ultimately famine may result. Famine typically occurs in areas that also suffer from poverty, civil unrest, or war. Desertification often helps to trigger a crisis, which is then made worse by poor food distribution and an inability of people to buy what is available. This is particularly true in Africa, where two-thirds of the continent is desert or drylands, and almost three-quarters of the extensive agricultural drylands are degraded to some degree.[74]

The stabilization of soil against water and wind erosion is diminished during desertification. Degraded land may cause downstream flooding, reduced water quality, sedimentation in rivers and lakes, and the silting of reservoirs and navigation channels. It can also cause dust storms that can exacerbate human health problems, including eye infections, respiratory illnesses, and allergies, and that can carry dust and its constituent organisms for thousands of miles (see "Introduced Species," chapter 2, page 47).

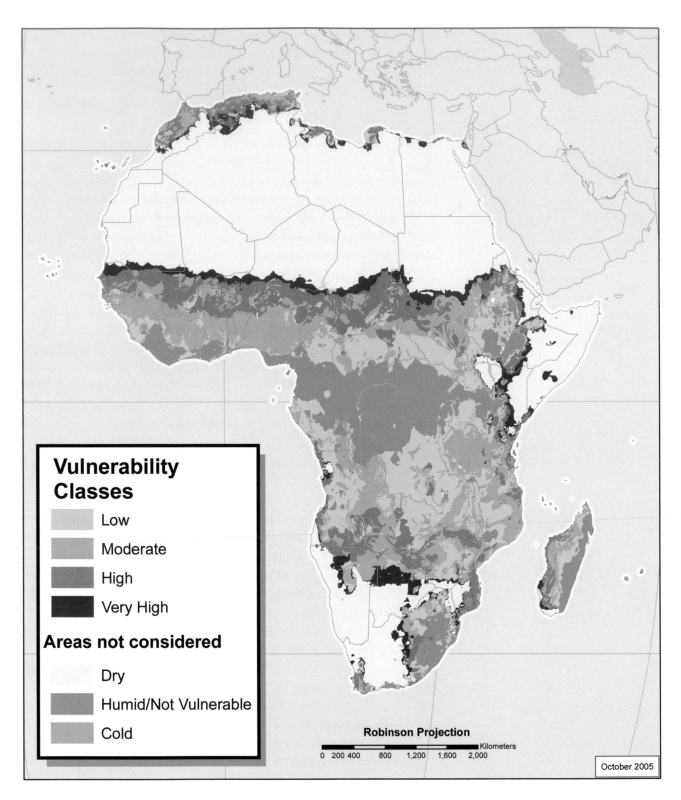

Vulnerability Classes

Low

Moderate

High

Very High

Areas not considered

Dry

Humid/Not Vulnerable

Cold

Robinson Projection

Kilometers

0 200 400 800 1,200 1,600 2,000

October 2005

Figure 3.15. Desertification Vulnerability in Africa, 2005. (Courtesy of Soil Survey Division, Natural Resources Conservation Service, U.S. Department of Agriculture.)

TABLE 3.2. PERCENTAGE OF TOTAL POPULATION LIVING IN CITIES, BY COUNTRY/REGION

REGION	1950	1970	1990	2010
United States	64	70	75	82
Japan	35	53	63	66
Europe	51	63	72	74
Central America and the Caribbean	38	52	64	70
Sub-Saharan Africa	12	19	28	40
China	12	17	27	45
World	29	36	43	51

Source: World Resources Institute, EarthTrends. 2006; available from earthtrends.wri.org/ [cited September 26, 2006].

Finally, critical habitat for both plant and animal species is lost as desertification proceeds. Loss of habitat has a range of consequences, including some economic ones. For example, in Africa, where desertification is currently having its greatest impact, ecotourism is being negatively affected in some areas.

Urbanization

Urbanization and population growth have been among the most unique features of the twentieth century. In 1700, only five cities in the world, all political capitals, were home to more than half a million people. By 1900, this number had climbed to forty-three cities. In 1950, only one city, New York, had more than ten million people. By 1975, there were five cities with populations of more than ten million, and in 2001, there were seventeen, with projections that this number will climb to twenty-one by the year 2015. By the year 2000, almost 50 percent of the world's people were urban dwellers.[76]

Cities can have significant benefits for the environment. They can attract people from rural areas where they may be doing more damage. Costa Rica, for example, is a conservation success story, as much because the Intel Corporation created thousands of jobs in San Jose as it is because land has been set aside in reserves. However, cities also gobble up land and take in ever increasing amounts of energy, water, and materials. They pump out commercial goods and services, along with pollutants and solid wastes. The impact of cities on the environment is wide ranging. Land-use changes and pollution associated with urbanization alter the goods and services that natural ecosystems provide. Plant and animal habitat is lost and some of ecosystems' stabilization functions are diminished. For example, urbanization often leads to increased erosion and reduced natural watershed control of floods. The filling in of wetlands for urban expansion eliminates their water cleansing function.[77] Many of these "lost" functions are costly, if not impossible, to replace.

Wetland Drainage

For hundreds of years in many places across the globe—for example, in the Netherlands, England, Germany, India, Burma, Vietnam, Thailand, the Philippines, Sudan, New Zealand, and the United States—people have drained wetlands to make new agricultural land. One estimate is that for the world as a whole, 10 million square kilometers (about 3.9 million square miles) of wetlands, an area about the size of Canada, have been drained during the twentieth century.[78]

In the lower forty-eight U.S. states, drainage has reduced the wetland area by about half—from 100 million hectares (about 247 million acres) to 53 million hectares (about 131 million acres). Much of the corn belt in the United States was created by the drainage of 17 million hectares (about 42 million acres), mostly in the twentieth century. In the South, the Mississippi River bottomlands were drained and eventually became important sites for growing rice and soybeans. In the Florida Everglades, another million hectares or so (about 2.5 million acres) were drained for agricultural use. And much of the Central Valley of California was also converted from wetlands to croplands and pastures.[79,80]

Wetland drainage has produced some of the world's most productive agricultural land, but at the expense of critical wildlife habitat, flood plains, and vitally important natural filtration systems for flowing waters.

Pollution

The pollution of rain and snowfall, air, water, and the land has diminished ecosystem goods and services in a variety of ways. The air pollutant ozone, for example, can reduce the growth of agricultural crops and plants of many different kinds in natural ecosystems. It is estimated that ground-level ozone levels in China are high enough to reduce crop yields nationwide by between 10 and 20 percent each year.[81] (The reader should note that ozone at ground level is a pollutant, damaging some plants, including crops, and causing flares of respiratory disease in people, while the same compound, O_3, in the stratosphere acts as a protective barrier, blocking harmful ultraviolet radiation from reaching the surface of Earth [see chapter 2, page 60].)

Pollution of rain and snowfall with sulfur and nitrogen compounds results in acid rain that damages plants and impoverishes soils. It also acidifies surface waters, killing plants and the animals that inhabit them. The nitrogen component of acid rain can act as a fertilizer to both land and water plants. In some estuaries, such as the Chesapeake Bay, the nitrogen inputs in acid rain are high enough to cause unwanted algal blooms that make the water unattractive for recreation and can lead to the creation of oxygen deficits in the water column associated with the decomposition of dead algae.[82] If the oxygen deficit is large enough, dramatic fish kills can result. Similar results can be caused by nitrogen pollution of estuaries from agricultural runoff and from point sources such as industrial complexes and sewage treatment plants.

Heavy metals spewing from smelters in such places as Sudbury, Ontario, have accumulated in soils downwind, causing the death of much of the plant life in the affected areas. Without the protection of vegetation cover, erosion has become a major problem at these sites.[83]

Dams and Water Diversions

Dams and water diversions have had major effects on the ability of aquatic ecosystems to provide goods and services. Dams have been built for several purposes, including extending irrigation, controlling floods, and generating electricity.

For dams, a slate of successes in food and energy production, and in flood control, has sometimes been overshadowed by environmental problems. One problem has been that dams change the natural flows of rivers and alter the quality of the aquatic habitat, resulting in species losses. A dam causes water to back up, flooding large areas of land and forming a reservoir, which destroys former plant and animal habitats. The natural beauty of the countryside is often negatively affected, and certain forms of wilderness recreation are compromised or made impossible.

In arid regions, the creation of reservoirs behind dams results in a greater evaporation of water, because the reservoirs have a larger surface area in contact with air than did the original rivers. As a result, serious water loss and increased salinity of the remaining water can occur. When the dammed water is used for irrigation in arid regions, there is always the risk of salinization, the process of various mineral salts accumulating in the soil. In rain-fed agriculture, precipitation that moves through the soil profile runs off to a river, carrying the salts away. Irrigation water, however, generally soaks into soils and does not run off the land into rivers. When the irrigation water evaporates, the salts remains behind and gradually accumulate. Salinization results in crop yield declines and, in extreme cases, renders the soil completely unfit for agriculture. This has occurred in some regions of the Central Valley of California, for example, once promoted as the "Fruit Basket of the World," where selenium salts have now reached high levels in irrigated agricultural soils.[84] By the end of the twentieth century, salinization had affected about 20 percent of the world's irrigated land.[85]

Dams may also encourage the spread of waterborne diseases, such as schistosomiasis, that may spread throughout local populations. Schistosomiasis is a tropical disease caused by a parasitic worm that can damage the liver, urinary tract, nervous system, and lungs. (See chapter 7, page 297, for a detailed discussion of schistosomiasis.)

Invasive Species

Centuries of human commerce and travel have led to a redistribution of Earth's biota. This process has accelerated through time, and today, invasive species are considered a major environmental issue. Invasive species compete with native species for food and habitat or may prey on them. They may also cause disease. By altering the food web and affecting ecosystem functions in a variety of ways, invasive species reduce the ability of ecosystems to deliver life-sustaining goods and services to people.

While invasive species are occasionally introduced into an area by natural means, people are usually responsible for the introductions, both with and without intent. For example, because it had attractive flowers, the Water Hyacinth was brought from South America to Florida. Over the years, this rapidly growing plant has crowded out native species and impeded boat traffic, clogging many of Florida's waterways. In 1990, the Amazonian Water Hyacinth population also exploded in Lake Victoria in East Africa. Bordered by Kenya, Uganda, and Tanzania, Lake Victoria is an essential source of water and fish protein for its surrounding human populations. A combination of invasive species and nutrient enrichment from land-use changes have transformed it from a clear, well-oxygenated lake with an incredible diversity of cichlid fishes (a large family of freshwater fish, some species of which are important food fish, e.g., tilapia, that live mostly in tropical areas of Africa and the Americas) to a murky, oxygen-depleted, weed-choked lake with markedly reduced fish diversity. For many experts, the changes due to eutrophication and invasive species have been so great that the ability of the lake to meet human needs is now threatened.[86]

CONCLUSION

All of us, regardless of where we live on this planet, depend completely on its ecosystems and on the services they provide, such as food, water, climate regulation, disease management, the breakdown of wastes and the recycling of nutrients, spiritual fulfillment, and aesthetic enjoyment. A central conclusion of the Millennium Ecosystem Assessment, a recent report by the United Nations of the status of Earth's ecosystems and the services they provide to us, is that over the past half century, humans have changed our planet's ecosystems more rapidly and extensively than in any comparable period of time in our history. Most of these changes were made to meet rapidly growing demands for food, fresh water, timber, fiber, and fuel. While the changes have clearly contributed to substantial gains in human well-being and economic development for some, many others have benefited little. In addition, the changes have resulted in a substantial and largely irreversible loss in the diversity of life on Earth and have had large costs in the form of the degradation of many ecosystem services and of an exacerbation of poverty for some groups of people. It is imperative that we understand much better than we do the makeup and functioning of the planet's ecosystems and how human activity threatens the services they provide, and it is essential that we do everything we can to preserve them.

◆ ◆ ◆

Suggested Readings

Biophilia, Edward O. Wilson. Harvard University Press, Cambridge, Massachusetts, 1984.

Breakfast of Biodiversity: The Truth about Rain Forest Destruction, John A. Vandermeer. Food First, Oakland, California, 1995.

The Dancing Bees, Karl von Frisch. Harcourt, Brace & World, Inc., New York, 1953.

Ecosystems and Human Well-being: Synthesis, Millennium Ecosystem Assessment. Island Press, Washington, D.C., 2005.

"The Great Race." *Economist*, 2002;364(8280).

Human Wildlife: The Life That Lives on Us, Robert Buckman. Johns Hopkins University Press, Baltimore, Maryland, 2003.

Microcosmos: Four Billion Years of Evolution, Lynn Margulis and Dorion Sagan. University of California Press, Berkeley, 1986.

Millennium Ecosystem Assessment, www.millenniumassessment.org.

Nature's Services: Societal Dependence on Natural Ecosystems, Gretchen C. Daily (editor). Island Press, Washington, D.C., 1997.

The New Economy of Nature: The Quest to Make Conservation Profitable, Gretchen C. Daily and Katherine Ellison. Island Press, Washington, D.C., 2002.

The Trees in My Forest, Bernd Heinrich. Harper Collins Publishers, New York, 2003.

Valuing Ecosystem Services: Toward Better Environmental Decision-Making, National Research Council. National Academies Press, Washington, D.C., 2004; see www.nap.edu.

The Work of Nature: How the Diversity of Life Sustains Us, Yvonne Baskin. Island Press, Washington, D.C., 1997.

CHAPTER 4

MEDICINES FROM NATURE

David J. Newman, John Kilama, Aaron Bernstein, and Eric Chivian

The library of life is burning and we do not even know the titles of the books.

—DR. GRO HARLEM BRUNDTLAND,
former Director-General of the World Health Organization and
former Prime Minister of Norway

While biodiversity makes possible the ecosystem services that keep us and all over living things on this planet alive, it also provides us with medicines that relieve our physical suffering and treat, and in some cases even cure, our diseases. Even with the advent of modern combinatorial chemistry, which can churn out thousands of synthetic chemicals in the hope that one or a few may have biological activity, pharmaceuticals derived from Nature remain a mainstay of medical practice today, as they have for millennia. In the United States, for example, half or more of the most prescribed medicines come from natural sources, either directly, or indirectly when these natural compounds serve as models or as chemical templates for new drugs.[1] And despite significant contributions in recent years from what is referred to as "rational drug design" (which is designing drugs based on a knowledge of the molecular target they are intended to interact with), the majority (116 out of 158) of new small-molecule drugs that were licensed by the U.S. Food and Drug Administration between 1998 and 2002 (or their equivalents in other countries) can be traced ultimately back to natural origins.[2] The developing world relies even more heavily on Nature for medicines, with a significant proportion of developing country residents, as high as 80 percent in a study sponsored by the World Health Organization, relying on medicines from natural sources.

Most literature on the subject of natural medicines tends to focus on how plants, particularly plants from tropical rainforests, have given us medicines such as quinine (for malaria) and pilocarpine (for glaucoma) and has made the argument that if we cut down and burn these forests, we will lose other highly useful medicines yet to be discovered, some from species that will be lost before they have even been identified. Given that a large number of such medicines have already been found, that fewer than 1 percent of known plants have been fully analyzed for their potential pharmacologic activity, that only a fraction of the world's total complement of species have been discovered to date (see chapter 1), and that we are losing species at a very rapid rate, 100 to 1,000 times or more above background levels, this warning is certainly justified. But the focus on higher plants in the tropics, while clearly of great importance, tends to obscure the vital contributions that other species, including animals and microbes, have made to medical treatment, both on land and in the oceans, both in the tropics and in other regions of the world.

In this chapter we look at some important examples of medicines we have obtained from Nature. We also consider the use of natural pesticides, because these indirectly contribute to human health, most significantly in developing countries, but also now increasingly in the industrialized world. As tempting as it is to cite pharmaceuticals as the principal reason for why we need to preserve biodiversity, because the benefits are so clearly identified, the reader should bear in mind that such an emphasis can result in our overlooking all the other ways that our health and our lives depend on other species and on the healthy functioning of natural ecosystems, subjects that are covered in other chapters of this book.

WHY NATURAL MEDICINES?

All organisms have developed a host of compounds to protect themselves against infections and other diseases, and all organisms interact, at least in part, with other members of their own species and with other species by means of chemicals. Some of these chemicals have evolved, such as the antimicrobial peptides present in an enormous variety of species (see chapter 5, page 198), to prevent infections by bacteria, fungi, and other organisms. Others have developed so that predators can instantly subdue their prey, such as the toxic peptides of cone snails, snakes, scorpions, and spiders. Still others serve to defend vulnerable species from being eaten, such as the potent alkaloids in the skin of some frogs and toads. Nature has been a combinatorial chemist for at least 3.5 billion years in manufacturing these and other chemicals and has tested and retested them in "field experiments" that have involved many millions of subjects, sometimes over millions of years, endlessly modifying their structures to best fit the functions for which they were intended. The compounds that did not work are no longer around or have assumed other functions.

The significance of this in our search for medicines in Nature is that in many cases "clinical trials" have already, in essence, been done. It is our task, if we are observant enough, to find those clues in plants, animals, and microbes that reveal the presence of potential human medicines, be they antibiotics that prevent microbes from developing resistance to them, or potent cancer chemotherapeutic agents that work by novel mechanisms. Because of the remarkable uniformity of all living things, particularly at the genetic and molecular level (see chapter 5 for further discussion of this point), these biochemical leads from other organisms can result in the discovery of important new drugs, some of which might never be discovered in the lab.[3]

THE HISTORY OF NATURAL PRODUCTS AS MEDICINES

With the exception of some specialized mixtures of minerals and metals, used predominately in Asia, most medicines prior to the middle of the twentieth century were derived from plant sources. The first written records come from Mesopotamia around 2600 B.C. Among the substances used were oils of various cedar (*Cedrus* spp.) and Cypress (*Cupressus sempevirens*) trees, Licorice (*Glycyrrhiza glabra*), myrrh (*Commiphora* spp.), and the Opium Poppy (*Papaver somniferum*), all of which are still in some use today for the treatment of various ailments. Egyptian medicine is believed to date from an even earlier period, about 2900 B.C., but the first known record is the *Ebers Papyrus* from approximately 1500 B.C., which describes some 700 drugs, mostly plant based, and includes formulas. The Chinese *Materia Medica* has been extensively documented over the centuries, with the first record containing 52 prescriptions (*Wu Shi Er Bing Fang*, 1100 B.C.), followed by 365 (*Shen Nong Herbal*, ~100 B.C.), and then 850 prescriptions (*Tang Herbal*, 659 A.D.). Similarly, documentation of drugs from Indian Ayruvedic medicine dates from about 1000 B.C. in the writings of Susruta and Charaka, among others. This system formed the basis for the primary text of Tibetan Medicine, *Gyo zhi*, written by Yuthog Yonten Gonpo in the eighth century A.D.

In the ancient Western world, the Greeks contributed substantially to the development of herbal drugs, with Theophrastus (~300 B.C.), Dioscorides (~100 A.D.), and Galen (130–200 A.D.) being the major influences. Except for some recording of this knowledge by monasteries in Western Europe during the early to mid period of the Middle Ages (the fifth to twelfth centuries), it was Arab scholars who were primarily responsible for preserving much of the Greek and Roman texts and for expanding them by including their own discoveries, as well as those from Chinese and Indian herbs that were virtually unknown to the Greco-Roman world. Two of these scholars stand out. The first was the great Persian physician and philosopher Avicenna (980–1037 A.D.), who contributed much to the sciences of pharmacy and medicine through such works as his *Canon Medicinae*, which attempted to integrate the medical teachings of Hippocrates and Galen with the biological insights of Aristotle and which served as a textbook in medical schools in Asia and Europe for several centuries. The second was Ibn al-Baytar (c. 1179–1248) from Andalusia, who described some 1,400 plants and their medicinal uses in two seminal books—*The Ultimate in Materia Medica* and *Simple Medicaments and Nutritional Items*.

The 600 or so medicinal plants catalogued by Dioscorides, a Greek physician, pharmacologist, and botanist who practiced in Rome during the reign of Nero, in his five-volume book *De Materia Medica* were central to the practice of medicine until the early Renaissance (i.e., for more than 1,500 years!), when interest in herbal treatments grew markedly. In 1597, John Gerard, curator of the Physic Garden of

Figure 4.1. From a Sixteenth-Century Chinese *Materia Medica*. This page describes the medicinal use of the Pomelo (*Citrus maxima* or *C. grandis*), a large citrus fruit, also called a shaddock, that gave birth to the grapefruit (a hybrid cross between a Pomelo and an orange). Scrapings from the inside of the rind have long been used in China to treat skin infections, rashes, and other skin inflammations, a practice that continues today. (Original held at the Harvard-Yenching Library of the Harvard College Library, Harvard University.)

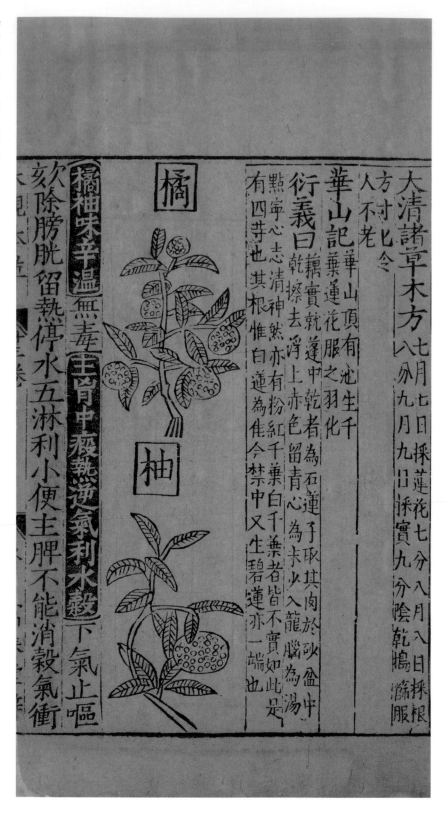

AVICENNA.

Ein Arabischer Arzt, zu Bochara geb. 980 und gestorben 1036.

the College of Physicians of London, published *The Herball*, or *Generall Historie of Plantes*, a massive volume containing 2,200 woodcut images of medicinal plants, which became the most influential text for prescribing medicines in the Western world for 200 years or more. In the nineteenth century, chemists learned how to extract compounds from plants by the use of solvents, distillation, and other means, and a rich period of isolating and identifying biologically active chemicals followed, especially for plant alkaloids. These included morphine in 1804, atropine in 1831, and cocaine in 1860.

Today, the extraction and identification of pharmaceuticals from Nature rely on such techniques as thin-layer chromatography and high-pressure liquid chromatography, techniques that allow separation of one compound from another in mixtures; mass spectrometry and nuclear magnetic resonance spectroscopy, which permit analysis of the three-dimensional structures of complex organic molecules; and other methods that can be used to test the biological activity of potential new medicines, for example, in cancer tissue cultures or in live animals.

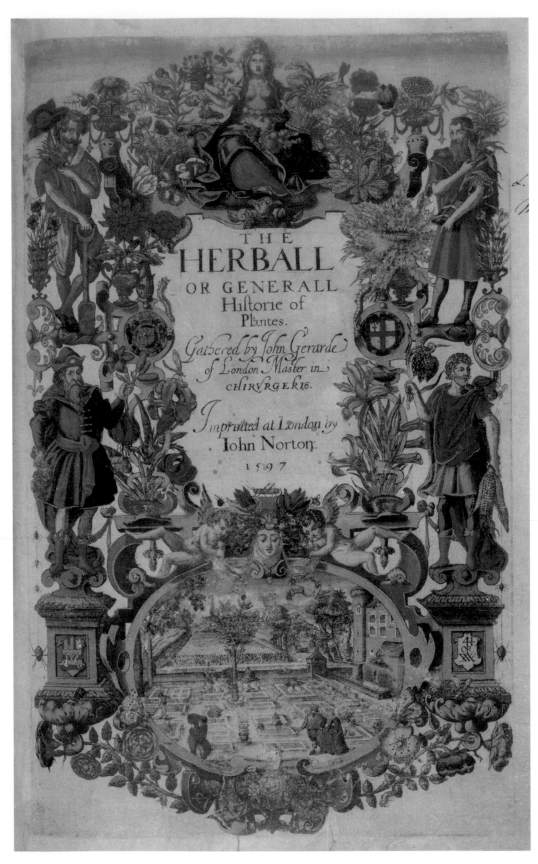

Figure 4.3. Hand-Colored Frontplate of Gerard's *Herball*, 1597. (Courtesy of Bodleian Library, University of Oxford, Reference L.1.15.Med.)

THE ROLE OF TRADITIONAL
MEDICINE IN DRUG DISCOVERY

Ethnobotany, that is, the scientific study of the use of plants by native cultures, including their use as medicines, can be said to have begun with Carl Linnaeus, who in the 1730s published *Flora Lapponica*, his detailed account of plant use by the Lappish, or Sami, people, living north of the Arctic Circle. These observations, like many made since then that draw on knowledge of the natural world gathered over many generations by indigenous peoples, have contributed significantly to the practice of medicine today.

The history of two modern pharmaceuticals—quinine and artemisinin—serve to illustrate our enormous debt to traditional medical healers.

Figure 4.4. Harvesting and Drying Cinchona Bark at the Tjinjroen Plantation in Java, 1882. (From J.C.B. Moens, De Kinacultuur in Azie. Ernst & Co., Batavia, 1882. Used with permission of the library of the Arnold Arboretum, Harvard University, Cambridge, MA.)

Quinine

The isolation of the antimalarial drug quinine from the bark of cinchona trees (e.g., *Cinchona officialinis*) was accomplished by the French chemists Pierre-Joseph Pelletier and Joseph-Bienaimé Caventou in 1820. The bark had long been used by indigenous peoples in the Amazon region for the treatment of fevers. Spanish Jesuit missionaries, after the conquest of the Inca Empire in Peru in the late sixteenth and early seventeenth century, learned of this use from the natives and found that the bark was effective in preventing and treating malaria. They brought this knowledge, along with the bark, back to Europe, where it became widely used and was often referred to as "Peruvian bark." With quinine as the model, chemists subsequently synthesized the antimalarial drugs chloroquine and mefloquine, and they have continued to modify the basic structure of quinine to produce even more effective agents, such as the new antimalarial bulaquine.

Artemisinin

The Sweet Wormwood plant (*Artemesia annua*) was also used as a treatment for fevers in China for more than 2,000 years (it is called *qing hao* in Chinese), but it was not until 1972 that the active compound artemisinin (*qing hao su*, which means the active principle of *qing hao*) was extracted and later identified as a potent antimalarial drug by Chinese scientists. This effort was part of a systematic examination at that time of indigenous plants in China as sources of new medicines. More soluble derivatives, artemether, artether, and artemotil, have been developed in recent years. These medicines, in combination with other antimalarials such as mefloquine, have proved highly effective in treating malaria, particularly the most deadly form caused by *Plasmodium falciparum* (see chapter 7 for a further discussion of malaria), which has become increasingly resistant to the first-line treatments—chloroquine and sulfadoxine-pyrimethamine—in Asia, South and Central America, and Africa. Given that malaria, despite intensive efforts by the world community, continues to kill between one and three million people each year, approximately three-fourths of whom are African children, and to cripple economies around the world, the importance of artemisinin and other effective antimalarials cannot be overstated.

Another possible use of artemisinin is in the treatment of cancer. Its antimalarial activity is thought to be due to its interaction with iron, present in very high concentrations in the malarial parasite. Since some cancer cells, particularly leukemia cells, also have high iron concentrations, they may also be killed by artemisinin, as has been demonstrated in some initial studies with cancer cells in tissue culture. The potential of artemisinin

Figure 4.5. The Sweet Wormwood Plant (*Artemisia annua*). (Photo by Scott Bauer, U.S. Department of Agriculture.)

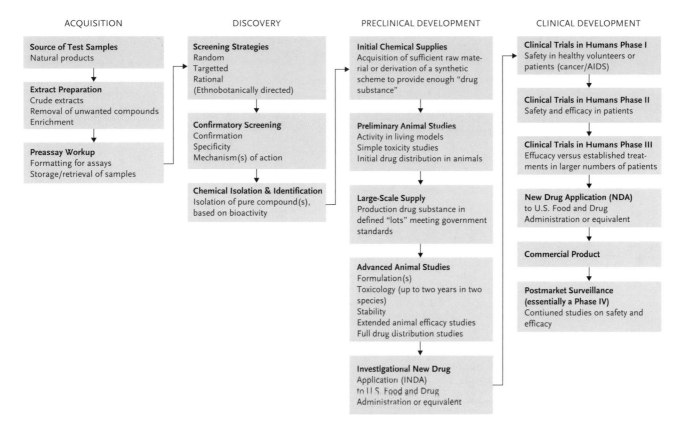

ACQUISITION	DISCOVERY	PRECLINICAL DEVELOPMENT	CLINICAL DEVELOPMENT
Source of Test Samples Natural products	**Screening Strategies** Random Targetted Rational (Ethnobotanically directed)	**Initial Chemical Supplies** Acquisition of sufficient raw material or derivation of a synthetic scheme to provide enough "drug substance"	**Clinical Trials in Humans Phase I** Safety in healthy volunteers or patients (cancer/AIDS)
Extract Preparation Crude extracts Removal of unwanted compounds Enrichment	**Confirmatory Screening** Confirmation Specificity Mechanism(s) of action	**Preliminary Animal Studies** Activity in living models Simple toxicity studies Initial drug distribution in animals	**Clinical Trials in Humans Phase II** Safety and efficacy in patients
Preassay Workup Formatting for assays Storage/retrieval of samples	**Chemical Isolation & Identification** Isolation of pure compound(s), based on bioactivity	**Large-Scale Supply** Production drug substance in defined "lots" meeting government standards	**Clinical Trials in Humans Phase III** Effucacy versus established treatments in larger numbers of patients
		Advanced Animal Studies Formulation(s) Toxicology (up to two years in two species) Stability Extended animal efficacy studies Full drug distribution studies	**New Drug Application (NDA)** to U.S. Food and Drug Administration or equivalent
		Investigational New Drug Application (INDA) to U.S. Food and Drug Administration or equivalent	**Commercial Product**
			Postmarket Surveillance (essentially a Phase IV) Contiued studies on safety and efficacy

Figure 4.6. Natural Product Drug Discovery and Development in the United States.

and its derivatives as cancer chemotherapeutic agents is being actively investigated in a variety of anticancer screens.[4,5]

The combination of a high demand for artemisinin-based antimalarials and limited commercial-scale production of *Artemesia annua* (in only a few locales in China and Vietnam) has left artemisinin-based therapies in short supply. The World Health Organization has stepped in to develop a plan to bolster production.

One possible solution to the supply problem may come from biotechnology. Scientists in California have recently produced the base structure of the chemical artemisinin in the bacterium *Escherichia coli*, and in yeast (*Saccharomyces cerevisiae*), by transferring the necessary genes from *Artemisia annua* into these microbes. For *E. coli* or yeast to become a viable source for artemisinin, the base structure would need to be modified, and the entire process would have to be scaled up to achieve commercial production levels.[6,7]

Traditional medicine, as practiced by indigenous people today, relies on its own version of "clinical trials," where natural products continue to be used only if they have been shown to be effective. These trials may take place over very long periods of time, sometimes over hundreds of years by generations of healers, and they lead to a vast and detailed knowledge of the medicinal properties of many natural substances. That is why many believe there is such enormous potential for finding new medicines among those used by traditional healers.

But there are also problems in using these leads for drug discovery. For one, there is the problem of diagnosis. In the absence of diagnostic tools such as blood tests,

X-rays, MRIs, and invasive techniques such as surgery, traditional healers must rely largely on a patient's history, on the physical exam, and on the external manifestations of disease, all of which can be unreliable. Superstition may also prevent accurate diagnosis and, along with the placebo effect, cloud objective evaluation of the success of treatment. Furthermore, some diseases, for example, those involving the elderly, such as Alzheimer's and most cancers, may be rare in some indigenous populations where life expectancy is short. Finally, knowledge may or may not have been faithfully transcribed from one generation to the next. Traditional medical practices in several parts of Asia, including China, Japan, Korea, and India, have been recorded in great detail over the centuries in written texts, in contrast to those in some other parts of the world, such as among South American Indians, where the passage of knowledge has been primarily by oral means. While these oral traditions may reflect very careful trials and observations, they are prone to errors as a result of unreliable transmission and anecdotal reports.

Nevertheless, indigenous healers have been critically important in the discovery of many new drugs. One study demonstrated that of 119 drugs (derived from some ninety plant species) currently in use in one or more countries, almost three-quarters were discovered by extracting the active chemicals from plants used in traditional medicines.[8]

Tragically, traditional healers now face a double threat—both from the loss of biodiversity that depletes the natural sources that make up their pharmacopoeia, and from encroachment by the outside world that may wipe out their cultures. In the first three-quarters of the twentieth century, more than ninety tribes have become "extinct" in Brazil alone. Scientists are racing to record the secrets these native healers hold before they and the plants and other species they use are gone.

SOUTH AMERICAN INDIGENOUS MEDICINES

As an illustration of indigenous medicines, we focus on a few examples from South America. If we were to cover as well the rich medical traditions of native populations in Africa, Asia, Pacific island nations, and in some other parts of the world, this section would be a book by itself.

Unha-de-gato *or Cat's Claw*

Unha-de-gato, the dried root bark of the tree Cat's Claw (*Uncaria tomentosa*) from the Amazon, is used extensively by indigenous peoples of the area, particularly the Campa, Amuesha, and Ashaninca tribes. Although found mainly in Peru, the plant is also fairly widely distributed throughout South America, in the forests of Bolivia, Brazil, Colombia, Guatemala, Honduras, Surinam, Trinidad, and Venezuela. Used as a contraceptive and as a treatment for the symptoms of arthritis, rheumatism, gastric ulcers, and wounds, its active agents have yet to be identified.

Jaborandi or Ruda-do-monte

This material is extracted from the leaves of the plant *Pilocarpus jaborandi*. Indians of northeast Brazil, including the Apinaje, have used it as an inducer of breast milk production and as a diuretic. The active principle, the alkaloid pilocarpine, was first isolated in Brazil in 1875, and from the early to the middle part of the twentieth century it was the drug of choice for the lowering of fluid pressure inside the eyeball (intraocular) in glaucoma, an eye disease that can lead to an irreversible loss of vision because of high intraocular pressures. The value of pilocarpine, present in other *Pilocarpus* species, as well, but found in the highest concentrations in *P. jaborandi*, reached an estimated US$40

Figure 4.7. Member of the Apinaje Community in Brazil Gathering Jaborandi Leaves. (© Michael Balick.)

million by 1980. By 1997, approximately 25,000 people were employed in gathering *P. jaborandi* in the wild, collecting as much as 1,200 tons of its leaves in the Maranhão, Piauí, and Pará states of Brazil. Such overharvesting was thought to be unsustainable, leading Merck Pharmaceuticals in the late 1990s to set up a 700-acre (about 283 hectares) plantation growing another *Pilocarpus* species, *P. microphyllus*, that produced approximately ten tons of leaves with high pilocarpine concentrations per acre each year.

Pau d'Arco *or* Lapacho

Pau d'arco, also known as *lapacho*, extracted from the bark of the tree *Tabebuia impetiginosa*, is known in the West as lapachol. The mixture had been used in the Amazon River basin for the treatment of several diseases, including general fevers, malaria, syphilis, and cutaneous infections, and in dry, rural, forested areas from Mexico to Paraguay for treating stomach disorders. Research in the 1970s at the National Cancer Institute showed lapachol to have antitumor activity in mice. However, clinical trials of the drug were halted due to unacceptable toxicity. Recently, a derivative of lapachol, beta-lapachone, obtained from another species of *Tabebuia*, *T. avellanedae*, has stimulated renewed interest in this class of compounds due to its activity against a range of cancer cell lines, including those of leukemia, breast, and prostate cancer, as well as in several multidrug-resistant cancers. The drug is now in early-stage (phase I/II) clinical trials in the United States.

A REVIEW OF SOME MEDICINES DERIVED FROM NATURE

The Terrestrial Environment

PLANTS

Even in modern times, plants are irreplaceable sources for the development of medicines. As stated above, the World Health Organization has concluded that a significant proportion of people in developing countries rely on traditional medicines, the great majority of which are derived from plants, and has recently begun cataloguing and evaluating the safety and efficacy of these remedies. Industrial countries also rely heavily on plant products for medical treatment. For example, an analysis of prescriptions dispensed from community pharmacies in the United States from 1959 to 1980 indicated that about 25 percent contained plant extracts, or active compounds, from higher plants.[9] Such compounds are not only useful as drugs in their own right, but may be even more useful as leads to other molecules, though synthetic in nature, that are based on them. There are many examples of such plant-based drugs in current use, some of which are given below.

Morphine

Morphine was isolated in 1804 by the German chemist Friedrich Wilhelm Adam Serturner from the Opium Poppy (*Papaver somniferum*), a plant that has been the source of pain-killing drugs for close to 5,000 years. The Sumerians in lower Mesopotamia are thought to have first cultivated opium poppies, which they called *Hul Gil* or "joy plants," around 3000 B.C. It is still the drug of choice in many cases of severe pain and in other conditions, including the agitation that may accompany preterminal states. By using the morphine structure as a model, chemists subsequently developed buprenorphine, a highly effective semisynthetic opiate with a significantly reduced potential for addiction and tolerance, but with a potency approximately 25–50 greater than morphine. As discussed in chapter 6, there are other natural products, for example, some derived from amphibians and from cone snails, that are even more effective painkillers than buprenorphine and that do not seem to result in addiction or tolerance at all.

Vinca Alkaloids

The Madagascar or Rosy Periwinkle (*Catharanthus roseus*, also known as *Vinca rosea*) has had a long history in folk medicine in various parts of the world, including during World War II, when it was used for treating diabetes. After the war, extracts of the plant were investigated by scientists in Canada, and in the United States at the Eli Lilly Company, for their utility in mediating glucose metabolism. It was found in test animals, however, that the extracts caused significant damage to their white blood cells. (Adipocytes, fat-storing cells, were also affected, possibly explaining the positive effect Madagascar Periwinkle extracts had on diabetes.) Prompted by the observations of white blood cell mortality, researchers began looking at how the Madagascar Periwinkle might be used in white blood cell cancers. Four compounds, known generically as the "*vinca* alkaloids," were isolated in the 1960s, and two of these have become some of the most effective chemotherapeutic agents available today. One, vincristine, also known as Oncovin, has revolutionized the treatment of childhood leukemia, turning it, when used together with other cancer medicines, into a disease that can be totally cured in many patients. The other, vinblastine (trade name Velban), has done the same for Hodgkin's disease, a cancer of the lymphatic system. There have been many attempts to develop even more effective agents, and to date, more than 500 potential compounds have been synthesized, with two semisynthetic derivatives, vinorelbine and vindesine, in clinical use.

Figure 4.8. Flowers and Pods of the Opium Poppy (*Papaver sominiferum*), with Opium Sap Dripping from a Cut Pod. (© Michael Balick.)

Aspirin

According to Hippocrates, the ancient Greeks used a concoction made from leaves of the willow tree (which would have contained

Figure 4.9. Madagascar or Rosy Periwinkle (*Catharanthus roseus*).
(Courtesy of U.S. National Tropical Botanical Garden.)

salicylic acid, a precursor of aspirin) as a painkiller during child-birth. In the mid-eighteenth century, a British Episcopal priest named Edmund Stone made careful observations about using the bark from the White Willow (*Salix alba vulgaris*) to treat fevers, a treatment suggested to him by the fact that the bark tasted bitter like that from the cinchona tree, the famous "Peruvian bark," known for its ability to reduce fevers. The chemical salicylic acid was extracted in the 1830s, both from the willow and also from a plant called "queen of the meadow" or Meadowsweet (*Spiraea ulmaria* or *Filipendula ulmaria*), which was well known to practitioners of folk medicine. Salicylic acid, it turns out, is widely distributed in plants, where it functions as a chemical defense against pathogens and as a precursor to other molecules used to defend against various environmental stresses. It subsequently became a popular treatment for fevers, pain, and the inflammation associated with arthritis and gout. In 1860, the German chemist Hermann Kolbe synthesized salicylic acid, and in 1898, Felix Hofmann, working at Friedrich Bayer's laboratory in Germany, added an acetyl group to make it less irritating to the stomach, and called his new product (acetylsalicylic acid) "aspirin" where the "a" referred to the acetyl group, and "spirin" to *Spiraea ulmaria*, the source of the salicylic acid at that time. Aspirin, the first modern drug to be synthesized, may be considered the foundation of today's pharmaceutical industry.

Aspirin has been taken by more people in the last 100 years than any other drug. In the United States alone, more than thirty billion aspirin tablets are consumed each year. But in spite of this widespread use, it was not until the 1970s that its mechanism of action began to be understood, when the British scientist John Vane discovered that aspirin blocks cyclooxygenase, an enzyme that helps produce substances called prostaglandins. Prostaglandins are chemical messengers released by cells that serve many functions, including stimulating nerves that carry pain signals, promoting the leakage of fluid from blood vessels into tissues that have been injured, and causing fevers to develop. By blocking prostaglandin synthesis, aspirin thus helps prevent the pain, swelling, and fever associated with injury and inflammation.

Prostaglandins also make blood platelets stick together, assisting in the formation of clots that can stop the flow of blood, an attribute that is helpful if you're bleeding, but potentially deadly if you're not. Because of aspirin's ability to prevent prostaglandin production, it can also help prevent the blood clots that trigger some heart attacks and strokes. The benefit of taking aspirin for the prevention of these conditions is particularly pronounced in people at high risk of having a heart attack or stroke, such as those who have already had one or those with atrial fibrillation or peripheral vascular disease. And aspirin may be protective against the development of cognitive decline and dementia in older individuals, perhaps because of its action in preventing the formation of small clots in the brain. Simply put, small doses of aspirin help save the lives of millions of people around the world each year, and it all started with the willow tree.[10]

The story of the calanolides is an example of how a combination of serendipity and systematic searching resulted in the discovery of an anti-HIV agent from a plant. It is also an example of how a highly promising new drug was almost lost to discovery, and how many others, present in rainforests and in other biologically diverse habitats, may be lost as well, if we continue to squander our natural resources.

In 1987, leaves and twigs from a South Asian tree, *Calophyllum lanigerum* (known locally as *Bintangor*), that is related to the rubber tree were collected by John Burley of Harvard University's Arnold Arboretum with support from the National Cancer Institute. Burley was in a rainforest thought to be the oldest on Earth, on the island of Borneo in the Malaysian state of Sarawak. The sample subsequently yielded a new compound, calanolide A, that demonstrated significant activity against HIV. Given this promising result, a return visit to find the original tree was made, but the tree was nowhere to be found (presumably it had been cut down), and samples from other *C. lanigerum* in the region failed to yield any of the original compound. Fortunately, another species of *Calophyllum*, *C. teysmannii*, was collected by Dr. Doel D. Soejarto of the University of Illinois at Chicago, and this species also produced an anti-HIV drug called calanolide B. While calanolide B is less potent than calanolide A, it can be produced in greater quantity. Furthermore, it can be obtained from the tree's sap, so it is not necessary to sacrifice the tree in order to produce a continuous, renewable supply of the drug. Of note, in 2004, another *Calophyllum* species from Mexico was reported to produce both calanolides A and B.[11]

Calanolide A was soon synthesized by a small U.S. company, MediChem Research Inc., and under a provision set forth by the National Cancer Institute for

Figure 4.10. *Calophyllum lanigerum*, Photographed by Doel D. Soejarto in Sarawak, March 16, 1996. (© D.D. Soejarto.)

licensure—that development of the drug had to involve the source country—Medi-Chem established a new company together with the state of Sarawak called Sarawak Medichem Pharmaceuticals. This joint venture has been held up as a model of successful benefit sharing by the U.N. Convention on Biological Diversity (see www.biodiv.org/programmes/socio-eco/benefit/case-studies.asp).

The calanolides belong to a class of anti-HIV agents called nonnucleoside reverse transcriptase inhibitors (NNRTIs), but in contrast to other drugs of this class, which generally share common structural elements and can thus induce cross-resistance that limits their effectiveness, calanolides A and B have unique structures and do not induce cross-resistance with other NNRTIs, such as nevirapine. Calanolide B is in preclinical trials in the United States, and calanolide A is approaching phase II clinical trials in combination therapy with other anti-HIV medications.

Sweet Clover (Melilotus Species)

Warfarin is the drug of choice for the long-term prevention and treatment of blood clots. Its ascent to occupy this position involved a series of improbable and, at times, bizarre circumstances that brought it from the cow pastures of North America to the medicine cabinets of millions of people around the world. The story begins with immigrant farmers who settled North Dakota and the Canadian province of Alberta around the turn of the twentieth century. The unforgiving climate there was unsuitable for raising traditional silage crops for their cattle, so these farmers were forced to cultivate melilots, or sweet clover (*Melilotus* spp.), which had been introduced to North America centuries earlier from Europe and Asia. (*Melilotus* species are now considered invasive species in North America.) The use of clover, although a nutritional success, came at a terrific cost. Cows began to die by the score, either from minor bumps or cuts that led to unstoppable bleeding, or from spontaneous internal bleeding. Publications about this hemorrhagic disease in cattle, which became known as "sweet clover disease," began to circulate in the early 1920s. Two veterinarians, Frank Schofield in Alberta and Lee Roderick in North Dakota, deduced that it was only when cattle ate spoiled sweet clover hay that they got the disease. However, it took many more years and a chance encounter with a despondent farmer before the molecule that was to become warfarin was discovered.

On a blisteringly cold afternoon in February 1933, a farmer named Ed Carlson showed up at Karl Paul Link's laboratory at the University of Wisconsin in Madison. He had trekked 190 miles in a blizzard with a dead heifer, a milk can filled with unclotted blood, and 100 pounds of spoiled sweet clover hay (the strength and endurance of farmers in this region was legendary) in hopes of obtaining advice about how to stem the epidemic of "sweet clover disease" in his cattle. Link, it turns out, had become interested in "sweet clover disease" the year before and had been working to develop a sweet clover plant that was low in coumarin, a compound in melilots that has a sweet smell, similar to vanilla, but a bitter taste that caused cows generally to avoid plants with high coumarin concentrations. Carlson's chance visit to Link's lab—he had actually planned to go to the nearby Agricultural Experiment Station, but it was closed—resulted in Link becoming more intensely focused on his coumarin research, and six years later, Harold Campbell, one of Link's colleagues, isolated the compound dicoumarol from spoiled sweet clover hay. Dicoumarol, formed when various molds, including some species of

Figure 4.11. Sweet Clover Field in Custer State Park, in the Black Hills of South Dakota. (© 2005 Gerald Brimacombe.)

Penicillium and *Aspergillus*, metabolize coumarin in sweet clover, inhibits the blood clotting process by interfering with the synthesis and metabolism of vitamin K. It has been successfully used as an anticoagulant since the early 1940s.

Warfarin is a synthetic compound derived from coumarin. Its development and use followed another strange turn of events. When Link was confined to a sanatorium after having been diagnosed with tuberculosis, he had a great deal of time on his hands and soon became an expert in the history of rodent control. Following his release, he resumed his work on coumarin, and in 1948 he patented warfarin, which was, and still is, an extremely effective rodenticide, killing rats and other rodents by causing them to bleed to death. Link then tried to interest physicians to use warfarin with their patients, but they were reluctant, given the compound's reputation as a rat killer, until it became widely known that a Navy captain had survived a suicide attempt with warfarin. Warfarin then rapidly supplanted dicoumarol and to this day is a mainstay in anticoagulant therapy.[12]

ANIMALS

The Medicinal Leech (Hirudo medicinalis)

Not much transpired in the world of anticoagulant therapy after the introduction of warfarin until the advent of lepirudin in the 1990s, the first major breakthrough for

Figure 4.12. Medicinal Leech (*Hirudo medicinalis*). Leeches are once again being used in medicine. (© Carl Peters, Biopharm Leeches.)

anticoagulation in more than forty years. Lepirudin relied on the discovery of yet another natural substance, this one from the saliva of an organism that had been familiar to medicine for centuries: the Medicinal Leech, *Hirudo medicinalis*. The leech's use in medicine was documented almost 3,000 years ago in the tombs of Egyptian pharaohs and is described in texts from India, China, and ancient Persia. In Western medicine, Medicinal Leech use peaked in the early to mid-nineteenth century. Blood-letting had become common in Europe by that point—about 100 million leeches were used annually in the 1830s, with a typical treatment course requiring a dozen or more leeches. Doctors themselves came to be referred to as "leeches." The demand for leeches grew so strong that some governments offered incentives for companies to bolster leech production, which prompted some enterprising individuals to wade fearlessly into marshes in the hope of having as many leeches latch on to them as possible.

In 1884, John Haycraft, working in Strasbourg, discovered that leeches were able to keep their victim's blood flowing by secreting hirudin, an anticoagulant, in their saliva. Hirudin works by binding to and blocking the blood protein thrombin, a key component in the clotting process. Though hirudin was identified at the turn of the twentieth century, it did not become available for use as a drug until 1994, when advances in biotechnology allowed for its mass production. At that time, the leech gene that encodes the hirudin protein was successfully inserted into yeast cells, which were then able to serve as miniature "factories" to churn out the recombinant protein (called "recombinant" because the leech gene was inserted into, or "recombined" with, the yeast genome in order to produce the protein) known as lepirudin, which is nearly identical to hirudin.[13]

Lepirudin has become an essential medicine for some patients whose immune systems destroy their own platelets as a consequence of prior treatment with heparin (another anticoagulant from natural sources, first isolated from dogs' livers in 1916 but not widely available until the 1930s). Part of the heparin molecule resembles a portion of the surface of platelets, and the immune system, in generating antibodies that target heparin, can in some individuals end up destroying platelets. In such cases, heparin, the traditional drug of choice for short-term anticoagulation, cannot be used, and lepirudin is often employed as one of very few possible alternatives.

In addition to hirudin, other contributions to modern medicine have been made by the Medicinal Leech. *H. medicinalis* has become an important ally of plastic surgeons around the world, particularly those who practice microvascular surgery, which is central to many operations, including those where severed body parts such as fingers are reattached. One of the most challenging aspects of such surgeries is the rejoining of ruptured veins. When veins are disrupted, drainage from the wound is inadequate, making the injured tissue prone to swelling, which can compromise blood flow and prevent healing. Leeches help drain excess fluid, allowing the tissues to reestablish their own blood supplies and to heal. The leech's anticoagulants may also facilitate blood flow to the site of the leech bite and, as a result, also to the surrounding tissues. Leeches have helped save thousands of severed fingers, toes, noses, and ears in recent years and have also been used increasingly in breast reconstructive therapy.

Figure 4.13. Three Leeches Being Used to Treat Degenerative Osteoarthrosis (a new term for osteoarthritis) of the Ankle Joint. (Photo by Andreas Michalsen, University of Duisburg-Essen, Germany.)

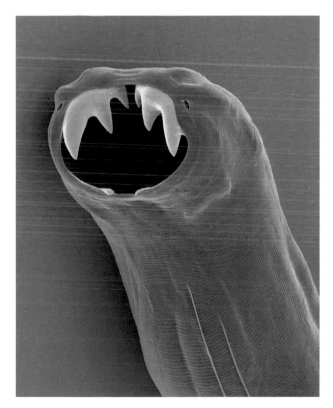

Figure 4.14. Scanning Electron Micrograph (Colorized) of Canine Hookworm (*Ancylostoma caninum*) Mouth. (© Dennis Kunkel Microscopy, Inc.)

The Medicinal Leech has, in addition, been found to be effective in treating the pain, and sometimes the inflammation, of osteoarthritis of the knee and other joints, perhaps due to the anesthetics, anti-inflammatory compounds, and complex cocktails of other bioactive chemicals that it injects with its saliva. What these compounds are and how they are working are not clear, but there is great interest in identifying them as potential new medicines in themselves and as leads to other effective synthetic compounds.[14]

While large numbers of leeches are being raised in various laboratories around the world, leech overharvesting in the nineteenth century, compounded in the twentieth by the draining of wetlands, which is prime leech habitat, has led to marked reductions in the wild of *Hirudo medicinalis* populations in Europe. To make matters worse, dwindling global amphibian populations have compromised the viability of newborn leeches, because they rely on amphibian eggs for some of their first meals. As a result of these conservation concerns, protections have been offered for the Medicinal Leech under endangered species laws in several European countries, and the leech has been listed in appendix II of the Convention on International Trade in Endangered Species of Wild Fauna and Flora (or CITES—see appendix B).

The Canine Hookworm (Ancylostoma caninum)

Phase II clinical trials are currently under way for use of the nematode anticoagulant peptide NAPc2 for the treatment of certain types of heart attacks. This potent medicine, which acts on a different portion of the coagulation cascade than other licensed medications, was not isolated from the saliva of the Canine Hookworm until the late 1990s, even though the anticoagulant properties of hookworm saliva had been recognized since the first years of the twentieth century.[15]

In 2003, NAPc2 showed additional promise when it increased the rates of survival in an experimental primate model of Ebola virus infection. The rationale behind NAPc2's use in Ebola patients was that it might prevent the life-threatening bleeding that can occur with Ebola infections by interfering with the virus's ability to manipulate blood coagulation. (See the section on Ebola in chapter 5.) *A. caninum* has also been the source of a substance that inhibits platelet aggregation, in much the same way that aspirin does.[16]

Pit Vipers

The renin–angiotensin–aldosterone system consists of a set of enzymes and a hormone that act in concert to maintain blood pressure in humans. Central to this system is an enzyme called the angiotensin-converting enzyme, or ACE, which serves to

Figure 4.15. The Pit Viper *Bothrops jararaca*.
(© Wolfgang Wüster.)

convert angiotensin I, with ten amino acids, into angiotensin II, an eight-amino-acid peptide. ACE also promotes the breakdown of bradykinin (a peptide that causes dilation of blood vessels and results in a drop in blood pressure) into inactive substances. Angiotensin II elevates blood pressure by causing blood vessels to constrict and the kidney to retain sodium and fluids. Thus, the action of ACE increases blood pressure, both by converting angiotensin I to angiotensin II and by inactivating bradykinin.

In 1949, the Brazilian scientist Mauricio Rocha e Silva discovered that bradykinin was produced in animals when the venom of the pit viper *Bothrops jararaca* was injected into their blood. In 1965, his student Sergio Ferreira found that the venom not only generated bradykinin in the exposed mammal but also enhanced its ability to cause a life-threatening plunge in blood pressure.[17] These insights, made possible through studies of *Bothrops*, led John Vane (who shared the Nobel Prize in Medicine for his discoveries of aspirin's inhibition of prostaglandins) to hypothesize that inhibiting ACE might be an effective means for treating human hypertension.

Subsequently, Miguel Ondetti and David Cushman and their co-workers at what was then the Squibb Pharmaceutical Company showed that the active molecule in *Bothrops* venom was a simple nine-amino-acid peptide now known as teprotide and that this peptide achieved its hypotensive effect by inhibiting ACE. Teprotide was then synthesized and investigated as a potential drug for hypertension, but its pharmacokinetic properties (i.e., how it was absorbed and metabolized) made it a poor choice. Subsequent investigation led to the synthesis in 1981 of captopril, the first ACE inhibitor that could be taken by mouth. Numerous ACE inhibitors are now available on the market and are among the most effective and best tolerated antihypertensive medications available. They may have never been discovered were it not for *Bothrops jararaca*.[18]

MICROBES

Although the development of medicines from natural products has historically focused on plants, microbes, as the most diverse organisms on Earth, are likely to become the most important sources for new pharmaceuticals in coming years. It is estimated that less than 1 percent of all microbial flora has been investigated to date, but even this figure is probably a significant overestimate, because the microorganisms present in most environments have barely been studied. The surface waters of the open ocean, for example, are believed to contain, on average, a total of more than 500,000 microbes per milliliter (more than 1.6 million per cubic inch).[19] And recent studies have demonstrated an extraordinary, and largely unexplored, diversity of microbial species in marine environments, particularly in the deep oceans, that may be 10 to 100 or more times greater than for that of any previously described microbial environment[20] (see chapter 1, page 14, for further discussion of microbial diversity).

Initially, scientists concentrated on certain microbes in soil that could be easily visualized and grown using culture media similar to those used in clinical microbiology laboratories. These were species belonging to the order *Actinomycetales*. However, once techniques became available in the 1990s to extract DNA from soil samples, the *Actinomycetales*, though initially thought to be the most abundant of all soil microbes, were, in fact, shown to constitute only a small proportion of its total microbial world. *Actinomycetales* do, however, remain critically important in the treatment of human disease, particularly in the development of antibiotics. Some of these will be discussed below, along with some other highly useful medicines derived from microbes.

Penicillin

Microbes were not considered to be important sources for medicines until 1928, when the Scots physician Alexander Fleming noticed that a mold, *Penicillium notatum*, that had contaminated one of his cultures of staphylococcus bacteria inhibited the

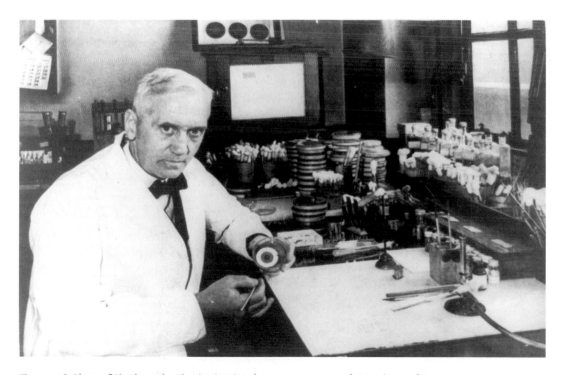

Figure 4.16. Photo of Sir Alexander Fleming in His Laboratory. (Courtesy of U.S. Library of Congress.)

Figure 4.17. Fleming's Original Petri Dish. (a) *Penicillium* mold. (b) Colonies of staphylococcus bacteria. Note how the colonies avoid the area around the mold. (© Alexander Fleming Laboratory Museum/St. Mary's NHS Trust.)

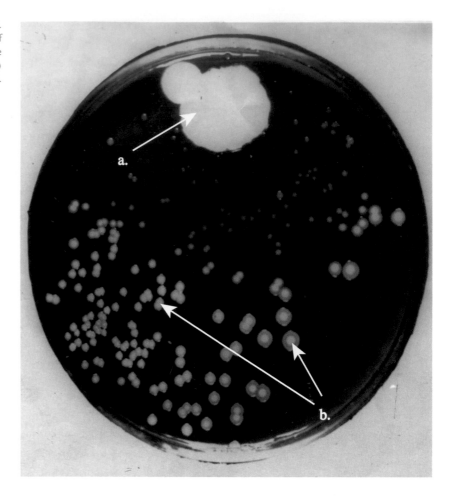

growth of those bacteria that were adjacent to it. He deduced that something from the mold must be killing them, and shortly thereafter he isolated penicillin. A decade later, thanks to the work of many scientists, including Howard Walter Florey, the systemic drug penicillin was developed, and over the next several years, it proved to be a remarkably effective antibiotic for millions of patients. Fleming was awarded the Nobel Prize in Medicine or Physiology in 1945. In the late 1940s, however, initial reports of resistance to penicillin, secondary to its destruction by bacteria, began to surface. Ever since, we have been in a continuous search to find new antibiotics that remain one step ahead of bacteria, organisms that have the ability to rapidly develop antibiotic resistance.

Over the last fifty years, tens of thousands of semisynthetic and synthetic beta-lactams, the chemical family to which penicillin as well as the cephalosporins belong, have been described. Cephalosporins, another group of antibiotics made by a fungus, *Cephalosporium acremonium*, were discovered in 1945 by Giuseppe Brotzu, a professor of hygiene from Cagliari, Sardinia, in a sample of seawater that was adjacent to a sewer drain. Brotzu had observed that young people who swam in the polluted water, in contrast to many others in the city, never came down with typhoid fever and surmised that something in the water must be protecting them. Following this hunch, he was able to culture *C. acremonium* in the water and isolate cephalosporins.[21] Almost all beta-lactams are now made by semisynthetic means from the basic penicillin and cephalosporin building blocks. Approximately thirty of these are still in use today.

The original penicillin-based antibiotics have been repeatedly modified over the years so that they can kill bacteria that became resistant to their predecessors. Early resistance to penicillins occurred because some bacteria developed the ability to cleave penicillin's beta-lactam ring by the action of the enzyme beta-lactamase, thus interfering with the antibiotic's ability to block bacterial cell wall manufacture. This spurred the development of beta-lactamase–resistant penicillins, such as methicillin. However, resistance to methicillin among *Staphylococcus aureus* populations soon appeared and has become, in some areas, quite common. Methicillin-resistant *S. aureus* (called MRSA) poses a major risk to human health, because very few antibiotics can treat it. One of these, vancomycin, also derived from a microbe, is described below.

The Aminoglycosides

Stimulated by the discovery of penicillin, Selman Waksman in the United States investigated a number of actinomycetes, tropical soil bacteria that are members of the *Actinomycetales*, to determine if they, too, contained antimicrobial compounds. It should be pointed out that in the middle 1940s, the actinomycetes, because they had aerial myceliae (the network of threadlike projections that are characteristic of fungi) were considered to be fungi like *Penicillium*, but subsequently it was shown that actinomycetes lacked nuclear membranes (fungi, being eukaryotes, possess nuclear membranes), and they were correctly classified as bacteria.

In 1944, Waksman and his co-workers reported the discovery of streptomycin, the first of a group of antibiotics called the aminoglycosides, isolated from the bacterium *Streptomyces griseus*. Streptomycin was highly effective against the bacterium that causes tuberculosis, *Mycobacterium tuberculosis*. However, resistance to streptomycin in *M. tuberculosis*, and in many other microbes, soon appeared, leading to the development of a host of semisynthetic variants, some of which are still widely used.[22]

The Glycopeptides

Glycopeptides, another important antibiotic class, of which vancomycin is the prototype, are the mainstay of treatment for multidrug-resistant *Staphylococcus aureus* infections. Vancomycin comes from a fungus, *Amycolatopsis orientalis*, which was originally found in the early 1950s in a soil sample from the jungles of Borneo. Once easily treated, infections from *Staphylococcus aureus* have in many cases become life-threatening, due to their evolution of resistance to the beta-lactams and to many other classes of antibiotics.

The Tetracyclines

From a screening program at Lederle Laboratories in the middle 1940s, the first of the tetracycline antibiotics, chlortetracycline, was discovered from the bacterium *Streptomyces aureofaciens*, another actinomycete. Although this material was not active against *Mycobacterium tuberculosis*, it was against a broad range of other microbes. In 1950, another tetracycline, oxytetracycline, was found in *Streptomyces rimosus*, and over the next two decades, thousands of derivatives of the base tetracycline structure were either synthesized or isolated from fermentation broths.[23]

Resistance to the tetracyclines, however, also developed rapidly, a result of the ability of resistant bacteria to remove the antibiotic via a pump, thereby reducing intracellular concentrations to ineffective levels. (This basic enzyme system, called the *tet*-efflux pump in bacteria, has also been found, in modified forms, in some tumor cells that develop resistance to certain chemotherapeutic agents.) New tetracyclines that are not recognized by the molecules of this pump system are currently being developed.

The Anthracyclines

Though the anthracyclines, derived from yet another *Streptomyces* species, and their many derivatives that have been synthesized over the last forty years, have activity against some Gram-positive bacteria and some yeasts, they have been used primarily to target cancer cells. Perhaps the best known is Adriamycin (doxorubicin), isolated from a variant of *Streptomyces peucetius* by Aurelio DiMarco in Italy in 1967. Despite having significant side effects (irreversible cardiac toxicity that limits the total lifetime dosage one can take), doxorubicin is still a prime treatment for breast and ovarian carcinomas.[24] Semisynthetic derivatives of the basic structure have been approved in the last decade, and still others are in clinical trials, with efforts being made to overcome some of the toxicity problems.

The Statins

The statins have become among the most, if not the most, prescribed medications in the world today, because of their ability to reduce an individual's risk of having a heart attack or stroke by about 25 percent and, even more remarkably, to cut that person's chance of dying from these diseases by almost as much.

Research has shown that high blood cholesterol levels, particularly the type of cholesterol known as low-density lipoprotein, increase these risks. A number of factors contribute to how much cholesterol people have circulating in their blood, but the two most important are how much cholesterol they eat and how much their livers produce. In most people, these contributions are roughly equivalent. The success of statins can be attributed to their ability to interfere with an enzyme, called HMG CoA reductase, that the liver relies on to manufacture cholesterol. (How this enzyme functions was demonstrated in baker's yeast, *Saccharomyces cerevisiae*, an organism whose immense value to medicine is covered in chapter 5.)

The first statin, called compactin (also called mevastatin), was isolated in 1976 from two different *Penicillium* species, *P. citrinum* and *P. brevicompactum*, by two groups of researchers working independently, one in England and the other in Japan. The drug showed promise in treating patients with a disorder known as familial hypercholesterolemia, an inherited condition in which blood cholesterol levels are dangerously high and can, in its most severe form, lead to heart attacks in children before they turn three years old. In 1987, Merck Pharmaceuticals patented and marketed lovastatin, a drug derived from yet another fungal species, *Aspergillus terreus*, that is basically a modification of the original compactin molecule. Since that time, several other statins have been developed.[25,26]

Current research on statins has investigated whether these drugs may have even broader clinical application. Animal experiments have demonstrated, for example, that statins can increase survival from severe bacterial infections that become systemic and are able to compromise blood flow to vital organs. And statins are being investigated for their ability to help lower the risk of developing dementia, but the jury is still out on this question.[27,28]

Given the value of these fungal compounds to people with heart disease and strokes, two of the most common diseases of humanity, and the promising preliminary results for treating other diseases, the statins must be considered among the most important medications in the world today.

Rapamycin

Originally found to be a potent antifungal drug, rapamycin was headed for the dustbin of medical history shortly after its isolation by the Indian scientist Suren Sehgal in 1972. He found that the molecule produced by a bacterium, *Streptomyces hygroscopicus*, harvested from the soils of Easter Island (known in the native tongue as *Rapa Nui*—hence the name *rapa*mycin), had a worrisome side effect—it suppressed the immune system. Twenty years later, this side effect would warrant its approval for use as a medicine.

Having been abandoned as an antifungal agent in the 1980s, rapamycin's development tracked along three separate, though related, paths. The first entailed its immunosuppressive activity. Scientists discovered that rapamycin could, in a unique way, block the activation of T-lymphocytes, the immune cells that can attack transplanted organs. (An additional benefit has come from this research—because of its unique cellular activity, rapamycin has also advanced our knowledge about how T-lymphocytes become activated.) Rapamycin is less toxic to the kidneys and to other organs than currently available immunosuppressive drugs and has, as a result, become an important part of antirejection therapy in kidney transplant patients.[29]

Figure 4.18. How Coronary Artery Stents Work. (© medmovie.com.)

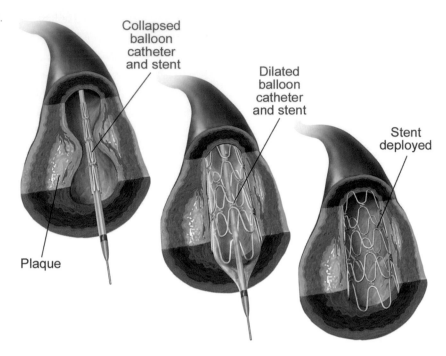

Rapamycin has also shown great promise as an anticancer agent, halting the growth of several tumors. At the present time, it is being investigated for the treatment of brain, lung, and endometrial cancers, as well as for leukemias and lymphomas.[30] A particularly intriguing study of some kidney transplant recipients, performed in 2005, showed that rapamycin could simultaneously prevent rejection of their transplanted kidneys and progression of their Kaposi's sarcomas.[31] Kaposi's sarcoma, a type of malignant tumor of the connective tissue, is now commonly seen in some HIV/AIDS patients. It has been associated with human herpesvirus-8, which, like HIV, can be transmitted sexually and is the result of weakened or suppressed immune systems (the immune response of kidney transplant patients is suppressed to prevent rejection of the transplanted kidney).[32]

The third area involves the use of rapamycin, also called sirolimus, in coronary artery stents, devices introduced in the late 1980s to keep coronary arteries, narrowed from cholesterol deposits, open. Stents are inserted as part of coronary angioplasty, a procedure performed by threading a balloon into a blocked coronary artery and inflating it to open the channel. The stents prop open the artery once the balloon has been removed. The early bare metal stents typically worked for a few months. Thereafter, 30–40 percent of patients would have a recurrence of the blockage, called a restenosis, as the cells lining their coronary arteries grew into the passage created by the stent. Given the prevalence of coronary artery disease, and its place atop the list of causes of death worldwide, finding a way to prevent the restenosis of coronary stents became a top priority. As described in chapter 6, the drug paclitaxel (Taxol) has been used with great effectiveness to coat polymer stents and to prevent the regrowth of endothelial cells. In a series of large clinical studies published in 2005 and 2006, polymer stents coated with paclitaxel (Taxus, Boston Scientific) or sirolimus (Cypher, Cordis, Johnson & Johnson) were compared. Some studies have indicated that the stents are equal in their ability to prevent restenosis and in their complication rates, whereas others showed that sirolimus-coated stents may be best. In either case, coating a stent with either of these drugs derived from natural products has been a major advance in the treatment of coronary artery disease.[33–35]

There has been some recent concern that these so-called drug-eluting stents may be subject to delayed thromboses, some of which occur as long as several months after implantation.[36] Proponents of the devices, however, say that the higher than antici-pated number of reported cases of thrombosis (development of blood clots) reflects the fact that the stents are being implanted in large numbers of high-risk patients, and that clotting can be avoided if anticoagulant medication is continued.

Information from Genome Sequencing of Microbes

Terrestrial microbes are the source for several other naturally occurring antimicro-bial and antitumor agents that are either in clinical use or in clinical trials. But what is potentially the most important development of the last few years is that some organ-isms, such as *Streptomyces coelicolor* (which could be considered to be the equivalent of the "laboratory mouse" for fermentation scientists) and *S. avermitilis* (an organism that produces avermectin, a potent, broad-spectrum pesticide that kills insects and mites and is also perhaps the most important of the veterinary antiworm medica-tions), have had their genomes sequenced. Both genomes are similar in organization. About 60 percent of each codes for essential life functions. Included in the remaining 40 percent are multiple gene clusters that control the bacteria's production of com-pounds called "secondary metabolites" (molecules made by cells that are not imme-diately essential to an organism's growth, development, or reproduction) such as antibiotics, which are usually produced in response to stress. With *S. coelicolor*, while only three antibiotics have thus far been identified, the gene sequences reveal that there are twenty-three or more distinct gene clusters that have the potential for pro-ducing others. In the case of *S. avermitilis*, only avermectin has been discovered, yet there are thirty such clusters. Sequencing the genes of microbes is beginning to uncover the existence of a vast, and still largely unknown, universe of microbial compounds potentially valuable to medicine.[37]

The Marine Environment

Although the diversity of life on land is enormous, that in the oceans may be even greater. As discussed in chapter 1, the marine environment not only contains twice as many phyla as on land, but also has seemingly limitless microbial diversity. Yet, until the 1970s, when safe underwater breathing devices and new ocean sampling technologies began to be developed, most marine environments remained largely unexplored.

Early marine explorations found that life in the oceans produced chemical com-pounds unknown in the terrestrial world, including unique chemicals that relied upon bromine, chlorine, and iodine and made use of previously unknown biochemical processes. These chemical novelties reflect a genetic distinctiveness among marine organisms that took shape over hundreds of millions of years before life ever colo-nized the land. Some of these are described below.

PLANTS

Given the importance of terrestrial plants as sources for human medicines, one would expect the same to be true for plants in the oceans. And, in fact, dozens of molecules from marine plant species, a few of which we examine here, are currently being investigated for their medicinal potential.[38]

While snorkeling near coral reefs off the island of O'ahu in Hawai'i in 1991, Mark Hamann and Paul Scheuer decided to harvest marine mollusks of the sea slug species *Elysia rufescens*. They had a hunch that since other coral reef mollusks had been found to contain unique compounds useful to medicine, there was the possibility that this species might as well. Their hunch proved right, at least in part. They did indeed find a group of unique bioactive compounds in *E. rufescens*, the most notable of which was a potent cancer chemotherapeutic agent, named kahalalide F, which is currently in phase II clinical trials for the treatment of melanoma, non-small-cell lung cancer, and liver cancer. But the mollusk *E. rufescens*, it turns out, does not manufacture kahalalide F; rather, it acquires it by eating plants—various species of green algae belonging to the genus *Bryopsis*. (*Bryopsis*, in turn, may be obtaining the kahalalides from a microbe.)[39]

Several species of red algae from the phylum Rhodophyceae, including those from the genera *Gigartina* and *Kappaphycus*, have been used to harvest molecules known as carrageenans, which have long been employed as thickening agents for foods and cosmetics. Experiments in the late 1980s conducted by Erik de Clercq and his colleagues at the Rega Institute for Medical Research in Belgium showed for the first time that carrageenans might also be useful as antiviral drugs, particularly for sexually transmitted HIV and herpes simplex viral infections (genital herpes infections, which affect about forty-five million people in the United States, can cause painful, and at times debilitating, ulcerations, and while they can be treated with

Figure 4.19. *Bryopsis* **Species**. (Courtesy of William Capman, Augsburg College, Minneapolis, MN.)

certain antiviral agents, they cannot be cured). Carrageenans may also be able to prevent transmission of the human papilloma virus, the cause of genital warts, and of the bacterium that causes gonorrhea, which can lead to infections of the reproductive tract in women and result in sterility.[40]

Two carrageenan molecules, lambda-carrageenan and kappa-carrageenan, have been mixed together in a gel formulation named Carraguard (developed by the Population Council) that is in a phase III clinical trial for HIV prevention in a group of more than 6,000 women in South Africa. Carraguard is applied inside the vagina prior to intercourse and is thought to work by binding either to the HIV virus or to cells infected with HIV, so that they cannot adhere to the cells lining the vaginal canal and cause infection. And, in contrast to many other compounds that are rapidly broken down in the acidic environment of the vagina, Carraguard is quite stable and may remain active for periods of 18 hours or more.[41]

Another molecule that has generated great interest is fucoidan, a complex string of sugars and sulfate molecules derived from some species of brown algae, originally isolated in 1913 by the Swedish scientist Harald Kylin. Fucoidan has shown promise as an anticoagulant,[42] cancer chemotherapeutic agent, contraceptive, and antimicrobial. In tissue culture studies, fucoidan was shown to kill T-cells infected with human T-cell leukemia virus type 1 (HTLV-1).[43] Leukemias are cancers of the white blood cells, and T-cell leukemia may develop when they become infected with HTLV-1 (see also chapter 7, page 317, for a discussion of HTLV-1). Fucoidan may help prevent the spread of tumor cells by blocking their attachment to the extracellular matrix, the web of sugar and protein molecules between cells that provides structure to tissues.

A similar mechanism may explain fucoidan's ability to block the attachment of a sperm to an egg (and thus its potential use as a contraceptive), as well as its ability to prevent herpes simplex and HIV infections in animal models. The combination of these two attributes in the same molecule—as a contraceptive and as a preventative for sexually transmitted viral diseases—has raised considerable interest in fucoidan's future as a medicine.[44,45]

ANIMALS

Marine animals also produce a host of unique and medically useful molecules.[46] One is bryostatin-1, a potent anticancer agent discovered in the bryozoan *Bugula neritina*. Bryozoans, or "moss animals," are tiny marine creatures that live in colonies held together by calcium carbonate that frequently attach to hard surfaces such as ship bottoms or pier pilings. They can encrust such surfaces or form lacelike or fanlike structures. Bryostatin-1, both a potent inhibitor of cancer cell growth and a potent activator of the immune system, is currently in clinical trials for various types of cancer, including leukemias and lymphomas. Bryostatin-1 may not be made by the bryozoan itself but by a commensal bacteria that lives inside it.[47]

Another example is trabectedin, developed by the Spanish company PharmaMar and licensed for eventual sales in the United States by Johnson & Johnson (pending FDA approval), a potent anticancer agent extracted from the Caribbean sea squirt *Ecteinascidia turbinata*. (Sea squirts, also known as ascidians or tunicates, are balloon-shaped animals that feed by siphoning water. Because they possess a primitive

Figure 4.20. The Bryozoan *Bugula neritina*. (Courtesy of San Francisco Bay: 2K, California Academy of Sciences.)

backbone and are thought to be a link between invertebrates and vertebrates, their genomes are being closely studied.) In early clinical trials, trabectedin has shown excellent results in the treatment of some of the more difficult to treat cancers, such as soft tissue sarcomas (rare malignant tumors that occur in such tissues as fat, tendon, and muscle) and breast cancer.[48–50] The mechanism of action for trabectedin is unlike that of any other available cancer chemotherapy in that it interferes with the DNA repair machinery in tumor cells. Finding new medicines that work by unique mechanisms, as was also the case with paclitaxel (see chapter 6), has the potential of leading to the discovery of whole new classes of therapeutic agents.

Yet a third compound from marine animals, from one living in deeper waters, is discodermolide from the Caribbean sponge *Discodermia dissoluta*. Discodermolide is currently in phase I clinical trials. It works by stabilizing the microtubules of the mitotic spindle, in much the same way that paclitaxel does, so that cancer cells cannot divide.[51] Finally, there is manoalide, a compound isolated from the sponge *Luffarriella variabilis* that has anti-inflammatory activity and has led to a family of similar compounds that have been synthesized, some of which have entered clinical trials.[52]

In other cases, medicines have been developed from coral reef organisms. One good example is the soft coral *Pseudopterogorgia elisabethae*, an animal found in the reefs off the Bahamas Islands that produces pseudopterosins, a class of anti-inflammatory agents used in skin creams (although, as with *Bugula neritina*, a commensal one-celled organism living in *P. elisabethae*, possibly a dinoflagellate, seems to be responsible for at least part of the manufacture of the compound). The weight of *P. elisabethae* required annually to meet commercial needs is in excess of 2,000 kilograms (2.2 tons), an amount that would quickly decimate populations of the soft coral animal. Through careful study of its reproduction and regrowth, it was discovered that natural populations of *P. elisabethae*, when carefully pruned, would fully regrow in less than eighteen months. Based on these findings, a program was created to manage coral reefs near Grand Bahama Island for the cultivation of *P. elisabethae*.

Figure 4.21. The Soft Coral *Pseudopterogorgia elisabethae*. (© 2005, Howard R. Lasker.)

Figure 4.22. Close-up of *P. elisabethae*. (© 2005, Howard R. Lasker.)

The success of this program, now in operation for more than twelve years, has convincingly demonstrated that the corals represent a significant sustainable resource when harvesting is scientifically managed and controlled, provided, of course, that the coastal marine ecosystems in which they live are preserved.[53]

Among the many compounds that have been developed from marine animals, a few in particular reveal how marine biodiversity loss can foreclose on the possibility of discovering new medicines. During an exploration of marine invertebrates in the waters off the coast of the Central Philippines, a sea squirt, identified as a species of *Diazona*, was collected and examined as part of a National Cancer Institute project. The animal

Figure 4.23. A Species of the Sea Squirt, Genus *Diazona*. (Courtesy of Paddy Ryan/www.ryanphotographic.com.)

contained an exciting new class of anticancer agents, called the diazonamides, which are potent inhibitors of cancer cell growth, achieving this effect by a mechanism of action that was at the time unknown to researchers. Multiple recollection attempts were made without success, and it was initially thought the organism, along with the promising diazonamides it contained, was lost. Eventually, the *Diazona* species was found deep within coral reef caves, and in recent years, one of the diazonamides, diazonamide A, has been synthesized. But the initial inability to find another of the specific *Diazona* species prevented early development of these compounds as anticancer drugs.[54]

The same thing happened with curacin A. This compound, which has a mechanism of action similar to that of paclitaxel, was discovered in blue-green algae (cyanophytes) in 1994 off the coast of the Caribbean island Curaçoa. Luckily, its discoverer was able to culture the original source, because when he returned to the same area a short time later, it had been developed and the original site no longer existed. The stories of the diazonamides and curacin A are the marine equivalents of what happened on land with calanolide A, described above.[55]

The Sponge Cryptotethya crypta

In the autumn of 1945, Werner Bergmann found a previously undescribed sponge, which came to be named *Cryptotethya crypta*, while swimming in the waters off

Elliott Key, Florida. Bergmann and his colleague Robert Feeney discovered two remarkable compounds in *C. crypta*, one they named spongouridine, the other spongothymidine. What was remarkable about these compounds was that they contained the sugar arabinose in place of the usual ribose or deoxyribose as the sugar component of each nucleoside, chemicals that form nucleotides, the building blocks of RNA and DNA. It had been believed prior to Bergmann and Feeney's discovery that only ribose and deoxyribose sugars would have any biological activity.

Scientists had tried to manipulate the other major component of nucleosides, the "bases," in the hope of making new drugs. After the discovery of spongouridine and spongothymidine, they began to focus instead on the sugars, and in the last few decades, these efforts have led to the discovery of critically important new medicines. One is cytarabine, or Ara-C, patterned directly after spongouridine and spongothymidine. Produced in 1960, Ara-C is a key component of combination chemotherapy for acute leukemias and lymphomas. Another is azidothymidine or AZT, synthesized in 1964. AZT, belonging to a group of drugs called the nucleoside reverse transcriptase inhibitors, was the first antiviral medication approved for the treatment of HIV/AIDS (see also section on primate research in chapter 6, page 240) and is still an essential part of HIV/AIDS therapy and of preventing the transmission of the virus from mother to child. Still other drugs, such as acyclovir, the treatment of choice for herpes simplex virus infections, and newer HIV/AIDS medications, owe their inspiration to the discovery of the novel compounds from the sponge *Cryptotethya crypta*.[56,57]

MARINE MICROBES

Since the 1970s, the search for new pharmaceuticals in the oceans has focused on macroscopic plants and animals, in particular, those found in coral reefs. These investigations continue today, as has been described above, with many studies demonstrating the enormous value of these marine resources to human medicine. What has been especially exciting in recent years is the discovery in the marine environment of another world that holds immense potential for new medicines with novel structures and activities. William Fenical's group at the Scripps Institute of Oceanography in La Jolla, California, is one that has pioneered research in the medicinal potential of marine microbes.[58] They have looked into compounds derived from microbes that live symbiotically with marine invertebrates and marine plants, as well as those that are free living.

As mentioned above, ordinary seawater contains more than 1.6 million microbes per cubic inch (100,000 per milliliter) and an astounding level of microbial diversity.[59] Levels of microbial diversity in ocean bottoms may be even greater. Like soils on land, they ultimately become the repositories for the settling of all marine organic matter, which provides an unusually rich environment for microbes.

Until recently, microbiologists did not appreciate the diversity of marine microbes because they relied on culture techniques that originated in the nineteenth century to isolate and identify them. What is clear today is that these old methods failed to find their targets because the cultures lacked the proper ingredients to support the growth of marine microbes, as they did not reflect the environments in which these microorganisms evolved. With new cultures based on careful analyses of the nutritional requirements of marine bacteria and fungi and new molecular biology methods that analyze genetic material, researchers have begun to give us a glimpse into the enormous biodiversity of the marine microbial world.

To ensure further discovery, access to even the most remote ocean environments needs to be possible, and most important, these undeveloped marine resources must be carefully preserved. Over the past five to ten years, the field of marine microbial drug discovery has been clearly established as one of the most exciting frontiers in natural drug discovery, with more than 100 papers published in the scientific literature each year.

Almost all of the examples of marine pharmaceuticals mentioned above (with the exception of discodermolide) have come from shallow waters, yet deep-water environments contain organisms belonging to plant, animal, and microbial families that also offer great promise to human medicine, some of which have never been seen before. These include species that live in very hot water, for example, those that live in the so-called "deep smokers" of volcanic vents, and those that reside in very cold environments, such as the "cold methane seeps." Accessing such deep-water organisms, however, remains technically difficult, and as a result, the oceans at depths greater than about 50 meters (165 feet) remain largely unexplored. Dredging or trawling can be employed, but both lead to extensive seafloor habitat damage and the problem of nonselective sampling. Manned submersibles or remotely operated vehicles can also be used, but the high cost of these methods precludes their being extensively used for routine collecting.

One example of recently identified, deep-water organisms that have potential use as medicines are a major new group of actinomycetes in deep ocean sediments. As mentioned above, this group of Gram-positive bacteria is the most prolific source for antibiotics used for human infections, such as the aminoglycosides, tetracyclines, and anthracyclines. Prior to 2002, only one marine species in the order *Actinomycetales* had been described, and it was common to assume that other marine actinomycetes must have come from the land. However, Fenical's group showed that this new marine actinomycete, found at depths of 1,100 meters (3,660 feet) in tropical oceans, belonged to a completely new genus, *Salinispora*. They also discovered that this *Salinispora* species contained a unique proteasome inhibitor (the proteasome is an organelle within mammalian cells that recycles proteins the cell no longer needs) that, like a few others found in some fungi and bacteria and also manufactured synthetically, can result in the death of tumor cells. The proteasome inhibitor from the *Salinispora* bacterium is about to enter phase I clinical trials as an antitumor agent.[60]

A theme that is becoming more and more apparent in marine drug discovery, as well as that on land, is that often compounds that are thought to be made by plants and animals are actually being produced by microbes that reside within them. One such case, in addition to those mentioned above, is a set of cyclic peptides thought to have been made by the sea squirt *Lissoclinum patella* but discovered instead to be manufactured by a cyanobacterium, *Prochloron*, an organism that completely infiltrates the sea squirt. The same may be true for some sponges. Evidence is accruing that *Porifera* sponges from both shallow and deep waters, for example, contain microbes that may be involved in the production of pharmacologically active molecules, originally isolated from the sponges. An example is manzamine A, a potential antimalarial and tuberculosis treatment, isolated from a *Porifera* sponge that was collected at a depth of 150 feet off of Indonesia but made by actinomycetes living commensally within it.[61]

Such examples make clear that marine microbes may be sources not only of the enormous array of novel pharmaceutical compounds that had been attributed to them, but also of others that had been thought to be produced by larger organisms.

HERBAL MEDICINES IN INDUSTRIALIZED COUNTRIES

No discussion of natural products would be complete without mentioning herbal medicines. Several are in wide use in the United States, Germany, and other industrialized countries. What distinguishes these "medicines," designated along with vitamins and minerals as "dietary supplements" in the United States, from those approved for therapeutic use (e.g., by such agencies as the U.S. Food and Drug Administration) is that (1) they often contain a mixture of active substances, and it is generally not clear what specific compound, or combination of compounds, is producing the observed, or hoped for, therapeutic effect; (2) their preparation is not strictly standardized or monitored, which means that when one takes *Echinacea*, for example, the contents of the pill may vary from one preparation to the next; and (3) the active compounds in the preparation tend not to be modified to enhance their pharmacologic effect, as is the case for many of the natural products that have been approved for use as drugs.

Three of the most popular and best-studied herbal medicines—*Echinacea* St. John's Wort, and Saw Palmetto—are briefly discussed below.

Echinacea

Echinacea extracts used for medicinal purposes typically come from one of a few species: *E. purpurea*, *E. angustifolia*, or *E. pallida*. These species of narrow-leafed, purple-flowered perennials are native to the North American prairie, where they were originally harvested by several Native American tribes for use as medicines, especially for the treatment of venomous bites and stings. The plant and its use as a medicine were introduced to Europeans during the eighteenth century. Currently available extracts may come from any one, or any combination, of the roots, stems, and/or leaves of these three species, and active molecules may be extracted with water or alcohol. The variability in *Echinacea* preparations makes it difficult to know exactly what one is taking and whether a therapeutic benefit can be ascribed to the preparation.[62]

The most widely used *Echinacea* species is *E. purpurea*, in part because of its relative ease of cultivation when compared to other species, and also because of a case of mistaken identity. In the late 1930s, Gerhard Madaus, the founder of one of the leading German manufacturers of *Echinacea* products, came to Chicago in search of *E. angustifolia* seeds, because *E. angustifolia* extracts were the best studied at the time. Though the bag of seeds he brought home may have been labeled *E. angustifolia*, it

really was *E. purpurea*. Undeterred, Madaus created an industry around *E. purpurea* extracts and developed the popular *Echinacea* preparation known as Echinacin.[63]

Each year, worldwide sales of *Echinacea*-based products total hundreds of millions of dollars. Most people take them in the hope that they will either prevent or shorten the duration of the common cold. Considerable research has been conducted to determine whether *Echinacea* extracts can boost the immune system and affect the course of such infections. To date, the results are, at best, mixed.

While well over a dozen clinical trials have been completed, most have serious design flaws. One recent study, however, did use a rigorous, double-blind, randomized, placebo-controlled format to help ensure that any difference between people who took *Echinacea* (three different extracts from *E. angustifolia* were used) and those who took placebo would reflect the effects of the medicine. This study, published in the *New England Journal of Medicine* in 2005, found that those who received *Echinacea* were equally likely to develop an infection with a common cold-causing virus, and had equally severe symptoms if they did develop a cold, as those who received placebo. While critics have asserted that the doses used in the study may have been too small to observe a therapeutic effect, the doses were based on data from a recognized German expert.[64]

St. John's Wort (Hypericum perforatum)

Hypericum perforatum is a perennial plant native to most of Europe (except for those regions above the Arctic Circle), northern Africa, and western Asia. Its medicinal use dates back to ancient Greece and Rome, where it was written about by Hippocrates, Theophrastus, Dioscorides, Pliny the Elder, and Galen, who described that it was effective as, among other indications, a diuretic, a treatment for sciatica and hip pain, and a salve for burns and other skin wounds.

Figure 4.25. St. John's Wort (*Hypericum perforatum*). (© 2006 Steven Foster.)

The medicinal use of *H. perforatum* has evolved over time. During the Middle Ages, the plant was bestowed with a power to ward off evil spirits and was given the name *fuga demonum*, "the devil's scourge." The Latin name *Hypericum* in fact derives from the Greek *hyperikon*, which can be translated as "above an apparition," suggesting that the Greeks, too, might have believed the plant could drive away demons. Yet no mention of this use has been found in their writings. Linnaeus added the species name *perforatum* in 1753 as the plant's leaves appear to have small holes, or perforations. The precise origins of the name St. John's Wort are not known, but the association of St. John with *H. perforatum* may be a result of the plant blooming (at least in most of Europe) sometime around June 24, the presumed birthday of St. John.

The first clear description of the use of *H. perforatum* to treat psychiatric conditions comes from the writings of the sixteenth-century Swiss physician Paracelsus. Today, extracts from *H. perforatum* are one of the most widely used treatments for mild to moderate depression. How St. John's Wort works as an antidepressant is not known, though it has been found to affect levels of several neurotransmitters in the brain, including serotonin, dopamine, and norepinephrine. Unlike many other herbal medicines, the production of St. John's Wort has been somewhat standardized so that each preparation contains similar amounts of the two main constituents—hypericin and hyperforin—thought to be the therapeutic ingredients.

The effectiveness of St. John's Wort has been studied in more than thirty clinical trials. Overall, it appears to be as effective as prescription antidepressant medications, such as the selective serotonin reuptake inhibitors, or SSRIs, for treating mild to moderate depression, but is less effective than prescribed antidepressants for more severe, or long-standing, depression.[65,66]

Because this is a book about biodiversity, it must also be mentioned that *H. perforatum* is considered an invasive species in most places where it grows. It has been cited as a threat to the extinction of several native plant species, including the Small Purple Pea (*Swainsona recta*) in Australia and an orchid, the Small White Lady's Slipper (*Cypripedium candidum*) in Canada.

Saw Palmetto (Serenoa repens)

This plant, native to southeastern United States and the West Indies, is a dwarf palm, with fans 50–100 centimeters (about 20 to 40 inches) long, barbed leaflets, and a stone fruit about the size of a cherry that has a deep purple color when ripe. Native Americans who lived where *S. repens* grew, notably the Seminoles, used the plant to make flour and its fruit to make beverages. (The fruits are evidently an acquired taste, with their flavor having been likened to "rotten cheese steeped in tobacco.")

Native Americans used Saw Palmetto as a medicine as well, at least since the early 1700s, to treat problems ranging from impotence and urinary tract disorders to coughs and general inflammation. Europeans who settled in Florida, where the plant is most abundant, did not use Saw Palmetto medicinally until they began to notice that their livestock fared better when they ate it. That was not until well into the nineteenth century.[67]

Saw Palmetto is a first-line treatment, particularly in Europe, for men with a condition known as benign prostatic hypertrophy (BPH), a noncancerous enlargement of the prostate gland that affects more than 50 percent of men older than sixty years of age. More than twenty randomized clinical trials have shown the extracts to relieve symptoms of BPH, including the retention of urine and decreased urine flow, with fewer side effects than conventionally used medicines.[68]

POTENTIAL MEDICINES IN FOODS

Although we do not consider foods as medicines, many things we eat contain substances that are just as potent and may be just as beneficial to our health as prescribed medications. Most fruits and vegetables, for example, contain compounds (in addition to vitamins) such as plant sterols and the antioxidant flavonoids that are thought to prevent some disease states. These include such flavonoids as the catechin-polyphenols in green tea, which are said to be protective against some cancers and cardiovascular disease, and the flavan-3-ols and the procyanidins in cocoa that may have cardiovascular health benefits. Resveratrol, which belongs to the class of antibiotic compounds called the phytoalexins, produced by plants as defenses against disease, is found in peanuts and some berries such as mulberries but is present in the highest concentrations in the skin of red grapes. It is said to be the component in red wine that explains "the French paradox," the fact that there is a lower than expected incidence of cardiovascular diseases among French people, despite their high-saturated-fat diets.[69] Cruciferous vegetables, such as broccoli, cauliflower, and cabbage, contain isothiocyanates and glucosinolates (among other compounds) that may help prevent certain cancers and contribute to the healthy functioning of the immune system. Oats contain beta-glucan, which may help reduce cholesterol in the blood and prevent heart disease. All of these compounds require additional, carefully controlled studies to determine whether the health benefits ascribed to them are as significant as claimed.[70]

One type of fat in foods, known as omega-3 because of its chemical structure, deserves special mention here. Omega-3 fatty acids come in two forms—long and short. Short-chain omega-3 fatty acids come from plants, and in particular from walnuts and the seed oil of the flax plant *Linum usitatissimum*. Longer chain omega-3 fats are found in fish, especially fatty fish such as salmon, tuna, herring, mackerel, anchovies, and sardines. Two recent studies have shown that they are also found in meat from livestock, including beef, lamb, pork, and poultry,[71] with grass-fed beef having significantly higher concentrations in their meat than grain-fed, feedlot beef, high enough to qualify as a food source for omega-3 fatty acids, according to Food Standards Australia New Zealand (an independent agency that works with the two country's governments to set food standards).[72]

The evidence is now overwhelming that increasing consumption of long-chain omega-3 fatty acids provides protection against death from heart disease. In several clinical studies, omega-3 fatty acids derived from fish oils have been shown to prevent dangerous cardiac arrhythmias, such as fatal ventricular fibrillation (so-called "sudden cardiac death") in some 45 percent of people at risk. These fats have also been shown to reduce the incidence of stroke in men and women. In addition, research has shown that consuming more fatty fish may lower high blood pressure.[73]

The benefits of omega-3 fatty acids to health are not limited to the cardiovascular system. Fish consumption has also been shown to markedly decrease the risk of developing Alzheimer's disease and slow the progression of dementia.[74]

It is generally believed, though debate continues, that the short-chained omega-3 fatty acids derived from plants do not exert the same effects as their long-chain, animal-derived counterparts. And although humans are able to convert short-chain omega-3 fatty acids to long-chain forms, the conversion percentages are small, so eating fish, and perhaps also grass-fed beef, may still be best way to obtain these heart healthy omega-3s.

Given the extreme pressures upon world fish stocks (see sections in chapter 2, page 44, and chapter 8, page 367), many have voiced concern about the harvesting of wild fish for use in the production of fish oil supplements. It turns out that fish acquire their long-chain omega-3 fatty acids from the chloroplasts of marine algae that they eat. In an effort to produce large amounts of long-chain fatty acid supplements that are free of pollutants such as PCBs (which are found in some wild and farmed fish and may be present in concentrated form in some fish oil supplements), while at the same time protecting wild fish stocks, marine algae farms have been developed commercially.

NATURAL PRODUCTS AS INSECTICIDES AND FUNGICIDES

An area of natural product development that has enormous ramifications for human health, but that is not usually considered in such discussions, is the use of natural compounds as insecticides and fungicides. Without adequate supplies of food, people will be malnourished, and no treatment, including that by the most effective drugs derived from natural or synthetic sources, will keep them healthy. Thus, an extremely important use of natural products, from crude extracts to purified compounds and their derivatives, is in agriculture, particularly in developing countries, where the use of expensive synthetic agents is often not feasible. In this section, we consider such compounds from plants, animals, and microbes, both on land and in the oceans.

Insecticides

PYRETHROIDS

Plants have to defend themselves against herbivores that eat them and pathogens that threaten their survival and have, as a result, developed a host of chemicals, including salicylates, for these purposes. It is not surprising, then, that plants produce compounds that are useful insecticides. Among the first to be used and most successful are the pyrethroids. Used since the nineteenth century as insecticides, the toxic constituents of pyrethrum flowers from the plant *Chrysanthemum cinerariaefolium* are referred to as pyrethrins. *C. cinerariaefolium* is native to the Dalmatian Mountains in Croatia, which was the original source for pyrethrins, but the major producers now are Kenya, Uganda, Rwanda, and Australia.

Pyrethrins are composed of six closely related compounds that together account for their insecticidal activity. Chemists have modified pyrethrin structures to produce other compounds, known collectively as the "pyrethroids," in order to improve their properties as insecticides. In contrast to pyrethrins, pyrethroids are stable when exposed to sunlight, are almost insoluble in water, and form breakdown products that are relatively nontoxic. As a result, they are less poisonous to those applying them (and also to other mammals and to birds) and, while being more stable in the environment than pyrethrins, have a lower risk of environmental contamination. Pyrethroids have proven to be highly useful, broad-spectrum insecticides, effective against a wide variety of insect pests—including moth larvae that attack many crops (especially *Heliothis* and *Spodoptera* species on cotton), certain pests of forest trees, and the eggs, larvae, and adults of many species of *Coleoptera* (beetles), *Diptera* (flies), and *Heteroptera* (a large group of 25,000 species of insects) that cause damage to a wide variety of economically important crops, including cotton, soybeans, apples, peaches, strawberries, tomatoes, and rape seed (used for making canola oil).[75]

CARBAMATES

Biologically active carbamates have been used since the seventeenth century in a region of southeast Nigeria, where the Effiks, who inhabited the area, used Calabar Bean seeds from a tree later known as *Physostigma venenosum* to establish the guilt of prisoners. Those accused were forced to swallow a milky potion of crushed seeds in water, and if they survived, they were declared innocent. The toxic alkaloid physostigmine was isolated from Calabar Beans in 1862; it was synthesized in 1935. Since that time, a large number of other compounds have also been synthesized, based on the action of physostigmine, and they have proven to be toxic to many insect species. Physostigmine and its synthetic derivatives cause paralysis and death in insects (and humans) by inhibiting acetylcholinesterase (the enzyme found at the junction of nerve endings and muscles, which breaks down the neurotransmitter acetylcholine. If the enzyme is inhibited, this recycling does not occur, and the muscle ceases to function).[76]

NICOTINE

In the late seventeenth century, water extracts of tobacco leaves (from both the plants *Nicotiana tabacum* and *Nicotiana rustica*), containing nicotine and two closely related compounds, nornicotine and anabasine, were used to control sucking insects. The insecticidal activity of all three alkaloids is due to their action as nerve poisons (see discussion of nicotinic acetylcholine receptors in "Cone Snails," chapter 6,

Figure 4.26. Calabar Bean Tree. (© 1995–2004 Missouri Botanical Garden, ridgwaydb.mobot. org/mobot/rarebooks.)

Leguminosae.

Physostigma venenosum Balfour.

page 265). Today, pure nicotine sulfate, applied by slow-release techniques, is used rather than extracts of the tobacco leaves.

Nicotine never achieved the prominence of other insecticides due to its greater expense, extreme toxicity to mammals, and limited insecticidal spectrum. However, many synthetic derivatives were made using nicotine's basic structure, and a number of these are now in general use (of note, the chemical epibatidine, discussed in chapter 6 under "Amphibians," is closely related to nicotine).[76]

ROTENONE

Preparations from the roots of plants from the genera *Derris, Lonchocarpus,* and *Tephrosia*, which are now known to contain the chemical rotenone, were used as commercial insecticides in the 1930s to control beetles, aphids, weevils, and many other insect pests of various truck crops. Rotenone dusts have also been used against animal parasites such as fleas, lice, and ticks, and by indigenous people throughout the tropics as fish poisons. Rotenone is a metabolic poison, causing a shutdown of energy-producing cellular respiration.[76]

NEEM

Native to India and Burma, the Neem tree is a member of the mahogany family Meliaceae and is known as the Margosa Tree or Indian Lilac (*Azadirachta indica*). The ability of the Neem to control insects was first recognized during locust plagues, when it was observed that locusts would swarm Neem trees but would leave without feeding on them. Extracts from the seeds and leaves have yielded potent insect control products that contain at least four major active compounds (including chemicals known as terpenes, azadirachtin, meliantriol, and salannin) and perhaps twenty minor ones. These are powerful inhibitors of insect feeding and also interfere with insects' ability to breed and metamorphose into adults. Neem products are in wide use as agricultural insecticides.[76]

Although Neem products appear to have little or no toxicity for warm-blooded animals, there are no comprehensive toxicology data available for humans, despite the Neem tree having been used for generations of people in India for its antiseptic properties.

NEREISTOXIN

Animals and microbes, including some marine organisms, have also been important sources for insecticides. The marine flat worm *Lumbrineris brevicirra*, initially used widely as bait for fishing in Lake Victoria, was found to contain an insecticidal poison, called nereistoxin. There are now a family of synthetic agents derived from nereistoxin (known by their trade names Cartap, Bensultap, and Thiocyclam) that are active, both as contact and as ingested poisons, against a broad range of sucking and leaf-biting insects, particularly *Coleoptera* (beetles) and *Lepidoptera* (butterflies and moths), killing adults as well as their eggs and larvae.[76]

Figure 4.27. Seeds and Leaves from a Neem Tree (*Azadirachta indica*). (© Gerald D. Carr.)

AVERMECTINS

The avermectins, originally isolated from a Japanese soil sample in 1976, are a mixture of natural products produced by the soil microbe *Streptomyces avermitilis*. They are effective in controlling the Fire Ant (*Solenopsis inpicta*) and the German Cockroach (*Blattella germanica*) both in agricultural and in indoor urban settings. The mixture, sold as Abamectin, is also effective against leafminers (bugs that burrow between the upper and lower surfaces of leafs), psyllids (also known as jumping plant lice, these are insects that suck on a plant's sugar supplies), and *Lepidoptera* when used on some food crops and ornamentals. Because the genome of *S. avermitilis* has been sequenced, the gene clusters responsible for producing the avermectins have been identified and used to develop improved compounds, such as doramectins.[76]

I n the production and storage of food, fungi have played critically important roles in the history of humankind, not only serving as foods themselves (e.g., mushrooms and truffles) but also as the agents of fermentation for producing such products as bread, beer, wine, soy sauce, and cheese. They have also been the causative agents of crop damage and, once crops have been harvested, of food spoilage. Based on the belief that some fungi might produce agents that would protect them against other, more noxious fungi, and that these agents might become useful fungicides, scientists began to look for such compounds.

STROBILURINS

Strobilurins were first identified as the result of a program begun in late 1976 aimed at discovering new antibiotic agents from the Basidiomycetes (a class of fungi). Studies of the pigments and toxins of the mushroom *Strobilurus tenacellus* yielded two compounds, strobilurin A and B, both of which exhibited potent antifungal activity, a result of their ability to inhibit fungal respiration. Synthetic variants of strobilurin A were found to be even more potent antifungal agents than the natural compound and were highly effective at low concentrations for the treatment of various fungal pathogens on plants, including Wheat Powdery Mildew (*Erysiphe graminis*), Wheat Brown Rust (*Puccinia recondita*), Rice Blast Disease (*Pyricularia oryzae*), Barley Net Blotch (*Drechslera teres*), Grapevine Downy Mildew (*Plasmopara viticola*), and Late Potato Blight (*Phytophthora infestans*).

Strobilurins represent one of the most significant modern innovations in crop protection. There has been a virtual explosion of research in this field, with more than

Figure 4.28. The Mushroom *Strobilurus tenacellus*. (Photo Pietro Curti, AMINT President, www. amint.it.)

500 international patent applications from more than twenty companies and research institutes published, more than 3,000 strobilurin analogues synthesized, and several strobilurin-derived products on the market. There is also interest in a group of related compounds that show similar activity called the oudemansins, isolated from the mushroom *Oudemansiella mucida*.[77]

CONCLUSION

Over the past fifteen years or so, most pharmaceutical companies around the world have relied on combinatorial chemistry, rather than on compounds from Nature, as leads for new drug discovery. The same has been true for most agrochemical companies. This is not surprising, given the promise, widely accepted by these companies, that with combinatorial chemistry techniques one can produce literally millions of novel compounds synthetically, combining chemical building blocks in innumerable different ways.

Indeed, the early experiments in producing libraries of biologically active compounds were initially quite successful, and these initial successes led to massive investment in combinatorial chemistry as the predominant method for providing chemical leads for future work. Another major factor in such a focus, no doubt, was the recognition that any compound produced in this manner was potentially an entirely new one and therefore, if active, patentable. By contrast, there was the belief, although not based on fact, that natural products either could not be patented, or if one were found to have novel activity, intellectual property rights for it could not be claimed.

In retrospect, however, what was not realized was that these combinatorial chemical libraries were based mostly on structures found in Nature, or on those that had already demonstrated biological activity. A recent analysis of the original sources for all drugs approved worldwide during the time frame 1981 to 2002 has concluded that none of these (1,031 in all) could be unequivocally traced to a totally synthetic, combinatorial source. And of the more than 300 compounds from this group that were in phase I, phase II, or phase III clinical trials for cancer treatment, only one could be called a true discovery resulting from combinatorial chemistry alone.

What was also found was that a significant number of the medicines in late clinical trials, or those entering clinical use, had been optimized by combinatorial chemical modifications. These include drugs that were active against methicillin-resistant *Staphylococcus aureus* (MRSA) infections and those that inhibited cholesterol synthesis in ways that differed from the statins. What this study demonstrated was that Nature still provides the best chemical leads for biologically active compounds that have medicinal value, but that combinatorial chemistry techniques may be able to modify these structures to find medicines that may be even more effective than the parent compounds.

We have endeavored in this chapter to demonstrate the extraordinary chemical richness of plants, animals, and microbes on land and in the oceans that hold clues for the discovery of new medicines, clues that we have barely begun to explore, particularly in the microbial world. Human activity is increasingly threatening this richness, and with it, our ability to cure many diseases and relieve the enormous human suffering they cause. We need to greatly expand our investigations of natural compounds before the species that make them are lost, in order to understand and to harness their enormous potential for human medicine. But we need to do so ethically, with deep care and respect for the organisms we are studying, and sustainably, so that our investigations do not threaten their survival.

◆ ◆ ◆

SUGGESTED READINGS

The Elusive Magic Bullet: The Search for the Perfect Drug, John Mann. Oxford University Press, Oxford, U.K., 1999.

The Healing Forest: Medicinal and Toxic Plants of the Northwest Amazonia, Volume 2, *Historical, Ethno- and Economic Botany*, Dioscorides Press, Portland, Oregon, 1990.

Medicinal Resources of the Tropical Forest: Biodiversity and Its Importance to Human Health, Michael J. Balick, Elaine Elisabetsky, Sarah A. Laird (editors). Columbia University Press, New York, 1996.

Microbe Hunters, Paul de Kruif. Pocket Books, New York, 1964.

Murder, Magic, and Mystery, John Mann. Oxford University Press, Oxford, U.K., 2000.

The Natural History of Medicinal Plants, Judith Sumner. Tiber Press, New York, 2000.

Natural Products Branch of the National Cancer Institute, dtp.nci.nih.gov/branches/npb/index.html.

Plants, People, and Culture: The Science of Ethnobotany, Michael J. Balick and Paul Alan Cox. Scientific American Library, New York, 1996.

CHAPTER 5

BIODIVERSITY AND BIOMEDICAL RESEARCH

Eric Chivian, Aaron Bernstein, and Joshua P. Rosenthal

Nature is trying very hard to make us succeed, but Nature does not depend on us. We are not the only experiment.

—BUCKMINSTER FULLER, Interview for the *Minneapolis Tribune*

Biomedical research has always relied on other species — animals, plants, and microbes — to help us understand human physiology and treat human disease. From the bacterium *Escherichia coli*, one hundredth the thickness of a human hair, to an eleven-foot-tall, 1,300 pound (591-kilogram) male Polar Bear; from the Fruit Fly (*Drosophila melanogaster*), which has a life span of only several weeks, to chimpanzees, which, like us, can live for decades, these and numerous other species have brought medicine into the modern era of antibiotics, vaccines, cancer therapy, organ transplantation, and open heart surgery.

Some species possess easy-to-study anatomical structures that make them especially useful as laboratory subjects, such as the enormous axons in squid that conduct electrical impulses from nerve cells to other cells, or the unusually large eggs of the African Clawed Frog (*Xenopus laevis*). Others, such as some species of bears or the Spiny Dogfish Shark (*Squalus acanthias*), have evolved physiological processes so unique that they offer us clues, which might not otherwise be discovered, to the healthy functioning of the human body or to the treatment of disease. Still others, because they are easy to keep in the laboratory, reproduce rapidly and in large numbers, and are able to produce unique strains of genetically uniform individuals, have become the workhorses of biomedical experimentation. We owe an enormous debt that we can never repay to these species, to the countless mice, rats, Guinea Pigs, hamsters, rabbits, Zebrafish, and Fruit Flies, as well as to the myriad dogs, cats, monkeys, sheep, pigs, frogs, and horses, all of which have been sacrificed so that we could expand our knowledge of human health and disease.

(left)
Scanning Electron Micrograph of the Fruit Fly *Drosophila melanogaster*. The image has been colorized. (© Dennis Kunkel Microscopy, Inc.)

While evolution has resulted in significant differences between humans and other life forms, such as the ability (as far as we know) to engage in abstract thinking, Nature has a striking uniformity at the molecular, cellular, tissue, organ, and organ system level that allows us to use a wide variety of other organisms to better understand ourselves.

The underpinnings of this uniformity become clear when we compare our own genetic makeup to that of other organisms. We share about half of our estimated 25,000 genes with both the Fruit Fly (*Drosophila melanogaster*) and the microscopic roundworm *Caenorhabditis elegans*, and considerably more with the Laboratory, or Common House, Mouse (*Mus musculus*). We even share more than 1,000 genes with yeasts, unicellular organisms that, like human cells, have nuclei, and a few hundred genes with bacteria, which do not. This core set of a few hundred genes, believed to be universal to all living things, encodes the information necessary for such basic life functions as DNA replication, the production of proteins from RNA, energy metabolism, and the synthesis of compounds called nucleotide triphosphates, such as adenosine triphosphate, or ATP, the energy currency for all life on this planet. The universality of these genes provides evidence that all organisms evolved from a common ancestor, which most likely had this core set of genes around three billion years ago.[1]

For medical research this is highly significant because, for example, about two-thirds of human mutations known to be linked to certain types of cancer and developmental abnormalities; to some cardiovascular, endocrine, and immune system disorders; and to some diseases, such as diabetes, have corresponding genes in the Fruit Fly. In mammals such as the mouse, the percentage is even greater. As a result, we can study in these other organisms biochemical and physiological processes controlled by the genes we share and can arrive at insights about human health and disease that would be very difficult, and in most cases essentially impossible, to achieve otherwise.

Most of the species used in biomedical research are abundant in Nature, and neither they nor the groups of organisms they belong to are threatened with extinction. We include these species in this chapter to demonstrate the invaluable information they contain for human medicine and to make the point that we must attempt to preserve all other species on this planet, the majority of which have not yet even been identified, for they may be similar treasure troves of medical knowledge. Other species used in biomedical research do come from groups of organisms that are threatened, such as nonhuman primates, amphibians, sharks, bears, horseshoe crabs, gymnosperms, and cone snails. Many species in these groups, which are examined in chapter 6, are Endangered or Critically Endangered, and some have become extinct. When a species is lost, it takes with it the anatomical, physiological, biochemical, and behavioral lessons that it, and perhaps it alone, possesses, lessons that are the result of millions, if not hundreds of millions, of years of evolutionary experiments.

Because medicine today would still be in the Dark Ages without the knowledge we have gained from studying other organisms, the current crisis of biodiversity loss represents nothing less than an enormous threat for biomedical research, the full magnitude of which we can now only guess.

A BRIEF HISTORY OF
BIOMEDICAL RESEARCH

The first documentation in Western medicine of the use of other organisms to comprehend the analogous structures and functions of the human body was in Greece about 2,500 years ago (~450 B.C.), when Alcmaeon cut the optic nerve in a living animal and noted that it became blind. A few decades later in Greece, Hippocrates, considered the "father of medicine" in the Western world, studied the act of swallowing in living pigs and directly observed the beating of the heart, noting how the upper chambers, the atria, alternated their contractions with the lower ones, the ventricles. The strict prohibition against all forms of human dissection at this point in history forced the need for animal experimentation to obtain knowledge of human anatomy and physiology. Early research on animals continued in ancient Alexandria (the seat of learning in the Western world for hundreds of years after Alexander the Great founded it in 332 B.C.), where, for example, two Greek physicians, Herophilus and Eristratos, were able to distinguish the functional differences between nerves that carried sensory information, nerves that stimulated muscles, and tendons that regulated muscle contraction. The work of Galen (129–199 A.D.), the Greek physician to the Roman emperor Marcus Aurelius and perhaps the most influential physician of all time, was particularly important. Galen made many astute observations about human physiology, particularly about the heart, lungs, brain, and spinal cord, by studying animals. One of his enduring legacies was his treatise *De Anatomicis Administrationibus* (On Anatomical Procedures) that contains the first known documentation of precise scientific techniques for animal experimentation.[2]

There is little documentation about Western medicine in general, and medical research in particular, during the Middle Ages (roughly from the fourth to the fifteenth centuries), but it is widely believed that little was accomplished in these areas then, because scientific inquiry, with its emphasis on direct observation in the pursuit of truth rather than on faith, was seen as a threat to Christianity and was thus severely constrained by the political power of the Church.

The same, however, was not true during this period in the Islamic world or in China and India, where science and medicine continued to flourish.

The establishment of universities in Europe beginning at the end of the eleventh century rekindled an interest in science and medicine, as did the ongoing writings of Islamic scholars, who kept the treatises of Greek and Roman scientists and physicians alive by translating them into Arabic. But it was not until the sixteenth century, when Galen's work was rediscovered, that there was a revival in applying scientific methods to the study of anatomy. In 1543, Andreas Vesalius, a Belgian working at the University of Padua in Italy, produced his masterpiece on human anatomy, *De Humani Corporis Fabrica*, relying on the illegal dissection of human corpses (obtained from fresh graves), supplemented with insights gained from vivisection (the dissection of living animals). Almost 100 years later, the British physician William Harvey, also working in Padua, elucidated the circulation of blood in humans using animal models, a seminal demonstration that such experiments not only could provide valuable anatomical knowledge, but could also be used to understand human physiology. Harvey's most famous treatise, *Exercitatio Anatomica de Motu Cordis et Sanguinis in Animalibus* (or An Anatomical Exercise on the Motion of the Heart and Blood in Animals), published in 1628, reports on studies of sheep, dogs, pigs, pigeons, chicks, toads, frogs, serpents, small fishes,

Figure 5.1. Front Page of Galen's *De Anatomicis Administrationibus Libri Novem* (S. Colinaeus, Paris, 1531). (From the Boston Medical Library in the Francis A. Countway Library of Medicine.)

crabs, shrimp (whose transparency made it possible to see the heart beating directly), snails, slugs, scallops, crayfish, bees, wasps, hornets, and flies.

In the first part of the nineteenth century, the center of Western biomedical research moved to France. François Magendie, considered to be the founder of modern physiology, performed pioneering animal experiments that demonstrated that bodily functions resulted from an interplay of various organs. And his student Claude Bernard, with the publication in 1865 of *Introduction à l'Étude de la Médecine Experimentale*, established, based on his animal experiments, one of the most fundamental principles of modern science—that to study one variable, one had to hold all others constant.

BOX 5.1

Islamic Medicine in the Middle Ages

From the eighth through the thirteenth century, the Golden Age for Islamic science and medicine, prominent physicians in the Eastern Caliphate of Baghdad and the Western Caliphate of Cordoba made detailed observations about the functioning of the human body in health and disease and wrote sophisticated medical texts that were influential in the practice of medicine for several centuries (see references a–c). Some of the medical scholars that stand out include the following:

NINTH- THROUGH ELEVENTH-CENTURY BAGHDAD CALIPHATE

- Hunain Ibn-Ishaq (also known as Johannitius), who wrote "Ten Dissertations on the Eye," the oldest systematic treatise on ophthalmology.

- Abu-Bakr Muhammad Ibn-Zaharia (al Rhazes), the greatest Arab clinician of his time, who wrote *Liber Continens*, an encyclopedia of medical practice and treatment, which included *Liber de pestilential*, a treatise on smallpox and chickenpox that is considered a masterpiece in the annals of clinical medicine.

- Abu-Ali Husain Ibn-Abdallah Ibn-Sina (Avicenna), who integrated the medical doctrines of Hippocrates and Galen with the biological concepts of Aristotle. His *al-Quanun* (Canon Medicinae), one of the most influential medical books of all time, served as a textbook in medical schools of the Western world for centuries (see also chapter 4, page 119 and Figure 4.2).

TENTH- THROUGH TWELFTH-CENTURY CORDOBA CALIPHATE

Cordoba at the beginning of the tenth century, according to Maria Rose Menocal, was one of the most sophisticated and cultured cities the world has ever known.[d] In addition to its 900 baths, paved and oil-lamp-lit streets, and clean water brought in by aqueducts, it was the city of books and the center of great learning in science, medicine, art, religion, philosophy, and literature. The caliph's library alone was said to contain 400,000 volumes, and there were an additional 200,000 volumes in scores of other libraries around the city, at a time when the largest library in Christian Europe had perhaps a total of 400 manuscripts. In addition, Cordoba was a place where Christians, Jews, and Muslims lived together peacefully and prospered, perhaps more than at any other time in history, before or since. Some of the luminaries in Cordoba during this time include:

- Abu-Al-Quasim Khalaf Ibn-Abbas Al-Zahrawi (Albucasis), who wrote *al-Tasrif* (The Method), the most important work on medieval surgical practices, used in Europe until the seventeenth century.

- Ibn-Rushid (Averroes), known for the *Colliget*, or the Collection, an encyclopedia on medicine in the Galenic tradition.

- Moses Maimonides or Moshe ben Maimon, also referred to as the Rambam and sometimes as "the Eagle" (and known in Arabic as Abu-Imran Musa Ibn-Maimon), the great Jewish Rabbi, philosopher, and physician who wrote a number of medical texts, the most famous of which was his *Medical Aphorisms* (written in Arabic and titled *Fusul Musa*), used widely by physicians for more than 500 years. Though born in Cordoba, he fled with his family as a teenager, ending up eventually in Cairo, where he was chief physician to the Sultan of Egypt.[e]

The discovery of ether anesthesia in the 1840s and, later in the century, the development of sterile surgical techniques and of the sciences of bacteriology and immunology resulted in an explosion of animal experimentation that has lasted until the present time. To illustrate one example, consider the work of the French chemist

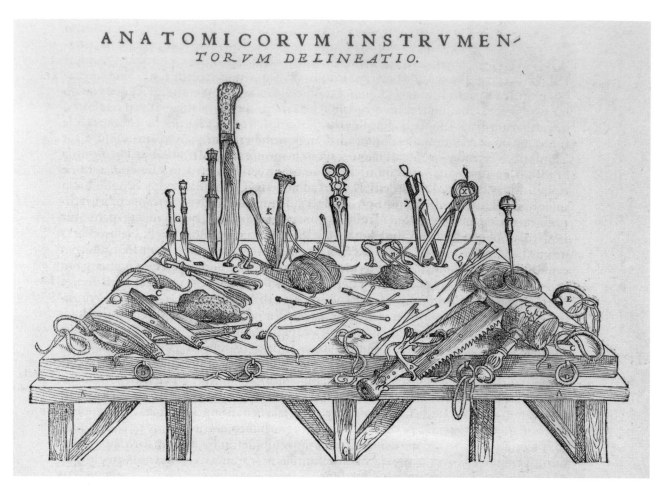

Figure 5.2. Figure of Vesalius's Dissecting Table with His Instruments. (From *De Humani Corporis Fabrica Libri Septem* Ex officina Ionnis Oporini, Basileae, 1543; Boston Medical Library in the Francis A. Countway Library of Medicine.)

Louis Pasteur. In the 1880s, believing that such diseases as cholera and anthrax were caused by microorganisms and were not the result of various imbalances in the body, as had been widely believed since the time of Hippocrates, Pasteur isolated a microbe from the intestines of chickens suffering from cholera, grew it in culture, and then, using the culture, produced cholera in healthy chickens and rabbits. He did the same with anthrax, using rabbits and Guinea Pigs (*Cavia porcellus*). Pasteur noticed that animals repeatedly exposed to these cultures developed a resistance to their effects, and as a result, he began to experiment with weakened cultures, discovering that he could induce immunity with these early vaccines. These animal experiments, which could never have been done on humans, resulted in the discoveries that cholera and anthrax were caused by microorganisms, and that vaccines could be made either to prevent these diseases from developing or, in the case of cholera, to reduce its severity and duration of symptoms.

One important outcome of Pasteur's findings—that microorganisms caused infections—was the work of the British surgeon Joseph Lister. Lister developed the practice of sterilizing surgical instruments, sutures, and wound dressings. Prior to Lister's sterile techniques, almost half of patients who had major surgery died from infection.

Some of the historical milestones in human medicine that have relied on experimentation with other species are noted in figure 5.4, box 5.2, and table 5.1. It should

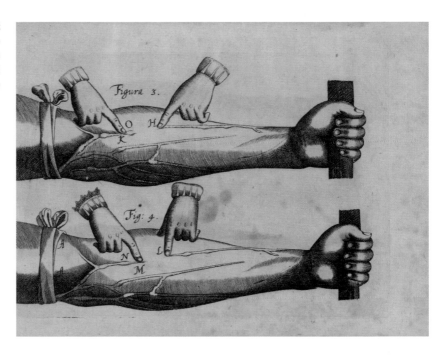

Figure 5.3. **Figure of Harvey's Experiment with the Valves in Veins.** (From *Exercitatio Anatomica de Motu Codis et Sanguinis in Animalibus* [Francofvrti: Sumptibus Gvilielmi Fitzeri, 1628], Boston Medical Library in the Francis A. Countway Library of Medicine.)

also be noted that all medications and vaccines, before being approved for use in humans, are first tested for safety in animals (see figure 4.6 in chapter 4) and that veterinary medicine is totally dependent on animal research for the development of vaccines and of effective treatments for diseases and injuries in pets, domestic animals, and wildlife (see table 5.2).

Although we do not consider plants in detail in this chapter, focusing instead on animals and a few key microbial species, the reader should understand that plants, too, have provided critically important insights for biomedical research. For a recent example, the mustard plant (*Arabidopsis thaliana*), also known as Thale Cress, has become an important model for studying why the bacterium *Pseudomonas aeruginosa* is so virulent and difficult to treat in people.[3] And in 2005, a team of researchers at Purdue University in Indiana made a startling discovery. They observed that an *Arabidopsis* plant, which had inherited two mutant copies of a gene, had the same appearance as a wild-type plant, when it should have looked different. The explanation was that *Arabidopsis* may carry a backup copy of at least some of its genes, which can be called into action when the need arises, a finding that may mean that such backup copies also exist in other organisms as well, including ourselves.[4]

THE ROLE OF ANIMALS AND MICROBES IN BIOMEDICAL RESEARCH

In this chapter, we look at the contributions that some other species have made to our understanding of how the human body works when it is healthy and when it is not. We begin with a brief review of some key areas of biomedical research—genetics, the regeneration of tissues and organs (including a discussion

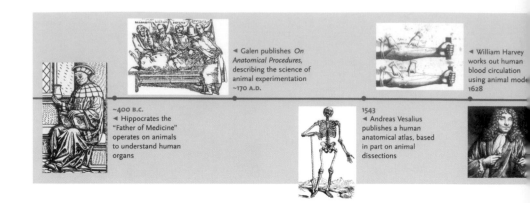

Figure 5.4. Time Line of Some Historical Medical Milestones from 400 b.c. to 1885 a.d. That Rely on Research with Animals, Plants, and Microbes. (Design by Tong Mei Chan.)

of stem cell research and of neurogenesis [the growth of new nerve cells]), and how human immunity, particularly what is called innate immunity, works—all of which demonstrate how dependent human biomedical research in these fields is on a wide variety of other organisms. We could have chosen many other domains of research to illustrate this dependency. As mentioned above, most of the species used in this research are not endangered, but are discussed to provide examples of the critically important medical information that other species, some of which are threatened with extinction, possess.

Genetics

Research on genetics began long before anyone knew what a gene was. Based on their observations that individual date palm trees possessed only male or only female reproductive structures, the ancient Babylonians and Assyrians began practicing artificial pollination from the time of King Hammurabi, around 2000 B.C. Thousands of years earlier, archaeological evidence shows that Native Americans in Mexico began to cultivate the predecessor of corn or maize.[5] In these early genetic experiments, and in those to follow in cultivated plants and in domesticated animals over the millennia, people began to observe that various traits in one generation were passed on to succeeding ones.

Hippocrates, Aristotle, and Plato—from the mid-fifth to the early fourth centuries B.C.—wrote about the inheritance of human traits, including some that seemed dominant (i.e., showed up more frequently) over others in offspring. Beginning with these Greek philosophers and lasting until the eighteenth and even until the nineteenth centuries, a debate raged about whether new life was preformed as miniature complete organisms in the egg or sperm, or whether it developed from components supplied by one or both parents. Animal breeding experiments by the French mathematician Pierre-Louis Maupertuis in the eighteenth century, in which he noted that first-generation hybrids had characteristics of both parents, led him to conclude that individuals were formed from the "seminal fluid" of each parent.

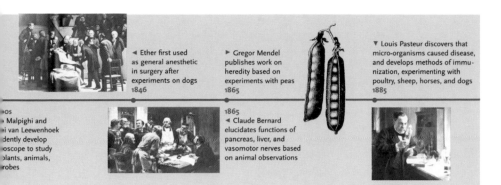

Several prominent figures in the history of science and medicine during the seventeenth, eighteenth, and nineteenth centuries such as Robert Hooke, Caspar Friedrich Wolff, Matthias Schleiden, Theodor Schwann, and others contributed to our understanding of the role of cells, cellular division, and nuclei in the passing on of traits from parents to offspring. They used a wide variety of organisms in their work, including insects, sponges, bryozoans (small aquatic organisms that affix themselves to rocks and form colonies that resemble mosses), foraminifera (single-celled microscopic organisms that have a hard calcium carbonate shell with holes through which filaments project), chickens, pigs, and toads. The Cork Oak tree (*Quercus suber*) was particularly important because it was the appearance of its cork tissue under a microscope that led to Hooke's discovery of plant cells, which he called "cells" because the boxlike structures reminded him of the cells that monks occupied in monasteries.

But it was not until the elegant experiments by the Augustinian abbot Gregor Mendel from Brno (in what is now the Czech Republic) that genetics was firmly established on scientific grounds. Studying hybrids in Sweet Peas (*Lathyrus odoratus*), Mendel was able to show mathematically that traits are passed from one generation to the next as discrete units (what we now know as genes), that each parent supplies to its offspring an equal number of these units, and that the traits in the offspring are the result of how these units combine, with some being dominant and others recessive.

Mendel's seminal 1865 paper "Experiments in Plant Hybridization," published only six years after Darwin's *Origin of Species*, marks the beginning of modern genetics. Unfortunately, it seems that Darwin did not know about this work, or if he did (this question is a subject of some controversy), he did not understand its significance, because Mendel's insights could have helped him explain how selected traits were passed on to succeeding generations. It was not until the twentieth century that the scientific concepts of evolution and genetics were successfully integrated.

Reviewing the genetic discoveries of the past 100 years is beyond the scope of this book (see the paper by Lorentz and colleagues[6] for a review). Instead, we concentrate on the history of a few of the key model organisms, including the Laboratory, or Common House, Mouse, the Fruit Fly, Zebrafish, the bacterium *E. coli*, Baker's Yeast, and the microscopic roundworm *C. elegans*. Studies of these organisms and their genetics have contributed greatly to our knowledge about human health and disease.[7]

BOX 5.2.

Important Biomedical Research Advances from Diverse Species over the Past 125 Years

- Around 1881: Agar first used to culture bacteria. Although the microbiologist Robert Koch has been credited with the first use of agar, derived from any one of several marine red alga species (genera include *Euchema*, *Gelidium*, *Gracilaria*, *Hypnea*, *Gigartina*, and *Macrocystis*), the idea to do so came from Fanny Eilshemius, the wife of one of Koch's colleagues, who had been using it in her puddings. Agar plates remain to this day the mainstay of bacterial culture around the world. A Japanese legend dated to 1658 tells of the first appreciation of agar's ability to solidify when an innkeeper in Japan by the name of Minoya Tarazaemon prepared a warm-weather seaweed soup in winter that, when disposed of on snow, turned gelatinous.

- 1883: Phagocytes, the first line of defense in innate immunity (see text), are discovered in bipannaria (the first stage of starfish larvae). Nobel Laureate Ilya Mechnikov was able to make this discovery because of bipannaria's translucent body cavity that enabled him to directly visualize a phagocyte eating a splinter he had inserted into the organism.

- 1905: Discovery of sex chromosomes in the Yellow Mealworm Beetle and various Hemiptera (a large order of insects that includes various species of beetle and bugs). Nettie Stevens showed that differences in the composition of chromosomes in male and female Yellow Mealworm Beetles (*Tenebrio molitor*) were easily identifiable. Edmund B. Wilson demonstrated similar differences with many species of butterflies. These experiments led to a recognition that an organism's sex depends on the presence of X and Y chromosomes.

- 1909: Discovery of giant axons in squid. L. W. Williams describes the squid giant axon in his book *The Anatomy of the Common Squid, Loligo pealei*. In it he wrote: "The very size of the nerve processes has prevented their discovery, since it is well-nigh impossible to believe that such a large structure can be a nerve fibre." This axon was ideally suited to the experiments of K. C. Cole, who in 1937 made the first measurements of voltage changes underlying a nerve impulse, as well as further experiments by Alan Lloyd Hodgkin and Andrew Fielding Huxley (for which they won the Nobel Prize, sharing it with John Eccles, in 1963), and others that further clarified how nerve cells transmit information.

- 1931: Discovery of genetic transposition. In her studies on corn chromosomes, Barbara McClintock showed that genes are transferred between paired chromosomes during cell division, a process essential to promoting genetic diversity among offspring. For this pioneering work, she won the Nobel Prize in 1983.

- 1962: Green fluorescent protein (GFP) isolated from the sea sponge *Aequorea victoria*. GFP has been a powerful tool in molecular biology. Scientists can detect the presence or absence of a gene of interest by associating it with the gene that encodes this fluorescent marker. Since the discovery of GFP, more than thirty other fluorescent proteins, all derived from Nature, have been discovered.

- 1964: Mitochondrial DNA from *Neurospora crassa*, a red bread mold, identified and characterized. Edward Reich and David Luck used this organism to uncover that mitochondria, the main fuel source of eukaryotic cells, not only possess DNA, but that this DNA is maternally inherited. *Neurospora* had also been used earlier in the century by George Wells Beadle and Edward Lawrie Tatum to establish that each gene encodes just one protein, for which they shared the 1958 Nobel Prize with Joshua Lederberg.

- 1984: A complete map of neurons in the brain of *C. elegans* is completed, comprising 302 neurons and approximately 8,000 synapses (connections between neurons). This first comprehensive blueprint of an organism's brain has yielded an exceptional opportunity to observe how genes and environment interplay in the course of neural development.

- 1997: Sheep clone produced from adult sheep cells.

- 2004: Adenovirus used to reverse the bleeding tendencies of dogs and mice with hemophilia by introducing a normal form of the gene defective in hemophilia. Viruses have emerged as the most promising means of administering gene therapy, because they have a built-in apparatus for delivering genes to cells, and they have easily manipulable genomes that can be designed to carry a normal copy of a defective gene without causing disease.

TABLE 5.1. SOME MAJOR MEDICAL DEVELOPMENTS DEPENDENT ON ANIMAL RESEARCH

Medications

Anesthetics used in surgical procedures

Antibiotics, including penicillin

Anticoagulants such as warfarin

Antidepressants

Antiretroviral drugs for HIV

Asthma medicines

High-blood-pressure and heart-failure drugs

Insulin for diabetes

Leukemia chemotherapies

Painkillers such as ibuprofen

And most other human medicines (which are tested first on animals for toxicity—see chapter 4)

Other Therapies/Medical Diagnostic Tools

Blood transfusions

Breast cancer treatments and other cancer chemotherapy

Cardiac catheterization

Cardiac pacemakers

CT or "CAT" scans

Electrocardiograms (EKG)

Heart and lung bypass machines for open heart surgery

Intravenous feeding

Kidney dialysis

Organ transplantation for corneas, heart valves, hearts, kidneys, and bone marrow

Tests for infectious disease, including HIV

Vaccines

Haemophilus influenzae B

Hepatitis B

Measles

Meningitis

Polio

Tetanus

Whooping cough

THE LABORATORY, OR COMMON HOUSE, MOUSE (*MUS MUSCULUS*)

Researchers who independently reported their rediscovery of Mendel's laws (the mechanisms of inheriting dominant and recessive traits) in 1900 had worked, like Mendel, with higher plants as their experimental subjects. The question immediately

TABLE 5.2. SOME ADVANCES IN VETERINARY MEDICINE FROM ANIMAL RESEARCH

Vaccines against distemper, rabies, anthrax, tetanus, foot-and-mouth disease, lungworm, rinderpest, and infectious hepatitis

Treatment of animal parasites

Orthopedic surgery for horses and for hip dysplasia in dogs

Radiation therapy and chemotherapy for cancer in dogs

Identification and prevention of brucellosis and tuberculosis in cattle

Treatment of feline leukemia

Improved pet nutrition

Figure 5.5. Thale Cress (*Arabidopsis thaliana*). The *Arabidopsis* plant has become a key model organism in biomedical research. (© Gary Peter, University of Florida.)

arose whether these laws also applied to animals, and the answer was not long in coming. By 1902, Lucien Cuénot in France had demonstrated that Mendel's laws explained the inheritance of coat colors in mice, and William Bateson and Edith Saunders in England had shown they applied to the inheritance of comb characteristics in chickens. Seven years later, when two scientific breakthroughs were achieved—Ernest Edward Tyzzer's finding that mice inherited resistance to the growth of transplanted tumors, and Clarence Cook Little's developing the first inbred mouse strain—mouse genetics started on a course that it was to follow for the next century. Thus began the widespread and enormously fruitful application of mouse genetics to an analysis of mammalian physiology, biochemistry, and pathology.

Two research topics largely dominated the first fifty years of mouse genetics. One was the study of genetic factors that determined susceptibility to transplanted tumors. The other was an analysis of the genetic basis for differences in the incidence of spontaneous tumors, work that eventually led to the discovery of the role of retroviruses in transforming normal cells into cancerous ones. These two lines of research provided

BOX 5.3

CONCERNS ABOUT THE USE OF ANIMALS IN RESEARCH

While the use of animals is widely accepted as being essential to biomedical research, some believe that this practice should not be allowed under any circumstances. They argue that animals are sentient beings (i.e., capable of experiencing feelings) and that it is morally wrong to subject them to the pain and suffering involved in experimental procedures or to the distress related to confining them under unnatural conditions. Those species most similar to humans, and others that generate the deepest emotional and aesthetic attachments, have aroused the greatest concerns. These arguments may at times be bolstered by examples that adequate alternatives to the use of animals in research exist—for example, in employing epidemiological investigations, autopsy findings, careful clinical trials and observation, and human tissue and cell culture studies or by evidence that one cannot extrapolate from the findings in animals to human beings.[a] But the history of modern medicine is largely a history of advances resulting from insights that have come from animal research, advances that otherwise would not have been possible. And, of course, the same is true for veterinary medicine.

In the Western world, from the time of Galen almost to the present, animals were generally considered to lack a rational soul and were treated as mere objects. It was not until the early nineteenth century in Great Britain that there were organized efforts to protect the welfare of animals. The British Society for the Prevention of Cruelty to Animals was established in 1824. A New York chapter opened in 1863, with other chapters to follow over the years in Boston and Philadelphia. Great Britain was also the first to pass legislation to protect animals in research. The Cruelty to Animals Act, passed in 1876, is still in force today and has been supplemented by the Animals (Scientific Procedures) Act of 1986, which sets up some of the strictest guidelines in the world, requiring, for example, that the costs (in terms of potential animal suffering) be less than the research's potential benefits. The first law in the U.S. that protected animals employed in research was not passed until 1966—the Animal Welfare Act, which requires that minimal standards for the care of animals be met and that analgesia or anesthesia be used when the animals are subjected to painful experiments, unless this would interfere with the experiment.[b] At the present time, many countries have similar laws that govern research with animal subjects.

We do not treat this critically important topic further in this chapter. Rather, we begin with the assumption that the judicious use of animals in biomedical research is an ethical imperative, because it contributes immeasurably to reducing pain, suffering, and the loss of life in human beings and is essential for ensuring that people are receiving safe and effective treatments, such as medications and vaccines. Physicians need to be able to look into the eyes of their patients, and into those of people who love them, and be able to say unequivocally that medical science has done everything in its power to help them.

We strongly believe that all efforts should be made to guarantee that research animals are bred in labs in order to protect wild populations, that they be treated humanely and with great care and respect for their welfare, that unnecessary experiments be strictly prohibited (e.g., those on cosmetics), and that research involving animals proceed only after alternative means have been fully considered and deemed inadequate. Animal welfare advocates have contributed immeasurably to focusing much needed attention on these complex issues and to making sure that humane and sustainable practices in animal research are practiced more widely.

> QUÄLE NIE EIN TIER ZUM SCHERZ, DENN ES FÜHLT WIE DU DEN SCHMERZ.
>
> ("NEVER TORMENT AN ANIMAL FOR FUN, BECAUSE IT FEELS THE PAIN LIKE YOU,"
>
> A COMMON GERMAN TRANSLATION OF THE MORAL FROM AESOP'S FABLE OF "THE BOYS AND THE FROGS.")

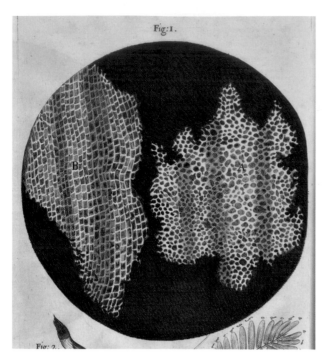

Figure 5.6. Figure of Cork Cells, Drawn by Robert Hooke. (From Robert Hooke, *Micrographia*. Jo. Martyn, London, 1665. From Boston Medical Library in the Francis A. Countway Library of Medicine.)

the original motivation for establishing other inbred mouse strains, now numbering more than 300, that are used around the world.

The period since 1980 has seen major advances in genetic technology, especially in our ability to engineer the genome (the sum total of the genetic information present in an organism), one gene at a time. In late 1980 and early 1981, six laboratories independently showed that rabbit DNA injected into mouse egg cells could become incorporated into their chromosomes. The resulting offspring carried an entirely new gene, and this gene was fully functional. Thus, it was shown that DNA from one group of mammals—rabbits—could function properly in another—mice—despite these species having been separated by about seventy-five million years of evolution.

In 1990, it became possible to replace an existing gene with an altered copy that had been rendered nonfunctional. This ability to "knockout" a gene function quickly led to a flood of experiments testing the function of specific genes in mammalian physiology, such as the role of the gene that produces the cystic fibrosis transmembrane regulator (known as CFTR), the protein that is defective in cystic fibrosis. (See also discussion of the Dogfish Shark salt gland in chapter 6, page 275.) At the present time, knockout mutations in mice number in the thousands.

The identification of the complete mouse genome in 2002 was a milestone event in biomedical research. Evidence shows that humans and mice both descended from a common ancestor, a small mammal that lived approximately 125 million years ago. Despite this long period of evolutionary divergence, only about 300 genes out of the mouse's estimated total of 25,000 have no known counterparts in the human genome.[8] Furthermore, some sequences of DNA in the mouse are so similar to those in the human that the two cannot be distinguished.

Knowing both the mouse and the human genomes allows researchers to begin to decipher the role of specific human genes. By finding a gene's counterpart in the mouse, they can genetically engineer a mouse strain that lacks this gene and learn what the missing gene does from the structure or function that has been altered or lost.[9,10]

There are now more than twenty-five million *Mus musculus* mice in laboratories around the world, comprising more than 90 percent of all mammals used in research. Many human diseases occur naturally in mice or can be genetically engineered in them. Some mouse strains, for example, have a greater susceptibility for developing tumors; others are obese; still others develop high blood pressure. By understanding the relationship between genetic alterations in laboratory mice and the defects and diseases that result, we may better understand how human diseases develop at a molecular, genetic, and cellular level and be able to find effective treatments for them.

Because mice are mammals like we are, most people are able to grasp that understanding how their genes work can contribute to an understanding of our own. It may be more difficult for many to make this same connection with organisms that are not mammals, such as Zebrafish, or with those which are not even vertebrates, such as Fruit Flies or the microscopic, though still multicellular, roundworms *C. elegans*. And it may be more difficult still to appreciate how one-celled (though nucleated) organisms such as

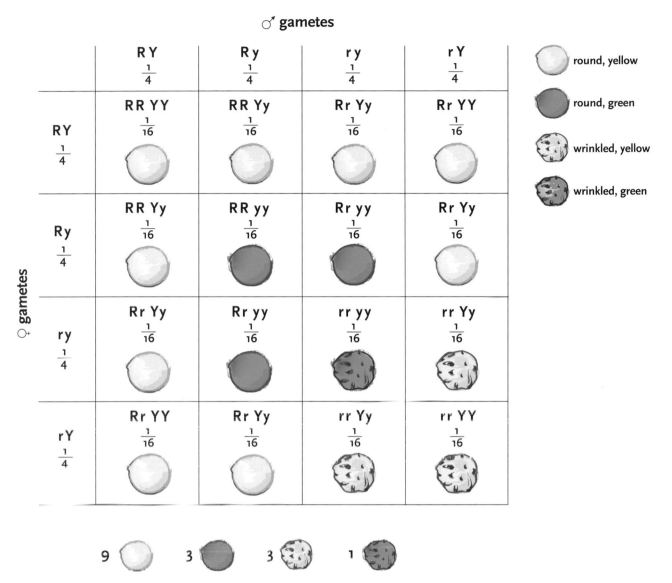

Figure 5.7. Diagram of Mendel's Dihybrid Sweet Pea Cross. It was experiments like this one where Mendel interbred peas with dominant and recessive traits (in this case, the dominant traits were "round" and "yellow," and the recessive were "wrinkled" and "green") that allowed him to determine the basic principles of how traits are inherited. (Illustration by Elles Gianocostas.)

Baker's Yeast or bacteria such as *Escherichia coli*, which do not even have a nucleus, can teach us much that is useful about ourselves. But in fact, all these species have been central to genetic research, with each possessing a different, and critically important, piece of the puzzle about the structure and function of the human genome. As in other areas of biomedical research, genetics relies on knowledge obtained from many species.

ESCHERICHIA COLI

Although lacking a nucleus and having a genome only about one thousandth the size of the typical mammalian genome, the ubiquitous bacterium *Escherichia coli*, which resides (in addition to numerous other locations) in the human intestine and is necessary for its healthy functioning (see also chapter 3, page 81 for a discussion of *E. coli*), has been essential to our understanding of some of life's most fundamental processes. These include how

Figure 5.8. Drawing of *Eomaia scansoria*, a Small Mammal Considered to Be the Common Ancestor of Both Mice and Humans. (Reconstruction artwork by Mark A. Klinger/Carnegie Museum of Natural History.)

DNA copies itself, how genes turn on and off, how DNA makes RNA, and how RNA makes proteins. All other genetic research has relied on these initial insights, learned from *E. coli*. More is known about the molecular biology of *E. coli* than about that of any other organism, yet there is still a great deal that is not understood, despite more than 100 years of research, such as the functions of many of the proteins encoded by its genome.

Because of the simplicity of *E. coli*'s genome relative to that of other laboratory organisms, it is an excellent model for genetic research. For example, it is being used to determine how other organisms cope with conditions that lead to an unraveling of their proteins, the so-called "heat-shock" response. By studying how *E. coli* decides which proteins to destroy because they cannot be salvaged, and which ones to refold, scientists may learn how similar molecular decisions are made in humans, in such protein-folding diseases as Alzheimer's disease, Huntington's disease (an inherited, degenerative, fatal neurological disease characterized by involuntary jerking movements and memory loss), and Creutzfeldt-Jakob disease (a rapidly fatal neurological disease, thought to be the human form of "mad cow" disease).[11,12] Also, some genes originally discovered in *E. coli* have been shown to have important counterparts in other organisms, including humans. For example, the *recQ* gene in *E. coli* has five homologues (homologues are genes that are related to each other by descent from a common ancestor) in humans, including one responsible for Werner's syndrome (which leads to premature aging), one for Bloom's syndrome (characterized by a high number of chromosomal recombinations), and one for Rothmund-Thomson syndrome (associated with chromosomal

TABLE 5.3. SOME PLANTS, ANIMALS, AND MICROBES WITH KNOWN (OR ALMOST COMPLETELY KNOWN) GENOME SEQUENCES

SPECIES[a]	NUMBER OF GENES (ROUGH ESTIMATES)	YEAR SEQUENCE WAS COMPLETED	COMMON NAME
Anopheles gambiae	15,100	2002	Mosquito
Arabidopsis thaliana	25,800	2000	Thale (or Mouse-Ear) Cress
Caenorhabditis elegans	20,400	1998	roundworm
Drosophila melanogaster	14,000	2000	Common Fruit Fly
Escherichia coli	4,400	2003	intestinal bacterium
Homo sapiens	25,000	2001	Human
Mus musculus	25,900	2002	Common House Mouse
Neurospora crassa	10,000	2003	bread mold
Pan troglodytes	25,000	2005	Chimpanzee
Rattus norvegicus	22,000	2004	Brown Rat
Saccharomyces cerevisiae	6,000+	1997	Baker's Yeast

[a]Nearly 300 species have had their genomes decoded, and more than 700 others are in various stages of analysis.

Sources: Cogent database (maine.cbi.ac.uk) and Entrez Genome Database (www.ncbi.nlm.nih.gov/Genomes/).

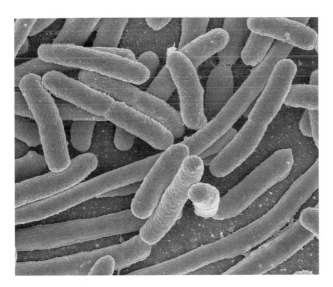

Figure 5.9. Scanning Electron Micrograph of *E. coli*. (Courtesy of National Institute of Allergy and Infectious Disease, National Institutes of Health.)

instability), all three of which predispose to human cancers when they are mutated.[13]

THE BACTERIUM *THERMUS AQUATICUS*

The bacterium *Thermus aquaticus* has also been a key organism for genetic research. First isolated by Thomas Brock and Hudson Freeze in 1966 from a hot spring in Yellowstone National Park, this extremophile (see chapter 1, page 11) grows at temperatures of up to 79 degrees Celsius (more than 174 degrees Fahrenheit!). What makes *T. aquaticus* able to thrive under conditions that would kill most other microbes (the process of pasteurization, e.g., designed to sterilize such liquids as cow's milk for human consumption, is typically done at temperatures between 63 and 72 degrees Celsius) are its uniquely designed proteins that can withstand high temperatures. One of these, the enzyme Taq polymerase, which *T. aquaticus* uses to replicate its own DNA, has led to the development of the polymerase chain reaction (PCR), which is a process employed by scientists to generate millions of copies of a particular piece of DNA. PCR has revolutionized molecular biology through vastly extending its capacity to identify, manipulate, and reproduce DNA. By making the cloning of genes possible, for example, PCR has paved the way for the determination of all the genomes that have been sequenced thus far. It has also become an important tool for diagnosing many infectious agents, such as the hepatitis C virus, the tuberculosis bacterium, and *Chlamydia trachomatis* (a leading cause of blindness worldwide), and for screening newborns for some inherited diseases, such as cystic fibrosis. And, as is well known, it has been used to identify individuals from minute traces of their genetic material that they have left behind (e.g., in cells from the inside of their mouths present on a licked postage stamp, or in sperm present in semen). In coming years, PCR will undoubtedly become even more widely used as a diagnostic tool, for it provides rapid and highly reliable results.[14]

BAKER'S YEAST (*SACCHAROMYCES CEREVISIAE*)

Baker's Yeast (*Saccharomyces cerevisiae*), a one-celled, free-living fungus used to make bread dough rise and to brew beer and other alcoholic beverages, and its cousin, *Saccharomyces pombe*, are ideal laboratory subjects. Despite at least one billion years of evolution separating yeasts and humans, still almost one-third of the more than 6,000 genes in the yeast genome have equivalent genes in humans. In fact, portions of our genome are so similar to that of yeasts that more than seventy human genes are able to repair various genetic mutations in yeasts.[7] In addition, almost 40 percent of yeast proteins are similar to mammalian proteins,[15] and some fifty genes in the yeast *S. pombe* are counterparts of human genes known to be involved in various human diseases, half of which are cancers.[16] Because yeasts are among the simplest and most ancient organisms containing nuclei, this suggests that many human diseases are caused by disruptions of fundamental cellular processes that have remained largely unchanged for hundreds of millions of years and that are likely to be found in all other organisms possessing nuclei.

Yeasts have been the workhorses of eukaryotic genetic research for decades. In the late 1950s, for example, transfer RNA, or tRNA (the molecule responsible for the ultimate translation of information encoded in DNA into the manufacture of proteins), was discovered in a yeast by Robert Holley.[17] However, yeast's most important contribution has been to reveal the workings of how eukaryotic cells make copies of themselves by cell division. As frequently as every 90 minutes, yeasts reproduce by "budding." This process involves genes that regulate and enable such key cellular processes as the duplication of chromosomes, checking the accuracy of that duplication (so as to ensure that DNA mutations do not get passed on to offspring), and finally cell division itself. Studies with yeasts have demonstrated which genes regulate this reproductive cell cycle and how they work together. These studies have also begun to shed light on how cancer cells develop in humans, as all human cancers are thought to involve one or more defects in this cycle.[18]

THE ROUNDWORM *CAENORHABDITIS ELEGANS*

The genome of the microscopic nematode or roundworm *Caenorhabditis elegans*, only one millimeter (0.04 inches) long, was sequenced in 1998. As a multicellular organism with about 19,000 genes (more than three times the number of genes found in yeast), it shares approximately 40 percent of these with humans. *C. elegans* can teach us about the human genome in ways that bacteria and yeast cannot, such as about the control of embryonic development. It has also helped us understand two basic processes found in all multicellular animals.

The first process involves the events that occur when a cell must die, either as a part of normal development or because it has been irreparably injured. *C. elegans* is a key model for studies of this "programmed cell death," also known as apoptosis. Because the organism is transparent, it is possible (under a microscope) to observe its cells dividing, differentiating, and developing into a complete organism—from the stage of a fertilized egg to a mature adult, with each worm eventually having exactly 959 cells. In comparison, humans have trillions of cells. Programmed cell death is essential for normal development in all embryos, fetuses, and adults. When tadpoles metamorphose to become frogs, for instance, some tadpole tissue has to be eliminated, so that tissues for the frog can develop. The same is true in humans, who lose the cells that make up the webbing between our fingers and toes during fetal development (testimony to our common origins with amphibians). In rare instances, this webbing may persist, in whole or in part, in the human newborn, a condition called syndactyly. The balance between cell division and cell death has to be tightly regulated and exquisitely timed in all multicellular organisms, so that the right number, right type, and right organization of cells develop. Studies with *C. elegans* have identified the specific genes that control apoptosis, a finding that ultimately may facilitate better treatments for diseases characterized by excessive apoptosis, such as degenerative diseases of the nervous system, like Alzheimer's disease, Parkinson's disease, and Huntington's disease, as well as for those where apoptosis is reduced, such as in autoimmune diseases and cancers.[19]

> YOU HAVE MADE YOUR WAY FROM WORM TO MAN, AND MUCH IN YOU IS STILL WORM.
>
> —Friedrich Nietzsche, *Thus Spake Zarathustra*

Research on *C. elegans*'s ability to enter into a metabolically slowed down state, known as the "dauer" phase, has been equally rich. *C. elegans* larvae, which feed on soil bacteria, can enter this hibernation-like state when food supplies are low. Compared to their normal life span of two to three weeks when food is abundant, worms in the "dauer" phase can live up to eight times longer. Several genes regulate what state the nematode is in, but one in particular, discovered by Gary Ruvkun and his colleagues at the Massachusetts General Hospital in Boston, the *daf-2* gene, has been the subject of much attention. Worms with a defective *daf-2* gene do not enter the "dauer" phase, and yet they can still live significantly longer than normal worms. While looking at the human genome for genes similar to *daf-2* in *C. elegans*, Ruvkun made a startling discovery: the *daf-2* gene closely resembles the human gene that codes for the insulin receptor, the protein that in the presence of insulin tells a cell to take up glucose from the bloodstream. This observation suggested that one way *daf-2*–deficient *C. elegans* may achieve their longer life span was by slowing down their consumption of glucose.[20]

Another *C. elegans* gene, called *ctl-1*, is also involved in the dauer phase. It has been shown to promote the destruction of free radicals, highly reactive, cell-damaging molecules that form, for

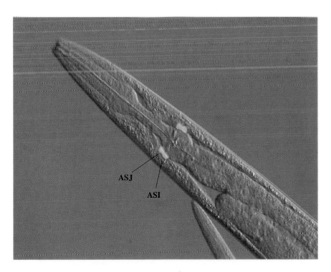

Figure 5.10. Photomicrograph of *C. elegans*. The cells of *C. elegans* can be easily visualized under a microscope, making it possible to study them directly in real time. In this image, two neurons labeled ASJ and ASI, which monitor external levels of glucose (two corresponding unlabeled neurons are on the other side), are stained green because they carry a green fluorescent protein transgene (see box 5.2). (Photo by Weiqing Li, Gary Ruvkun Lab.)

example, as a result of the metabolism of glucose and other sugars. By studying the *daf-2* and *ctl-1* genes in *C. elegans*, we may be able to better understand and treat human diabetes, especially type 2 diabetes, which is a problem of insulin receptor dysfunction, secondary to too much insulin rather than too little.[21] We may also develop better insights into human aging, adding to our knowledge about the role of free radicals in the aging process and complementing research in rats, monkeys, and humans that has examined how caloric restriction may prolong life span.[22]

C. elegans has also been a prime model for investigating how to use "RNA interference" (RNAi), a powerful genetic mechanism first identified in petunias, to turn off disease-causing genes. Cells may have evolved RNAi machinery as a defense against some viruses that possess double-stranded RNA. The mechanism of RNAi entails a cell's designing a piece of RNA so that it binds to a section of abnormal RNA produced by a disease-causing gene. In so doing, the RNA effectively makes it impossible for the defective gene to express its disease-causing potential.[23]

The use of *C. elegans* with RNAi techniques has been promising on a number of fronts, including cancer research. For example, scientists have found several genes, mutated in human cancers, that have counterparts in *C. elegans* and that can be silenced using RNAi. Because these genes function in the worms and in ourselves in similar ways, the ability to turn them on or off in *C. elegans* is proving to be a powerful research tool, one, for example, that has led to the discovery of dozens of other genes that may be involved in cancer formation and to the prospect of RNAi-based cancer therapies.[23]

The technique has also been used successfully in mice to treat diseases ranging from hepatitis B and influenza to non-small-cell lung cancer. And it is being investigated for the treatment of age-related wet macular degeneration in people, a condition that results in a deterioration of part of the retina.[24] Macular degeneration is the leading cause of blindness in the elderly in the United States (see also the discussion of macular degeneration in the section on sharks in chapter 6). RNAi has the potential to be one of the most significant medical breakthroughs of the twenty-first century.

Figure 5.11. Petunias (*Petunia hybrida*). An accidental finding during an experiment to create petunias with a deeper purple color, by introducing double-stranded RNA pigment genes into their cells, led to the discovery of RNAi. (© 2002 Alia Luria.)

THE COMMON FRUIT FLY
(*DROSOPHILA MELANOGASTER*)

Common Fruit Flies (*Drosophila melanogaster*; see the opening figure for this chapter), which seem to generate spontaneously in our kitchens when fruit is left out to rot, have been, along with mice, one of the most important model organisms for studying genetics for almost the past 100 years. The fly's genome, published in the year 2000, contains, like that of *C. elegans*, a surprising similarity to our own. For example, of the genes known to be mutated or deleted in human diseases, at least 60 percent are found in *Drosophila*. If one looks only at such genes associated with human cancers, the proportion that are similar in the Fruit Fly climbs to 68 percent.[25]

Beginning in 1910, Thomas Hunt Morgan at Columbia University began his seminal studies with *Drosophila*, which were to be continued by his students Alfred Henry Sturtevant, Calvin Blackman Bridges, and Hermann Joseph Muller over the next three decades. Morgan discovered that some traits in *Drosophila*, such as having white eyes (instead of the usual red ones), were sex-linked (i.e., their genes were carried by one of the sex chromosomes—usually the X chromosome—but not the other). From his knowledge of the X and Y sex chromosomes derived from other insect studies, Morgan determined that the gene for white eyes was located on the X chromosome, paving the way for an understanding of sex-linked traits in humans. Morgan's students greatly expanded his work with *Drosophila*, showing for the first time that genes were located on chromosomes and were arranged linearly like pearls on a necklace, that genes for specific traits could be found in precise and fixed locations on the chromosome, and that gene mutations could be induced by X-rays. These early insights derived from studies of *Drosophila*, insights that we now take for granted but that were revolutionary in their day, have formed much of the basis of our understanding of chromosomes and genes.[26]

The foundation for modern genome research, that is, determining the entire genetic sequence of an organism, as opposed to working on specific parts of the sequence, also

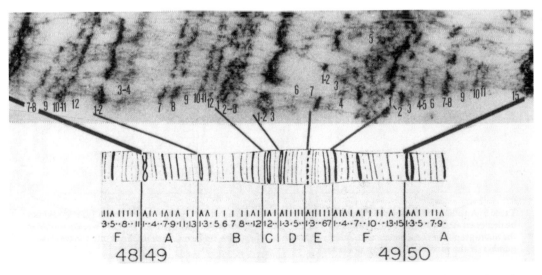

Figure 5.12. Section 49 of Polytene Chromosome 2R of a *Drosophila*. These chromosomes are found in the salivary glands of some flies, including *Drosophila*. They are long and thick, made up of a large number of parallel chromosome fibers. The dense bands mark the locations of specific genes, making it possible to investigate their functions. (© Anja Saura, University of Helsinki.)

BOX 5.4

SOME MOUSE MODELS OF GENETIC DISEASES IN HUMANS

- Down Syndrome: One of the most common genetic birth defects in humans, occurring in one out of every 800 to 1,000 live births, Down syndrome results from an extra copy of chromosome 21, an abnormality known as trisomy. The Ts65Dn mouse, developed at the Jackson Laboratory in Maine, mimics trisomy 21 in humans, exhibiting many of the same behavioral, learning, and physiological defects, including mental deficits, small size, obesity, hydrocephalus (a condition where too much fluid accumulates in the brain), and impaired immunity. The Ts65Dn mouse is the best research model we have for Down syndrome.

- Cystic Fibrosis (CF): The *Cftr* knockout mouse has helped advance research on cystic fibrosis, the most common fatal genetic disease in the United States today, occurring in approximately 1 in every 3,300 white births, 1 in every 9,500 Hispanic-American births, 1 in every 15,300 African-American births, and 1 in every 32,000 Asian-American births. Scientists now know that CF is almost always caused by a defect in the gene that encodes the information necessary to produce CFTR, a protein that regulates the passage of chloride and other substances in and out of cells. Studies with the *Cftr* knockout mouse have shown that there is an inability to fight certain bacteria in the lung, perhaps because of the presence of thick mucus characteristic of the disease, resulting in potentially life-threatening infections. These mice have become models for developing new approaches to correct the CF defect and cure the disease.

- Cancer: The *p53* knockout mouse has a disabled *Trp53* tumor suppressor gene that makes it highly susceptible to various cancers, including lymphomas and osteosarcomas. The mouse has emerged as an important model for Li-Fraumeni syndrome, a rare genetic syndrome in people in which individuals are predisposed to developing many different types of cancer.

- Type 1 Diabetes: This autoimmune disease, also known as juvenile diabetes, or insulin-dependent diabetes mellitus (IDDM), accounts for up to 10 percent of diabetes cases. Nonobese diabetic (NOD) mice are enabling researchers to identify IDDM susceptibility genes and disease mechanisms in people.

- Muscular Dystrophy: The Dmd^{mdx} mouse is a model for Duchenne muscular dystrophy, a rare human neuromuscular disorder in young males that is inherited as a sex-linked recessive trait and results in progressive muscle degeneration.

- Ovarian Tumors: The mouse models known as SWR and SWXJ provide excellent research platforms for studying the genetic basis of a type of tumor called the ovarian granulose cell tumor, a very serious malignancy in young girls and post-menopausal women.

Adapted from National Human Genome Research Institute, Background on Mouse as a Model Organism. 2005; available from www.genome.gov/10005834 [cited August 30, 2007].

came from initial experiments with *Drosophila*. David S. Hogness and his colleagues at Stanford University used *Drosophila* in 1974 to produce the first cloned segment of chromosomal DNA. This was the first step in making a complete map of the *Drosophila* genome and in the development of the techniques used in the Human Genome Project.[25]

Drosophila continues to be a mainstay of genetic research, facilitated by its short life span and the ease of maintaining large numbers of individuals. In 1983, the homeotic or "hox" genes, which determine the basic body plan of animals, were first discovered in the Fruit Fly. And so was the gene that encodes the Toll receptor, which led to the discovery of Toll receptors in mice and in humans and to a quantum leap in our understanding of innate immune systems[27] (see also "The Immune System," below). *Drosophila* have also been essential to identifying genes involved in the generation of circadian rhythms (the biological clocks for various cycles in animals) and in aging, aggression, learning, and memory.[28]

Finally, the role of the *p53* gene, which produces a protein essential for protecting cells from developing cancers, has been worked out in experiments with *Drosophila* (as well as in the Laboratory Mouse—see box 5.4). The p53 protein has the ability to detect DNA damage in a cell about to undergo cell division and to "direct" that cell to activate its DNA repair proteins. If the DNA is too damaged to be fixed, the p53 protein "directs" the cell to undergo apoptosis. Mutations in the *p53* gene are the most common of those found in human cancers. By studying uncontrolled division in *Drosophila* cells that have an inactive *p53* gene, we may begin to uncover some of the genetic mechanisms involved in the development of human cancers.[29]

ZEBRAFISH (*DANIO RERIO*)

As vertebrates, fish are genetically closer to humans than are *Drosophila*, *C. elegans*, Baker's Yeast, and *E. coli* and can therefore be used, as these other organisms cannot be, as models for some diseases, such as those involving bone formation. Fish are especially important to our understanding of vertebrate evolution because they have the largest number of known species and the greatest population sizes of any vertebrate. The Zebrafish (*Danio rerio*), originally from rivers in Southeast Asia, has long been a favorite for those keeping tropical fish. In recent years, they have also emerged as major laboratory animals in genetic research, with more than 1,000 researchers in some 250 labs around the world studying them. They offer some important advantages over mice in that they are less expensive to keep in a lab, and they reproduce in greater numbers. A female Zebrafish, for example, will generally produce dozens to hundreds of eggs at a time, thus being able to create larger populations of specific mutant strains for study.

Also, in contrast to the mouse, the fertilization of eggs in Zebrafish takes place outside of the animal, and because their young are transparent, Zebrafish organs can be observed and manipulated at every stage of development. Furthermore, early development is extremely rapid, with fertilized eggs becoming small, swimming, feeding fish with all their organs formed in just five days. Though studies with Zebrafish began only about thirty years ago, what they have taught us in this short period of time about the mechanisms involved in the formation and the function of various tissues such as muscle, cartilage, bone, and skin, and of various organs such as the heart, eye, brain, and kidney, has been truly remarkable. In addition, studies

Figure 5.13. Male and Female Zebrafish (*Danio rerio*). The larger female is on the top. (© Ralf Dahm, Max Planck Institute for Developmental Biology, Germany.)

Figure 5.14. Stages in the Development of a Zebrafish Embryo. The transparent Zebrafish embryo makes it possible to study in detail the stages involved in the formation of various organs. (From P. Haffter et al., The identification of genes with unique and essential functions in the development of the Zebrafish (*Danio rerio*). *Development*, 1996;123:1–36. © The Company of Biologists.)

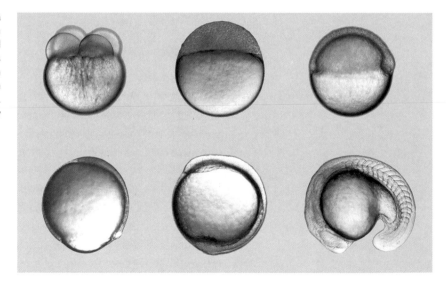

with Zebrafish mutants have broadened our knowledge of many diseases, such as thalassemia, an inherited anemia resulting from defects in the production of hemoglobin; osteogenesis imperfecta, an inherited disease in which patients have fragile bones; and porphyria, a genetic disorder resulting in the accumulation of porphyrins, natural compounds that are the main precursors of heme (an essential component of hemoglobin and other proteins involved in cellular metabolism), which in high concentrations becomes toxic to various organ systems, including the nervous system.[30]

Zebrafish have been especially important for studying the development of the human heart. More than fifty strains of Zebrafish have been isolated that model various developmental and functional heart abnormalities. These Zebrafish mutants have phenotypes similar to the human syndromes and contain mutations in the same genes that are affected in the human inherited diseases.[31]

The Regeneration of Cells, Tissues, and Organs

Since the elegant regeneration experiments with the freshwater polyp hydra by the Swiss scientist Abraham Trembly in 1744, who demonstrated that small pieces of the animal could grow into complete organisms, there has been a great deal of speculation and scientific interest about whether it might be possible for humans to regrow injured or diseased body parts. In fact, we do regrow cells and tissues all the time, replacing cells that are sloughed off from our skin, the inside of our mouths, and the lining of our intestines. We are also constantly replenishing our red and white blood cells and replacing injured liver cells, and recent research (which we shall discuss below) strongly suggests that we are also regenerating nerve cells in our brains (a process called neurogenesis) on a regular basis.

But the prospect of replacing heart tissue after a heart attack or nerve cells and their connections in the spinal cord after spinal injury has intrigued and defied investigators for most of the past century.

REGENERATION RESEARCH

All regenerating tissues require control mechanisms to regulate the stages of the regeneration process and to guide the behavior of the cells involved. In general terms, in order for regeneration to happen, one or both of two different processes must occur. The first involves the persistence of undifferentiated cells, called stem cells, that are pluripotent or totipotent. If a cell can differentiate only into certain types of tissues it is said to be *pluripotent*; if it has the potential to develop into all the tissues in that organism, it is said to be *totipotent*. Stem cells can serve as needed to replace more differentiated cells that have become damaged. A good example of this process in humans can be found in our skin, where stem cells that sit at the skin's bottom layer reproduce to supply new cells to the upper layers where they are routinely lost. The second process entails cells that have already differentiated into a specific cell type, such as a hair cell, blood cell, or liver cell, to revert to a less differentiated state (i.e., become a type of stem cell), enabling them to become other types of cells. This process is known as *de*differentiation.

Regeneration research is a young science, and the specific genetic and molecular mechanisms that govern such processes as just described, which mediate, for example, the regrowth of limbs in salamanders or of heart tissue in Zebrafish and mice, are only just beginning to be mapped out. One of the difficulties in uncovering these mechanisms has been that some of the best regeneration subjects, like the hydra, planaria (minute freshwater flatworms), and urodele amphibians (salamanders and newts), are species for which mutant strains are not readily available for study, nor do they have well-described genetics. Nevertheless, these organisms continue to provide valuable information about regeneration that complements work done on mice and Zebrafish, organisms that have had their genomes sequenced.

Hydra

The hydra, a member of the Cnidaria, an ancient group of invertebrate organisms that evolved around 700 million years ago, has long been a favorite of high school biology classes and a popular model for regeneration research. In evolutionary terms, cnidarians were the first animals to have a defined body axis—that is, distinct tops and bottoms, fronts and backs—and the first to possess nervous systems. Visible to the naked eye, hydras live in freshwater systems and feed by paralyzing their prey with stinging cells located on their tentacles. They have two tissue layers, an outer layer called the ectoderm and an inner one called the endoderm, separated by a noncellular gelatinous layer. The species that has been most widely used in research is *Hydra vulgaris*.

The ability of a hydra to regenerate is without peer. When a hydra is cut into small fragments, each fragment can develop into a complete organism in as short a time as two to three days. Even individual cells are able to reaggregate with others and regenerate into whole animals. If one were to cut a hydra in half, the cells nearest the cut are somehow able to determine where they are located in the organism; the bottom half will, amazingly, grow a new head, and the top half a new foot. If cells are taken exclusively from the head or the foot, a complete hydra does not form, demonstrating that these cells have reached a terminal state of differentiation, in contrast to cells in the body column, which continue to function as stem cells, capable of differentiating into different types of cells.[32]

Although much remains to be discovered, current research suggests that from the moment a hydra is cut, its cells begin a complex and intricate cascade of molecular events.[33] These involve specific peptides binding to receptor sites that initiate regeneration, and the activation of regulatory and orienting genes (the latter providing information about body axis—e.g., where the organism's head should be) that control the regeneration process. It is not known whether the genes that encode the peptides and receptors that are responsible for hydra regeneration are also found in vertebrates. Until recently, understanding the genetics of regeneration in hydra has been hampered by an inability to manipulate gene

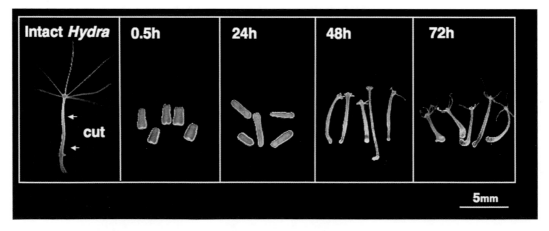

Figure 5.15. Head and Foot Regeneration of Hydra. Cutting sections from a hydra's column results in each piece regenerating a functioning new head and new foot after 72 hours. (© Toshitaka Fujisawa, National Institute of Genetics, Japan.)

activity. However, in the past few years, researchers have had greater success, notably through the use of RNAi techniques (see section on *C. elegans*, above) and in observing their effects through the use of fluorescent proteins to mark genes.[32]

Tantalizing molecular clues suggest that ancient genetic and molecular mechanisms that regulate regeneration in hydra have been conserved for hundreds of millions of years of evolution. If so, hydra, given their deep evolutionary roots, may be an ideal organism to clarify some of the fundamental pathways of regeneration. Such information may be of tremendous value in trying to sort out our own capacity to regenerate, which in humans largely disappears by the time one is born.

Planaria

Regeneration in planaria—freshwater, free-living, cross-eyed flatworms that have also captivated generations of high school students—was first described by the German scientist Peter Simon Pallas in 1766. Like hydras, planaria have extraordinary regenerative powers. Thomas Hunt Morgan, before he began the work for which he is known on *Drosophila*, demonstrated that if one cut a planarium into pieces, a fragment as small as 1/279th of the total worm could still regenerate into a complete organism. Several other investigators continued Morgan's work, but until quite recently, planaria have been largely ignored in studies on development and regeneration, because they did not lend themselves easily to genetic and molecular approaches. But as interest in stem cell research has steadily increased, planaria, which are among the simplest organisms that have three tissue layers, bilateral symmetry, and distinct organs, and that regenerate by mechanisms more like those of higher animals than of hydras, are once again receiving attention.

There are thousands of species of planaria, but the one that has become a laboratory favorite is *Schmidtea mediterranea*, a species that has one strain that reproduces

Figure 5.16. Drawings of Regenerating Planarium. These are the original drawings of Thomas Hunt Morgan, showing the stages of regeneration of a planarium (*Planaria maculata*) cut in half. (From T.H. Morgan, Experimental studies of the regeneration of *Planaria maculata*. Archiv für Entwicklungsmechanik der Organismen, 1898;7:364–397.)

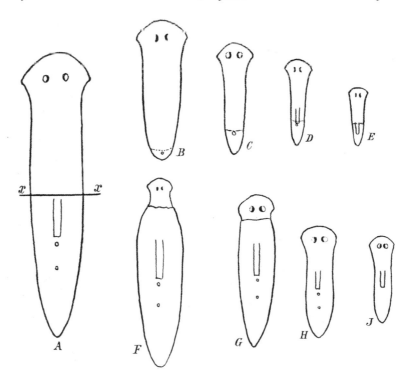

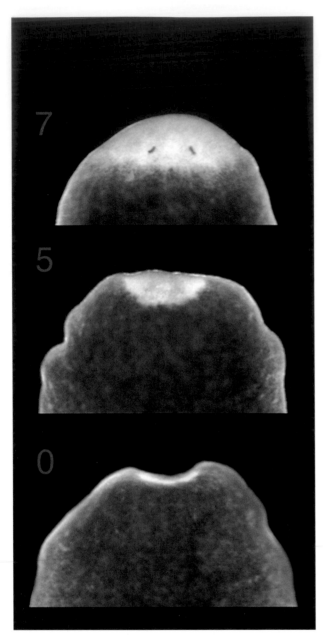

Figure 5.17. Formation of the Blastema in Planaria after Decapitation. After five days, the wound has been filled in by stem cells called neoblasts that make up the blastema. After an additional two days, specialized organs like the eyes begin to form. (© Alejandro Sanchez Alvarado, University of Utah School of Medicine.)

only sexually and another that is exclusively asexual (reproducing by dividing in two). Both strains have remarkable regeneration abilities.

When *S. mediterranea* is injured, stem cells called neoblasts, which are widely distributed in adult tissues, begin to proliferate and migrate to the wound site, forming a specialized grouping of cells called a blastema, from which the regenerated tissues originate.[34] While the mechanisms that govern the proliferation of the neoblasts and the regeneration process itself have not yet been determined, *S. mediterranea* and the other planarian species that have been studied have been shown to possess all the major gene families found in higher animals,[35] as well as numerous genes that are closely related to those in mammals.[34] As a result, planaria are likely to become increasingly important in our efforts to understand the fundamental mechanisms of how stem cells and the regeneration process work.[36]

Zebrafish

Hydra, astonishingly, regenerate without producing new cells. The stem cells in their body columns differentiate into all the tissues required to form complete animals. By contrast, Zebrafish, like the champions of vertebrate regeneration, salamanders and newts, regenerate body parts through the production of new cells and the formation of a blastema, a process that seems to have first evolved in early flatworms such as planaria.

In the past several years, research on regeneration and on stem cells in Zebrafish has flourished, driven by a detailed knowledge of their genetics, by a growing understanding of the molecular pathways involved in Zebrafish regeneration, and by the relative ease of manipulating them in laboratory experiments. Zebrafish are able to regenerate multiple body structures—fins, retinas, scales, and portions of their hearts and spinal cords.[37] Experiments on fins have concentrated on the formation and growth of the blastema and have demonstrated that regeneration does not occur without epidermal cells (the epidermis is the most superficial cell layer) being present at the wound site or without an intact nerve supply, indicating that there may be growth factors released by these tissues.[38] These experiments have uncovered some of the genes involved, which have been shown to have human counterparts,[39] in wound healing, the formation of the blastema, and fin regeneration.

Zebrafish can also fully regenerate as much as one-fifth of their heart tissue within two months after that amount has been removed.[31,40] However, with specific mutations, Zebrafish will form scar tissue rather than cardiac muscle when their hearts have been damaged, just as human heart cells do after a heart attack. This process of scar formation seems to be inhibited by activation of genes involved in heart regeneration.[40] Identifying these genes may shed light on how to prevent the injured human heart, or other tissues, from scarring, as well as how to initiate a regenerative response.

Mammals generally cannot regenerate appendages, with the exception of members of the deer family, the Cervidae (which includes deer, moose, caribou, and elk), which can regenerate their antlers each year after they are shed.[41] Mice can regrow the tips of their foretoes when they are amputated past the last joint,[42] and in some rare cases, young human children, and even some adults, have been able to regrow cosmetically perfect and fully functional fingertips when accidental amputations have occurred after the last joint.[43] While they cannot in general replace appendages, however, mammals can routinely replenish blood cells and other cells in the liver and other organs, as mentioned above. Among vertebrates, the ability to regenerate organs in the adult seems mostly to have ended with the evolution of urodele amphibians (because amphibians are addressed in chapter 6, we reserve the discussion of amphibian regeneration for that chapter).

A mutant mouse, called the MRL mouse, produced by interbreeding other mutant strains, was originally of great interest to researchers because of its large size and its inability to regulate its immune cells. With age, some of the immune cells (lymphocytes in their lymph nodes and spleen) in MRL mice reproduced uncontrollably, making the strain a possible model for immune system research, particularly for work on autoimmune diseases like systemic lupus erythematosis. The MRL mouse also has the capacity to regrow a variety of tissues after injury with perfect fidelity and without scarring. This ability was first discovered after the mice completely healed (within 30 days) 2 mm (~0.08 inch) holes that had been surgically punched in their ears for identification. The punched holes in other lab mice, by contrast, remained open and developed scars at their margins. The healed skin of the MRL mouse appeared normal to the naked eye and, in addition, looked identical to the tissue it had replaced when examined under the microscope.[44]

This observation has led to a spate of other research with the MRL mouse. Among the most exciting are studies that have examined the mouse's regeneration of heart muscle and parts of its spinal cord after injury. In both sets of experiments, normal tissue regrows without scarring. It seems it is this ability of the MRL mouse to inhibit scar formation, or to break it down once it begins to form, that is a basic feature of its ability to regenerate tissues.

In studying the MRL mouse and the Zebrafish, one comes to the intriguing conclusion that the ability to regenerate tissues and organs may be conserved in the human genome but that it is inhibited. If we are able to decipher the genetic, molecular, and cellular events responsible for the MRL mouse's unique regenerative ability, we may be able to discover how to release this potential in human tissues and organs destroyed by injury or disease.[45]

STEM CELL RESEARCH

Regeneration research in essence encompasses stem cell research, because the process of regeneration involves the activity of undifferentiated stem cells, or the dedifferentiation back into stem cells, of cells that have already become specialized. The focus, however, in most stem cell research is on transplanting cells rather than on

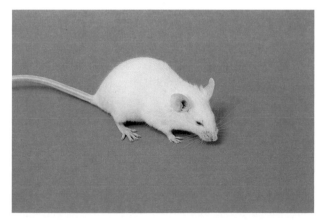

Figure 5.18. MRL Mouse. (Courtesy of the Jackson Laboratory.)

promoting the replacement by the organism's own cells of its injured or diseased parts. It is thought by many that stem cells have the potential of treating a wide variety of human ailments—damaged organs, such as heart muscle after heart attacks or brain tissue after strokes; degenerative diseases such as Parkinson's disease and amyotrophic lateral sclerosis (ALS or Lou Gehrig's disease); and diseases where the specialized cells are absent or fail to function, such as in type 1 diabetes. Stem cells are also being touted as possible solutions to the shortage of organs available for transplantation. Perhaps no other area of medical research has generated more excitement, and more controversy, than stem cell research.

Research with experimental animals has laid the groundwork for stem cell research in humans. Animal models have been essential for pioneering the techniques used to isolate stem cells in people and have also offered a first glimpse at how these cells are signaled into action and then grow and differentiate.[46] This work complements ongoing studies with stem cells taken from human embryos, and with those that are normally present in human adults—for example, in their intestines, skin, testes, bone marrow, and other organs.

Animal stem cell experimentation has moved in several directions that hold great promise for human medicine. We look briefly at research on two diseases: Parkinson's disease and type 1 diabetes.

Parkinson's Disease

Parkinson's disease, the most common neurodegenerative movement disorder in people, afflicts approximately one million people in the United States; about 40,000 people develop the disease each year.[47] Data on how many people worldwide have Parkinson's are sparse, but at least four million, and quite possibly millions more, have the disease. The observation that Parkinson's involves a progressive degeneration of specific nerve cells that produce a chemical called dopamine, and that these cells are located in a region of the midbrain called the substantia nigra, was initially made in rabbits and was later confirmed in nonhuman primates and in humans (research on Parkinson's involving nonhuman primates is discussed in chapter 6). This finding led to treatment for Parkinson's patients with a medication called levodopa or L-dopa, a compound that restored brain dopamine levels and that is still the treatment of choice.

In the past two decades, studies in animals and in people have demonstrated that transplanted nerve cell tissue can heal injuries to the spinal cord and the brain that have long been thought to be irreparable. Since the late 1980s, for example, hundreds of patients with Parkinson's disease have had transplants with human fetal nerve cells. These cells were able to make dopamine and, in some cases, to reduce the debilitating symptoms of the disease.[48] One of the major difficulties with current medicines for Parkinson's disease, including L-dopa, is that they lose their effectiveness when patients take them over long periods of time, and that they can often lead to the appearance of debilitating involuntary movements. As a result, these early encouraging trials with embryonic nerve cell transplantation generated a great deal of interest. The results, however, have been inconsistent, and younger patients, who have Parkinson's more rarely, show more clinical benefit than older ones.[49] Practical difficulties have limited research in this area, not the least of which is the problem of obtaining human embryonic tissue, a result of the current heated debate about the ethics of its use.[50]

Attempts to better understand the potential for such transplantation have been carried out using undifferentiated rodent embryonic stem cells. After implantation into the brains of rats that had the equivalent of Parkinson's disease, these cells both proliferated and fully differentiated into dopamine-producing nerve cells, resulting in a gradual and sustained improvement in the motor functions of these "parkinsonian" animals. One of the most exciting and unexpected findings in this research was that only about 10 percent of the implanted cells became functional neurons; the remainder seemed to play a role in preventing diseased cells from dying.[51,52] Such an experiment has recently been successfully repeated in Cynomolgus Monkeys, long-tailed, Old World monkeys of Southeast Asia and Indonesia, also called the Long-Tailed, or the Crab-Eating, Macaque (*Macaca fascicularis*). This work raises the intriguing possibility that embryonic stem cell transplants in human brains could not only result in a replacement of diseased or dead neurons in patient's with Parkinson's disease, strokes, and perhaps even Alzheimer's, but could also potentially arrest the degenerative processes.

Mouse embryonic stem cells have also been shown to repair damaged nerve cells in the spinal cord following injury,[53] and in the brain following a stroke, and to myelinate axons in the brain and spinal cord in myelin-deficient rats (myelin, a substance consisting of fat and protein, coats the projections of neurons and increases the transmission speed of their electric signals). These experiments with mice and rats demonstrate the enormous potential for embryonic stem cell transplantation to treat a range of human neurological conditions, including spinal cord injuries and strokes.[54,55]

Type 1 Diabetes

Like Parkinson's disease, where a specialized cell (the dopamine secreting nerve cell) is dysfunctional, type 1 diabetes (previously known as juvenile onset diabetes) is a disorder involving the destruction of a specific cell type in the pancreas, the beta cell, that secretes insulin, a protein that controls the amount of sugar present in blood. Because both disorders involve single cell types that are defective, they are potentially well suited to stem cell transplantation.

The limited availability of human pancreases for transplantation has pushed researchers to look for other ways to meet the enormous clinical demand. Two approaches have thus far yielded disappointing results. The first, growing human beta cells that have been forced to replicate in tissue cultures has failed because these cells tend to lose their ability to produce insulin.[56] The second, involving transplantation of human embryonic stem cells, has also been problematic because, although these cells do differentiate into insulin-producing cells, they do so only rarely and, even when they do, they do not produce a sufficient quantity of insulin.[57]

Again, it has been necessary to rely on laboratory animals to answer basic questions about how cells capable of producing insulin might be successfully transplanted into humans to treat type 1 diabetes. Mice, rats, and the African Clawed Frog (*Xenopus laevis*) have revealed the role of the *Pdx1* gene, which regulates the expression of multiple genes involved in the production of insulin and in the transport of glucose. Through the introduction of the *Pdx1* gene, fetal liver cells from both mice and humans have been transformed into insulin-secreting cells. And when transplanted into mice having the equivalent of type 1 diabetes, these cells restored normal function for prolonged periods.[58] These early cell transplantation experiments in mice offer some hope for eventually curing type 1 diabetes, a disease that afflicts roughly 5.3 million people world-wide, including almost 400,000 children, and that causes untold human suffering.[59]

The potential application of stem cells extends well beyond these two diseases. Mice, Zebrafish, and other laboratory animals are being used to investigate the possibilities of stem cell transplantation for treating a wide variety of other human maladies, including various blood disorders, heart attacks, and diseases of the liver.

NEUROGENESIS

Despite laboratory evidence as early as the 1960s that showed that new neurons were being continuously regenerated in adult mammalian brains, this capability, called neurogenesis, was not widely accepted until the late 1990s. Until then, the conventional dogma—that neurogenesis in mammals was restricted to the embryonic period or, for some nerve cells, to no later than the time just after birth—was considered incontrovertible. Indeed, when one of the editors of this book (E.C.) was in medical school during the late 1960s, he was very confidently taught that we are born with all the nerve cells we are ever going to have and that we spend the rest of our lives, at first slowly and then more rapidly, losing them. Happily, this does not seem to be the case. How was this discovered?

Research done in the 1960s by Joseph Altman showed for the first time that new neurons were being formed in the brains of adult rats and cats. Given the strong belief at that time that this could not be happening, however, his ground-breaking work was mostly dismissed. Research on neurogenesis continued over the next decade in various fish models but had little impact in changing prevailing beliefs.

It was not until the seminal and elegant work with songbirds by Fernando Nottebohm and his group at Rockefeller University was first reported in the late 1970s that the possibility of neurogenesis in the central nervous systems of warm-blooded vertebrates began to be taken seriously. Nottebohm and his colleagues noticed that the regions of the brain that made up the song system in canaries were larger in males than in females (male canaries do most of the singing and sing far more complex songs than do females), and that these regions in male brains changed in size depending on the time of year—they were largest in the spring at the start of the breeding season. They discovered, further, that the male hormone testosterone, if given to adult female canaries, resulted in their singing more, and in an enlargement of their brain song centers, particularly the region called the "high vocal center." Initially, it was speculated that these changes in volume might be the result of changes in the number of synapses or connections between neurons, but it became clear after several years of radioactive labeling experiments that new neurons were indeed being produced, that they came from stem cells that were located in a region of the brain called the lateral ventricles, that they migrated to the song centers of canary brains where they replaced dying neurons, that these new neurons joined existing neuronal circuitry, and that they, too, were eventually replaced. While this research did not set out to study neurogenesis, but rather the song centers in the brains of some songbirds, it nevertheless has spurred further neurogenesis research and ultimately a more detailed understanding of the human brain.[60,61]

Currently, work by Elizabeth Gould at Princeton University and others has established that neurogenesis occurs in the dentate gyrus (a region of the hippocampus, the area of the brain most actively involved in the establishment of new memories) of adult rats and mice and of primates—tree shrews, marmosets, and macaques.[62,63] And Peter Eriksson and his colleagues from Göteborg, Sweden, have conclusively demonstrated that neurogenesis occurs in the human hippocampus throughout adulthood.[64]

Further investigation has identified some of the controls on hippocampal neurogenesis. Estrogen stimulates the production of new neurons in the dentate gyrus in rats, as does living in more complex artificial environments (e.g., cages containing mazes for rats and mice) or in enriched natural ones, such as Black-Capped Chickadees (*Parus atricapillus*) living in the wild as opposed to in captivity. Glucocorticoids, such as cortisone, which are released during stressful experiences, decrease neurogenesis in this brain region. It has also been shown that prenatal stress in pregnant rhesus monkeys and rats has lasting effects on their offspring, reducing neurogenesis in later life.[65]

The implications of this work for human medicine, begun in rats and songbirds, and developed in half a dozen other species, cannot be overstated. It has changed the way we think about learning and memory and raised the possibility of halting or even reversing some of the devastating effects of some degenerative conditions such as Alzheimer's. It has altered our outlook on the possibility of repairing damaged central nervous system tissue in victims of stroke or head injury, not from grafts of embryonic stem cells, but from mobilizing their own neuronal stem cells, present in their brains, both in the young and in adults, to carry out the repair. It has highlighted the importance of reducing stress during the prenatal and early postnatal periods, as stress at those times may lower the number of new neurons formed in the human brain well into adulthood. And it has affected our understanding of how stress, hormones, and the complexity of one's environment, by altering rates of neurogenesis in the hippocampus, may affect one's capacity for learning and for creating new memories. Knowledge about neurogenesis in the adult human brain is still at a very rudimentary level, and no one can now estimate how and when neurogenesis may be employed to solve human ills. But, given that as recently as twenty years ago it was widely believed that old neurons could not be replaced at all by new ones in adult mammalian brains, much less in human ones, we should allow for the enormously exciting possibility that someday we may be able to harness our own neurogenic potential to treat diseases that cause great suffering, and to revitalize brains that, as a result of injury, disease, or aging, are beginning to fail.

The Immune System

All plants and animals live within a sea of microbes that envelop their exterior surfaces and line the walls of internal organ systems that communicate with the outside world, such as the gastrointestinal tracts and lungs of animals. Almost all of these microbes live without causing their hosts any harm, and some even provide life-sustaining services, such as the conversion of unusable nitrogen gas into usable forms by nitrogen-fixing bacteria on the roots of certain plants, or the production of vitamin K by bacteria that live in the human intestine (see box 3.1 on microbial ecosystems in chapter 3). Only very few microbes are pathogens and consider their host organisms as food, initiating local infections or causing a variety of systemic infectious diseases. Of course, when organisms die, they all become food for a great number of different species of microorganisms, with the process of putrefaction beginning immediately.

Plants and animals have evolved multilayered defensive systems to protect them against attack by microbial pathogens. While ancient Greek physicians recognized some aspects of these defenses in humans, for example, that those who survived bouts of the plague developed resistance to becoming infected again, and while Pasteur (see "A Brief History of Biomedical Research," above) and others had some understanding of immunization more than 100 years ago, it has been only in the last few decades that researchers have begun to unravel the molecular intricacies of the immune system. Many species, from the Fruit Fly and the Cecropia Moth (*Hyalophora cecropia*) to lampreys and the Domestic Pig (*Sus domestica*), have contributed to these insights.

There are two main components of the immune system—the ancient innate immune system, which all organisms possess, and the adaptive immune system, which is found only in higher vertebrates, having first evolved around 450 million years ago in the first sharks, skates, and rays (known collectively as the elasmobranchs), the earliest known vertebrates to have jaws. Adaptive immunity is a highly complex system in which the foreign markers of pathogens are recognized, and highly specific chemical and cellular responses are mobilized to destroy them. These responses can be "remembered" by immune system cells known as lymphocytes and can be triggered again by future exposures to the same pathogens. This capacity explains the ability of vaccines to protect us from infectious diseases. Adaptive immunity may have allowed early elasmobranchs to venture into new environments where they were exposed to pathogens possibly unfamiliar to their innate systems. Humans have both systems, as do all other vertebrates that have evolved since the elasmobranchs.

In this section, we discuss only innate immunity, examining some of its most basic components and how they interact with adaptive systems. We consider hagfish and lampreys, both of which, while possessing only innate immune systems, demonstrate some primitive features of adaptive immunity. (Chapter 6 discusses sharks, which possess all the requisite parts of adaptive systems.) By this examination, we show how research on the ancient immune systems of these organisms has led to a better understand of the origins and functioning of our own.

INNATE IMMUNITY

All organisms, from the simplest one-celled archaea to the most complex higher primates and marine mammals, possess some form of innate immunity that protects them from infections. Although originally thought to be quite primitive, responding only nonspecifically to microbial assaults, the innate immune system has been shown, in multicellular organisms, to be highly complex and pathogen specific and under the control of a sophisticated set of receptors. The term "innate immunity" has generally been applied only to multicellular organisms, but in this section we use the term to include any response, including those by single-celled organisms, that fights pathogens. All life, including single-cell eukaryotes such as paramecia and single-cell prokaryotes such as bacteria and archaea, must have the ability to produce antimicrobial peptides, or other antimicrobial compounds such as the potent antibiotics made by some species of the bacterial genus *Streptomyces* (see the discussion of microbes in chapter 4, page 136), in order to survive.

Innate immunity is designed to generate an immediate response, in the first minutes to hours to days following an infectious challenge. By contrast, the mechanisms of adaptive immunity do not become fully mobilized for several days, and it is believed that in most instances they are never activated, because innate immune defenses are sufficiently effective in terminating the great majority of infections before they can take hold and cause disease.

Receptors

What are the components of the innate immune system? First there are the receptors. A variety of receptors called pattern recognition receptors (PRRs) have evolved to recognize specific molecular structures that have been highly conserved by evolution in pathogenic microbes. These structures are known as pathogen-associated molecular patterns (or PAMPs), and they are characteristic for specific pathogen groups. For example, there are receptors for the lipopolysaccharides (abbreviated as LPS) of Gram-negative bacteria such as *Salmonella* (see box 4.1 in chapter 4 for an explanation of Gram staining); the glycolipids of mycobacteria, such as the mycobacterium that causes tuberculosis; the lipoteichoic acids of Gram-positive bacteria, such as *Pneumococcus* that can cause pneumonia; the mannans of yeasts; and the double-stranded RNAs of viruses (produced by most viruses at some phase in their replication).[66] Because some pathogens can present multiple PAMPs, each of which can be identified by a distinct receptor, there are potential built-in redundancies in the system.

The seminal discovery in *Drosophila* in 1996 of the Toll receptor led to an appreciation of the complexity and specificity of innate immune receptors. The Toll receptor, which had been known to regulate dorsoventral (back vs. front) orientation in the developing fly, was also found to be essential for recognizing and defending the Fruit Fly against the fungus *Aspergillus fumigatus*. The subsequent discovery in mice and in humans of a Toll like receptor, called TLR4, and of its ability to recognize LPS from Gram-negative bacteria demonstrated the central role that these and other receptors are thought to play in innate immunity.[27] To date a total of thirteen mammalian TLR subtypes, each identifying a specific PAMP, have been identified (ten in humans), and there is evidence that collectively Toll like receptors (it is assumed that more await discovery) may be able to recognize most, if not all, common pathogenic viruses, fungi, bacteria, and protozoa.[67]

Cellular and Chemical Response

After a pathogen has been identified by the appropriate receptor or receptors, a variety of cellular and chemical responses are set in motion. These include the arrival of tissue macrophages, which produce high levels of chemicals called cytokines or chemokines, which serve to direct blood flow to the infection and to regulate the traffic of the body's cellular defenses; neutrophils, amoeba-like cells that, along with the macrophages, engulf and destroy invading microbes; eosinophils, which play a specific role in some parasitic infections such as malaria; and "natural killer" (NK) cells, which are thought to help destroy virus-infected cells (and cancer cells) and to produce proteins that cause microbial pathogens to break apart. NK cells may also kill dendritic cells, which serve to present the pathogen's antigens (specific surface proteins) to lymphocytes, a process that serves to trigger the antibody and cellular responses of the adaptive immune system.[68] If the innate system has been successful in terminating the infection, the adaptive system is not needed.

Antimicrobial Peptides

As a part of the innate immune response, a vast array of compounds is released into infected areas and into the internal compartments of macrophages and neutrophils containing the microbes that have been ingested. These include enzymes such as lysozyme and proteases, both of which break apart microbes; complement proteins that attack microbial membranes; toxic metabolites (e.g., reactive molecules containing oxygen or nitrogen); and perhaps the most ancient of all defenses, the antimicrobial peptides.[68]

The first published work on antimicrobial peptides was done by Hans Boman at the University of Stockholm in the 1980s. By studying compounds produced by the pupae of a giant silkworm moth, the Cecropia Moth (*Hyalophora cecropia*), that had been exposed to bacteria, Boman and his colleagues isolated two antibacterial peptides, which they named cecropin A and cecropin B, that destroyed bacteria by poking holes in their cell walls.[69] Since this discovery, more than 800 other antimicrobial peptides have been identified (for a catalogue of them, see www.bbcm.univ.trieste.it/ [click on Services and then Antimicrobial Sequences Database] or aps.unmc.edu/AP/main.php). They have been isolated from bacteria and single-celled protozoa, from invertebrates such as shrimp, from the blood cells of all vertebrates and the epithelia of all animals that have been studied, and from various tissues in plants. They are thought to be present in all living things.[70]

Some antimicrobial peptides that have been widely studied are drosocin in the Fruit Fly, which attacks primarily fungi; magainins in the skin of frogs (see section on amphibians in chapter 6, page 204) that have very broad-spectrum antibacterial activity; and defensins in rabbits, cows, insects, horseshoe crabs, pigs, and humans (in our intestines) that demonstrate potent antibacterial, antifungal, and antiviral activity against a wide variety of microorganisms.[71] Organisms tend to produce antimicrobial peptides that are designed to attack those microbes they generally encounter. For example, aquatic frogs,

Figure 5.19. Cecropia Moth (*Hyalophora cecropia*). With a wingspan of five to six inches, the Cecropia Moth is the largest moth in North America. (Photo by Scott H. Hale.)

such as the African Clawed Frog (*Xenopus laevis*), produce peptides that are particularly active against some species of Gram-negative bacteria and fungi (as well as a wide range of protozoa) that flourish in the damp places where they live. Frogs that live on land in dry niches, by contrast, produce antimicrobial peptides that target species of Gram-positive bacteria and fungi that have adapted to drier conditions. The Scots Pine produces a completely novel peptide that targets fungi infecting its roots.[72]

While pathogenic microbes are generally able to develop resistance rapidly to commonly used antibiotics, creating a looming crisis for clinical medicine today, the same has not been the case for antimicrobial peptides. They have remained effective in fighting pathogenic microbes for hundreds of millions of years. How has this been accomplished? A few explanations have been offered. For one, when an organism releases antimicrobial peptides in response to an invading microbe, it generally releases a diverse cocktail of them, each with a different structure and each potentially lethal to the microbe. For the microbe to develop resistance to this defensive assault, it would have to do so for all the peptides at once, a highly unlikely event. In addition, antimicrobial peptides strike at vital microbial structures. Bacterial membranes are structurally distinct from the membranes of multicellular organisms (notably, the outermost layer is predominantly composed of negatively charged lipid-containing molecules). Antibacterial peptides specifically target these distinctive molecular structures, punching holes in them and causing bacteria to break apart. To develop resistance against this type of attack, bacteria would have to redesign the basic building blocks of their membranes, yet another highly unlikely event. Antimicrobial peptides can also interfere with essential molecular pathways.[73,74] For example, a peptide from the European Sap-Sucking Bug (*Pyrrhocoris apterus*) kills both Gram-positive and Gram-negative bacteria by disrupting proteins (called DnaK proteins) that are responsible for preserving the structure of the bacteria's so-called housekeeping enzymes, which are necessary for survival.

The ability of antimicrobial peptides to stave off resistance has catalyzed a wave of research designed to develop effective and safe antibiotics for fighting various pathogenic agents in people. Although difficulties have been encountered in these searches, the field is still in its infancy, and it is clear that these substances, which were first discovered in the Cecropia Moth, hold enormous promise for the treatment of infectious disease.

THE IMMUNE SYSTEM OF JAWLESS FISH—THE AGNATHANS

The oldest known survivors of the agnathans, or jawless fish, which share a common ancestor with sharks and other elasmobranchs, are the hagfish and the lampreys. However, unlike the elasmobranchs and the vertebrates that have evolved after them, all of which have adaptive immune systems, hagfish and lampreys possess innate immunity alone. They do not seem to be at a loss, however. Their innate immune systems have served to protect them, as well as all invertebrates, plants, and microbes, from infections for hundreds of millions of years. As the evolutionary bridges to the development of adaptive systems, hagfish and lampreys offer unique windows into the origins and the basic functioning of the human immune system.

Atlantic Hagfish (*Myxine glutinosa*), as scavengers that eat dead and decaying marine organisms, are exposed to very high concentrations of microorganisms.

Figure 5.20. Atlantic Hagfish (*Myxine glutinosa*). (© Illustration by Jacqueline A. Mahannah.)

Precisely how they contend immunologically with these microbes remains poorly understood, but some clues have recently been found. Three potent, broad-spectrum antimicrobial peptides, all members of the cathelicidin family, have been identified in Atlantic Hagfish. These peptides kill a variety of Gram-positive and Gram-negative bacteria. Hagfish cathelicidins are produced by lymphocyte-like cells in their intestines that are positioned to respond to microbes attempting to pass through their thin intestinal epithelial walls. Because lymphocytes are the primary cells involved in adaptive immunity, this finding, along with the fact that cathelicidins have been shown to trigger adaptive immune responses in mammals, suggests that the Atlantic Hagfish, as well as the two other hagfish species, the Pacific Hagfish (*Eptatretus stoutii*) and the Inshore Hagfish (*Eptatretus burgeri*), may possess precursors of adaptive immunity and be living links between innate and adaptive systems. Cathelicidins may also play a role in human immunity. Some people who lack cathelicidins in their saliva, for example, seem to be more prone to periodontal disease.[75]

Hagfish are of research interest, in addition, because they produce a defensive mucus slime, which contains high concentrations of flexible microscopic fibers that are being used to understand the mechanics of how cells keep their shape and how they move.[76]

In some regions of the Pacific, such as off the coast of Japan, Pacific Hagfish populations have been depleted by overfishing (mostly to satisfy the demand for leather products made from hagfish skins), and hagfish fishing has moved to the United States, especially to New England, where, for example, twelve million pounds of Atlantic Hagfish were landed in Massachusetts and Maine in the year 2000. There are no regulations governing this harvesting along the East Coast of the United States, and there are growing concerns that some hagfish populations may be threatened.[77] Not only are hagfish important research models, but because they are major scavengers of marine carrion, much like vultures and crows on land, they are likely to contribute to the recycling of nutrients in the oceans.[78]

Several species of lamprey, including the Spotted Lamprey (*Petromyzon marinus* L.), American Brook Lamprey (*Lampetra appendix*), and the Northern Brook Lamprey (*Ichthyomyzon fossor*), have also been shown to possess primitive components of adaptive immunity. They, too, have lymphocyte-like cells, with receptors capable of producing cellular responses similar to those seen in adaptive immunity. Like lymphocytes, these cells are also more sensitive to radiation than are other blood cell types and are able to aggregate and divide in response to stimulation by microbial surface antigens.[79]

CONCLUSION

This chapter, like chapter 4, summarizes some of the contributions that other species—plants, animals, and microbes—have made to human medicine. We selected three key areas of biomedical research for examination—genetics, regeneration, and the development of the immune system. But we could have selected many other equally important areas of research, for example, how human cancers develop, the process of aging, or the genesis of cardiovascular disease, to demonstrate the central roles that other species have played in our understanding of these conditions.

Some might say that it is not necessary for us to preserve organisms in the wild, that we can produce all the variants we could possibly ever need in the lab, with selective breeding and genetic modification. Certainly our experience with developing ever-increasing numbers of mouse strains illustrates the importance of such efforts. Others might say that, like Noah, we should simply preserve a few individuals of each species that is threatened in seed banks or zoos or botanical gardens or aquaria so that we will have them for the future, for breeding or for whatever information we might need from them. And again, there is great merit in these attempts at preservation.

But because we know so little about what species exist and about what lessons are to be learned even from those species we have identified, because Nature has been figuring out how organisms can best survive and protect themselves against diseases for hundreds of millions of years, and because whatever organisms we can save in artificial environments represent only a minute sample of the genetic diversity of their species, there is no substitute for preserving species and the ecosystems in which they live in their natural states, for them, and for us.

◆ ◆ ◆

Suggested Readings

"Beyond E. coli: The Role of Biodiversity in Biomedical Research," Joshua P. Rosenthal and Trent Preszler. In *Conservation Medicine: Ecological Health in Practice*, A.A. Aguirre et al. (editors). Oxford University Press, New York, 2002.

Howard Hughes Medical Institute Model Organisms, www.hhmi.org/genesweshare/e300.html.

Making PCR: A Story of Biotechnology, P. Rainbow. University of Chicago Press, Chicago, 1996.

Medicine's 10 Greatest Discoveries, Meyer Friedman and Gerald W. Friedland. Yale University Press, New Haven, Connecticut, 1998.

Microbe Hunters, Paul de Kruif. Harcourt, Brace & Company, New York, 1926 (new edition 1996).

Model Organisms in Biomedical Research, www.nih.gov/science/models/.

Winterworld: The Ingenuity of Animal Survival, Bernd Heinrich. HarperCollins, New York, 2003.

CHAPTER 6

THREATENED GROUPS OF ORGANISMS VALUABLE TO MEDICINE

Eric Chivian and Aaron Bernstein

Learn from the beasts the physic [the art of medicine] of the fields.

—ALEXANDER POPE

In this chapter, we look at seven groups of threatened organisms, three from the ocean and four from the land—amphibians, bears, nonhuman primates, gymnosperms (a group of plants that includes the conifers, cycads, and the Ginkgo Tree), cone snails, sharks, and horseshoe crabs—all of which are critically important to human medicine. These case studies are really at the heart of this book, for they provide specific examples of what we are now losing, and what we will be losing to a much greater degree, when we degrade the natural world, and serve to demonstrate concretely some of the many ways that human health depends on biodiversity.

There are other plants and animals we could have chosen to make these same points. We chose these organisms because most people are familiar with them, because a great deal is known about their contributions to medicine, and because they are among the most threatened organisms on Earth.

We do not mean to imply in any way that the most important reason for these organisms' existence is our use of them. All have lived on this planet for millions, and some for many hundreds of millions, of years longer than we have, evolving over these enormous spans of time into the magnificent organisms they are today. Their lives are no less sacred than our own.

But this is a human-dominated planet, and our actions are threatening their existence. And it is the belief of those of us who wrote this book that, until people around the world begin to realize that we have no other choice but to protect the animals and plants described in this chapter, and the countless millions of other species on Earth, because our health and our lives depend on them, we will not do so.

(left)
Polar Bear with Her Two Cubs. (Photo by Steve Amstrup, U.S. Fish and Wildlife Service.)

AMPHIBIANS

The Amphibian Extinction Crisis

The first amphibians, having evolved from early fishes that moved onto the land during the Devonian Period some 350 million years ago, are the ancestors of all living terrestrial vertebrates, including humans. The name "amphibian" comes from the Greek *amphis*, meaning "double," and *bios*, meaning "life," because their life cycles frequently involve breeding in aquatic environments, with their eggs and larvae developing in water, but with adults spending their lives mostly on land. This double exposure may explain, in part, their extreme sensitivity to environmental hazards.

Living amphibians are divided among three orders—Anura, composed of frogs and toads and comprising by far the largest number of known amphibian species; Caudata (also called Urodela), made up of salamanders and newts; and the Gymnophiona (also called the Apoda), consisting of the caecilians—little-known, legless amphibians that resemble giant earthworms. Of all the major groups of organisms on Earth, amphibians are, with primates, the most threatened, according to the International Union for the Conservation of Nature and Natural Resources (IUCN) 2006 Red List of Threatened Species. Nearly one-third—1,811 out of 5,918 amphibian species—are threatened with extinction, with all three orders well represented in this count, and 122 species are believed to have already gone extinct in the past few decades.[1] In addition, because almost one-quarter (22.5 percent) of the total number of known amphibian species are considered to be "data deficient" when it comes to their conservation status, the number of threatened amphibian species is likely to be even greater than 1,811. As with other groups of organisms, new species of amphibians are constantly being discovered, and as of early 2007, the total number of species had grown to 6,155 (see www.amphibiaweb.org). The conservation status, however, of most of the newly discovered species has yet to be determined.

Even in comparison to other groups of species at great risk, such as birds and mammals (where some 12 percent and 23 percent, respectively, are considered threatened), the threat to amphibians stands out. Many in the scientific community have referred to it as the "Amphibian Extinction Crisis."[2] Major international scientific concern about the status of amphibians began in earnest in 1989, when reports of significant declines in different parts of the world were shared at the first World Congress of Herpetology, held in the United Kingdom. It was noted then that these declines had begun as early as the 1970s in places such as the western United States, Puerto Rico, and northeastern Australia. And research from the 1980s, demonstrating that some 40 percent of amphibian species had vanished from one site in Costa Rica alone, and that such sudden disappearances had also occurred in mountainous areas of Ecuador and Venezuela, including in many places that were thought to be pristine and far from human habitation, further raised international concern about amphibian survival.[3] There is no indication from the fossil record that amphibian extinctions had occurred at this high rate in the past. Indeed, major amphibian groups alive today are well represented in the geological record and appear to have survived largely intact through the main prehuman extinction events of the past few hundred million years.[3]

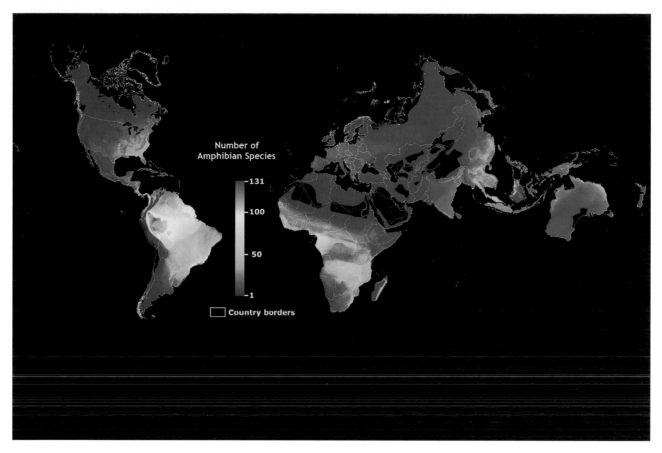

Figure 6.1. Fuller Projection of Global Amphibian Diversity. This map was created using the Environmental Systems Research Institute's ArcGIS software, which allows the GIS data to be plotted onto the Fuller Projection. It shows the global distribution of amphibian species. Note the marked concentration of amphibians in South America, particularly in the Amazon Basin. (Map and land mass arrangement created by Clinton Jenkins, 2005.)

With the exception of the oceans and the poles, amphibians, especially frogs and toads, are found throughout the world, living in a variety of habitats from deserts to subpolar regions, at sea level and at mountain snow lines, from the ground to the highest treetops. Salamanders are confined to the Northern Hemisphere, with the exception of the Plethodontidae, the lungless salamanders (which breathe through their skins), whose range extends into Central and South America. North America has the greatest salamander diversity of any continent, with nine of the ten existing families represented, but Central and South America together have the largest number of salamander species, all of which are plethodontids. For the mole salamanders, the Ambystomatidae, all of which have lungs, the United States has the greatest diversity of any country. Caecilians are found only in the tropics. (See figure 6.1 for the global distribution of amphibian species and figure 2.1 in chapter 2, which shows Buckminster Fuller's global projection map, called the Dymaxion Map, from which figure 6.1 has been derived.)

THREATS TO AMPHIBIAN SURVIVAL

Amphibians are threatened by many factors that are the direct result of human activities—the degradation, destruction, and fragmentation of their habitats;

overexploitation; the introduction of alien species; increased exposure to ultraviolet B radiation; pollutants of many types; global climate change; and various infections. Generally in the past, the scientific literature discussed each of these separately, with the implication that one can explain the precipitous and widespread losses of amphibian populations in different parts of the world by looking at individual factors alone. Increasingly, however, such discussions have begun to describe the declines as a function of multiple factors acting together, and by studying the interactions among these factors, we may better understand amphibian vulnerability. Those species that appear to be the most vulnerable—species that live near streams in mountainous tropical environments[3]—merit particular attention for such study.

One theory about why amphibians are so endangered is that some species, in particular, some tropical species, have evolved to live within narrow ranges of environmental conditions. When these conditions, such as temperature and humidity, change even by modest amounts so that they are outside the ranges of adaptation, a collapse of the exposed amphibians' vital systems (e.g., their immune system) can result, and they become vulnerable to other threats, such as infectious diseases.[4] The events that appear to be the cause of death in these situations, such as infections by chytrid fungi (see below), may instead be the ultimate manifestation of a cascade of events, such as those that can be seen in some people with HIV/AIDS who succumb to opportunistic fungal infections when their immune systems have been sufficiently weakened. In this section, we look at each of the factors that are thought to threaten amphibian populations individually, but the reader needs to bear in mind that, in general, they are not acting in isolation.

Loss of Habitat

Because many amphibians occupy two distinct habitats during their lifetimes, water and land, and can be affected by alterations in either one, they are potentially more vulnerable to habitat loss than most other organisms. As a result, many of the major alterations to Earth's landscape—including the clearing of forests, draining of wetlands, conversion of land for agriculture or development, and the building of roads that cut off access to aquatic breeding areas—may all affect amphibian populations. Frogs and toads with extremely small ranges, such as *Oreophrynella weiassipuensis* (which does not have a common English name) that inhabits only a single mountain near the Guyana–Brazil border, are particularly at risk.[5]

Salamanders have been the object of many habitat studies. Some have shown that old-growth forests contain three to six times more salamanders than second-growth forests.[6] This is most likely the result of old-growth forests' ability to maintain moist and cool habitats, such as in the many fallen, decaying logs that are present, which salamanders depend on for their survival. Clear-cutting is especially destructive to salamanders, with one study demonstrating that salamander catches were five times higher in mature forest stands than in those that were clear-cut. This study concluded that clear-cutting in U.S. national forests causes the loss of some fourteen million salamanders each year.[7]

Overexploitation

Humans have eaten frogs for millennia, particularly larger species such as bullfrogs that have meaty hind legs. Most of the current trade supplies European markets, especially France and Belgium, where frogs' legs have long been sought-after delicacies. Tens of millions of frogs are sacrificed each year, with most of these being taken from such countries as Indonesia, Malaysia, and Bangladesh, sometimes illegally. India banned the export of frogs in 1987, after determining that the profits from such export were less than the costs of importing pesticides to substitute for the pest-control services provided by the frogs. Asian Bullfrogs (*Kaloula pulchra*) are the main species being exploited, but other species are also taken.[8] To what degree this over-exploitation of some Asian frog species contributes to their being threatened has not been studied. The same is true for the question of how losing millions of frogs such as Asian Bullfrogs, a major insect predator, will affect insects that attack rice fields in Asia, or mosquitoes that carry major human infectious diseases such as malaria and Japanese encephalitis.

Introduction of Alien Species

Amphibian declines and extinctions have also been ascribed to the introduction of a variety of species, including fish, other amphibians, and crayfish. Stocking mountain lakes with Rainbow Trout (*Oncorhynchus mykiss*) and Brook Trout (*Salvelinus fontinalis*) for sport fishing in the California Sierra Nevadas, for example, was found to cause a major decline in the Mountain Yellow-Legged Frog (*Rana muscosa*), whose eggs and tadpoles were eaten by the fish. Removal of the trout resulted in a rebound in *R. muscosa* populations.[9,10] In addition to feeding on native amphibian species, other introduced species, such as the North American Bullfrog (*Rana catesbeiana*) may carry infectious diseases, like chytridiomycosis (see below), which may spread and threaten native populations. And finally, introduced amphibian species have been known to hybridize with native species or outcompete them, in both cases threatening survival.[11]

Ultraviolet B Radiation Exposure

The amount of ultraviolet B radiation (UVB) reaching the surface of Earth in recent decades has increased as a result of human-caused depletion of the stratospheric ozone layer (see section on UVB in chapter 2, page 60). UVB is potentially harmful to all life because it can cause mutations in DNA, and experiments in amphibians have shown that exposure to UVB can cause increases in embryo and larval mortality, skin damage, and behavioral changes and can impair growth and development.[12] Some field studies have shown UVB to decrease the viability of some amphibian species' eggs, but others have not. Why the results of these studies vary may be explained, at least in part, by the variability in the amount of UVB the amphibians are exposed to, because this amount depends on cloud cover, altitude (there is greater exposure at higher altitudes), and the amount of dissolved carbon in the aquatic environments (which blocks out UVB) where amphibians lay their eggs. Amphibians also vary in their ability to deal with UVB exposure—some may sustain greater damage than others given the

Figure 6.2. Western Toad. The decline of the Western Toad (*Bufo boreas*) may be due to a combination of factors, including climate change, increased ultraviolet B radiation, and fungal infections. (Photo by Christopher W. Brown, U.S. Geological Survey.)

same dose. The mixed data on the effects of UVB have caused some to minimize its importance in amphibian declines. However, because UVB damages DNA, and good research has documented that UVB can affect amphibian reproductive success, dismissing UVB as a contributor to amphibian declines is a hard case to make.

The strongest case for the effects of UVB on amphibian populations in the wild may be that of the Western Toad (*Bufo boreas*), a once widely distributed species in the western United States but now rare in much of its original range. Andrew Blaustein, Joseph Kiesecker, and their colleagues from Oregon State University have shown that the hatching success of Western Toads in the Pacific Northwest of the United States is less when embryos develop in shallow ponds unshielded from UVB radiation. This effect has also been found in Cascades Frogs (*Rana cascadae*), Long-Toed Salamanders (*Ambystoma macrodactylum*), and Northwestern Salamanders (*Ambystoma gracile*). As is the case with most amphibian die-offs, however, other potential factors, such as climate change and fungal infections, are also likely to have contributed to the loss of Western Toads. What may be happening is that lower water levels in the shallow lakes and ponds of the Pacific Northwest where *B. boreas* breeds, the result of climate-change—mediated reductions in precipitation, have led to greater UVB exposures for *B. boreas* embryos (the deeper the water is, the more it acts to filter out UVB), and consequent lethal infections caused by the fungus *Saprolegnia ferax*. Perhaps the UVB has weakened the embryos' immune response, allowing *S. ferax* infections to take hold.[13–15]

Pollution

Their residence in aquatic environments and their permeable skins make amphibians vulnerable to a wide range of pollutants. Many studies, for example, have demonstrated that acid rain has a detrimental effect on reproduction in some amphibian species.[16] The acidification of water may be a particular problem for species that breed in vernal ponds (temporary ponds that form in the spring), because they are more likely to be filled by precipitation (rather than by groundwater springs) and have less buffering capacity than do permanent ponds. Acidification also increases the solubility of some toxic metals, such as lead, aluminum, and mercury, the last of which has been found to bioaccumulate in some amphibian species, such as in the Northern Two-Lined Salamander (*Eurycea bislineata bislineata*) in Maine.[17]

The Cricket Frog (*Acris crepitans*), once the most common amphibian in the state of Illinois, has undergone a marked population decline in recent decades; it is no longer found in Canada and is rarely seen in the Upper Midwest region of the United States. By studying museum specimens collected over the past 150 years, scientists were able to link an increased incidence in hermaphroditism (i.e., having both male and female sex organs) in Cricket Frogs, both temporally and geographically, with the presence of organochlorine compounds such as PCBs and DDT. They hypothesized that these compounds may have acted to disrupt sexual development and reproduction.[18] Cricket Frogs may also have suffered from periods of drought and cold winters. Unlike some closely related species in these areas, which survive the winters by burrowing underground, submerging under water, or producing antifreeze, Cricket Frogs spend the winter in shallow, wet areas at the edges of wetlands. When it is too dry or too cold, they cannot survive at the northern and western extremes of their ranges, areas where much of their population losses have been found.

Pollutants may also be a cause of limb deformities in some frogs, making them easier prey for such predators as herons and egrets. These deformities, found in some sixty species of frogs, salamanders, and toads in forty-six states in the United States, have been linked in some cases to infection by a trematode worm (*Ribeiroia ondatrae*).[19] Exposure to agricultural chemicals may make these infections more likely to occur.

Two of the agricultural chemicals that may pose the greatest danger for amphibians are the herbicides glyphosate (the active ingredient in Roundup) and atrazine. Glyphosate is now the most commonly used pesticide in the United States, with a total of about 40 million kilograms (around 44 million tons) applied on 20 million acres in 1999.[20] Roundup is highly toxic, when applied at manufacturer's recommended concentrations, to three species of amphibians, Leopard Frogs (*Rana pipiens*), American Toads (*Bufo americanus*), and Gray Tree Frogs (*Hyla versicolor*), killing 96–100 percent of the larvae in test ponds after three weeks, and 68–86 percent of the juveniles on land.[21] Given the rapid, worldwide increase in the use of this chemical to grow some genetically modified crops, glyphosate's demonstrated toxicity to some amphibians, in the midst of a growing amphibian extinction crisis, should raise serious international concern.

Atrazine is the second most commonly used herbicide in the United States and also one of the most used worldwide. At concentrations that were 400 times lower than levels found in rainwater in some parts of the U.S. Midwest, and thousands of times lower than those that can be present in agricultural runoff (less than one part per billion [1 ppb], or 1 millionth of a gram of atrazine in a liter of water), atrazine was found to promote hermaphroditism and impair sexual development in both African Clawed Frogs (*Xenopus laevis*) and Northern Leopard Frogs (*Rana pipiens*).[22,23] And at somewhat higher concentrations, 4 ppb, which is still only 1 ppb more than the maximum allowable level of atrazine in U.S. drinking water,[24] atrazine resulted in embryo and larval mortality in Streamside Salamanders (*Ambystoma barbouri*).[25] When atrazine was mixed with minute concentrations (0.1 parts per billion) of eight other pesticides to replicate conditions found in a typical Nebraska corn field, where multiple herbicides, fungicides, and insecticides are generally used in combination, more than one-third of the Northern Leopard Frog tadpoles exposed to the mixture died.[26] These findings have been challenged by the agrochemical company Syngenta, the primary producer of atrazine. But given that there may be no places in the United States where atrazine levels are less than those shown to be toxic to *X. laevis* and *R. pipiens*, and that the herbicide may affect pregnant women by causing their fetuses to grow more slowly or by predisposing them to early deliveries,[27] the U.S. Environmental Protection Agency's granting permission for the continued use of atrazine, and the review by its Scientific Advisory Panel, especially in light of a ban on its use in France, Germany, Italy, Sweden, and Norway, should be reconsidered.

Global Climate Change

In the case of the Western Toad described above, and other amphibians found in the Cascade Mountains of the Pacific Northwest, reduced precipitation resulted in shallower pond and lake depths where these amphibians lay their eggs and may have contributed to high rates of mortality. This climate effect was thought to be a product of an intense El Niño event (a normal cycle of the Pacific Ocean in which water warms near the Equator every three to seven years and disrupts weather patterns over a broad expanse of countries bordering the Pacific and beyond). Mounting evidence suggests that global warming, by further warming sea-surface temperatures around the world, more so in eastern equatorial regions than in western ones, is likely to create more El Niño conditions on average and more extreme precipitation patterns (floods and droughts).[28,29]

Climate change may also affect amphibians by changing the timing of seasons and, as a result, when a species will migrate and reproduce (see section on global climate change in chapter 2). In Britain, for example, the spawning time of the Common Water Frog (*Rana esculenta*, also known as the Edible Frog because of the use of its legs in cooking) and the breeding pond arrival time for the Common Newt (*Trituris vulgaris*) have come significantly earlier over the seventeen-year period between 1978 and 1995. Such alterations may threaten populations by disrupting their relationships with other species on which they depend, for example, those that constitute their food supply or their predators or competitors, all of which may change the timing of their own biological cycles, but by differing amounts.[11,30]

The impacts of global climate change on amphibians have been perhaps more thoroughly explored in the cloud forests of Central America, particularly the Monteverde Cloud Forest Preserve in Costa Rica (which contains 10,500 hectares, or about 26,000 acres), than anywhere else. Here, all forested regions have warmed significantly—from 1975 to 2000, by 0.18 degrees Celsius (0.32 degrees Fahrenheit), on average, per decade. The warming has resulted in a raised height for cloud formation (also caused by deforestation in the lowlands surrounding the mountains); shifts in the habitat occupied by populations of birds, reptiles, and amphibians, with some species moving up the mountains to occupy new ranges that meet their biological needs (some organisms that already live at the very top of the mountains have nowhere to go); increased cloud cover from larger amounts of water vapor in the

Figure 6.3. Male Golden Toad (*Bufo periglenes*). The Golden Toad, whose range was limited to the cloud forests of Monteverde, would gather by the hundreds in pools during mating season. In 1989, only a single male Golden Toad seeking a mate was found. The species is now believed to be extinct, because no individuals have been seen since that time. (© J.W. Raich/Iowa State University.)

atmosphere secondary to the warming and to particulate air pollution from upwind urban populations; and alterations in daytime and nighttime temperatures. These climate-induced changes are thought to have created favorable conditions for the pathogenic fungus *Batrachochytrium dendrobatidis* to thrive, contributing to the deaths of some amphibian species in Monteverde, and perhaps in other parts of Central and South America.[31] The Golden Toad (*Bufo periglenes*) was last seen in 1989 and is presumed extinct (see figure 6.3). Its bright orange image has become an icon for amphibian extinctions.[32]

The Harlequin Frog (*Atelopus varius*) was thought to be extinct, with reports that it had not been seen since 1996. But in 2003, at a site in Costa Rica within its known geographical range, three live *A. varius* adults were found. Other *Atelopus* species once thought to be extinct have also been found.[33,34] Although such rediscoveries tend to instill hope that these species may survive, the reality is that they are still considered to be Critically Endangered and perilously close to extinction.

Figure 6.4. Female Harlequin Frog (*Atelopus varius*) Showing Warning Colors. The Harlequin Frog, thought to be extinct, was rediscovered in Costa Rica in 2003. As a group, however, more frogs of the genus *Atelopus* are thought to have gone extinct than those belonging to any other amphibian genus. Several members of this genus contain toxic alkaloids, like the tetrodotoxins, that have potential human medical uses (see page 214). (© Michael and Patricia Fogden, www.fogdenphotos.com.)

Of some 113 species of *Atelopus* frogs that live in the tropics of Central and South America, at least thirty have been missing from all their known habitats for periods of eight years or more and are presumed to be extinct. Only ten *Atelopus* species have stable populations.[35] All of the extinctions may involve the chytrid fungus *B. dendrobatidis*. An important threshold seems to have been passed in the late 1980s when mass mortality among frogs and toads of the American tropics seems to have begun.[31]

Infection

Chytrids are ubiquitous fungi that are found in aquatic habitats and moist soils. They are believed to be among the most ancient of the fungi; the oldest fossils of fungi that have been found are chytrid-like organisms. Those species that are parasitic infect mainly plants, algae, protists, and invertebrates. The chytrid *Batrachochytrium dendrobatidis*, which has been implicated in the deaths of amphibians around the world from the disease chytridiomycosis, an infection of the most superficial layers of the amphibian skin, is the only chytrid known to parasitize vertebrates. First discovered in 1998 in dead amphibians in Australia and Panama, *B. dendrobatidis* has been linked to the decline of some ninety-three amphibian species worldwide, with compelling data to suggest that the fungus was the immediate cause of these declines (see www.jcu.edu.au/school/phtm/PHTM/frogs/chyglob.htm). The fungus is lethal to amphibians most likely because, as it grows on skin, it interferes with the skin's vital functions, including respiration and the exchange of fluids. The fungus can also grow in the mouths of tadpoles and obstruct their feeding.

The sudden appearance of lethal chytridiomycosis from *B. dendrobatidis* in amphibians over the past ten to twenty years has suggested to some that it is an emerging infectious disease, invading new habitats and exposing amphibians that have little or no immunity to it.[36,37] Supporting this theory is the observation that outbreaks of chytridiomycosis seem to affect amphibians as if the outbreaks were spreading from one area to the next over time.[38] One hypothesis is that *B. dendrobatidis* is being introduced by some amphibian species that carry the fungus without showing signs of disease, such as the African Clawed Frog (*Xenopus laevis*, used extensively in biomedical research [see chapter 5, page 193] and as an early pregnancy test [female frogs injected with the urine of pregnant women ovulate])[39] and the American Bullfrog (*Rana catesbeiana*), which is used as a food source.[40] They are thought to be among the agents for spreading the disease to other amphibians in new habitats. People, including researchers, have also been implicated in the spread of the pathogen, carrying it into previously *B. dendrobatidis*-free environments. A recent study has shown that in one area where there had been no chytridiomycosis, El Copé in Panama, the fungus had an extremely rapid spread over a four-month period in 2004, during which time amphibian abundance decreased by more than 50 percent, and thirty-eight frog species (some 57 percent of the total number at this site) were infected. Temperature and rainfall patterns at the site during this epidemic were apparently similar to those found in long-term records.[38]

As with other contributors to amphibian declines, how much chytrid fungi have affected amphibian populations remains controversial, because some species, and in some instances even different populations of the same species, may not be affected when the fungi infect them. Some antimicrobial peptides found in frogs' skin, such as the magainins (in *X. laevis*) and the dermaseptins (in the South and Central American tree frog *Phyllomedusa sauvagei*), show activity against *B. dendrobatidis* (see section below on antimicrobial peptides, page 215) and may play some role in protecting these and other species from the fungus.[41] Evidence that some amphibian species such as the Australian Eungella Day Frog (*Taudactylus eungellensis*), which had experienced significant population declines thought to be due to *B. dendrobatidis*, but which are now found to be healthy in the wild while still being infected, suggests that some individuals may have greater immunity than others, or that the fungus may have more or less deadly strains.[42] Furthermore, evidence of infection by *B. dendrobatidis* has been found in one study in a significant proportion (7 percent) of healthy amphibians in eastern North America collected since the 1960s.[43] For some chytridiomycosis epidemics, then, the fungi may already be present, and other factors, such as habitat loss, climate change, and pollutants, may have been the triggers.

ECOSYSTEM EFFECTS OF AMPHIBIAN LOSS

With the loss of amphibians will come serious consequences for the terrestrial and aquatic ecosystems they inhabit. In eastern U.S. forests, for example, salamanders are thought to be the most abundant of all vertebrates, both in terms of numbers and biomass,[7,44] and they are critically important in such ecosystem functions as the decomposition of organic matter and nutrient recycling.[45] They also are important predators of prey too small for birds and mammals and are themselves an important food source for many other organisms, including birds, snakes, and mammals.[44] And larval amphibians can be important herbivores in aquatic ecosystems.

Amphibians are major predators of insects, and their loss is likely to have large effects on pest management in agroecosystems and on the emergence and spread of some vector-borne diseases, particularly those carried by mosquitoes. These and other ecosystem consequences of the escalating Amphibian Extinction Crisis must be given the highest research priority, because frogs, toads, salamanders, newts, and caecilians are rapidly disappearing before our eyes.

Potential Medicines from Amphibians

Because amphibians are among the most threatened of all major groups of organisms on Earth, and because they possess an enormous variety of biologically active compounds, some of which could become important new medicines, there has been an intense race against time to identify these compounds and to save as many amphibian species as possible before they disappear.

To defend themselves against predators and infections, amphibians have developed a vast array of compounds that they release when stressed or injured, often as complex cocktails, from what are called "granular glands" in their skins. This section describes some of these compounds, isolated from more than 500 amphibian species,[46] that may have important medical uses. We organize them into the following categories: alkaloid toxins, antimicrobial peptides, other bioactive peptides such as bradykinins, and novel compounds or mixtures of compounds, such as a substance known as "frog glue."

THE ALKALOID TOXINS

Tropical frogs belonging to the family Dendrobatidae are commonly referred to as poison-dart or poison-arrow frogs. This designation, however, is not strictly correct, since only three species, all belonging to the genus *Phyllobates* and all found in the Choco region of western Colombia in South America, have been used to coat blow gun darts, employed by native tribes to hunt birds and mammals, and only these three contain the highly toxic batrachotoxin alkaloids in sufficient concentrations to be used as dart poisons (see section below on batrachotoxin, page 220). Of the other frog species in the family Dendrobatidae, only eighty or so that belong to the genera *Dendrobates*, *Minyobates*, *Epipedobates*, and *Phyllobates* contain alkaloids. None of these frogs is as toxic as the three true poison-dart frogs; their alkaloids are either much less toxic to begin with or are present in lower concentrations. But these compounds still serve as powerful deterrents to predation, because they have a bitter taste and cause unpleasant effects in the mouths of predators. Some of the dendrobatid frogs are threatened by habitat destruction or by disease, and as a result, restrictions to their collection have been widely adopted under the Convention on International Trade in Endangered Species of Wild Fauna and Flora (or CITES—see appendix B). Nevertheless, by using samples that were, in most cases, collected before these restrictions were put into place, researchers have identified more than 800 alkaloids from more than fifty species of these frogs.[47]

Dendrobatid frogs raised in captivity, it turns out, cannot synthesize alkaloids by themselves. Rather, they must acquire the building blocks for these compounds from their diets, in large part from alkaloid-containing ants, mites, beetles, and millipedes.[48,49] After ingestion, the alkaloids are stored in the granular glands of the amphibian skin and, in at least one case, can be made more toxic through chemical tinkering by the frog.[50] Other amphibians also possess such alkaloids in their skins—*Mantella* frogs from Madagascar, *Melanophryniscus* toads from South America, and *Pseudophryne* frogs from Australia (these Australian frogs, in contrast to the dendrobatids, actually do synthesize one class of their alkaloids). Toxic alkaloids have also been found among the highly endangered Harlequin Frogs of the genus *Atelopus* (see the discussion above on widespread extinctions of *Atelopus* frogs), including tetrodotoxin, chiriquitoxin, and more recently, zetekitoxin AB, a potent sodium-channel blocker (see page 220), found in the Panamanian Golden Frog (*Atelopus zeteki*),[51] which is nearly extinct in the wild.[2]

Numerous alkaloids have been identified, such as the pumiliotoxins and epibatidine, that have potential as, or may serve as models for, important new medicines.

The Pumiliotoxins

The first two pumiliotoxins were isolated from the Panamanian Poison Frog (*Dendrobates pumilio*). Nearly 100 pumiliotoxins and many other compounds similar in structure, namely, the allopumiliotoxins and the homopumiliotoxins, have been discovered in frog skin extracts. They have been considered as possible medicines for strengthening the contraction of the heart, due to their effects on the flow of sodium and calcium across membranes.[52] To date, however, this has not been pursued because of their toxicity. Synthetic modification of these compounds, however, could create effective derivatives with lower toxicity.

Figure 6.5. *Epipedobates tricolor.* (Courtesy of www.jjphoto.dk.)

Figure 6.5. *Epipedobates tricolor.* (Courtesy of www.jjphoto.dk.)

Epibatidine

In 1976, John Daly at the National Institutes of Health isolated from skin extracts of the Ecuadorian Poison Frog (*Epipedobates tricolor*) a small amount of a novel alkaloid that was later named epibatidine, a compound that turned out to have highly potent analgesic properties in mice. At the time, it was not possible to collect more extract because of collection restrictions, and it was not until several years later when new techniques became available, that the molecular structure of epibatidine was defined and the compound synthesized. The site of action was found to be nicotinic receptors (one of the two classes of receptors of the neurotransmitter acetylcholine, found throughout the central and peripheral nervous systems). Epibatidine, representing a whole new class of analgesics, was found to have remarkable pain-killing potency— approximately 200 times that of morphine—and, in addition, it did not lead to the development of tolerance (the state in which, over time, higher doses of a medication must be used to achieve the same effect, routinely seen with opiates in the treatment of pain).[53]

The structure of epibatidine has been used as the starting point for the development of a large series of related compounds, modeled on the portions of the epibatidine molecule that were thought to be responsible for its pharmacological effect. One of these was the compound that Abbott Laboratories called ABT-594. ABT-594 went through phase II clinical trials for pain control, but it was withdrawn in 2003 because of undesirable side effects. Although ABT-594 itself was not successful, subsequent attempts to develop other painkillers based on epibatidine have shown promise,[54] and epibatidine and other nicotinic receptor-binding compounds have become important research tools to help understand the structure and function of these critically important nervous system receptors.

ANTIMICROBIAL PEPTIDES

The first amphibian skin peptide that was demonstrated to have antibacterial activity was bombinin, isolated in the late 1960s from the European Yellow-Bellied Toad

(*Bombina variegate*). But research to find other antimicrobials from amphibians did not begin in earnest until Michael Zasloff reported in the late 1980s a new class of antimicrobial peptides, the magainins, found in the African Clawed Frog (*Xenopus laevis*; see figure 6.9 for a photo of *Xenopus laevis* and page 222 for a discussion about its importance to biomedical research). During the same period, working at the University of Rome "La Sapienza," Vittorio Erspamer was isolating a large number of bioactive peptides from frogs of the genus *Phyllomedusa*. In the last two decades, more than 200 antimicrobial peptides from a total of six "structural families" of peptides have been found in the skin of a wide variety of frogs and toads, and numerous analogues based on these naturally occurring peptides have been synthesized.[55] The diversity and potency of these compounds should come as no surprise, given the great variety of habitats in which amphibians live, the great diversity of pathogenic microbes found in these habitats, and the marked vulnerability of amphibians to skin infections, given that their skin functions not only as a protective barrier but also as a regulator of the passage of both water and electrolytes. We describe here the magainins from *Xenopus* and antimicrobials from the frog *Phyllomedusa sauvagei*.

Magainins

The magainins found in the skin and the gastrointestinal tract of *Xenopus laevis* are broad-spectrum antimicrobial agents that have an affinity for bacteria, binding to acidic phospholipids (fatty-acid–containing compounds that are major components of all biological membranes) on the surfaces of bacterial membranes. Magainins kill both Gram-negative and Gram-positive bacteria by making their membranes more permeable, thereby causing them to break apart. Remarkably, they also are lethal to some pathogenic fungi and protozoa, but they do not kill mammalian cells, with the possible exception of certain cancer cells that have been harvested from humans and mice.[56]

A synthetic derivative of magainin called pexiganan has successfully completed phase III clinical trials for the treatment of infected diabetic foot ulcers, demonstrating both efficacy and safety in treating infection and promoting wound healing. However, its commercial development has been delayed because of requests by the U.S. Food and Drug Administration (FDA) for additional studies.[57] Pexiganan is also being investigated for the treatment of other bacterial infections, including those leading to septic shock.[58]

Dermaseptins

South and Central American leaf frogs of the genus *Phyllomedusa* produce skin secretions that contain, according to Vittorio Erspamer, a "huge factory and store-house of a variety of (biologically) active peptides."[59] Among these are antimicrobial compounds called the dermaseptins, five of which have been isolated from the skin of the frog *Phyllomedusa sauvagei*, which inhabits the Chaco (dry prairie) regions of Argentina, Brazil, Bolivia, and Paraguay.

In the lab, dermaseptins are obtained from live frogs (as is also now the case for other amphibian skin compounds) by gently squeezing skin glands on their backs, washing off the secretions, and isolating the molecules. The dermaseptins have

Figure 6.6. *Phyllomedusa sauvagei*, the Waxy Monkey Tree Frog, a South and Central American Leaf Frog Found in Argentina, Brazil, Bolivia, and Paraguay. (Courtesy of Johannes Otto Foerst, Bamberg, Germany.)

broad-spectrum activity against both bacteria and fungi and are thought to play a role in defending the vulnerable skin of *P. sauvagei* against pathogenic microbial invasion.[60] Of particular interest is that they were among the first vertebrate peptides to demonstrate lethal activity against some species of fungi, such as *Aspergillus*, *Candida albicans*, and *Cryptococcus neoformans*, that can cause life-threatening infections in people with weakened immune systems, such as those with HIV/AIDS.[61]

Also isolated from *P. sauvagei* are two other novel classes of antimicrobial peptides, the dermatoxins and the phylloxins. In contrast to the dermaseptins, these are highly selective antibiotics that act against only some Gram-positive and Gram-negative bacteria, as well as against mollicutes (a group of bacteria, also called mycoplasmas, that do not have cell walls).[62,63]

As antibiotic resistance among pathogenic bacteria becomes more widespread, the need to find new antibiotics to treat the infections they cause takes on much greater urgency, especially for microbes that are resistant to multiple antibiotics, such as methicillin-resistant *Staphylococcus aureus* (MRSA) and new strains of the bacterium that causes tuberculosis, some of which have been newly classified as "extensively drug-resistant" or XDR, because they cannot be treated either with first-line drugs, such as isoniazid and rifampicin, or with three or more classes of second-line drugs.[64,65] Amphibians may not only contain antibiotics, such as the magainins and dermaseptins, that are capable of treating these bacteria, but also may provide unique models for new synthetic antibiotics or for new antibiotic treatment strategies designed to overcome antibiotic resistance. Furthermore, magainins, dermaseptins, and other amphibian antimicrobial peptides are critically important for us to study because they target microbial *membranes* and therefore are not likely to induce significant antimicrobial resistance—bacteria cannot change their cell membranes to thwart such an attack without interfering with the essential functions of the many proteins that sit on, or in, their membranes and that rely on the membrane's integrity to function normally.[66] Contrast this with many commercially available antibiotics, such as penicillin, or with its derivatives such as methicillin, that work by disrupting bacterial cell *walls* (bacteria have both cell membranes and cell walls), structures that bacteria can easily mutate to make themselves resistant to antibiotics.

Frogs seem to release multiple antimicrobial peptides simultaneously, which suggests that such combination therapy may provide for them the most complete protection against the broad diversity of pathogenic microbes they encounter across their large geographic ranges, and that such release has probably evolved as an effective strategy to prevent the development of antibiotic resistance.[60,67] Of particular significance is the surprising observation that each frog species—even closely related species—produces peptides that are characteristic of its species, and that each peptide exhibits a specific range of antimicrobial activity. The peptide "cocktails" of frogs represent a potential antibiotic library of enormous proportions. We have a great deal to learn about these peptides and about the unique antimicrobial defenses that amphibians employ. (See also chapter 5 and the discussion of antimicrobial peptides, page 198.)

OTHER BIOACTIVE PEPTIDES

In addition to antimicrobials, amphibians secrete a large number of other bioactive peptides from their skin granular glands that are thought to protect them against predators. Several hundred of these have been characterized and classified into families according to their structures, such as the bombesins, tachykinins, caeruleins, and bradykinins.[68]

Of these, the bradykinins and their related compounds have generated the greatest interest. These include maximakinin and kinestatin from the Chinese Large-Webbed Bell Toad (*Bombina maxima*); phyllokinin from *Phyllomedusa* spp.; Leu[8]-bradykinin from the North American Pickerel Frog (*Rana palustris*); tryptophyllin-1 from the Mexican Leaf Frog (*Pachymedusa dacnicolor*); and bradykinin itself and its structural variant, Thr[6]-bradykinin, from the Oriental Fire-Bellied Toad (*Bombina orientalis*).

Bradykinins have multiple actions. They induce contraction of various smooth muscles, including those in the mammalian intestine and uterus. Their presence makes pain receptors more sensitive when cells are injured. But of greatest importance, at least for human medicine, is that bradykinins cause dilation of blood vessels and a consequent drop in blood pressure. Some frog peptides, such as maximakinin, are fifty times more potent than bradykinin in binding to bradykinin receptors on mammalian

Figure 6.7. Oriental Fire-Bellied Toad (*Bombina orientalis*). The toad *Bombina orientalis*, found in eastern China and North Korea, produces peptides in its skin that may become leads for new human medicines. (© Michael and Patricia Fogden, www.fogdenphotos.com.)

arterial smooth muscle. Others, such as kinestatin, are highly potent bradykinin receptor antagonists, and are not only more potent than bradykinin, such as the compound phyllokinin, but also produce more prolonged dilation of blood vessels and lower blood pressure. Still others, such as a tryptophyllin from *Pachymedusa dacnicolor* called PdT-1, causes dilation of arteries, but it is not clear whether it is doing so by acting at the bradykinin receptor or at another, as yet unidentified receptor site. All of these compounds could lead to important new medications for high blood pressure.[69]

OTHER NOVEL COMPOUNDS

The Australian frog *Notaden bennetti* spends as much as nine months of the year living underground in dried mud, emerging only during torrential rains. Once on the surface, it is vulnerable to various predators, including biting insects, and it defends itself by secreting a sticky, protein-based material from glands in its skin that acts as a pressure-sensitive adhesive. This "frog glue" hardens in seconds and sticks well even when wet.[70] For repairing various human (and other animal) tissues, there is a need for a strong, flexible, and porous adhesive. Synthetic glues, such as the cyanoacrylates (the main ingredient of Superglue), are certainly strong enough for such repairs, but they also are generally toxic and brittle. Moreover, they tend to form impervious barriers that do not allow exchange of gases, fluids, nutrients, and electrolytes, or the infiltration of cells, all of which are necessary to promote healing. Most biological glues, on the other hand, such as those based on fibrin (a sticky protein in blood plasma) or on albumin (a common protein found in blood), are not strong enough to repair tissues that are subject to strong shear forces, such as those that are present following tears of the meniscus cartilage in the human knee. Meniscal tears are common injuries, frequently sustained in sports such as skiing or football, and are difficult to repair successfully.

The glue from *N. bennetti* skin has been tested in sheep, where it held together experimental meniscal cartilage tears until collagen, the main component of cartilage, was able to heal the wound.[71] As a result of this initial success, there is great interest in pursuing the possibility of using "frog glue" in human surgical repairs.

Amphibians in Biomedical Research

Few organisms have as rich a history in biomedical research as do amphibians. In the seventeenth and eighteenth centuries, they became the models for understanding how electricity works in the nervous system; in the nineteenth century, they were used to help clarify how organisms develop from the very earliest stages of life; and in the twentieth century and today in the twenty-first century, they have become a major resource for investigating the processes of regeneration.

ELECTRICAL CONDUCTION IN THE NERVOUS SYSTEM

In 1791, Luigi Galvani, a distinguished professor at the University of Bologna, published *De Viribus Electricitatis in Motu Musculari Commentarius* (Commentary on the Effect of Electricity on Muscular Motion) that described a series of experiments he had performed over the previous ten years on the effects of electricity on frog muscles. He demonstrated that applying electric currents to a nerve in the legs of dead frogs could make them twitch. To modern minds, such a result might seem entirely predictable, but at the time it caused nothing short of a sensation and led to countless frog-collecting expeditions solely because people wanted to see for themselves what Galvani had seen.

Galvani's seminal experiments on frogs were followed by those of several of his countrymen, including Alessandro Volta, Leopoldo Nobili, and Carlo Matteucci, all of which ultimately led to a detailed understanding of how muscles respond to electrical stimuli generated by the nervous system. They were also an important step that led to the development of the electrocardiogram, a device that measures the electrical activity of the heart and that relies on a device bearing Volta's name, the voltmeter.

FROG SKIN AND MOLECULAR SCIENCE

The peoples of Central and South America have long lived with poison-dart frogs (see page 214) and have known for centuries that they produce lethal toxins in their skin.

Figure 6.8. Golden Poison Frog (*Phyllobates terribilis*), One of the Most Poisonous Organisms on Earth. (© Mark Moffett/Minden Pictures.)

In the 1960s, John Daly began work on these chemicals. One of his first discoveries, originally isolated from the skin of a frog that lives in western Colombia, *Phyllobates aurotaenia*, was the most potent toxin of them all. He named it batrachotoxin.

Batrachotoxin is one of the most toxic substances on Earth—only two-tenths of one microgram (roughly equivalent in weight to three grains of salt) is a lethal dose for humans. Only a few species of frogs carry batrachotoxin (it turns out that four species of birds [the *Ifrita kowaldi*, and three species from the genus *Pithoui*] and beetles from the genus *Choresine*, all of which are from Papua New Guinea, also possess the toxin), but none contains more than the Golden Poison Frog (*Phyllobates terribilis*).[72] This diminutive animal, no more than two inches in length, packs a deadly wallop, containing enough batrachotoxin to kill 100 or more adult humans, and as many as 20,000 mice.

Batrachotoxin acts on ion channels that allow the passage of sodium ions across membranes (see figure 6.37 and page 264 for a discussion of ion channels). It is the ability of our cells to regulate the movement of sodium (as well as that of other ions) that keeps us alive. Batrachotoxin kills by incapacitating the sodium channels of nerve cells so that muscles no longer contract, including those of the heart.

Part of what makes batrachotoxin so toxic is its remarkable strength in binding to sodium channels. While the Emberá and Choco Indians use the toxin to coat their darts, scientists have used it to explore the structure and function of sodium channels[73,74] and to better understand how the nervous system and how muscles, including heart muscle, work. While no medicines derived from batrachotoxin are currently being investigated, the compound has been used to test the action of various anesthetic, anti-arrhythmic, and anticonvulsant drugs on sodium channels, thereby leading to safer, more effective medications.[75]

REGENERATION

Amphibians of the order Urodela, which includes newts and salamanders, unlike nearly all other vertebrates (with the notable exceptions of Zebrafish and the lab-bred MRL mouse described in chapter 5), are capable as adults of regrowing tissues, such as heart muscle and nerve tissue in the spinal cord, and even organs, such as complete limbs, jaws, and tails.[76,77]

The first serious examination of urodele regeneration was made by the Italian priest and scientist Lazzaro Spallanzani, who published *An Essay on Animal Reproduction* in 1768, describing his experiments with limb, tail, and jaw regeneration in aquatic salamanders. A great deal has been learned since Spallanzani's early experiments. After a urodele's limb is severed, the cells at the limb's stump do a most extraordinary thing: They revert from being specialized cells to more generalized, dedifferentiated cells that are capable of reproducing the limb that has been lost (see "Regeneration Research" in chapter 5, page 187). In a urodele, following the loss of a limb, an entirely new one, almost indistinguishable from the original, can be grown in approximately three months' time. Injured spinal cords have been shown to heal in even less time. Studies with adult Eastern Spotted Newts (*Notaphthalmus viridescens*) demonstrate that in as little as four weeks, enough of the spinal cord has regrown that the newts recover their ability to swim, although complete regeneration of the spinal cord structure and all its connections may take as long as two years.[78]

Many genes have been identified that control this process, for example, the same genes that have been identified in Zebrafish known as *fgf20* and *hsp60* and that are also present in people,[79] although application of this knowledge has yet to be realized. What these findings have made clear is that regeneration is an evolutionarily ancient process and that it occurs in different ways in different species.[77] Of all the organisms known to be capable of regeneration, with the exception of the MRL mouse, urodele amphibians are the most closely related to us, and thus they may offer our best hope for understanding how we may be able to unlock our own latent ability to regenerate lost cells, tissues, and perhaps even organs.[80,81]

EARLY EMBRYONIC DEVELOPMENT

Amphibians have been at the center of research into the very first stages of animal development since the early nineteenth century. In the 1820s, the Italian scientist Mauro Ruconi and German scientist Karl Ernst von Baer used the embryos of various frog species, including the Common European Frog (*Rana temporaria*) and the Edible Frog (*Rana esculenta*), to investigate how these organisms (and, as it turns out, all animals) progress from single cells to a ball of cells, having a central cavity, known as a blastula. By the end of the nineteenth century, at least a dozen other amphibian species—notably the Fire-Bellied Toad (*Bombina bombina*)—would be used to figure out the complex developmental processes that occur during these earliest stages of embryonic life and to provide insights into how the blastula develops into the many different tissues that comprise the fully formed organism.

In the early twentieth century, amphibians contributed to nearly all the major advances made in the field of experimental embryology. In a legendary set of experiments, Hans Spemann and Hilde Mangold made use of blastulas from two salamander species, *Triton taeniatus* and *Triton cristatus*, to show that the differentiation of cells into, for example, neurons depends upon their being stimulated by signals from neighboring cells. They were able to make this discovery because they could transplant portions of the blastula from one species onto that from another, while maintaining the viability of the hybrid. Because the blastula cells of the two species were, most conveniently, different colors, they could see clearly where the transplanted cells had migrated in the developing organism. The Dutch scientist Pieter Nieuwkoop expanded on these findings using *Rana pipiens* and *Ambystoma punctatum*, among other amphibians. His studies involved drastic manipulations to the blastula, such as cutting it in half, rotating one part 180 degrees, and then reattaching the two halves. Such transformations of the blastula, which led to a better understanding of the mechanisms of early cellular differentiation, were not possible with other organisms.

Today, we still rely on amphibians for studies of early embryonic development. The African Clawed Frog (*Xenopus laevis*) has become an essential research model for scientists around the world, because it (like the other amphibian species mentioned above) produces relatively large and easily manipulable embryos that heal well and can be studied outside of the organism. What makes *Xenopus* stand out is that it is able to spawn year-round, in contrast to most other amphibian species used in biological research that are seasonal breeders. Research with *Xenopus* has continued to deepen our understanding of how animals take the very first steps in development. A so-called "fate map" has been developed for *Xenopus* that traces the course of each cell through its developmental transitions. This map has led to a more complete appreciation of such fundamental developmental processes as the development of a body plan (i.e., determining left from right and top from bottom).[82]

Figure 6.9. The African Clawed Frog (*Xenopus laevis*). (© Michael Redmer/Visuals Unlimited.)

FROZEN FROGS

For centuries, people have dreamed of being frozen so that they might be thawed at some point in the future. Some have even paid high prices to be stored, postmortem, in this way, hoping that scientists will someday figure out how to bring them back to life. While frogs cannot return to being alive after dying, frozen or not, at least five frog species—the Wood Frog (*Rana sylvatica*), Gray Tree Frog (*Hyla versicolor*), Spring Peeper (*Pseudacris crucifer*), Chorus Frog (*Pseudacris triseriata*), and Cope's Tree Frog (*Hyla chrysoscelis*)—can survive after being frozen solid. In the case of a Wood Frog, upon its first exposure to ice in the fall, it undergoes a remarkable transition, worthy of a science fiction novel, to a dormant state in which its heart ceases to beat for up to several weeks and the water that surrounds its cells turns to ice. The contact with ice first sets off a modified fight-or-flight response (the body's way of preparing for acute stress, which includes an increased heart rate, dilated pupils, and mobilized energy stores) that yields an enormous outpouring of sugar into its bloodstream (as much as 4,500 milligrams per deciliter has been recorded—more than ten times the level needed to diagnose diabetes mellitus in humans, and more than enough to kill us), in addition to other substances, that together act as antifreeze. These substances are taken up by the Wood Frog's cells, while at the same time proteins are released into its blood that promotes the formation of ice. Thus, the cells are protected—if ice crystals were to form inside the cells rather than in the extracellular spaces, they would be torn apart. Come spring, Wood Frogs reverse the process, though they do so from the inside out. In a somewhat inconceivable turn of events, despite warmer temperatures outside, they manage to thaw their brains and hearts first.[83] The frog's ability to survive freezing has drawn the attention of many, including those involved in organ transplantation, who are trying to apply some of what the frogs do to prolong the viability of organs for eventual transplantation.

BEARS

Human Threats to Bears

Nine species of bears are listed on the IUCN's 2006 Red List of Threatened Animal Species, including the Polar Bear (*Ursus maritimus*), the Giant Panda (*Alluropoda melanoleuca*), and the Asiatic Black Bear (*Ursus thibetanus*). In 2005, the Polar Bear Specialist Group of the IUCN Species Survival Commission reviewed the status of the Polar Bear and decided to list it as a Vulnerable species, increasing its degree of threat from Lower Risk, given the projected loss of habitat resulting from global climate change. And in 2006, the U.S. Fish and Wildlife Service began a review to consider whether Polar Bears should also be protected under the Endangered Species Act, a decision that is expected in early 2008. Influencing this decision will be a series of studies released in September 2007 by the U.S. Geological Survey predicting that two-thirds of the world's Polar Bears will be lost by 2050 because of melting summer sea ice in the Arctic secondary to global warming. These predictions are based on middle-of-the-road warming projections.[84]

Like amphibians, bears are threatened by many factors resulting from human activity. They are at risk because of habitat degradation and destruction, with human settlement encroaching upon their diminishing ranges and with the ecosystems in which they live being altered, such as by deforestation, beyond their ability to adapt. They are also killed in large numbers for their body parts, which command high prices in Asian black markets, such as in South Korea, Japan, Thailand, Taiwan, and China, where they are sold as traditional medicines used to treat a broad range of human ailments. Gall bladders from some bear species have been known to fetch more than their weight in gold.[85] In 2002, CITES ordered countries to report on their trade in bear parts and on inhumane practices, present in China, Korea, and Vietnam, such as keeping bears in small cages and milking their gall bladders for bile, often through open wounds.

Polar Bears are at great risk because they face a variety of environmental assaults. Some scientists predict that they, the largest carnivores on land (with males reaching 2.5 meters [more than 8 feet] in height from nose to tail, and up to 11 feet if one measures them standing on their hind legs, and weighing as much as 600 kilograms [more than 1,300 pounds]), will be extinct in the wild by the end of the twenty-first century, and perhaps even earlier, due to disappearance of the Arctic sea ice upon which they depend.[86] Like other bears, Polar Bears are threatened by habitat loss. They are also killed by hunters, both for their meat (for human consumption and for dog food) and to make rugs and clothes, and by trophy hunters, who pay $20,000 or more for the chance to shoot a Polar Bear. While such hunting is tightly regulated by the United States and Canada (though banned in Norway), there is concern that overhunting may be threatening Polar Bear populations in some areas, such as Baffin Bay (between Canada and Greenland), parts of the Chukchi Sea (north of the Bering Strait, between Siberia and Alaska), and parts of Russia.[87] Polar Bears are especially threatened by the accumulation of high concentrations of pollutants in their tissues and by global warming. In addition, because of increasing oil and gas exploration and transport in the Arctic, they may be exposed to a greater number of oil spills into the marine environment and to habitat alteration from increased shipping.

POLLUTANTS

Organochlorines—for example, PCBs; polybrominated diphenyl ethers (flame retardants known as PBDEs, most of which originate from the United States); perfluoroalkyl substances (e.g., perfluorooctanoic acid and perfluorooctane sulfonate, called, respectively, PFOA and PFOS), which are used in fire-fighting foams, stain repellents, lubricants, and the manufacture of Teflon; and some pesticides—accumulate in the fatty tissues of marine organisms and become progressively more concentrated up the food chain.[88] These compounds are being found in increasing concentrations in the Arctic, carried there, it is believed, by currents and northbound winds, largely from industrialized regions of North America and Europe, and perhaps also by the excreta of birds such as Northern Fulmars (*Fulmarus glacialis*, a gull-like bird of the North Atlantic).[89] Because Polar Bears are at the top of the marine food chain, they tend to be exposed to very high levels of these pollutants. PBDE153, for example, a PBDE congener (there are 209 different congeners, or molecular forms, of PBDE), was seventy-one times more concentrated in Polar Bears than it was in Ringed Seals, their main food, biomagnifying to the same degree as PCB congener PCB194.[90] The persistence of organochlorines in Polar Bears, however, varies—some are metabolized in the bears' livers into harmless substances, while others reach high levels in vital organs during the long periods of fasting that accompany denning.

Especially high levels of pollutants have been found in Polar Bears living in and near the region of the Kara Sea, into which drain several rivers carrying industrial wastes from western Siberia.[91] Recent studies indicate that PCB exposures (and perhaps those from other toxic substances as well) are already having significant biological impacts—including (1) immune system defects, as evidenced by reduced levels of antibodies in populations of Canadian and Svalbard Polar Bears (the Svalbard islands lie to the east of Greenland and just west of the Kara Sea, above the Arctic Circle),[92] (2) impaired bone mineral formation (in East Greenland Polar Bear populations) from higher blood concentrations of organochlorines,[93] and (3) effects on hormonal systems (in Svalbard bears), with levels of cortisol, progesterone, estrogen, testosterone, and thyroid hormone changing in association with exposure to toxic chemicals.[94][96] In laboratory studies, alteration of these hormones can lead to disturbed growth patterns, reproductive failure, and weakened immune systems that leave the bears at greater risk of disease and death. Pollutants such as PBDEs may also be causing increased rates of hermaphroditism that seem to be present among Svalbard Polar Bears.[90]

GLOBAL CLIMATE CHANGE

Of all the threats to Polar Bear survival, however, global warming is the greatest. Over the past three decades, the thickness of Arctic sea ice has been reduced by approximately 15 to 20 percent as a result of such warming, with some areas showing reductions of up to 40 percent. Predictions are that the pace of this melting will quicken in coming years, as Arctic temperatures are expected to rise by as much as 4 to 7 degrees Celsius (7–13 degrees Fahrenheit), about twice the amount that the planet as a whole is expected to warm, by the year 2100.[97] In Siberia and Alaska, temperatures have already warmed by 2 to 3 degrees Celsius in the past fifty years, melting permafrost and placing great stresses on wildlife, human populations, and ecosystems in these areas. The melting of Arctic ice reduces Polar Bears' access to their main prey, Ringed Seals (*Phoca hispida*) and Bearded Seals (*Erignathus barbatus*), which can then more easily avoid capture, because there is more open water for them to surface for air.[84] Such conditions have already resulted in nutritionally stressed Polar Bears in the most southern-based populations, such as those that live in Canada's western Hudson Bay, leading to higher mortality, particularly for cubs, and lower reproductive success for females. Populations have fallen by 22 percent from 1987 to 2004 in this region, the first ever documented decline.[98] Further, hungry bears are more likely to seek alternative food sources, increasing the incidence of human–bear interactions that may lead to their being killed. And while Polar Bears can swim for many miles—to hunt for seals and to move from one ice sheet to another—the rapidly increasing areas of open water in the Arctic may be causing the observed increase in the number of drowned Polar Bears.[99]

Other species, such as Ringed Seals (*Phoca hispida*), which depend on intact ice sheets for foraging and for giving birth to and rearing their young (see page 68 in section on Global Climate Change in chapter 2), and some migratory bird populations are also thought to be at risk from the melting of Arctic ice. And warming may expose Ringed Seal pups to greater levels of predation, as the roofs of their snow dens are more likely to collapse. The loss of Ringed Seals, a major food source for Polar Bears, will further compromise their survival.[100]

The loss of Polar Bears, as well as other denning bears, not only deprives us of these magnificent creatures but also results in the loss of potential medicines and prime biomedical research models for understanding and treating several human diseases that cause much human suffering and large numbers of deaths worldwide, such as osteoporosis, kidney failure, and diabetes types 1 and 2.

Figure 6.10. Mother and Cub Polar Bears on Ice Floes Separated by Large Areas of Open Water. Open water reduces polar bears' ability to hunt seals and leads to drownings. (© 2002 Tracey Dixon, www.trp.dundee.ac.uk/~spitz.)

Figure 6.11. Ringed Seal Pup. The Ringed Seal (*Phoca hispida*), made up of five subspecies, is the most numerous and widely distributed marine mammal in the Arctic. Like the bears, which depend on these seals for food, Ringed Seal populations are under pressure from overharvesting, pollution, and the melting of Arctic ice from global warming. (© B&C Alexander.)

Ursodeoxycholic Acid: A Medicine from Bears

The origins of ursodeoxycholic acid trace back to traditional Chinese medicine centuries ago when bear bile (bile is a liquid produced by the liver that is stored in the gallbladder and is released into the intestine, aiding in the digestion of fats) was made into a powder, known as *yutan*, to treat diseases of the liver and gallbladder. Not until the twentieth century and a multinational effort did the active pharmacological substance in *yutan*, ursodeoxycholic acid, become known.

In 1900, expeditions to the tundra of Greenland yielded samples of Polar Bear bile that were delivered to the lab of a Swedish researcher, Olof Hammarsten, a biochemist at the University of Uppsala. In 1901 he published a paper on the composition of Polar Bear bile in which he identified what we now know to be ursodeoxycholic acid (UDCA). Twenty-five years later UDCA was crystallized (this time using a bile sample from a Black Bear, *Ursus americanus*) and its chemical structure revealed. UDCA in humans, and in other vertebrates, typically constitutes less than 5 percent of total bile acids, quantities that are insufficient to do what it does in bears: help them maximize absorption of their high fat diets and ensure that they have adequate fat stores during periods of dormancy. At the same time, UDCA helps prevent gallstones from forming in bears when they den.

UDCA is currently used in several human diseases, including in the prevention of complications from the thickening of bile that can occur in pregnancy[101] and in some premature infants who receive their nutrition intravenously. It can also help dissolve certain kinds of gallstones. Its most important use, however, is in patients with primary biliary cirrhosis (PBC), a disease that can lead to destruction of the liver from inflammation of its internal bile ducts. UDCA not only alleviates the symptoms of intense skin itching associated with this disease but also may be the only medicine capable of prolonging survival in patients with PBC.[102] PBC usually affects women between the ages of thirty and sixty and has a poor prognosis in the absence of therapy. Patients who have symptoms of the disease at the time of diagnosis live, on average, only seven and a half years more unless they receive a liver transplant.

Denning Bears and Biomedical Research

During times when food is scarce—winter in northern temperate regions, summer in the Arctic—bears enter a period of three to five months or longer of reduced metabolic activity called "denning" during which they are in an inactive, lethargic state, and do not eat, drink, urinate, or defecate. In contrast to some rodents, such as woodchucks, which truly "hibernate" during the winter (i.e., dramatically drop their metabolic rates and body temperatures and are unresponsive to stimuli), "denning" bears essentially maintain their normal temperature and are able to rapidly become fully alert and responsive.[103]

The remarkable physiological processes that allow bears to survive the extreme privations of "denning" are not found in any other animal. Understanding these

Figure 6.12. Mother Black Bear (*Ursus americanus*) Denning with Cubs. (© Photo by Gary Alt.)

processes, which involve the recycling of essentially all of their body wastes, and identifying the substances that mediate them may lead to new insights for treating a number of human diseases.

OSTEOPOROSIS

Unlike all other mammals, including humans, who lose bone mass when they fast, are inactive, or are not bearing weight for significant periods of time, denning bears, despite several months of these bone-loss–causing behaviors, do not. In fact, they lay down new bone.[104] By contrast, a person who is bed-ridden for a five-month period increases the risk of losing one-fourth to one-third of his or her bone mass, a condition called osteoporosis, or "porous bone." Even pregnant bears, after they have delivered and nursed as many as five cubs (which takes a great toll on body calcium stores), maintain normal bone mass while denning.[105] Pregnant Polar Bears can den for as long as nine months. The reason bears are able to achieve this seemingly impossible biological feat is that they are able to recycle the calcium that is lost from bone back into forming new bone.[104]

Ralph Nelson and his laboratory at the Carle Foundation Hospital and the University of Illinois College of Medicine at Urbana-Champaign have isolated an extract from denning bears that inhibits the activity of osteoclasts, the cells that are involved in the normal process of bone breakdown, and stimulates cells called osteoblasts that form new bone, and others called fibroblasts that form new cartilage. This extract has also been shown to reverse bone loss in rats that had their ovaries removed, simulating a postmenopausal state in humans.

Osteoporosis is a major public health problem worldwide, particularly among postmenopausal women, the inactive elderly, and paralyzed and bed-ridden patients. Although calcium and vitamin D supplements and regular physical exercise reduce the rate of bone loss, and some medications can act to inhibit bone breakdown and stimulate bone formation, osteoporosis afflicts ten million people in the United States alone, with an additional thirty-four million people who have reduced bone mass, a condition called osteopenia, at risk for the disease. It causes 1.5 million bone fractures and 70,000 deaths each year and costs the U.S. economy more than $18 billion (in 2002 dollars),[106] and the world economy more than $130 billion annually in direct health care costs and lost productivity.[107] Worldwide, there are an estimated 740,000 deaths each year from hip fractures (1990 data), most of which are the result of osteoporosis, and by the year 2050, estimates are that there will be more than 6 million hip fractures globally each year from osteoporosis.[108,109] Understanding why denning bears do not get osteoporosis in spite of their having so many of the major risk factors, and isolating the compounds responsible could lead to new treatments and preventive measures for this disease.[110]

RENAL DISEASE

Denning bears do not urinate for five months or more, and yet they do not suffer toxicity from the buildup of urinary wastes. Humans unable to rid themselves of these wastes for periods lasting only a few days cannot survive. While patients with kidney failure can partially reduce their blood levels of urea, the main urinary waste, by restricting their dietary protein intake (it is the breakdown of proteins that produces urea), ultimately almost all of them progress to end-stage renal disease (ESRD), for which the only treatment is dialysis (using filtering machines that act as external kidneys to rid their blood of urea) or kidney transplantation. In the United States, more than 80,000 people with ESRD die annually (2003 figures), with a cost to the economy, for treatment and lost productivity, for both public and private expenditures, of more than $27 billion (in 2003).[111] Renal failure is also a major public health problem worldwide, with an estimated 1.5 million people receiving treatment for ESRD in 2001,[112] with middle-of-the-road estimates that this figure could climb by 2030 to somewhere between seven and fourteen million.[113]

Denning bears form urine, but it is completely reabsorbed by their bladders back into their blood streams,[114] and the urea is recycled back into amino acids that form new proteins.[115,116] Ralph Nelson and his team have isolated an extract from the blood of denning bears that has been shown, when tested in nondenning bears and in guinea pigs, to stimulate such urea recycling.

If we could increase urea recycling in humans using insights and medicines derived from denning bears, not only might we be able to find treatments for ESRD, but we might also be able to help large populations of starving people around the world reduce their protein wasting, one of starvation's most devastating and lethal effects, by stimulating recycling of their urea back into protein.

DIABETES TYPES 1 AND 2 AND OBESITY

The unique energy metabolism of fats and carbohydrates in denning bears may also hold clues for more effective treatments for diabetes types 1 and 2 and for obesity.

Denning Black Bears have low blood levels of insulin, as do people who suffer from type 1 diabetes (also called juvenile-onset diabetes) who are unable to produce enough insulin to control the amount of sugar in their blood.[117] But unlike type 1 diabetics, denning bears do not develop the consequences of inadequate insulin—high blood sugar concentrations, dehydration, and a condition called ketoacidosis.[118] Ketoacidosis is a toxic state where circulating levels of ketones, products of fat metabolism, are high. Studies have confirmed that denning Black Bears keep their blood glucose levels normal despite not having enough circulating insulin, because their cells have greater sensitivity to the effects of insulin.[119] Ketoacidosis also does not occur, because free fatty acids, rather than being metabolized into ketones, are instead recycled back into triglycerides.[118] Within these complex pathways in the denning Black Bear, there may be found new approaches for treating type 1 diabetes mellitus.

Free-ranging wild Polar Bears, by contrast, have been shown to be insulin-resistant, with fat bears (e.g., those ready for denning) having higher concentrations of insulin and higher levels of insulin resistance, than thinner bears.[120] In humans, insulin resistance is also correlated with obesity. With the recent epidemic of obesity, particularly in countries such as the United States, where about one-third of all adults were considered obese in a study done in 2004,[121] there is an associated epidemic of type 2 diabetes, an illness characterized by insulin resistance, with or without impaired insulin secretion.

Type 2 diabetes affects an estimated 15.7 million people in the United States, almost 6 percent of the population, with some 800,000 new cases diagnosed each year. It was either the underlying cause of death, or contributed to the death, of some 224,000 people in the year 2002.[111,122] The American Diabetic Association has estimated that diabetes, both type 1 and 2, cost the U.S. economy around $91 billion in the year 2001. Worldwide in 2002, according to the World Health Organization, there were some 150 million cases of type 2 diabetes, although other estimates (e.g., by the International Diabetes Foundation) have put this figure as high as 194 million (2005 estimate). Almost all diabetes cases are type 2 diabetes, in both developed and developing countries alike.

Under normal conditions in people, fat and carbohydrates compete as fuels for energy metabolism, with the contributions that each makes primarily determined by the availability of carbohydrates and by insulin levels. Following a carbohydrate-rich meal, the level of insulin rises, promoting the storage and metabolism of carbohydrates and restricting the release and subsequent metabolism of fat molecules from body stores. Conversely, when the dietary supply of carbohydrates is scarce or absent and circulating insulin concentrations are low, the use of fat for energy is enhanced, while that of carbohydrates is reduced. In type 2 diabetes this regulation is disturbed—increased concentrations of insulin no longer suppress the release and metabolism of fat, and cells become insensitive to insulin's effects on carbohydrate uptake. As a result, type 2 diabetics have high circulating blood lipids (fats) as well as high blood sugar concentrations and are at greater risk for atherosclerosis (the build up of cholesterol-laden plaques in the walls of arteries), heart attacks, and strokes.

Despite insulin resistance and obesity, Polar Bears do not show any signs of altered energy metabolism and continue to tightly regulate the release and metabolism of body fat for energy production, and they do not develop type 2 diabetes.[120] Clarifying the mechanisms involved in energy metabolism in Polar Bears, particularly the regulation of lipid mobilization and use, should provide greater insights into the pathogenesis of type 2 diabetes and into more effective treatments for this disease.

One possible treatment may involve employing a substance in denning bears' blood that has been shown to trigger the transition from a period of ravenous appetite, during the three months prior to denning, when bears consume several times their usual number of calories per day, to a period of fasting, just prior to denning, when bears are no longer hungry and stop eating. In the fasting state, bears begin urea recycling and start losing fat tissue, although they retain their lean body mass.[105,115] Such a medicine might be effective in treating obesity, as well as type 2 diabetes associated with it.

OTHER MEDICAL CONDITIONS

An additional marvel of the metabolism of denning bears that may have implications for human health is that they do not become deficient in essential fatty acids despite fasting for long periods of time. These fatty acids, linoleic and alpha-linolenic acids, cannot be synthesized by humans (or, it is believed, by bears) and need to be obtained from dietary sources. Denning bears are thought to be able to mobilize these fatty acids from fat stores in precisely the amounts needed for metabolic processes, even during pregnancy and lactation.[120] If we can learn how they accomplish this, we may have a better understanding of states of essential fatty acid deficiency in humans, such as those that are seen in chronic malnutrition.

PRIMATES

Primates Endangered

Of all the threatened species on Earth, no group evokes as much human concern and as much heartbreak as the primates, the mammalian order that includes our own species, and perhaps none illustrates more strikingly and more tragically the magnitude of what we will lose if we do not halt the present extinction crisis.

Figure 6.13. Black-and-White Ruffed Lemur (*Varecia variegata*). This lemur, like all lemur species, is native only to Madagascar. It grows to about 10 pounds and can live for twenty years. It has the second loudest call of any nonhuman primate; howler monkeys are the loudest. It is classified by the IUCN as Critically Endangered. (© Noel Rowe.)

Figure 6.14. Golden-Bellied or Yellow-Breasted Capuchin (*Cebus xanthosternus*). This New-World monkey is now found only in the Atlantic forest of southern Bahia, Brazil. It is Critically Endangered. (© Noel Rowe.)

Primates are subdivided into two major groups or suborders—the Strepsirhines, also called the Prosimians, the first primates to evolve, which includes the lemurs, lorises, and tarsiers (although some have placed the tarsiers in their own distinct suborder); and the Haplorhines, also called the Anthropoids. The suborder Haplorhines is in turn broken down into three groups: New-World Monkeys, which are found in tropical forest environments in Central and South America and primarily live in trees, such as squirrel and spider monkeys, capuchin monkeys, marmosets, and tamarins; Old-World Monkeys, which are found primarily in Africa and South and East Asia, such as baboons, mandrills, colobus monkeys, and macaques (including the Rhesus Monkey); and our superfamily, the Homonoidea, which includes gibbons—the lesser apes—and the great apes: orangutans, gorillas, Chimpanzees, Bonobos, and ourselves. Out of a total of 358 primate species (this number, which continues to grow as new species of primates are discovered, is the subject of ongoing debate, with estimates ranging from 233 to 374),[123] an alarming 114—almost one-third—are considered to be threatened, and of these, more than half are considered Endangered or Critically Endangered, according to the IUCN Red List. (Since 1990, ten new monkey species have been identified in Brazil alone;[124] in recent years, several new lemur species have been found;[125] in 2004, a new species of macaque named the Arunachal Macaque [*Macaca munzala*] was discovered, living in the remote mountains of the state of Arunachal in northeastern India;[126] and in 2005, two teams of scientists reported finding a new species of mangabey that they named the Highland Mangabey [*Lophocebus kipunji*] in the highlands and mountains of southern Tanzania [and shortly thereafter stated that they thought it was likely that it was Critically Endangered].)[127] Tragically, all primate groups are well represented in these threatened categories. The consensus among biologists who study primates is that this is just the beginning of a great wave of primate extinctions in coming years.

LESSER APES

Gibbons

The Homonoidea are particularly at risk, with all groups threatened by human activity (*Homo sapiens* excepted). Seven of the twelve species of gibbons are listed as threatened by the IUCN, but although they are the most species-rich of all the apes, gibbons are also the least studied.[128] Gibbons are threatened primarily because of a loss of their forest habitat, especially in Java, where more than

Figure 6.15. Celebes Macaque (*Macaca nigra*). These Old-World monkeys, also called Celebes Crested Macaques, or Crested Black Macaques or Black "Apes," live in rainforests in the northeast of the Indonesian island of Sulawesi (formerly called Celebes), as well as on smaller neighboring islands. They live in groups, usually led by a dominant female, that number up to twenty-five animals. They are considered Endangered as a result of habitat destruction, and also because they are often killed by people, both for bushmeat and when they raid farms to find food. (© Noel Rowe.)

90 percent of the forests have been cleared.[129] Two species—the Javan, or Silvery, or Moloch Gibbon (*Hylobates moloch*) and the Eastern Black-Crested Gibbon (*Nomascus concolor*)—are listed as Critically Endangered.

THE ASIAN GREAT APES: ORANGUTANS

The two species of orangutans—the Borneo Orangutan (*Pongo pygmaeus*) and the Sumatran Orangutan (*Pongo abelii*)—are also in grave danger, with the Sumatran species being classified as Critically Endangered. Logging (much of it illegal), gold mining, the establishment of plantations and villages, agriculture, industrial development, and the road building that accompanies all of these activities have destroyed, despoiled, or fragmented the forests that orangutans depend on for their survival.[130] In addition, forest fires, deliberately set to clear the land (often for palm oil plantations), threaten orangutans. In 1997 and 1998, such fires spread rapidly through Indonesian forests, fueled by a severe drought (perhaps as a result of global climate change exacerbating an exceptionally strong El Niño during those years) and killed large numbers of orangutans—from the direct effects of the flames and smoke; from starvation, as their forest habitat, along with their fruit dominated food supply contained within it, burned, and from frightened villagers and plantation workers who shot them in the hundreds as the orangutans fled the burning forests onto farmland. As a result of these fires, the orangutan population in Borneo is estimated to have plummeted from 23,000 in 1996 to 15,400 in 1998.[130] And finally, while hunting orangutans and keeping them as pets are both illegal, poaching is still common, and adults are sometimes killed so that the young, which are adorable and remarkably trusting and playful animals, can be kept as pets or occasionally raised as substitute children by childless couples. Some 420,000 Bornean and 380,000 Sumatran orangutans are thought to have populated Earth about 10,000 years ago. Today, there are fewer than 15,000 and 12,000, respectively.[131]

THE AFRICAN GREAT APES

The survival of our nearest relatives in the animal kingdom—gorillas, Chimpanzees, and Bonobos—is also at stake.[130] The loss of these creatures, so close in appearance and behavior to ourselves, touches many of us as if they were one of our own.

Scientists estimate that there are somewhere between 280,000 to just more than 400,000 African great apes alive at the present time,[132] out of a population that is thought to have numbered in the millions as recently as a century ago. A recent survey by Peter Walsh from Princeton University and his team has determined that even in the intact, pristine forests of Gabon, populations of gorillas and Chimpanzees have

Figure 6.16. Javan Gibbon (*Hylobates moloch*). The Javan Gibbon, also known as the Silvery or Moloch Gibbon, or in Indonesian as *owa jawa*, is found only in tropical rainforests in the western and west central parts of the island of Java in Indonesia. It is estimated that there are fewer than 2,000 individuals in the wild, the result of a loss of 98 percent of their natural habitat from deforestation for logging, farming, and human settlement, and from the capture of infants (with the associated killing of the mothers) for the illegal pet trade. Javan Gibbons are listed as Critically Endangered. (© Erni Thetford.)

Figure 6.17. Borneo Orangutans (*Pongo pygmaeus*)—Mother, Father, and Son. The Borneo Orangutan (in the Malay language, *orang-utan* means "man of the forest") is found in tropical, swamp, and mountain forests on the island of Borneo in Indonesia, and in Malaysia. Human activity has greatly reduced the numbers of *P. pygmaeus* in the wild, and it is now classified as Endangered by the IUCN. (© Karl Ammann, karlammann. com.)

fallen by more than 50 percent in the past twenty years. Based on these figures, the 23-member survey team, representing a broad range of research institutions and conservation organizations, predicted that gorilla and Chimpanzee populations would crash by an additional 80 percent in the next thirty-three years, causing them to recommend that these apes' threat statuses be changed immediately from Endangered, where it is now, to Critically Endangered.[133]

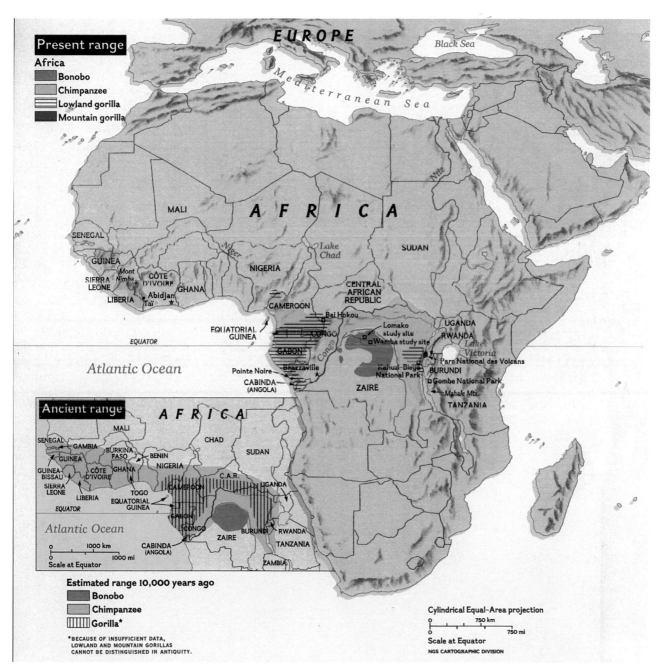

Figure 6.18. Map Comparing Ancient and Present-Day Great Ape Ranges in Africa. Ten thousand years ago, Chimpanzees are thought to have occupied a broad unbroken area of West–Central Africa, stretching from Tanzania to Senegal; gorillas were distributed in a continuous band from Tanzania to Angola, and Bonobos were found in a large area in what is now the Democratic Republic of Congo. Today, the ranges and numbers of each of the Great Apes have been markedly reduced and broken up into population islands, separated from one another, where various subspecies have evolved. (© National Geographic Image Collection.)

Gorillas

Although scientists continue to debate the details of how many gorilla species and subspecies there are, most recognize two distinct species—the one most often seen in zoos, *Gorilla gorilla*, and *Gorilla beringei*. *G. gorilla* lives in lowland forests of equatorial West Africa and likely has two subspecies: *Gorilla gorilla gorilla* and *Gorilla gorilla diehli*. *G. beringei* lives in eastern Central Africa and has one subspecies, *Gorilla beringei graueri* that lives in lowland areas, and a second, *Gorilla beringei beringei*, that lives in the mountains. Although precise figures for gorilla populations are difficult to obtain, the best recent estimates of population size put the number of western gorillas at around 95,000 and eastern gorillas at around 17,500, with Mountain Gorillas (*G. beringei beringei*) numbering only a scant 600 to 700 or so (this estimate includes what some primatologists have included as a separate subspecies, the Bwindi gorilla).[132] These numbers are several years old, however, and may be significant overestimates,

> IF WE DON'T DO SOMETHING RADICAL, GORILLAS AND CHIMPANZEES WILL BE EFFECTIVELY EXTINCT FROM WESTERN EQUATORIAL AFRICA WITHIN THE NEXT 10 YEARS.... PEOPLE WILL SAY IN 10–20 YEARS, THESE GUYS—THEY KNEW THIS WAS COMING AND THEY DIDN'T DO ANYTHING.
>
> —Peter Walsh

Figure 6.19. Mountain Gorilla (*Gorilla beringei beringei*). Mountain Gorillas have been one of the most endangered animals in the world due to habitat loss, poaching, and war, but conservation efforts have begun to reverse this trend somewhat. (© Martin Harvey.)

at least for the lowland gorillas (*G. beringei graueri*), given recent observations of significant population declines. But Mountain Gorillas seem to be doing well, with their populations steadily climbing.[134] These increases give hope that with sufficient protection, which the Mountain Gorillas have received, conservation measures can be successful.

Gorilla populations are dwindling under the strain of several human activities. First among these is forest habitat destruction, driven by commercial logging, and increasing rates of forest conversion—to cropland, to pasture for livestock, and to sites for the extraction of minerals and petroleum—all of which come with the construction of new roads that further erode forest integrity. Commercial enterprises bring with them large numbers of outside workers, increasing pressures on the local food supply that can increase demand for primate, including gorilla, meat. As mentioned in chapter 2 (see page 43), the hunting of gorillas and other bushmeat has been going on for millennia. One reason that this hunting now poses such a major threat to gorilla survival is that modern hunters have more sophisticated weapons. Most important, however, they also have access to previously impenetrable sections of forests, thanks to new roads. This has resulted in an explosion of bushmeat commerce, particularly in the big cities of Central Africa. In these markets, primate meat (and that of other exotic animals, such as elephants) commands high prices, whether it is bought as food or for use in traditional medicines, some of which purportedly endow

Figure 6.20. Murdered Gorilla Family. Because gorillas are often together as families, it is not difficult to slaughter them for bushmeat, all at the same time. (© Karl Ammann, karlammann.com.)

consumers with extra strength, sexual potency, and magical powers. In addition, a depletion of fish in the ocean off the coast of West Africa, the result of overexploitation by heavily subsidized European Union fleets, is thought to be forcing West Africans, deprived of this major protein source, to hunt more land animals, including primates, for food.[135]

War and conflict also has the potential to threaten gorilla and other primate populations. The 1994 Rwandan genocide, for example, resulted in hundreds of thousands of refugees (both Tutsis and Hutus), who fled through gorilla-forested habitat, felling trees for fuel and poaching animals along the way, including gorillas.[136] They also laid thousands of landmines, which endanger wildlife as well as people. Recent strife in the Democratic Republic of Congo (DRC) has seen heavily armed soldiers, encamped in forested areas where gorillas are found, hunting them for food. But despite these threats, one community of eastern Lowland Gorillas in the DRC has stabilized[137] and may even be growing, largely, it is believed, because of the courageous efforts of guards in the Kahuzi-Biega National Park, who have successfully driven off rebel armies and poachers. And Mountain Gorillas in the Virunga Volcanoes Park seem to have largely weathered the Rwandan Civil War.[134,138]

A final hazard to gorillas (and to Chimpanzees) comes from the Ebola virus. While the dynamics of the disease caused by this virus, Ebola hemorrhagic fever, are still not completely understood, the finding of symptomless Ebola infections in three different bat species—*Hypsignathus monstrosus*, *Epomops franqueti*, and *Myonycteris torquata*—all of which have ranges that overlap where the outbreaks among great apes and humans have occurred, suggests that they may be serving, perhaps along with other bats and other animal species, as Ebola virus reservoirs in the wild.[139] Ebola virus is thought to be capable of spreading between monkeys and apes, among great apes, and from apes to people.[140] In one recent study, Ebola infections were found to be the apparent cause of a reduction in gorilla populations by 50 percent between 2002 and 2003, and in Chimpanzee populations by almost 90 percent during the same period, in an area of the Republic of Congo,[141] making Ebola a possible rival of hunting as a threat to apes in this region.

Chimpanzees and Bonobos

There are two species of chimpanzees—*Pan troglodytes*, the most familiar species known simply as the Chimpanzee, and the Bonobo, *Pan paniscus*, formerly known as the Pygmy or Gracile Chimpanzee. In this chapter, we will refer to *P. troglodytes* as Chimpanzees, with a capital C, and *P. paniscus* as Bonobos. As with gorillas, appraising the size of Chimpanzee and Bonobo populations has proven difficult, but no one doubts that both have dropped significantly in recent decades and that both are in great peril. The IUCN lists them both, perhaps too optimistically, as Endangered.

As many as four subspecies of *Pan troglodytes* may live in the wild, with a total population of between 150,000 and 250,000 in western, central, and eastern Africa. There are no recognized subspecies of Bonobos, which are thought to number between 20,000 and 50,000.[132] The same forces that threaten the survival of gorillas threaten Chimpanzees—loss of forest habitat, bushmeat hunting, and infectious diseases such as Ebola. In addition, there was a report of an anthrax epidemic in the Tai

National Park in the Côte d'Ivoire that killed at least six Chimpanzees of the subspecies *Pan troglodytes versus* in late 2001 and early 2002.[142] While it is unclear how these Chimpanzees became infected, it seems likely that the bacterium that causes anthrax (*Bacillus anthracis*), like the viruses that cause Ebola and HIV (see chapter 7), may spread to humans following exposure to infected apes and other primates hunted for bushmeat. Bonobos, found only in the DRC, are also particularly vulnerable to hunting, including in Salonga Park, the only area where Bonobos are protected, at least in principle. Here, rebel gangs, heavily armed and hungry, have settled. While Chimpanzees are highly threatened by hunting, habitat loss, and infectious disease, most primatologists give them the best chance of any great ape of surviving the twenty-first century in the wild.[132]

Primates and Biomedical Research

Our biological similarity to primates is startling. Our DNA, for example, is almost identical with that of the Chimpanzee, differing by a mere 1.3 percent (corresponding figures for Old-World monkeys are 8 percent, and for New-World monkeys, 15 percent). (To compare the DNA sequence from a human gene with that from a chimp, visit the Silver Project: Ape Genome Sequencing [sayer.lab.nig.ac.jp/~silver/] and look, e.g., at a gene known as *CHRM2*, which codes for a particular type of nerve cell receptor. The sequences are identical for hundreds and hundreds of nucleotides in a row.) This closeness, which extends to their anatomy, physiology, and behavior, makes them critically important, and sometimes irreplaceable, models for some biomedical research. It also makes them central figures in the increasingly confrontational debate about the use of such animals as research subjects. The central question that must be asked when considering the ethics of experimentation on nonhuman primates (or, for that matter, on any animal) is whether there are diseases or other medical conditions for which nonhuman primates must be used as research subjects because no effective alternative means to animal experimentation are available, and because these diseases either do not occur in other organisms or, if they do, their manifestations are too different from those in humans to be useful (see box 5.3, which addresses concerns about the use of animals in research, in chapter 5).

Figure 6.22. Older Bonobo (*Pan paniscus*) Males. (© Karl Ammann, karlammann.com.)

This section explores three areas of research on nonhuman primates that are vital to human medicine: infectious diseases and the development of vaccines, neurological disorders, and behavioral disorders. Others could have been discussed here as well, such as reproductive disorders and in vitro fertilization, the development of birth control pills, aging, sickle cell anemia and the development of hydroxyurea therapy (the only known preventative measure for this debilitating disease), prevention of respiratory distress syndrome in premature infants through the use of surfactant (a soapy lubricant that facilitates lung expansion), or various forms of cancer, including prostate cancer, colon cancer, and leukemia. But the three areas covered below provide a useful overview to this large and important field of biomedical research.

INFECTIOUS DISEASES AND THE DEVELOPMENT OF VACCINES

Infectious diseases and the development of vaccines were the subjects of some of the earliest research on nonhuman primates, which began in the late 1800s with Pasteur's work on rabies and that of others on smallpox. About twenty years later, in 1909, Karl Landsteiner and Erwin Popper isolated poliovirus by injecting infected spinal cord from those who had died from poliomyelitis into Rhesus Monkeys, baboons, and chimpanzees, work that won them the Nobel Prize. Nonhuman primates were used extensively over the next several decades in developing a vaccine against polio.

Jonas Salk developed a polio vaccine using tissue cultures from monkey kidneys, and Albert Sabin used monkeys and Chimpanzees in his search for attenuated strains of naturally occurring polio virus, which eventually led to the highly effective oral polio vaccine. In addition to polio, other human infectious diseases such as yellow fever, measles, and rubella (German measles) have all depended on primate research for advances in understanding and preventing them.

Hepatitis C

Approximately nine million people in the United States and Europe and 200 million worldwide are infected with hepatitis C virus. Those using shared needles to inject drugs are at greatest risk of infection. Since the development of a screening test to detect antibodies to the virus in blood in the 1990s, hepatitis C is now rarely spread by blood transfusions. However, in some parts of the developing world, the reuse of contaminated medical equipment and inadequate blood donor screening contribute to high rates of hepatitis C infection.[143]

As many as one-fourth of those initially infected are able to rid their bodies of the virus, but most people develop a chronic infection, and of these, about 10 to 20 percent progress after a few decades to cirrhosis of the liver (the replacement of normal liver tissue by scarring as a result of injury or disease), and some to liver cancer or liver failure.[144] Current medicines are expensive, have significant side effects, and do not cure hepatitis C infections. The enormous variation in strains of the virus—it has 11 genotypes and 100 subtypes—and the ability of the virus to easily mutate have thus far thwarted the development of a vaccine.[145] As a result, hepatitis C infection has remained a leading cause for the need for liver transplantation worldwide.

Chimpanzees are the only known animals other than humans that can be infected with the virus, so much of what in known about the molecular biology of hepatitis C infections and the immune responses to it comes from studies involving Chimpanzees.[145] Several promising approaches have emerged from preliminary investigations of vaccines in Chimpanzee subjects,[146,147] but to date, no effective vaccine is available.

Hepatitis B

The hepatitis B virus is transmitted in the same ways as the hepatitis C virus—by contact with the blood or body fluids of an infected person. Only about 10 percent of those infected develop a chronic infection, but they are at great risk for cirrhosis of the liver and liver cancer.[148] Hepatitis B infection is an enormous global public health problem, with 400 million people chronically infected and around 500,000 deaths globally per year.[149]

A vaccine, developed using nonhuman primates (White-Moustached Marmosets [*Saguinus mystax*], Grivet Monkeys [*Cercopithecus aethiops*], and Chimpanzees), has been available for more than twenty years, but it is insufficiently used, especially in the developing world, and 10 percent or so of adults given the vaccine fail to respond. Hepatitis B viruses are found in a variety of mammals and birds, but only Chimpanzees and Rhesus Monkeys, as far as we know, can be infected with the types of hepatitis B that infect humans.[148] Furthermore, Chimpanzees show evidence of liver damage following infection by hepatitis B, and they demonstrate circulating antibodies and cellular immune responses similar to those found in humans.[150] New vaccines are being tested in Chimpanzees. Chimps have also proven invaluable for learning how the immune system can conquer the infection without harming infected liver cells.[151]

Malaria

According to the World Health Organization, approximately 300 to 500 million people contract malaria annually, and as many as one to three million people die each year from the disease, most of whom are children. Given the persistence of malaria as one of the most widespread and lethal infectious diseases in the world, concerns about the public health and environmental impacts of the toxic pesticides used to control the mosquito vectors of the disease, and the fact that these vectors and the malarial parasites themselves have both developed increasing resistance to the chemicals we have used in our attempts to control them, the need for a malaria vaccine remains a top global public health priority.

The infectious agents are protozoa from the genus *Plasmodium*, and they cause disease in a wide variety of mammals, including rodents and birds, in addition to primates. However, of the more than 100 *Plasmodium* species, only four infect humans (*P. vivax*, *P. falciparum*, *P. ovale*, and *P. malariae*), and for the two that cause the most disease, *P. vivax* and *P. falciparum*, nonhuman primates are the best research subjects.[148]

Much of the effort to develop a vaccine has focused on the most deadly malarial organism, *P. falciparum*. The successful sequencing of *P. falciparum*'s genome in 2002 has greatly facilitated this quest, making it more likely that a vaccine will be found that can both prevent the emergence and spread of malaria and reduce the severity of disease and risk of death in infected individuals. While other organisms such as yeasts, birds, mice, and rabbits have all been essential for malaria vaccine research, with mice and rabbits serving as the initial organisms for testing vaccines, nonhuman primates, especially owl monkeys (*Aotus* spp.), squirrel monkeys (*Saimiri* spp.), and Rhesus Monkeys (*Macaca mulatta*, also called Rhesus Macaques), remain the most important research models for evaluating vaccine effectiveness. The most promising vaccine for *falciparum* malaria, known as RTS,S/AS02A, which is being tested in phase II clinical trials in parts of Africa, has relied upon Rhesus Monkeys throughout its twenty-year development to improve the vaccine's efficacy.[152–154]

Ebola and Marburg

Perhaps no infectious agent has conjured up more images of terror in the human imagination than the Ebola virus. This has been, in part, the result of fictional accounts about Ebola in the popular literature and in movies, where its legendary abilities, such as its capacity to liquefy internal organs (this does not occur), have been touted. But the main reasons that Ebola virus, and the related Marburg virus, evoke such fear are that (1) diseases caused by these viruses seem to emerge out of nowhere and then just as quickly to disappear, only to reemerge at a later time; (2) their life cycles remain poorly understood, despite the recent identification of three bat species that may serve as reservoir hosts (see page 304); (3) no cure or treatment exists; and (4) the diseases are rapidly and catastrophically fatal, with mortality rates at times approaching 90 percent.

Marburg outbreaks were first described in 1967 in Germany and Yugoslavia, and then at various sites in Africa, such as South Africa in 1975 and Kenya in 1980 and again in 1987. The most recent and most deadly Marburg epidemic was in Angola in 2004–2005, with a reported total of 374 cases (both suspected or confirmed) and 329 deaths (almost an 88 percent mortality) according to the World Health Organization.[155] Ebola first emerged, as far as is known, in African rainforests both in the Sudan and in the Democratic Republic of Congo (DRC) in 1972. More recently, there have been scattered outbreaks in Uganda (2000–2002), Gabon (2001–2003), and the DRC (also 2001–2003). There have been a total of some 1,850 cases of Ebola worldwide, with a cumulative mortality of approximately 65 percent according to World Health Organization figures.[156]

In 1994, the first documented outbreak of Ebola occurred in nonhuman primates in the Côte d'Ivoire, killing eight Chimpanzees. The lesions present in the internal organs of these chimps resembled those seen in monkeys that had been experimentally infected with Ebola, and laboratory tests confirmed the presence of a subtype of the Ebola virus in one of the dead chimps.[157] The same subtype was identified in the blood of a researcher who developed a brief illness, characterized by high fever, vomiting and diarrhea, a rash, and temporary confusion and memory loss, presumably after she had become infected while performing a necropsy (an animal autopsy) on one of the dead chimps.[158] This case is thought to demonstrate the capacity of the virus to be transmitted from Chimpanzees to humans. In late 2001 and early 2002, there were fifty deaths from Ebola in Gabon, occurring predominantly in areas where Ebola outbreaks were, as has been mentioned above, most widespread among Chimpanzees and gorillas.[159]

Much has been learned from studies with Rhesus Monkeys, as well as with other nonhuman primates infected with the Ebola virus. These studies have shown that the virus can be transmitted through several routes: by direct contact with the skin of infected individuals, by exposure to their body fluids (where high concentrations of the virus are present), via aerosols from sneezing and coughing, and by eating infected meat.[160,161] Other key information has been obtained about the different subtypes of the Ebola virus and about the ability for Ebola to be freely transmitted among primates—from monkeys to apes and from apes to humans. There is, at present, no evidence that humans have infected nonhuman primates with Ebola or with Marburg.

Because infections with Ebola and Marburg viruses progress so rapidly, with little time for the body to mount an immune response, and because there are no effective antiviral treatments, an aggressive search has been under way for an Ebola and a Marburg vaccine. This search has yielded promising results, with the development in 2005 of two vaccines, one for Ebola and one for Marburg, both of which were tested in Cynomolgus Macaques (*Macaca fascicularis*) and shown to be 100 percent effective in preventing infection following exposure to the viruses.[162]

Ebola, one of the most deadly infectious diseases on Earth, and one that has put several species of nonhuman primates on the brink of extinction and decimated villages in several African countries, illustrates how nonhuman primates are essential for understanding the emergence and spread of the disease and for developing safe and effective vaccines, both for humans and for nonhuman primates alike.

Rotavirus

Worldwide, rotavirus causes some 138 million cases of gastroenteritis (an infection of the gastrointestinal tract that produces vomiting and diarrhea) each year, with most cases occurring in children younger than five years of age, and more than 600,000 deaths annually (more than one child a minute) from the dehydration that results from the illness.[163] In India alone, 100,000 children die every year from rotavirus infections.[164] Rotaviruses cause more infections of the gastrointestinal tract in children than any other agent, with more than 90 percent of children having been exposed by age three.[165]

Nonhuman primates seem to be similarly infected. At the Yerkes National Primate Research Center in Atlanta, Georgia, a majority of the Chimpanzees, and Old-World monkey species such as mangabeys (*Cercocebus* spp.), Pigtail Monkeys (*Macaca nemestrina*), and Rhesus Monkeys (*Macaca mulatta*), harbor antibodies to rotaviruses (and to noroviruses, which cause similar diseases). Researchers believe the same is true in the wild.[166] Infected nonhuman primates have provided a wealth of knowledge about how rotaviruses produce symptoms and have been essential for the development of a rotavirus vaccine. Two rotaviruses, RRV and SA11, for example, originally isolated from monkeys, have become the main strains used in laboratory research, and a new rotavirus oral vaccine, which relied on studies in nonhuman primates to establish safety and efficacy, was approved for use in the United States in early 2006.[167] The vaccine has been shown to prevent life-threatening vomiting and diarrhea from rotavirus infections and has the potential to curtail this leading cause of childhood mortality in the world.

HIV/AIDS

The human infectious disease that may best illustrate the need for nonhuman primates as research models is the human immunodeficiency virus, or HIV. Since the HIV pandemic was first recognized in 1981, more than sixty-five million people have been infected worldwide, and of these, there have been some twenty-five million deaths.[168] Each year, there are almost five million new infections and approximately three million deaths. Almost two-thirds of all people currently living with HIV live in sub-Saharan Africa, as do more than three-quarters of the world's infected women; the infection rate in some areas, such as in Botswana, may be close to 40 percent.[169] Such high rates of infection have created a generation of millions of orphans in Africa and other regions whose parents have died from HIV/AIDS. And because HIV can be transmitted from a mother to her child—during pregnancy, childbirth (most often), or nursing—and because large numbers of pregnant women are infected, countless infants and children, at least in the developing world (such transmission has been markedly reduced in industrialized countries), have become infected as well.[168]

In the absence of adequate medical therapy, almost all who are infected with either of the two major types of the HIV virus (HIV-1 and HIV-2) will eventually develop acquired immune deficiency syndrome, or AIDS. This syndrome develops because HIV infects and destroys cellular components central to our immune systems, especially cells known as "helper T-cells." The principal job of these cells is to selectively activate other parts of the immune system best able to eradicate pathogens they encounter. As HIV spreads through the population of helper T-cells, the immune system can no longer respond adequately to infections, and the patient becomes susceptible to so-called opportunistic infections by organisms that are easily conquered by normal immune systems. The immunodeficiency also promotes the development of several forms of cancer, including lymphomas and Kaposi's sarcoma, the latter a type of skin cancer caused by a herpesvirus.

As with the infectious diseases described above, our hopes for better treatment and prevention of HIV/AIDS rest largely in our ability to study the infection in nonhuman primates. Nonhuman primates are required subjects, in the field as well as in the lab, if we are to investigate a host of questions about the biology of HIV/AIDS. Some of these areas of investigation are summarized below.

Strong evidence shows that HIV-1, the strain primarily responsible for the HIV/AIDS pandemic, came from a related virus, SIVcpz, carried by a subspecies of the Chimpanzee, *Pan troglodytes troglodytes*, from West–Central Africa (for details, see "Species Exploitation and the Consumption of Bushmeat" in chapter 7). While people generally die from HIV-1 infection in the absence of medical treatment, Chimpanzees seem to carry their SIVcpz viruses without showing any obvious signs of infection, making them critically important research models in the wild that are naturally infected but do not develop illness from the infection.[170,171]

Laboratory experiments with a variety of Old- and New-World monkeys, including baboons (*Papio* spp.), macaques (*Macaca* spp.), owl monkeys (*Aotus* spp.), and mangabeys (*Cercocebus* spp.), have made it clear that some nonhuman primates have a built-in means to protect themselves from being infected with retroviruses (viruses that are able to transcribe and integrate their RNA genomes into the host's DNA genome—HIV is a retrovirus). One example comes from studies of Old-World monkeys at Harvard's Dana-Farber Cancer Institute. Researchers there found that an intracellular protein called TRIM5 alpha prevented HIV-1 infection in these animals (humans have their own version of TRIM5-alpha, but, unfortunately, it does not prevent HIV infection). Further research revealed that this protein blocks the virus from replicating its genome after it has entered a cell. Understanding the mechanisms of TRIM5-alpha protection in some nonhuman primates may yield new prospects for how to treat or even prevent HIV infection in humans.[172]

Presumably humans have been exposed to simian immunodeficiency viruses (SIVs) from apes and monkeys for millennia. Why has the HIV/AIDS epidemic emerged only in the latter half of the twentieth century? Answering this question, which involves a better understanding of the dynamics of SIV transmission from nonhuman primates to people, and of what factors allow this transmission to take place, may also help us abate a growing number of other emerging infectious viral diseases that have entered the human population from wildlife, such as hantavirus pulmonary syndrome, SARS, and Nipah virus infections.

Developing Treatments and a Vaccine for HIV/AIDS

The first effective HIV medication, AZT (zidovudine), has been extensively studied in several primate species, including Cynomolgus Macaques and Rhesus Monkeys, to establish safety and efficacy. And Pigtailed Monkeys , infected with an SIV, were essential to establishing the safety of AZT in humans when used to prevent the transmission of the virus during childbirth (AZT can reduce the risk of transmission from mother to child by 67 percent).[173] Primates have also been invaluable for the development of combination drug therapy for HIV, now the mainstay of treatment, as well as for investigations looking into how the virus becomes resistant to antiretroviral medications.[174]

Over the past twenty years, the race to develop a vaccine for HIV/AIDS has become a major focus of research. Many different strategies have been tried to make an effective vaccine, and primate models, in particular, the Rhesus Monkey, have been invaluable to testing candidate vaccines.[175]

NEUROLOGICAL DISORDERS

Of all the organs in our bodies, the human brain is the most unique when compared to that in all other animals, with the exception of our fellow primates. While we can learn a great deal about it from various other animal models, for example, about our visual systems from studying cats, about human neurodegenerative diseases from those reproduced in transgenic mice, and even about the genetics of human aggression from investigations with fruit flies, ultimately the search for a deeper appreciation of the intricacies of our brains and how they work mandates the use of nonhuman primates. This is because they share with us a similar anatomy and a complexity of organization on a cellular and molecular level that makes them the most suitable research models for the study of such areas as the circuitry in our brains and the function of different regions; human sensory and motor capabilities; human perception, cognition, memory, reasoning, and the development of language; and human neurological disorders.[176] In this section, we look only at neurological disorders, focusing on two major diseases that have been covered from different perspectives in other sections of this book: Parkinson's disease and Alzheimer's disease.

Parkinson's Disease

When it was discovered in the 1980s that those who used a synthetic form of heroin contaminated with a substance known as MPTP developed a disorder nearly indistinguishable from Parkinson's disease, and when it was shown that MPTP produced a disease that closely resembled Parkinson's in Rhesus Monkeys, marmosets (*Callithrix* spp.), and baboons, MPTP-treated nonhuman primates became the prime research models for the disease.[177] Although mice, dogs, cats, sheep, rats, rabbits, and goldfish (among other species) have all been exposed to MPTP to determine whether they would make suitable Parkinson's disease models, none of them reproduced the biochemical defects and neuropathological lesions (the loss of dopamine-containing cells in a region of the brain called the substantia nigra) or the motor and behavioral abnormalities (including low-frequency tremors while at rest) characteristic of the disease in humans. Nonhuman primates, however, do.[148]

Nonhuman primates have also become superb subjects for testing new therapies. As was mentioned in chapter 5, long-term use of L-dopa eventually results in less control of symptoms and in the development of spontaneous, uncontrolled movements called dyskinesias. MPTP-exposed primates treated with L-dopa develop these same dyskinesias and can, as a result, be studied to determine how these movements origi-nate, how the prevent them, and how to treat them when they occur.[148] Nonhuman primates also provide insights into the cognitive, behavioral, and emotional deficits of Parkinson's disease and are essential for evaluating new therapies, including stem cell transplantation, that hold promise for treating this disease.

Alzheimer's Disease

Alzheimer's disease is a progressive degenerative disease of the human brain result-ing in dementia that afflicts approximately 4.5 million people in the United States (a figure that is expected to rise to 10 million by 2025) and more than 28 million world-wide. The cost of dementia to the world economy each year has been estimated, in a 2005 study by the Alzheimer's Association, at $156 billion. Alzheimer's is uniformly fatal, with people living only about five years more after the initial diagnosis is made. In the United States in 2003, approximately 63,000 died from Alzheimer's, and the cost of caring for those with Alzheimer's is estimated at $50 billion each year.[178]

While there is no animal model that closely mimics Alzheimer's disease, aged Rhesus Monkeys (those in their mid to late 20s—Rhesus Monkeys can live to thirty-five years of age or more) do begin to show cognitive and memory disturbances, a loss of brain nerve cells, and brain lesions, all of which are similar to, but less severe than, those seen in Alzheimer's patients.[179] The characteristic lesions in the brains of Alzheimer's patients are called plaques. They contain clumps of a protein substance called beta-amyloid, which in Rhesus Monkeys has an amino acid sequence that is identical to that in humans.[148]

Alzheimer's disease is associated with a destruction of brain cells that commu-nicate via acetylcholine, a neurotransmitter that is involved in the laying down and retrieval of memories (among many other roles). Drugs currently in use that lead to increased levels of this neurotransmitter in the brains of Alzheimer's patients appear to show a minimal benefit in some patients. Aged Rhesus Monkeys provide the best model for testing the safety and effectiveness of such new drugs.[180] Rhesus Monkeys and other nonhuman primates are also being used in attempts to develop vaccines that would prevent the buildup of beta-amyloid, which is thought by some researchers to be the underlying process that causes Alzheimer's disease.[181,182]

BEHAVIORAL DISORDERS

Because primates have large, highly complex brains similar to ours, and behaviors that are the closest of all other animals to our own, they provide an important window into our own behavior and into its emotional, social, physiological, and anatomical underpinnings. Research areas include primate social systems and their relevance to our own; aggression; motivational states such as hunger, thirst, sexual excitement, and addictive drives; the effect of hormones on behavior; gene environment interac-tions in behavior; and, of course, studies on the primate equivalent of human psychi-atric states, such as depression and anxiety.[176] One area of tremendous importance to human development is that of mother–infant interactions. Harry Harlow pioneered this field of research in the late 1950s and 1960s and conducted experiments with

Rhesus Monkeys (which generated a great deal of controversy because of ethical concerns). He demonstrated how maternal deprivation and separation can produce psychologically and socially impaired offspring, which developed symptoms much like those seen in human depression, and how maternal physical contact was essential for normal development.[183]

Behavioral research has been conducted in natural habitats, in more controlled settings such as zoos, and in the laboratory. In this section we very briefly cover the critically important work that has been done in natural settings, in part to illustrate one aspect of the depth of our loss when we lose primates in the wild.

Some of the earliest systematic observations of nonhuman primates in the wild were made by C. R. Carpenter and his colleagues on gibbons in the 1930s. Along with Carpenter, Sherwood Washburn in the United States and Kenji Imanishi in Japan were instrumental in developing widespread interest in studying the behavior of nonhuman primates in the field for the purpose of understanding the origins of human behavior, and in establishing such interdisciplinary studies as central to modern anthropological research. A host of key comparative studies followed, including those by Irven Devore[184] and Stuart Altmann[185] on baboons. More recent work has included that by Dian Fossey on gorillas[186] and Biruté Mary Galdikas on orangutans.[187] The work on gorillas has investigated such questions as why the Mountain Gorilla (*Gorilla beringei beringei*) shows aggressive, and sometimes violent, behavior in male–male interactions when different groups come across each other, while the Western Gorilla (*Gorilla gorilla*), a closely related species, interacts peacefully and even intermingles when such groups come in contact. Given the enormous problems of human aggression and violence in communities, and how dangerous our own violent impulses have become in a world filled with terrorism and weapons of mass destruction, understanding some of the origins of this behavior may have great value. If gorillas go extinct in the wild, as appears increasingly likely in the next several decades, these studies will not be possible.

The same holds true for orangutans, which are at great risk for extinction. Studies done in 2003 demonstrated that some behaviors, such as bedtime rituals and sexual practices, differed from one group to another, presumably because these behaviors were learned from members in each of the groups.[188] While there is debate about whether these differences can be ascribed to cultural learning, because some see "culture" as a distinctly human characteristic (see Editors' Note), it is clear that more research in the wild is necessary to determine whether the differences observed can be explained by environmental differences that have not yet been discovered. Because orangutan populations are disappearing at a very rapid rate, this research may soon become impossible.

Of all the field studies with primates that have been undertaken, none perhaps has had the importance and the broad public appeal as those of Jane Goodall with Chimpanzees. This work is of particular significance because molecular analysis and comparative morphology has shown that humans are closest to the Chimpanzee and to Bonobos, having shared a common ancestor approximately five to seven million years ago. The work of Dr. Goodall has revealed that Chimpanzees have highly complex societies; experience deep, humanlike emotions; use tools; and conduct wars, work that has taught us a great deal about human behavior and human social systems. It has also called attention to the plight of Chimpanzees resulting from human activity and to that of all other primates.[189]

Gymnosperms

Endangered Gymnosperms

Every person is likely to have seen a gymnosperm, because this group of plants includes some of the most common species of trees alive today, such as pines and spruces. All told, some 980 species of gymnosperms have been identified from the tropics to the poles, in environments ranging from mountaintops to the fringes of the Arctic Circle. What unites the gymnosperms are their seeds, which are exposed or "naked" (*gymno* means "naked" in Greek; *sperm* means "seed"), that is, lacking a covering, such as a fruit, to surround them. The gymnosperms share other traits, as well—their evolutionary roots are some of the oldest of any plant in existence. Gymnosperms evolved roughly 370 million years ago and were the first plants with seeds. They gave rise to their evolutionary cousins, the angiosperms, or flowering plants, now the most numerous group of plants having seeds, with around 250,000 known species.

Several orders comprise the gymnosperms: the gnetophytes, which include the *Ephedra* species that have been the source of several medicines, including the widely used decongestant pseudoephedrine (see below for a discussion of medicines from *Ephedra*); the cycads, the most ancient and most endangered group of gymnosperms (all species are listed in CITES, and more than half are listed as threatened on the 2006 IUCN Red List), which produce potent neurotoxins that can induce neurodegenerative disease; and the Ginkgo Tree, *Ginkgo biloba*, which is the only member of its family.

Ginkgo biloba has changed remarkably little since it first evolved around 200 million years ago, making it a tree with one of the oldest evolutionary roots in existence.[190] Ginkgos, which may live for more than 2,000 years, have dodged extinction at several points throughout their history, including the Cretaceous extinction event sixty-five million years ago. And had it not been for its cultivation by monks in Japan and China, and by botanists in Europe and North America, the tree would likely not have survived into the present, because few wild stands remain. In spite of these efforts, *Ginkgo biloba* is still listed as Endangered on the IUCN Red List. It has been

Figure 6.23. Ginkgo Tree and Leaves. (From Philipp Franz von Siebold (1796–1866), *Flora japonica; sive, Plantae quas in imperio japonico collegit, descripsit, ex parte in ipsis locis Lugduni Batavorum.* 1835–1870. © President and Fellows of Harvard College, Archives of the Arnold Arboretum.)

a source of medicines for many hundreds of years in China, and extracts have become widely used around the world in the past decade. These uses are discussed below.

The gymnosperms also include the conifers, a group of around 600 tree species, including the pines, spruces, cedars, and cypresses. Conifer species are of vital importance to humanity not only because they are the most widely used timber for home construction and for paper pulp, but also because some species have yielded compounds that have been developed into medicines. One prominent example, a drug

Taxol (paclitaxel), derived from the Pacific Yew Tree (*Taxus brevifolia*), is described in detail below. Among the conifers can be found the oldest and tallest trees in the world: The oldest is the Bristlecone Pine (*Pinus longaeva*), which lives in harsh, dry conditions at more than 10,000 feet elevation in the White Mountains of California (one specimen is estimated to be well more than 4,700 years old); the tallest is the California Redwood (*Sequoia sempervivens*), one of which was discovered in 2006 in California's Redwood National Park to be 379.1 feet high.

With the exception of a few small islands, conifers live everywhere humans do. They also constitute the dominant species in the great boreal forests of the northern latitudes of North America, Europe, and Asia, which make up about one-third of all forested regions of the world. Yet, despite this overall abundance, one in four species of conifers is threatened with extinction.

The greatest threat to conifers is unsustainable harvesting. Forests are cleared to provide wood for building and for producing paper and to make way for human settlements and agriculture. For many species, however, more specific dangers exist, including pests that attack trees and, especially for boreal species, global warming. Temperatures in boreal regions of North America, for example, have already risen 2 degrees Celsius on average (around 4 degrees Fahrenheit) in the past fifty years and are predicted to rise by as much as another 3 to 6.5 degrees Celsius (around 5 to 12 degrees Fahrenheit) more by the year 2100.[191] As mentioned in chapter 2, these warming temperatures will push the zone in which species can survive ever closer to the poles. It is predicted, for example, that boreal forests will have to shift northward by 100 kilometers or more (more than 62 miles) over the next century, creating serious threats to survival for many species, including the conifers.[191,192]

Take spruce trees in Alaska, for example. An epidemic of the North American Spruce Bark Beetle (*Dendroctonus rufipennis*) wiped out more than 90 percent of White and Lutz Spruce Trees across a 3.2-million acre span of the Kenai Peninsula and Copper River region of south central Alaska between 1987 and 2000 (only the lack of sufficient numbers of living tree hosts put an end to the epidemic). This represents the largest loss of trees ever recorded in North America due to a single pest.[193,194] The beetle is native to the regions where the epidemic occurred, raising the question: What caused it to attack otherwise healthy stands of old-growth spruce? The answer, researchers believe, has to do with a warming climate. Alaska, like other boreal regions, has experienced significant increases in mean temperatures over the past few decades. From 1972 to 1978, the average annual summer (May through August) temperatures recorded at the Homer Airport on the southern coast of the Kenai Peninsula was 49.42 degrees Fahrenheit, while for the period from 1992 to 1998, it was 51.65 degrees Fahrenheit, a rise of more than 2.2 degrees Fahrenheit.[195] Warmer winters have resulted in greater beetle survival, and warmer springs and summers have allowed them to reproduce yearly, as opposed to their normal cycle of once every two to three years. The explosion in beetle populations, combined with dryer conditions in the forests that reduced spruce tree sap production and in the process compromised the trees' defense against the beetles (the sap blocks the beetles' larval galleries in the wood and may also fight fungal diseases the beetles carry), is thought to have led to the massive spruce mortality in the Kenai.[196]

Warmer temperatures may also spell disaster for other conifer species under assault from pests. Whitebark Pine (*Pinus albicaulis*), found throughout mountainous

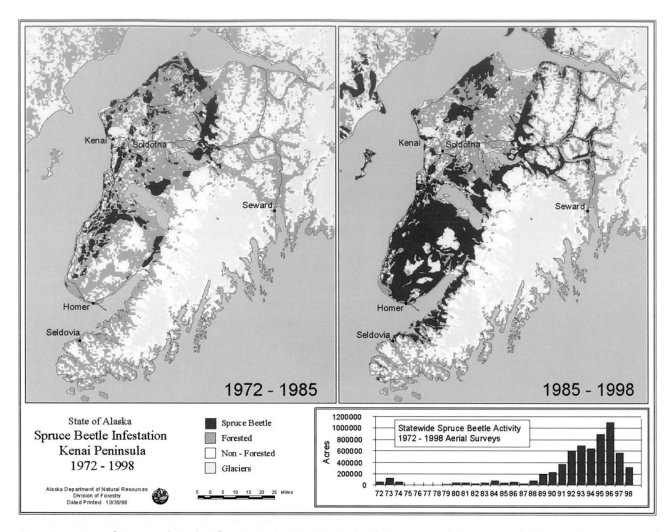

Figure 6.24. Maps of Spruce Bark Beetle Infestation in the Kenai Peninsula, Alaska, 1972–1998. Note the marked increase in the number of infested forested acres in the thirteen-year period from 1985 to 1998 compared to that from 1972 to 1985, thought to be related to significant warming in the region. Such infestations by Spruce Bark Beetles kill the trees and destroy the forest ecosystem, setting the stage for major forest fires. (Used with permission from the Alaska Department of Natural Resources.)

regions of the Pacific Northwest of the United States and Canada, is threatened with extinction according to the IUCN Red List. Not only has the fungus White Pine Blister Rust (*Cronartium ribicola*), introduced to North America on infected White Pine seedlings from Europe around 1910, killed about half of the Whitebark Pine trees in Glacier National Park and in the adjacent Bob Marshall Wilderness Complex in northwestern Montana (80 percent of the surviving half are also infected),[197–199] but the Mountain Pine Beetle (*Dendroctonus ponderosae*) has also become a major threat to the survival of the species. The beetle is present in about 16 percent of the one million acres that make up Yellowstone and Teton national parks and some surrounding areas.[200] Although indigenous to this territory, historically it has spared Whitebark Pine trees there, because the high elevations resulted in temperatures that were too cold for the beetle to survive. With warming temperatures, the Mountain Pine Beetle has now established itself in stands of Whitebark Pine.[201]

Whitebark Pine can be considered a keystone species because it supports many other trees, birds, and mammals. Its presence allows less hearty conifers such as the Subalpine Fir (*Abies lasiocarpa*) and Engelmann Spruce (*Picea engelmannii*) to grow in regions where they would not otherwise grow, and forests in regions where Whitebark Pine grows have more species where Whitebark Pine is the dominant tree than in those where it is not.[202] Some species, such as the endangered Grizzly Bear (*Ursus arctos horribilis*), depend upon nuts from the pine to survive. And while the Clark's Nutcracker (*Nucifraga columiana*) also depends on Whitebark Pine nuts for food, the pine depends almost exclusively on the Clark's Nutcracker for seed dispersal, making the survival of the two species highly dependent on each other.[201,203]

Hemlocks (also conifers) have a long history of being threatened by human activity. In the nineteenth century, for example, to supply the tanning industry, as many as two-thirds of all the hemlocks in the Adirondack Mountains were stripped of their bark for tannins. Now hemlocks are under a new threat. Those growing in the eastern United States have been assaulted by a pest, the Hemlock Woolly Adelgid (*Adelges tsugae*), introduced from Japan in the 1950s.[204] Some parallels can be drawn between the Hemlock Woolly Adelgid and the Spruce Bark Beetle infestations. For example, both may become more destructive with a warming climate. Cold winter temperatures limit the range of the adelgid, and decrease population size,[205] so as temperatures rise, the insect may spread to new regions and increase its numbers.

Hemlock canopies supply unique conditions. More shaded, cool, and moist than their surroundings, they are the preferred habitats for a wide range of organisms. More than a dozen species, from lichens and ferns to warblers, salamanders, and toads, depend on hemlocks in warmer months.[206] And in winter, the densely needled hemlock branches provide protective cover from heavy snowfalls for a variety of others.

Hemlocks also have a prominent role in shaping the cycling of nutrients, especially nitrogen, in the ecosystems they sustain. It should come as no surprise, then, that with their loss, scientists have observed profound ecological changes: New hardwood tree species, such as Black Birch (*Betula lenta*), Red Oak (*Quercus rubra*), and Red Maple (*Acer rubrum*), have migrated into the niches left by their absence, as have new species of shrubs and birds. Even fish, which live in rivers and streams adjacent to hemlock stands, may be affected, perhaps as a result of warmer water temperatures secondary to the loss of shade.[207] Such ecosystem changes not only put species highly dependent on hemlocks in jeopardy of local extinctions, but also may have implications for human infectious diseases, for example, by reducing vertebrate diversity in the forests of middle Atlantic and New England states in the United States, thereby potentially increasing the risk that people in these areas will get Lyme disease. This topic and several others that illustrate the relationship between ecosystem change and the spread of human infectious diseases are explored in chapter 7.

Medicines from Gymnosperms

EPHEDRINE

This amine (amines are nitrogen-containing compounds derived from ammonia) was first isolated in Japan in 1887, and was reisolated in 1923 from the gymnosperm *Ephedra sinica* (*ma huang* in Chinese), a plant long used in traditional Chinese medicine. The genus *Ephedra* is made up of about forty shrubby plant species that live in arid and semiarid regions of the Northern Hemisphere. They belong to the Gnetophyta phylum, which is made up of plants that possess unique reproductive and vascular structures that have led some botanists to believe that it was early Gnetophyta species that evolved into the angiosperms. Ephedrine has formed the basis for the synthesis of beta-adrenergic agonists. Some of these compounds are widely employed today to treat asthma (e.g., the medicines albuterol and salmeterol) because they dilate the respiratory tract to make breathing easier. Others, such as the drug isoproterenol, are used to stimulate heart rate in patients with heart block or in those who have dangerously slow rates.[208]

GINKGO

Ginkgo biloba has survived as a species for millions of years. Part of this longevity may have to do with a dazzling variety of protective molecules the tree makes. These molecules shield its leaves from damaging ultraviolet radiation, guard against microbial infection, and ward off insects. Remarkably, the tree appears to have evolved an ability to change its output of these defenses in response to seasonal cues that indicate which potential threat is greatest at a given time of year.[209]

The Ginkgo Tree has been used for hundreds of years in China to treat scores of conditions, including asthma, diarrhea, skin rashes, cancer, and tuberculosis. The first documentation for the internal use of leaf extracts (the seeds had been used even earlier) appeared during the Ming dynasty in 1436 A.D. with the publication of *Dian nan ben cao* (Medicinal Plants in Southern Yunnan). At present, tens of millions of people in Europe and North America use Ginkgo leaf extracts. Not all chemicals in Ginkgo are benign, however. The compound 4-*O*-methylpyridoxine, or MPN for short, found in Ginkgo seeds, can cause what has been called "Gin-man" food poisoning if ingested, resulting in seizures and sometimes death, particularly in infants.[210]

When one consumes a pill of Ginkgo extract today, it should contain a standardized mixture of molecules, known as EGb 761, derived from the tree's leaves. Some of these molecules are potent antioxidants and have been shown—in both test tube experiments and animal models—to protect cells, particularly neurons, from harmful free radicals (free radicals are highly reactive atoms or molecules that can damage tissues and are thought to contribute to the aging process and to the development of some cancers). Other Ginkgo compounds have been shown in the brains of rodents to improve the efficiency by which cells use glucose and oxygen; to relax blood vessel walls, improving circulation; and to stave off the appearance and toxicity of beta-amyloid (a protein whose presence is associated with Alzheimer's disease). The discovery of these promising properties has led to a flurry of interest into whether Ginkgo extracts could improve various conditions, ranging from high-altitude sickness and tinnitus (a persistent ringing in the ears) to the treatment of asthma, depression, and impotence. At present, however, no clear-cut scientific evidence supports Ginkgo's ability to treat any of these conditions.[211]

However, for some people with dementia, Ginkgo extracts do seem to provide some benefit, delaying by as much as six months such symptoms as worsening of memory, slowed reaction times, and decreased attention span. This relief is similar in duration to that provided by prescription medications for dementia, but perhaps with fewer side effects.[212]

Ginkgo preparations are not limited to medicines. The Chinese and Japanese have for centuries used dried Ginkgo leaves as an insect repellant. Individual leaves have been used as bookmarks in Japan, for instance, in the belief that they deter booklice and silverfish that damage bindings and paper. One important piece of research on Ginkgo's insecticidal potential, done at Seoul National University in South Korea, has shown that specific extracts of the leaves can kill, at very low concentrations, a major rice pest, the Brown Plant Hopper[213] (see the discussion of the Brown Plant Hopper and rice production in Indonesia in chapter 8, page 340).

PACLITAXEL (TAXOL)

Probably the most significant drug discovered and developed through the U.S. National Cancer Institute's Natural Products Branch was paclitaxel, isolated in 1969 from the bark of the Pacific Yew Tree (*Taxus brevifolia*) by Dr. Monroe Wall (a student of Selman Waksman, the discoverer, as described in chapter 4, of the aminoglycoside group of antibiotics) as part of an extensive plant-screening program. The tree, prior to this discovery, had been routinely discarded during logging operations in old-growth forests of the Pacific Northwest region of the United States, because it was thought to have no commercial value. Today, sales of paclitaxel are more than US$1.5 billion, with a single treatment course costing in excess of US$10,000.[214,215]

In early clinical trials in 1989, paclitaxel was found to be effective for inducing remissions in cases of advanced ovarian cancers, cancers that had generally responded poorly to most other chemotherapies.[216] Since that time, it has been shown to have significant therapeutic benefit for several other forms of advanced malignancies as well, including lung and prostate cancers, malignant melanomas, lymphomas, and metastatic breast cancers.[217] Paclitaxel inhibits the proliferation of cancer cells by stabilizing the cellular protein tubulin, thereby blocking the disassembly of the mitotic spindle (a cellular scaffolding made of tubulin that appears during cell division and enables chromosomes to divide and move to their new daughter cells) and preventing cell division. Discovery of this mechanism of action, unique among cancer chemotherapeutic agents, has opened the door to the development of an entire new generation of related drugs.[218]

As mentioned in chapter 4, paclitaxel also works to inhibit the proliferation of smooth muscle cells that line arterial walls (called endothelial cells) and has been successfully employed as a coating for coronary artery stents (coronary arteries supply blood to the heart), preventing endothelial cells from growing over and

Figure 6.25. Pacific Yew Tree (*Taxus brevifolia*) Needles and Cones. (From Charles Sprague Sargent's *Silva of North America*, illustrated by Charles Edward Paxon, Vol. 10. Houghton, Mifflin & Co., Cambridge, 1896, Plate DXIV. Used with permission from the Harvard University Botanical Library.)

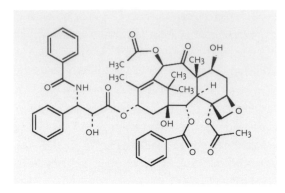

Figure 6.26. Chemical Structure of Paclitaxel (Taxol). The paclitaxel molecule is a highly complex, interlocking ring that would be nearly impossible to discover by synthetic means alone.

inside the stents, thereby opening blood flow through the arteries.[219] The use of paclitaxel-coated stents has not been an unqualified success, however. Evidence suggests that some patients' arteries will eventually become blocked again despite the placement of such stents, and, in a small fraction of patients, the stents may cause the formation of new blood clots, sometimes months or even a year or more after insertion. Although there remain some questions about the long-term safety of drug-coated stents, the most current and best research on this topic indicates that the clots that do form are unlikely to cause heart attacks or death.[220]

Originally in danger of short supply due to a limited number of Pacific Yew Trees, paclitaxel and other taxoids are now being produced by semi-synthetic conversions of precursor compounds that are found in yew tree needles, including from yew trees other than *Taxus brevifolia*, that are sustainably harvested in plantations in many parts of the world. In addition, scientists have discovered that it may be fungi that live symbiotically with the Pacific Yew that are producing the paclitaxel molecule, raising the possibility that taxoids may someday be produced without relying on yew trees at all.

The paclitaxel story illustrates the great importance of conserving natural resources, because this highly effective therapeutic agent was discovered only from a random screening of 35,000 plant samples. It also demonstrates how highly complex, naturally occurring, bioactive molecules such as paclitaxel are unlikely to be discovered by combinatorial chemistry alone, because the number of possible structures for this one molecule is so vast (on the order of 2^{11} possibilities), yet once they have been identified in Nature, they can serve as models for synthetic therapeutic agents that are as effective, or even more so, than the original natural product. By the end of 2005, there were more than 200 ongoing clinical or preclinical trials with paclitaxel and its derivatives, and paclitaxel was one of the top-selling cancer drugs in the United States.

Gymnosperms and Biomedical Research

Although the Ginkgo Tree may produce dozens of biologically active compounds, only ten or so have been the focus of scientific investigations. These include a group of five molecules, the ginkgolides, as well as a compound known as bilobalide.

Both bilobalide, and to a greater extent the ginkgolides, can interfere with the function of inhibitory neurotransmitter receptors in the central nervous system (i.e., those that decrease the ability of neurons to transmit signals between themselves). As a result, they have the potential to be valuable research tools for understanding how such receptors function in Alzheimer's disease, epilepsy, and depression, all of which involve significant inhibitory neurotransmitter activity.[221]

The ginkgolides have still further value to biomedical science. Some of them, particularly ginkgolide B, can reduce the production of a receptor found on the outer walls of mitochondria known as the peripheral benzodiazepine receptor (PBR). (Benzodiazepines, such as Valium, are used as muscle relaxants, anticonvulsants, anxiety relievers, and agents to induce sedation and sleep.) The PBR assists in the transport of cholesterol across the outer mitochondrial membrane, making it available for use in the production of various steroids within the mitochondria. These steroids include a subset known as the glucocorticoids, whose presence in hippocampal neurons at higher than normal levels can weaken these neurons, or even lead to their death. The ginkgolides are the only known pharmacological agents capable of reducing the production of the PBR, an ability that may someday be exploited in helping to improve memory. In addition, overactivity of the PBR may affect the spread of aggressive cancers, including some forms of breast cancer, and investigations are ongoing about the use of ginkgolides as therapy for tumors associated with PBR overexpression.[222,223]

CONE SNAILS

Endangered Cone Snails

Cone snails, a large genus of marine mollusks named *Conus*, number about 700 species. In the year 2004 alone, seven new species were identified.[224] The most familiar *Conus* species live in shallow waters, less than 20 meters deep (about 65 feet), inhabiting tropical coral reefs and soft bottom habitats such as those of mangroves. But there may be many others, not yet identified, that live in deeper waters. For hundreds of years they have captivated collectors, who have placed great value on the enormous beauty of the varied patterns on their shells (see the opening figure for chapter 4). As we discuss below, cone snails are also extremely valuable as sources for new medicines and in biomedical research.

No *Conus* species is listed as Critically Endangered or Extinct in the 2006 IUCN Red List of Threatened Species, and only four, all found off the coast of Angola, are classified as Vulnerable by IUCN criteria, that is, living within restricted areas and therefore being at risk for extinction as a result of human disturbances or natural catastrophes. However, there has never been a systematic assessment by the IUCN of threatened cone snail species, and no assessment at all has been done for more than ten years. The current listings, therefore, do not provide an accurate account of the magnitude of the threat to these creatures.

We can be sure that a great many cone snail species are threatened. For one, their primary habitats—coral reefs and mangroves—are being rapidly degraded and destroyed around the world. Many cone snail species live in regions where these habitats are among those in the poorest shape. In one study of 386 cone snail species, for example, almost 70 percent had more than half of their geographic ranges within areas where coral reefs were threatened.[225] An estimated 20 percent of the world's coral reefs are said to be so damaged that they are unlikely to recover, and an additional 50 percent are at risk of collapse[226] (see also chapter 2, page 35). For the world's mangroves, the situation is even worse, with an estimated 50 percent having been cleared for wood, development, and aquaculture.[227] In Southeast Asia, which harbors more than half the world's cone snail species, almost 90 percent of the reefs are threatened by human activities and almost 50 percent are highly

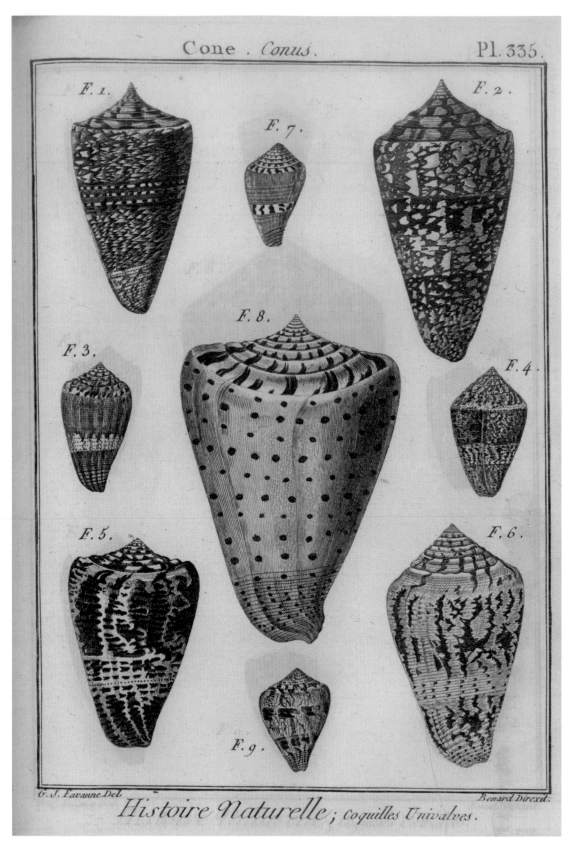

Figure 6.27. Cone Snail Species: Eighteenth-Century Print of Nine Species of Cone Snails. (In *Tableau encyclopédique et méthodique des trois règnes de la nature, vingt-unième partie /par le cit. Lamarck.* Chez Henri Agasse, Paris, 1798. From the collections of the Ernst Mayr Library, Museum of Comparative Zoology, Harvard University.)

threatened, and in the Philippines, which contains at least eight cone snail species found no where else, fully 97 percent of the reefs are threatened[228] and 60 percent of the mangroves have been destroyed.[227] Moreover, many cone snail species tend to be concentrated in narrow geographic ranges, putting them at a greater risk for extinction. The December 26, 2004, tsunami, by uprooting mangroves and damaging corals—smashing them with the force of the waves and with land-based debris, and covering them with pollution and silt—is likely to have further compromised cone snail habitat in Southeast Asia.

Direct exploitation also endangers cone snails. Just as they have been for centuries, cone snail shells are still today widely sold in thousands of curio shops and markets around the world. Although no precise figures are kept of the numbers involved, a good educated guess would suggest that millions of cone snails are being sacrificed each year to meet global demands. The exponential rise in research on cone snail toxins may be further contributing to declines in some populations, although most major *Conus* researchers have taken great care to ensure that wild populations are not threatened.[229]

The release of carbon dioxide from the burning of fossil fuels, the main cause of global warming, damages reefs in two ways. The first is from the direct effects of atmospheric carbon dioxide when it dissolves in seawater, causing it to become more acid and inhibiting calcification of coral skeletons (see section on acidification of the oceans in chapter 2, page 69). Reduced calcification reduces growth rates and weakens the structural integrity of the corals, imperiling their survival. The second is from the effect that warming air temperatures have on corals. As the lower atmosphere warms, it warms the oceans, and when sea-surface temperatures exceed local temperature maxima by as little as 1 degree Celsius (or about 2 degrees Fahrenheit) for more than a few days, the symbiotic algae that live in coral reef tissues leave or die, and the corals, which need these algae to supply them with nutrients, appear as if they had been bleached. Bleaching alone can kill the corals, or it can lead to an increased susceptibility to fatal infectious bacterial and fungal diseases.[230] Some corals seem to be more resilient than others and are able to survive bleaching and recover their functions, but it is not clear what distinguishes these from corals that are more sensitive. In recent years, coral bleaching has induced widespread mortality of corals and serious reef degradation. Protecting cone snails and other reef life requires dealing with global warming, the single greatest threat to coral reefs.

Further actions to protect reefs are needed, as well, such as establishing marine reserves, setting controls on coastal development and pollution, safeguarding mangroves and other reef-associated habitats from destruction, and banning destructive fishing practices, for example, those that make use of explosives such as dynamite, or poisons such as cyanide. Cone snail trade needs to be monitored and regulated to prevent cone snail populations from collapsing, perhaps by using CITES, much as has been done with trade in some species of coral.[228] Controls on the collection and trade of cone snails are in place in only a few countries such as Australia. The countries of Southeast Asia, where most cone snails are found, have no such controls. And although such regulations will not stop people in many countries from eating cone snails and selling their shells, they will help to reduce these practices and to protect some cone snail species that are endangered.

6/3/96

10/2/97

Medicines from Cone Snails

Cone snails defend themselves and paralyze their prey—worms, fish, and other mollusks—by injecting a cocktail of toxic peptides (small proteins) through a hollow, harpoon-like tooth.

Each of the estimated 700 species is thought to make as many as 100 to 200 distinct peptides, so there may be as many as 70,000 to 140,000 peptide toxins in all.[231] To put this in perspective, only about 10,000 plant alkaloids, which include some of our most useful medicines, such as morphine, vincristine, and pilocarpine, have been identified. Moreover, while other groups of poisonous animals, such as pit vipers, spiders, scorpions, and sea anemones, also produce peptide toxins, each of these animals makes only a handful of different types. The evolutionary explosion of *Conus* peptides is even more remarkable when one considers that they evolved around fifty million years ago, considerably more recently than snakes (about 125 million years ago), spiders and scorpions (close to 400 million years ago), or sea anemones (about 500 million years ago).

Figure 6.29. Black Band Disease in a Colony of *Montastrea annularis*, a Scleractinian (Stony) Coral in Caribbean Reefs. A number of different bacteria, the major component of which is the cyanobacterium *Phormidium corallyticum*, are present in the band. The white circular area at the apex is dead coral. Photo taken off Grand Cayman Island. ((c) Ray Hayes.)

In addition to being more numerous by orders of magnitude than the toxins of these other poisonous animals, cone snail toxins have a greater diversity of receptor binding sites (molecular structures on the surfaces of cells that activate different cellular processes), and for a given class of receptor, they tend to have greater selectivity. The enormous variety of such sites includes multiple subtypes of ion channels that regulate the flow of sodium, potassium, and calcium across cellular membranes, as well as numerous other receptors, including those that bind to compounds that function as neurotransmitters (chemicals that transmit messages between nerve cells), such as acetylcholine, serotonin, and norepinephrine.[232] It is this combination of exquisite selectivity and extraordinary diversity of binding sites that makes cone snail toxins among the most sought after natural compounds for biomedical research and for the development of new medicines.[233] More than 3,400 articles on these toxins have been published in the scientific literature since 1980 alone.[234]

Of the estimated 70,000 to 140,000 cone snail peptides, only around 100 have been characterized, and of the approximately 700

Figure 6.30. Close-up Photo of Cone Snail Harpoon Protruding from Proboscis. ((c) Clay Bryce.)

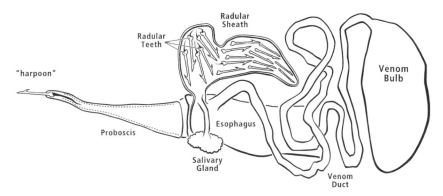

Figure 6.31. Drawing of Cone Snail Harpoon Anatomy. Individual harpoons form and are in various stages of assembly in the radular sheath, or "quiver." They work their way into the esophagus, where they are coated with venom, and then attach to the tip of the proboscis. (Courtesy of Baldomero M. Olivera.)

Conus obscurus

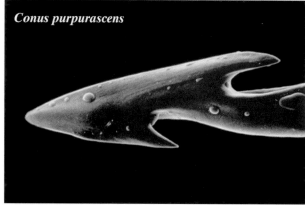

Conus purpurascens

Figure 6.32. Scanning Electron Micrographs of Harpoon Tips from *Conus obscurus* and *C. purpurascens*. Each *Conus* species makes a distinctive harpoon. (Courtesy of Baldomero M. Olivera.)

cone snail species, only six—*Conus geographus, C. magus, C. imperialis, C. purpurascens, C. radiatus,* and *C. striatus*—have been studied in any detail. But even though fewer than 1 percent of the likely total number of conopeptides have been investigated to date, several potential new medicines have already been identified. The most important of these are for treating pain.

PAIN MEDICATIONS

The standard treatment for severe chronic pain involves the use of opiates such as morphine. Opiates are highly effective when first given, but with continued use, they often result in addiction and in the development of tolerance, that is, the need over time to give higher and higher doses to achieve the same effect. Eventually, opiates may cease to be effective, or the dose required for pain relief may exceed dangerous levels.

One conopeptide, originally isolated by Baldomero Olivera's group at the University of Utah from the cone snail *C. magus,* blocks a specific type of calcium channel found on nerve cells that carry pain impulses to the brain. In its synthetic commercial form, ziconotide, which is identical both biologically and chemically to the natural product, has been shown to be safe and effective in animal models, and in a carefully controlled study, it has significantly, and in a few cases totally, relieved the pain of more than 50 percent of advanced cancer and AIDS patients tested, whose pain had been unresponsive to opiates.[235] There is evidence that ziconotide is as much as 1,000 times more potent than morphine.[236] What's more, it does not result in either addiction or tolerance. In 2004, the FDA approved the use of ziconotide (marketed as Prialt by the Elan Corporation) for patients whose pain no longer responded to opiates.

Three other pain medications derived from cone snail peptides—AMM336 from *Conus catus,* CGX1160 from *C. geographus,* and ACV1 from *C. victoriae*—have shown as much, and perhaps even more, potency in treating pain as ziconotide, and in addition, they appear to have larger therapeutic indices (i.e., the difference between the drug dose that produces pain relief and that which causes significant side effects). These potential new medicines, as well as at least four other conopeptide painkillers, are currently in clinical trials.[237] The use of conopeptides for pain therapy represents a watershed in the history of pain management, which has been centered on opiate-based medicines for centuries.

MEDICINES FOR OTHER MEDICAL CONDITIONS

Conopeptides may also be effective in diagnosing and treating an array of other medical conditions. For example, one conopeptide that blocks a specific type of neurotransmitter receptor, known as the *N*-methyl-d-aspartate, or NMDA, receptor, has

Figure 6.33. *Conus striatus* Harpooning a Fish. (Courtesy of Baldomero M. Olivera.)

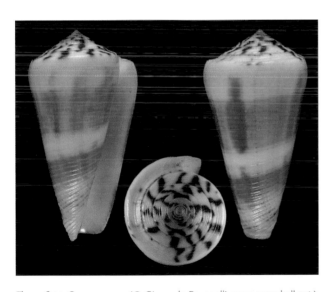

Figure 6.34. *Conus magus*. (© Giancarlo Paganelli, www.coneshell.net.)

been shown to protect neurons from cell death when there is inadequate circulation, such as during head injuries and strokes.[238] Conopeptides may, in addition, be able to prevent nerve cell death in some neurodegenerative diseases, such as amyotrophic lateral sclerosis (Lou Gehrig's disease), Alzheimer's disease, and Parkinson's disease.[239] And conopeptides may be a source for new antiepileptic medications. One NMDA receptor blocking conopeptide called conantokin-L, from the cone snail *C. lynceus*, for example, showed potent anticonvulsant activity in mice,[240] but its development by the company Cognetix was terminated because of toxicity. About 20 percent of the fifty million people worldwide with epilepsy continue to have seizures despite appropriate treatment, so the importance of finding new antiepileptic drugs, especially those that work by new mechanisms, as was demonstrated for conantokin-L, cannot be overstated.

Moreover, conopeptides may be useful in diagnosing Lambert-Eaton myasthenic syndrome (LEMS), an autoimmune neurological disease characterized by muscle weakness, fatigue, and symptoms such as mouth dryness and diminished sweating. These symptoms are caused by circulating antibodies, formed in response to certain cancers, such as small-cell carcinomas of the lung, that also attack nerve cells and cause them to malfunction. By being able to distinguish LEMS from other neurological disorders, conopeptides may provide an early warning for these cancers that are often hard to detect and difficult to treat.[241]

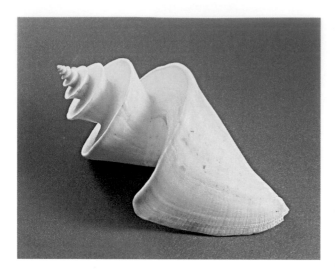

Figure 6.35. Japanese Wonder Shell (*Thatcheria mirabilis*). This turrid snail species is found only in deep waters, in contrast to other turrids that generally live in shallow, off-shore habitats. Like all members of the Turridae family, it paralyzes its prey with toxins delivered by a harpoon. (© Photo by Burt E. Vaughan, shells.tricity.wsu.edu.)

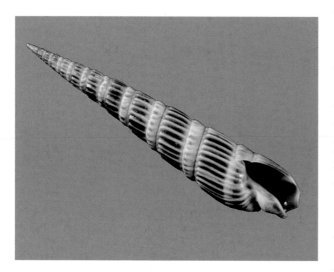

Figure 6.36. Dussumieri's Augur (*Terebra dussumieri*). This terebra snail species belongs to the family of venomous snails called the Terebridae. They fire harpoons coated with toxic peptides at their prey before eating them. (© Photo by Burt E. Vaughan, shells.tricity. wsu.edu.)

Conopeptides may also be effective in treating spasticity secondary to spinal cord injury, clinical depression, urinary incontinence, and cardiac arrhythmias; in preventing laryngospasm (a reflexive contraction of the larynx) when the larynx is being examined; and in creating functional, reversible brain lesions that mimic those that would be produced neurosurgically, in order to test their effects.[234] Of all the families of organisms on Earth, cone snails, the Conidae, may contain the largest and most clinically important pharmacopoeia of any in Nature. (To follow developments with cone snail peptides, see grimwade.biochem.unimelb. edu.au/cone/main.html.)

But there are also other families of venomous snails—the turrids (Turridae) and the terebras or auger snails (Terebridae)—that are relatives of the Conidae (all belong to the superfamily called the Coneacea) and that may rival, and possibly even exceed, them in their usefulness to human medicine. The turrids are the largest known family of marine gastropods (the class of mollusks that includes snails, slugs, and whelks), with perhaps more than 4,000 species. Like cone snails, they possess a venomous harpoon system that allows them to paralyze their prey, usually marine worms, before eating them. While we know very little about cone snail peptides, having characterized far less than 1 percent of them, we know next to nothing about the peptides in turrids and terebras.[242] The terebras number around 300 species and are warm water, sand-dwelling snails. They also harpoon their prey, again, mainly marine worms, paralyzing them with toxic peptides. If we lose these venomous snails through overexploitation and through our destruction of their habitats—coral reefs, mangroves, and other marine environments—we will have committed a self-destructive act of unparalleled folly.

Cone Snails and Biomedical Research

Understanding how a heart beats or a nerve transmits a sensation depends on understanding how cells are able to control the composition of fluids both within and outside their membranes. Cells are awash in a bath of ions (which are atoms or group of atoms that carry an electric charge), such as sodium, potassium, calcium, or chloride ions. A change in the concentration of any of these in intracellular or extracellular fluids can profoundly affect the way a cell behaves. Thus, the ability of a cell to regulate the flow of ions across its membranes is of paramount importance: whether or not a pancreatic islet cell releases insulin, a nerve cell releases a neurotransmitter, or an immune cell releases substances capable of killing bacteria—these, in addition to thousands of other cellular processes that keep us alive, all rely on the selective permeability of cell membranes to specific ions.

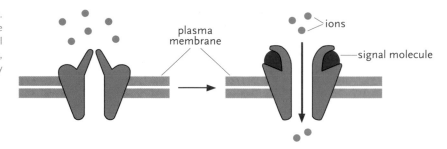

Figure 6.37. Ion-Channel–Linked Receptors. This schematic shows a closed membrane ion channel (left) and one opened by signal molecules that bind to specific receptors, allowing ions to pass (right). (© Drawing by Elles Gianocostas.)

Each type of ion channel has a distinctive molecular structure that enables it to control the passage of a specific ion. Scientists study these channels by using molecules that bind to them, so-called molecular probes, and that either enhance or inhibit their ability to function. Cone snail toxins, because of the enormous diversity of their ion channel targets and the remarkable potency and specificity in their binding ability, may be more important at the present time to ion channel research than molecules made by any other group of organisms.

Conopeptides have been used to advance our knowledge of many of the most fundamental processes of life, including skeletal muscle contraction, insulin secretion and the control of blood glucose, the workings of the retina, blood pressure regulation, and immune and kidney cell function, to name a few. But conopeptides have had their greatest impact in helping us understand the human nervous system.

Take the example of nicotinic acetylcholine (nACh) receptors that bind the molecule acetylcholine (they have been named "nicotinic" because they are also the site where nicotine molecules bind). These receptors activate ion channels that control the contraction of skeletal muscle, the activity of peripheral nerve endings of the autonomic nervous system (which regulate blood pressure, heart rate, and sweat glands, among many other processes), and the transmission of electrical impulses between some nerve cells in the brain. Each nACh receptor is composed of five subunits, and each of these five subunits, in turn, may be drawn from a pool of a dozen or so possible molecules, making the total number of available compositions for nicotinic acetylcholine receptors on the order of dozens. A recently discovered peptide from the snail *Conus bullatus* illustrates the exceptional ability of conopeptides to distinguish among the different nACh receptors. This peptide, only thirteen amino acids long, is 40,000 times more likely to bind to one form of an nACh receptor (described as "alpha-6/alpha-3/beta-2") than it is to another with almost the same composition ("alpha-4/beta-2").[243] Until conopeptides were used in research on nACh-activated ion channels in the brain, researchers could not distinguish between the various nACh receptor subtypes.[244] But conopeptides have now made possible the identification of fifteen different nACh receptors in the brain, yielding potential insights about the roles these receptors may play in Parkinson's disease, Alzheimer's disease, epilepsy, and alcoholism.[245–247]

Conopeptides have also been identified that bind to sodium, potassium, and calcium channels, as well as to ion channels controlled by the neurotransmitters serotonin and glutamate. Still other conopeptides attach themselves with great specificity to cell-surface receptors for neurotensin, vasopressin, epinephrine, and norepinephrine, all of which, upon stimulation, coordinate complex cellular responses.[248] With each passing month, new conopeptides are described in the scientific literature, each

Figure 6.38. The Cone Snail *Conus bullatus*. Once cone snails have detected their prey by siphoning and chemically analyzing the water around them, they extend their proboscis, fire a harpoon to subdue it, and envelop it to begin digestion. (© 2002 Charlotte M. Lloyd.)

one with its own unique characteristics. The fact that so much valuable biomedical information has been obtained by studying only 100 or so conopeptides in any depth, out of an estimated total of 70,000 to 140,000, demonstrates the vast, and still largely unexplored, potential these natural compounds have for illuminating some of the most fundamental aspects of how our bodies function in health and disease.

SHARKS

Shark Populations Threatened by Overharvesting

Sharks, and their relatives the rays and skates (known collectively as the elasmobranchs), are among the earliest vertebrates, having evolved in ancient seas around 400 to 450 million years ago. They include approximately 400 known species, but scientists do not know the exact number because of the difficulties involved in surveying the oceans. In recent decades, the exploitation of sharks has increased dramatically, with as many as seventy-three million sharks being caught each year.[249] Consequently, many species are now threatened. A recent study, using fishing records, of the northwest Atlantic Ocean has shown rapid declines in the populations of large coastal and open ocean (pelagic) sharks, with species such as the Scalloped Hammerhead (*Sphyrna lewini*), White Shark (*Carcharodon carcharias*), and Thresher Shark (*Alopias vulpinus*) down in number by more than 75 percent in the past fifteen years. The scientists who did this study concluded that "all recorded shark species, with the exception of makos, have declined by more than 50 percent in the past eight to fifteen years."[250] The same is true for other areas. In the Gulf of Mexico, for example, pelagic shark species have declined precipitously in recent decades, with species such as the Oceanic Whitetip Shark (*Carcharhinus longimanus*), which had been the most prevalent shark in the Gulf in the 1950s, dropping in population by 99 percent since then.[251] The IUCN lists twenty-three shark, skate, and ray species that are Endangered and nine that are Critically Endangered.

Sharks are more vulnerable to extinction than most other fish. For many, recovery may take decades after fishing pressure is reduced. This vulnerability and slow recovery is the result of sharks taking many years to reach sexual maturity, having long pregnancies, and producing relatively few young.[252] For example, the Spiny Dogfish Shark (*Squalus acanthias*) is not able to reproduce until it is ten to twenty years old, has a pregnancy that lasts as long as two years, and produces on average only six pups with each pregnancy.[253] The swordfish, by comparison, can produce a million eggs each year. The vertebrate with the most delayed sexual maturity may be the Dusky Shark (*Carcharhinus obscurus*), which is not able to reproduce until it reaches the age of twenty to twenty-five.[254]

Until recently, few people were interested in protecting sharks. For a very long time, sharks have been universally feared and maligned as vicious killers, thought to prey on people and to cause large numbers of deaths each year. In reality, many shark species are docile and rarely attack people. The two largest, for example—the Whale Shark and the Basking Shark—are harmless plankton eaters. In general, sharks that do attack may be mistaking people for their normal prey. There were sixty-one documented unprovoked shark attacks in 2004, with a total of seven fatalities worldwide, so for every human killed by a shark, approximately ten million sharks are killed by humans. Almost half of the attacks happen in North American waters, with most of these occurring in Florida.[255] The chance of dying from a shark attack, however, is exceedingly small, far less than that of being killed by a bee sting (90–100 deaths a year in the United States alone)[256] or lightning (average of eighty-four deaths per year in the United States alone from 1959 to 2003).[257]

People are doing more shark fishing in recent years for several reasons. First, shark meat has become increasingly sought after as a food, as populations of other traditional fisheries have collapsed. Sharks are eaten both as steaks and as "fish and chips" (in which dogfish shark meat is widely used) in places like the United Kingdom. Second, sharks are often caught incidentally (called bycatch) during fishing for swordfish and for tuna. Third, sharks may be killed for their teeth and jaws, because these can bring high prices as prized trophies—one Great White Shark jaw from South Africa, for example, was recently sold for US$50,000.[258] But the two main sources of shark overfishing in recent years have been the demand for shark fins and for shark cartilage.

For more than 2,000 years, Asian countries have traded in shark fins. Only in recent years, however, with the development of an affluent middle class in parts of Asia, particularly in China, has demand for shark fins increased to the extent that it now threatens many shark species. In Hong Kong, for instance, dealers imported more than six million kilos (13.2 million pounds, or 6,600 tons) of shark fins in 1995.[259] Shark fins are among the most highly valued of

I COULDN'T POSSIBLY WRITE *JAWS* TODAY. WE KNOW SO MUCH MORE ABOUT SHARKS—AND JUST AS IMPORTANT, ABOUT OUR POSITION AS THE SINGLE MOST CARELESS, VORACIOUS, OMNIVOROUS DESTROYER OF LIFE ON EARTH—THAT THE NOTION OF DEMONIZING A FISH STRIKES ME AS INSANE.

—Peter Benchley

Figure 6.39. Shark Fins Drying in the Sun, Prior to Being Sold. (© Adam Summers/Biomechanics Lab.)

marine fishery products; shark fin soup (considered by some to be a delicacy and to have medicinal and aphrodisiac properties) can sell for as much as US$200 a bowl (and some are eager to demonstrate that they can afford these prices), while top-quality fins can command US$700 or more per kilogram.[260] Consequently, thousands of vessels eagerly fish for sharks from Taiwan, Japan, and many other countries. To make the most efficient use of their vessel's limited refrigeration space, shark fin fisherman prefer to cut the valuable fins off and throw the much less valuable carcasses back into the ocean, leaving the sharks to die slow, painful deaths. This inhumane practice is finally being outlawed, with the sixty-three countries of the International Commission for the Conservation of Atlantic Tuna, for example, adopting in 2004 laws present in the United States, Canada, and several other countries that ban the landing of shark fins without their bodies attached.[261] This restriction has recently been extended to the eastern Pacific, but elsewhere the slaughter continues unabated, and monitoring and enforcement are often wanting.

Killing sharks for their cartilage is also contributing to the decimation of shark populations. The trade is driven by the belief that eating shark cartilage can cure cancer and other diseases, and this has fostered a brisk market for pills containing shark cartilage extract in the United States and Europe.

A collapse in shark populations has serious potential consequences. As the top, or "apex," predator of the open oceans, sharks, like other apex predators, are thought to help maintain the functioning of marine ecosystems and the diversity and structure of marine food webs, culling, for example, diseased and weak organisms from among their prey populations, such as Atlantic Herring (*Clupea harengus*) and Atlantic Mackerel (*Scomber scombrus*) in the case of Spiny Dogfish Sharks.[250,262] Sharply declining populations (by as much as 99 percent or more in some cases) of all eleven large shark species in coastal northwest Atlantic Ocean waters that prey upon other elasmobranches, such as rays, skates, and small sharks, are thought to have led to the destruction of the Bay Scallop (*Argopecten irradians*) fishery along parts of the Eastern Seaboard of the United States. With far fewer of their predators around, populations of some of these smaller elasmobranchs have exploded, like that of the Cownose Ray (*Rhinoptera bonasus*), which has wiped out century-old bay scallop beds in North Carolina.[263] This may be just one example of many to come of "trophic cascades" from plummeting large shark populations, where the loss of these apex predators will have major impacts on marine ecosystems and marine food production. It is hoped that we do not need to wait until more such impacts are discovered before better shark protection plans are implemented.

Potential Medicines from Sharks

Compounds found in sharks that are thought to have medicinal value fall into two categories—those contained in shark cartilage extracts, and the aminosterols, such as squalamine, a type of steroid.

SHARK CARTILAGE

Scientific studies on shark cartilage began in the early 1980s when Robert Langer and his colleagues at the Massachusetts Institute of Technology did research to determine if it contained substances that had anti-angiogenic activity (i.e., the ability to prevent the growth of new blood vessels). Langer, working with Judah Folkman and others, had originally found that cow cartilage contained such substances and that these substances stopped the growth of solid tumors that had been implanted into rabbits and mice. He then turned to sharks, because they provided a much larger source of cartilage. Their skeletons, in contrast to those in mammals, are composed entirely of cartilage. Langer used cartilage from Basking Sharks (*Cetorhinus maximus*; the second largest fish after Whale Sharks [*Rhincodon typus*] that, although reaching lengths of 40 feet or more, is harmless to people unless attacked; Basking Sharks are now rare because of a long history of exploitation). Basking Shark cartilage extract, like that from cows, was found to be a powerful anti-angiogenesis agent for tumors transplanted to rabbits, significantly inhibiting their growth.[264]

Since this experiment was performed, numerous other labs have looked into the anti-angiogenic and anticancer properties of shark cartilage, and there have been some promising results in animal models. As a result, many companies have sprung up and are selling enormous quantities of shark cartilage pills. Some of these companies have said that sharks do not get cancer and that it is shark cartilage that protects them. Two companies that have made such claims have been ordered by the U.S. Federal Trade Commission to stop these promotions, and one was fined US$1 million

Figure 6.40. Basking Shark (*Cetorhinus maximus*). (© Tony Sutton.)

for false advertising.[265] Let us look at the evidence about whether shark cartilage is an effective medicine.

Sharks, along with skates and rays, do, in fact, get several types of cancers, but there are no reliable data about whether their rates of developing cancers are similar to those in other animal groups or are unusually rare.[266] If their cancers were found to be rare, however, it might have little or nothing to do with whether or not they produced anticancer compounds in their tissues. Rather, it could be the result of sharks being largely pelagic, that is, they inhabit the open oceans, and are therefore exposed to lower concentrations of environmental carcinogens than they would be if they had spent most of their time in inland or coastal waters. Tumors in pelagic bony fish are generally rarer, for example, than they are in benthic (bottom-dwelling) bony fish that feed in polluted waterways.[266]

Very few studies (only five as of February, 2006) on the use of shark cartilage in human cancer subjects have been published in peer-reviewed journals. (See the National Cancer Institute's website, www.nci.nih.gov/cancertopics/pdq/cam/cartilage/HealthProfessional/page5.) Of these, the largest and most carefully controlled one, which studied incurable breast cancer patients and colorectal cancer patients, each randomly assigned to receive either shark cartilage or placebo, showed no difference in survival or in the quality of life between the two groups.[267] Another study used a shark cartilage extract called AE-941 (or Neovastat, made by Aeterna Laboratories in Quebec) and reported in a phase II trial in patients with renal cell carcinoma that those given more AE-941 had better survival rates than those given less. This was not, however, a randomized study, and it excluded from the analysis most of the patients who had originally been given the extract.[268]

Two randomized phase III clinical trials of AE-941 (in addition to chemotherapy and radiation therapy) have been approved by the FDA, one with patients with non-small-cell lung cancer, and the other in patients with metastatic renal cell carcinoma. The trial with renal cell carcinoma, although completed, was never reported in the peer-reviewed literature and so can be presumed to have shown no benefit for AE-941 in these patients.

Furthermore, although there have been claims that shark cartilage is effective in treating such disorders as psoriasis, macular degeneration, pain, osteoarthritis, and a number of other conditions, as Harvard Medical School's Consumer Health Information website concludes, "There is no scientific evidence to support its [shark cartilage] use for any medical condition."[269] Shark cartilage preparations are sold as dietary supplements and are therefore not strictly regulated as to strength, purity, or safety, so the consumer may be getting different quantities and qualities of active substances in different preparations, or none at all. The active ingredients in shark cartilage extracts (including AE-941) are presumably glycoproteins (i.e., proteins that have sugars attached to them), but neither they nor other compounds are identified in the published reports, so it is not clear what compounds are being given and in what concentrations. In addition, while claims have been made that the active ingredients in the extracts enter into the blood from the gastrointestinal tract, it is not clear whether they make it past the human stomach or, if they do, whether they are absorbed by the human intestine in sufficient quantities to be effective.

As with other therapeutic agents extracted from Nature, moves to identify and synthesize these compounds should be early and aggressive, so as to relieve the over-harvesting of the species or family of species from which they are obtained, especially if these organisms are threatened. In almost all cases, this can be accomplished. No such activity seems to be going on in shark cartilage companies.

Given the extreme pressure on shark populations, which have survived almost unchanged for 400 million years or more, and the importance of these apex predators to marine ecosystems, there should be extremely compelling evidence of the efficacy of shark cartilage extracts, or of the inability to identify or synthesize the active compounds in them, to justify the slaughter of hundreds of thousands of sharks. Until such efficacy is demonstrated, this chapter's authors believe the wide-scale harvesting of sharks by various companies for their cartilage is irresponsible and unethical. And until favorable results of carefully controlled clinical trials using shark cartilage to treat human cancers and other diseases are reported in peer-reviewed journals, the best advice to consumers contemplating taking these extracts is *caveat emptor.*

SQUALAMINE

Squalamine was first isolated in 1993 when researchers began looking for compounds with antimicrobial activity in Spiny Dogfish Sharks (*Squalus acanthias*).[270] This search was undertaken to see whether sharks, like some other organisms, including the African Clawed Frog (*Xenopus laevis*), pigs, mice, and humans, all of which produce potent antimicrobial peptides in their stomachs, did so as well. The thought was that sharks were likely to possess such compounds as part of

Figure 6.41. Spiny Dogfish Shark. (© Scott Boyd, Emerald Sea Photography.)

their innate immunity to complement their evolutionarily ancient adaptive immune systems (see discussion of innate immunity in chapter 5, page 196). Rather than peptides, however, a group of compounds called aminosterols, which had not been previously identified in other animals, was discovered. The most abundant of these was squalamine, present in all tissues of the dogfish shark but found in the greatest concentrations in the liver. Although initial experiments with squalamine and other aminosterols required harvesting them directly from a dogfish shark, these compounds were eventually made synthetically, and this made possible several lines of research.

Antimicrobial Activity

Early experiments showed that squalamine was a potent antibiotic for a variety of bacteria, as well as for some fungi and protozoa.[270] Other compounds patterned after squalamine were later manufactured, and these, too, were powerful broad-spectrum antibacterial agents as well as fungicides. Some even demonstrated the ability to kill methicillin-resistant *Staphylococcus aureus* and vancomycin-resistant *Enterococcus faecium*,[271] a bacterium that can cause bloodstream, urinary tract, skin, and abdominal infections and that has become deadly because of antibiotic resistance (see section in chapter 4 on vancomycin, page 139).

Figure 6.42. Wet Adult Macular Degeneration. (a) Photo of normal retina. (b) Photo of a retina with wet adult macular degeneration. (© Eye Centers of Louisville, D.B.A., Bennett & Bloom Eye Centers.)

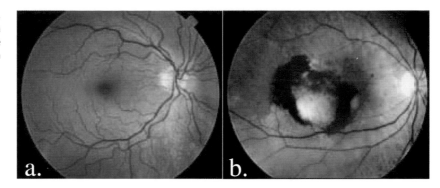

Normal Vision

Age-related Macular Degeneration

Figure 6.43. Wet Adult Macular Degeneration. (a) Normal vision. (b) What one with wet AMD would see. (Courtesy of the U.S. National Institutes of Health.)

Despite preliminary studies suggesting that it would be safe when given at doses sufficient to be effective as an antibiotic, squalamine's development as an antimicrobial, and that of related compounds, was interrupted by the subsequent discovery of side effects in experimental animals. These included a regression of blood vessels in chick embryos and in frog tadpoles, presumably because of an anti-angiogenic effect,[272] and a cessation of eating in mice, rats, monkeys, and dogs, because of an appetite suppressant effect. As a consequence of these unanticipated effects, the aminosterol drug development program of Genaera Pharmaceuticals, the company working on these compounds, was redirected toward developing medicines that took advantage of these and other properties of aminosterols—their anti-angiogenic, antitumor, appetite-suppressant, and antidiabetic activities.

Anti-angiogenesis Activity

Adult (also called age-related) macular degeneration (AMD) is the leading cause of blindness in the Western world, affecting between twenty and twenty-five million people worldwide, figures that are expected to triple in the next thirty to forty years.[273] AMD has two types—the "dry" type, which is common but mild, and the "wet" type, which, while accounting for only about 10–15 percent of all cases, is responsible for 90 percent of the severe loss of vision associated with AMD. Wet AMD results in the growth of new blood vessels in the retina and bleeding into the macula (a small part of the retina that can see fine detail), gradually resulting in its degeneration.

Based on studies in which squalamine was shown to inhibit the process of angiogenesis in the eyes of rats[274] and primates,[275] it was evaluated for safety and efficacy in people with wet AMD. After four months of squalamine treatment, most patients who received the medicine had no progression of their disease, and some had an improvement in vision.[276] Based on these encouraging initial results, a phase II randomized trial began in 2005 to evaluate squalamine in patients with wet AMD, both when used alone and in combination with Visudyne, an approved treatment for AMD. The results of this trial may help determine whether squalamine will help improve the vision of millions of people with this disease.

Antitumor Activity

Angiogenesis, or the growth of new blood vessels, has been shown to be essential for the growth of human tumors and for their metastasis (i.e., their ability to spread to distant sites throughout the body), so it was predicted that squalamine's ability to inhibit the growth of several types of transplanted tumors in mice and rats was the result of its potent anti-angiogenic activity. Several studies are under way to test whether squalamine, by itself or when used with other chemotherapeutic agents, shows antitumor activity. One is a phase I/IIA trial with advanced non-small-cell lung carcinoma, which has indicated that squalamine may improve survival when added to the standard regimen of carboplatin and paclitaxel (Taxol) for these patients.[277] In 2001, the FDA granted squalamine "orphan drug approval" (an approval designed to promote research and the development of drugs for diseases that affect fewer than 200,000 people in the United States) for the treatment of advanced ovarian cancer. Trials for other cancers are also being planned.

Appetite Suppressant Activity

During the extraction of squalamine from the liver of the dogfish shark, other structurally similar aminosterols were also discovered. One of these, designated as MSI-1436 by Magainin Pharmaceuticals (now a part of Genaera Pharmaceuticals), resulted in significant weight loss when tested in rodents, dogs, and monkeys.[278] Subsequent studies showed that MSI-1436, when given to mice and rats, resulted in major reductions in food and fluid intake, with consequent weight loss, but without dehydration or electrolyte imbalance. Unlike caloric deprivation, which causes animals to slow down metabolically, the reduction in food intake induced by MSI-1436 had no effect on their basal metabolic rates, their overall level of motor activity, or their behavior. MSI-1436, when administered to *ob/ob* and *db/db* mice (these are mice with genetic mutations that result in their becoming obese and developing diabetes), controlled their weight gain, preferentially increasing their metabolism of fat tissue and correcting their high blood sugar levels.[278] MSI-1436 appears to send the brain's feeding circuits a message that the body is storing too much energy as fat and that it can safely deplete these stored reserves. Given the epidemic of obesity and of type 2 diabetes that results from it in the United States and other countries (see also section on bears, above), and that many obese, diabetic patients require medication to help them suppress their appetites, in addition to regimens of dieting and exercise, an agent like MSI-1436 could become a breakthrough for the treatment of these conditions.[279]

Sharks and Biomedical Research

Sharks have been important research models primarily in two areas of investigation. The first takes advantage of their unique rectal salt-secreting gland as a model for fluid and electrolyte balance; the second involves study of their immune systems, because they have the most evolutionarily ancient adaptive immune system of any animal.

THE SALT GLAND OF SQUALUS ACANTIIIAS

The observation that the concentrations of various salts in the blood plasma of all vertebrates, including humans, mimic those of seawater led the Canadian biochemist Archibald Macallum to speculate that vertebrates largely retained the composition of seawater, which had been bathing their tissues when they first evolved, in the blood of their closed circulations. This was true for those animals that had lived in salt and fresh water, as well as those that had moved onto the land.[280] It is still the case today. A central challenge in these various environments has always been to maintain the right balance of water and salt for cells and organ systems to function. In addition to developing kidneys to regulate this balance, many animals, notably sharks, reptiles, and some marine birds, have developed specialized organs in other parts of their bodies designed specifically to manage salt concentrations in their blood.[281]

The salt gland of the Spiny Dogfish Shark (see figure 6.41) has become the prime model system for understanding how these salt-regulating organs work. Over the past forty-five years, mostly at the Mount Desert Island Biological Laboratory in Maine, research with the dogfish salt gland has led to a better understanding of the human kidney, and of some diuretics (medicines that increase the kidney's removal of water from the body, used to treat high blood pressure or congestive heart failure). The mechanism of action of furosemide, for example, one of the most important diuretics for the treatment of congestive heart failure, was partially worked out using the Spiny Dogfish Shark salt gland.[282]

The salt gland has also furthered our understanding of how chloride gets transported across membranes, which may be applicable to the human disease cystic fibrosis.[283] This devastating disease results from a mutation in the gene that encodes the cystic fibrosis transmembrane regulator (CFTR), a protein that controls chloride transport at the surface of secretory cells and that is present in, among other places, the respiratory tract and pancreas. Malfunction of these chloride channels results in markedly thickened secretions in both organs, which plug up ducts, resulting in a destruction of the pancreas by the patient's own digestive enzymes, and in blocked airways that make breathing difficult and that provide fertile ground for infections to take hold. The defective CFTR is also thought to cause higher salt concentrations at the surface of cells lining the respiratory system.[282] Both the thickened mucus and the saltier environment may interfere with the cystic fibrosis patient's ability to fight some life-threatening bacterial pneumonias, perhaps in part by blocking the activity of some naturally occurring antimicrobial peptides that line the inner surface of the lung.[284,285] The regulation of chloride transport in the salt gland of the Spiny Dogfish is similarly controlled by a CFTR-like gene, so studying chloride transport in the salt gland may yield important insights for treating cystic fibrosis.[283,286]

Polycystic kidney disease is another genetic disease and is quite common, afflicting as many as six million people worldwide.[287] Like cystic fibrosis, it may also involve an abnormality in chloride transport. In this condition, hundreds of fluid-filled cysts form in the kidney, and as they enlarge, through the process of chloride secretions into the cysts, they destroy normal kidney tissue and eventually compromise kidney function. A majority of patients with this disease develop kidney failure by the time they are in their seventies, making it one of the leading causes of end stage renal disease and of the need for dialysis.

Somatostatin, a hormone made by the pancreas, the gastrointestinal tract, the nervous system, and the thyroid, is known to inhibit secretory activity in the pancreas and in other organs. When it was discovered in nerves of the rectal gland of the Spiny Dogfish Shark and shown to inhibit chloride secretion there,[288] and when somatostatin receptors were discovered in the human kidney,[289] researchers decided to try somatostatin in patients with polycystic kidney disease.

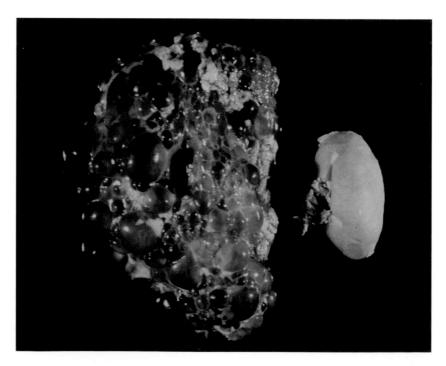

Figure 6.44. Polycystic Kidney Disease. A kidney with severe polycystic disease is on the left. A normal kidney is on the right. (© Polycystic Kidney Foundation.)

A preliminary study has shown that a derivative of somatostatin called octreotide is safe when given to these patients, and that it seemed to slow enlargement of their kidneys, presumably by preventing chloride secretion and growth of their cysts.[290]

IMMUNE SYSTEMS

Sharks have also been of great interest to scientists who study the origins and functions of the human immune system. Around 400–450 million years ago, they developed adaptive immune systems. In contrast to innate immune systems, which can be quite specific in their responses to some invading organisms, this more specialized and complex immunity can produce cells capable of attacking a much broader array of targets. Sharks, for example, are able to rearrange genes in seemingly limitless combinations in order to make antibodies that are able to recognize and bind to a multitude of targets on pathogens. The rearrangement of these genes is under the control of yet another set of genes known as the recombinase-activating genes, or RAGs, which are thought to have been originally transferred from bacteria into sharks at about the time that elasmobranchs first evolved. Studies of these first RAGs in living sharks, such as Nurse Sharks (*Ginglymostoma cirratum*), Bull Sharks (*Carcharhinus leucas*), and Sandbar Sharks (*Carcharhinus plumbeus*), have been instrumental in understanding how RAG genes function.[291]

Sharks were also the first organisms to possess a remarkable group of genes collectively known as the major histocompatibility complex, or MHC, present in all species, including humans, that evolved after sharks.[292] When cells become infected, MHC molecules capture a small piece of the invading microbe and display it on their surfaces, so that it becomes accessible to immune cells that pass by. As such, the MHC molecules serve as molecular signposts that contribute to the immune system's specificity. Not only does the surface of every cell in our bodies have many different MHC proteins, but each individual has a distinctive MHC repertoire.

The ability to distinguish one's own repertoire of MHC molecules, that is, to identify one's own cells from others, has great implications for organ transplantation, because transplanted cells have their own unique set of MHC molecules. At present, the success of a transplanted organ relies upon drugs that suppress the immune system so that one's body does not attack and reject it. Through studies of the shark immune system, we have learned a great deal about the molecular mechanisms involved in this rejection process, so that someday we may be better able to shut it down.

Because sharks were the first organisms to possess all the elements of the adaptive immune system, they exist as the template upon which all subsequent alterations to their basic, yet extraordinary, ability to defend themselves against disease have evolved. Indeed, they still use two kinds of antibodies that we are no longer capable of manufacturing.[293] What potential these creatures may still hold to further our knowledge about immunity is being rapidly depleted with the mass slaughter of sharks and the endangerment of shark species worldwide.[294]

HORSESHOE CRABS

Overexploited Horseshoe Crabs

Among the many fascinating organisms detailed in this book, the four species of horseshoe crab must be some of the most extraordinary. They have four eyes (as well as six other light-detecting organs, one of which can be found in the tail), six pairs of legs, and blood that turns a brilliant cobalt blue when exposed to air. Across its life span, each crab will molt more than two dozen times, a convenient adaptation to the problem of having a bevy of hitchhikers that latch onto its shell. All told, more than a dozen species—from *Bdelloura* flatworms (known as the "*Limulus* worm") and Atlantic Slipper Shells (*Crepidula fornicata*) to Blue Mussels (*Mytilus edulis*) and various bryozoans (encrusting colonial animals sometimes called seamoss)—have been found riding on horseshoe crabs, and that's not counting the microbes.

Horseshoe crabs also have an ancient history: The ancestors to the present-day species appeared sometime between 250 and 300 million years ago and miraculously survived the great Permian extinction that eliminated an estimated 95 percent of all marine species, including the organisms from which they had evolved, the trilobytes. A look at the crab's lineage reveals that it is much more a spider or scorpion than it is a crab. Belonging to the arthropod phylum, which includes the insects, spiders, and crustaceans, the four species of horseshoe crabs, because of their uniqueness, fall into their own class, the Merostomata, meaning "legs attached to the mouth." To the best of scientists' ability to tell, the crabs have remained essentially unchanged over hundreds of millions of years.[295]

Horseshoe crabs inhabit the Atlantic coast of North America and the shores of Southeast Asia, stretching from the Bay of Bengal to the seas southwest of Japan. While still abundant in many locales, populations have been entirely wiped out from others. On the whole, population census data are lacking. It is known, however, that these creatures take about a decade to reach maturity, and out of roughly

Figure 6.45. Horseshoe Crabs (*Limulus polyphemus*) Mating. There is one female under this pile of male horseshoe crabs. (© 1993 by Paul Erickson.)

90,000 eggs laid per female each breeding season, only around ten offspring survive, making the species vulnerable to overfishing. It is also known that over-harvesting in the Pacific has led commercial fisheries to exploit Atlantic stocks, where some data suggest that populations of the local species *Limulus polyphemus* may be in sharp decline, in spite of the stringent catch limits that have been imposed for some nesting beaches. The crabs historically were harvested for their shells, which were pulverized and used as fertilizer. Today they are used as bait for eel and whelk fisheries.

Ecologists have particular concern for the decline of *L. polyphemus* because their eggs, laid on beaches, provide the main source of food for millions of migratory shorebirds whose populations have fallen off markedly in recent years. Research by Guy Morrison at Carleton University in Canada, for example, has documented a 98 percent drop in the populations of the North American Red Knot (*Calidris canutus rufa*) since the 1980s. The rapid decline of Red Knot populations, which has put them on the verge of extinction in many places along their annual migratory route, from Tierra del Fuego at the southernmost tip of South America to the Arctic Circle, likely reflects the paucity of horseshoe crab eggs in the Delaware Bay that these birds rely upon to fuel their nearly 10,000 mile (16,000 km) journey.[296]

The horseshoe crab, in addition to its pivotal role in the ecology of migratory shorebirds, has exceptional value to human health, as will be presented below.

Horseshoe Crabs and New Medicines

For more than fifty years, the blood of horseshoe crabs has been known to be capable of killing bacteria, but it took decades for researchers to find the responsible antimicrobial peptides. Several classes of peptides have been identified, including the tachystatins, tachyplesins, and polyphemusins, all capable of killing a wide array of Gram positive and Gram-negative bacteria. Scientists have studied these compounds to understand their specific structure and function, and these studies have been applied to a more general understanding about how antimicrobial peptides work and to designing more effective antibiotic therapies.[297–299]

Had these peptides been the only contribution that horseshoe crabs made to the development of pharmaceuticals, they, as a group, would be no more distinguished than hundreds of other organisms that possess unique antimicrobial peptides. But this is not the case. Fortunately for us, the horseshoe crab's blood contains a plethora of other novel molecules, which have the potential to aid in the treatment of several other major diseases. A molecule called T140, for example, derived from one of the polyphemusins, polyphemusin II, locks onto a receptor that HIV viruses use to gain access into immune cells, thereby blocking their entry. Though clinical trials have yet to begin with T140, preclinical studies have shown that it inhibits replication of the HIV virus, at levels comparable to those for AZT.[300]

The receptor that T140 binds to is known as CXCR4, an extremely important molecule. In addition to its relationship with HIV, it guides immune cells to where they are needed. It also directs the movement of many other kinds of cells—blood

Figure 6.46. Harvesting Lysate from Horseshoe Crabs (*Limulus polyphemus*). (© 2005 Frans Lanting.)

cells and neurons among them—to their proper locations during embryonic develop-ment. Experiments with T140 are preliminary, but the molecule has shown promise in both preventing the spread of leukemia, prostate cancer, and breast cancer and as a possible treatment for rheumatoid arthritis.[301,302]

But by far the greatest gift that horseshoe crabs have given to humanity is not a pharmaceutical. In the late 1950s, Frederik Bang and Jack Levin, working at the Marine Biological Laboratory in Woods Hole, Massachusetts, found that cells in the crab's blood called amebocytes formed a clot whenever they encountered endotoxin, a substance found in the cell wall of Gram-negative bacteria. This discovery led to the *Limulus* amebocyte lysate (called LAL) test (a lysate refers to a solution containing the contents of cells whose membranes have burst open), which is very widely used to detect the presence of Gram-negative bacteria in medical devices and in inject-able solutions.[303] It is also being used as a screen to detect Gram-negative bacteria in body fluids, for example, in the cerebrospinal fluid of patients suspected of having Gram-negative bacterial meningitis. How good is this test at detecting endotoxin? The LAL assay is so sensitive that it can detect 1 picogram (one trillionth of a gram, or 0.000 000 000 001 grams—there are 454 grams in a pound) of bacterial endotoxin per milliliter of solution—which, according to Charles River Labs (one of the test's manufacturers), is roughly equivalent to finding one grain of sugar in an Olympic-sized swimming pool!

Horseshoe Crabs and Biomedical Research

Horseshoe crabs are an organism of choice for studying actin, a structural protein that can assemble into molecular motors, along with other proteins such as myosin, to contract muscles, alter a cell's shape or enable it to move, and allow sperm to fertilize eggs. Research on *Limulus* sperm, begun in the 1970s, has focused on the events that lead to entry of the sperm into the egg. When a sperm first encounters an egg, it triggers what is called an acrosomal reaction. The first event in this reaction is a release of enzymes from the sperm tip that digest a jelly-like substance that coats the egg; the last involves the sperm's DNA entering the core of the egg. By studying *Limulus* sperm, where it is possible to see the actin assembling itself, and where the steps of the acrosomal reaction appear to occur mostly one after the next instead of simultaneously as they do in other organisms, scientists have been able to better understand the process of egg fertilization and how actin makes it possible.[304]

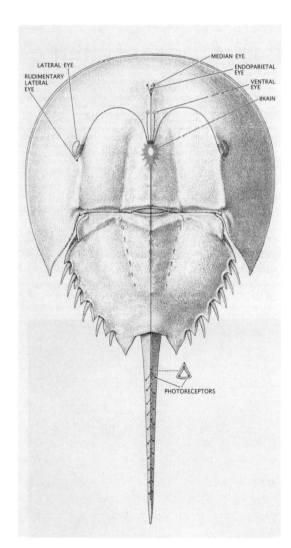

Figure 6.47. Eyes and Photoreceptors in the Horseshoe Crab (*Limulus polyphemus*). (From Robert B. Barlow, Jr., What the brain tells the eye. *Scientific American*, 1990;262(4). Image used with permission of Nelson H. Prentiss.)

Research labs around the world continue to rely on horseshoe crabs in other ways as well. Robert Barlow, a pioneer of *Limulus* research, has looked at how the crabs adapt their vision with cues from an internal clock, and how their eyes are fine-tuned to identify other crabs, work that has implications for human vision and for figuring out how our own biological clocks work.[305] Finally, because of its ancient origins, the innate immune system of horseshoe crabs has been studied to determine the evolution of the complement system, a group of some thirty circulating proteins that are used by crabs and by humans to kill bacteria.[306]

The eyes of the horseshoe crab *Limulus polyphemus* have proven to be an exceptional model to investigate human vision. Though each crab has ten eyes, the two lateral eyes (so-called because they sit along the sides of the crab) have received the most attention. The lateral eyes are compound eyes, meaning that they consist of many—about 500 to 1,000 in an adult *Limulus*—miniature, individual, light-detecting structures that work together to form the eye as a whole. Most arthropods, however, have compound eyes. What makes the lateral eyes of *Limulus* remarkable is that its optic nerve can be dissected down to individual fibers, making it possible to determine how input from one detector is integrated with those from others. In addition, the *Limulus* optic nerve can grow to be four inches in length, allowing for easier dissection, and it lies directly below the crab's shell, making it readily accessible.

The most famous discovery made in *Limulus* is of a phenomenon known as lateral inhibition, a fascinating mechanism that eyes use to exaggerate contrasts along the edges of objects, enabling crabs, and humans, to see these objects more clearly. Look at the areas in figure 6.47 at the junction of two bands and notice that a lighter appearing gray band appears even lighter in the region just next to an adjacent darker band, when compared to the rest of the stvripe. Each band is, in fact, exactly the same shade throughout, but our eye's circuitry fools us into believing that

Figure 6.48. Mach Bands Demonstrating Lateral Inhibition. (© Illustration by Elles Gianocostas.)

it is not. (To convince yourself of this, cover the picture so that only one band shows.) Lateral inhibition refers to the ability of one light-detecting cell to inhibit input from its neighbors, which accounts for why we see the exaggerated differences in shading in the diagram—the cells that see the lighter band act to inhibit those seeing the adjacent darker band, making it appear still darker and, as a result, making the lighter band appear lighter.

H. Keffer Hartline won the Nobel Prize in Physiology or Medicine in 1967 for his research on lateral inhibition, and on other visual processes, work made possible by studying the lateral eyes in *Limulus*.[307] Although Hartline was supposedly known for telling his students "to avoid vertebrates because they are too complicated, to avoid color vision because it is much too complicated, and to avoid the combination because it is impossible," he seems not to have entirely followed his own advice. After many years of research with *Limulus*, he began experimenting with vertebrate eyes, including those of frogs, relying on knowledge he first acquired from *Limulus*.

CONCLUSION

The organisms presented in this chapter are all remarkable in their own right—for their beauty and for their startling genetic, molecular, anatomical, biochemical, physiological, and behavioral complexity. However, they have become the focus of this chapter not for these reasons, but for what they show us about how they—and, indeed, how so many other organisms that could have readily been substituted for them—contribute to human health and to our well-being. Their scarcity, and humanity's role in causing it, is cause for alarm, for if we cannot protect these most visible, these most well-known creatures, we will be hard pressed to preserve all the others upon which our lives utterly depend.

◆ ◆ ◆

SUGGESTED READINGS

AMPHIBIANS

Amphibian (DK Eyewitness Books), Barry Clarke and Laura Buller. Dorling Kindersley Publishing, New York, 2005.

Amphibian Species of the World, research.amnh.org/herpetology/amphibia/index.php

Amphibians: The World of Frogs, Toads, Salamanders, and Newts, Robert Hofrichter (editor). Firefly Books, Buffalo, New York, 2000.

Checklist of Amphibian Species and Identification Guide, Northern Prairie Wildlife Research Center, www.npwrc.usgs.gov/resource/herps/amphibid/index.htm.

Declining Amphibian Populations Task Force, www.open.ac.uk/daptf/index.htm.

FrogLog Index, www.open.ac.uk/daptf/froglog/.

FrogWeb: Amphibian Declines and Malformations, www.frogweb.gov/.

BEARS

Bears: A Year in the Life, Matthias Breiter. Firefly Books, Buffalo, New York, 2005.

Bears (Wildlife Series), Daniel Wood. Whitecap Books, North Vancouver, British Columbia, 2005.

Black Bear Home Page, www.bear.org/Black/BB_Home.html.

Black Bears: A Natural History, Dave Taylor. Fitzhenry & Whiteside, Markham, Ontario, 2006.

Polar Bear International, www.polarbearsinternational.org/.

The World of the Polar Bear, Norbert Rosing. Firefly Books, Buffalo, New York, 2006.

PRIMATES

Eating Apes, Dale Peterson. University of California Press, Berkeley, 2003.

Gorillas in the Mist, Dian Fossey. Houghton Mifflin Company, New York, 1983 (2000 edition).

Great Ape Odyssey, Biruté Mary Galdikas (photographs by Karl Ammann). Harry N. Abrams, New York, 2005.

James and Other Apes, James Mollison. Chris Boot, London, 2004.

My Life with the Chimpanzees. Jane Goodall. Simon & Schuster, New York, 1996 (2002 paperback edition).

Primate Factsheets, pin.primate.wisc.edu/factsheets/.

Primate Photo Gallery, www.primates.com/.

Primates: Amazing World of Lemurs, Monkeys, and Apes, Barbara Sleeper (photographs by Art Wolfe). Chronicle Books, San Francisco, 1997.

A Primate's Memoir: A Neuroscientist's Unconventional Life among the Baboons, Robert M. Sapolsky. Simon & Schuster, New York, 2001.

Reflections of Eden: My Years with the Orangutans of Borneo, Biruté M.F. Galdikas. Hachette Book Group, Lebanon, Indiana, 1995.

Walker's Primates of the World, Ronald M. Nowak, Russell A. Mittermeier, Anthony B. Rylands, and William R. Konstant. Johns Hopkins University Press, Baltimore, Maryland, 1999.

GYMNOSPERMS

Ginkgo, nccam.nih.gov/health/ginkgo/.

Gymnosperms, www.biologie.uni-hamburg.de/b-online/library/knee/hcs300/gymno.htm.

Gymnosperms: A Reference Guide to the Gymnosperms of the World, Hubertus Nimsch. Balogh Scientific Books, Champaign, Illinois, 1995.

Origin and Evolution of Gymnosperms, Charles B. Beck (editor). Columbia University Press, New York, 1988.

CONE SNAILS

Cone Snails and Conotoxins Page, grimwade.biochem.unimelb.edu.au/cone/index1.html.

SHARKS

Biology of Sharks and Their Relatives, Jeffrey C. Carrier, John A. Musick, and Michael R. Heithaus (editors). CRC Press, Boca Raton, Florida, 2004.

The Encyclopedia of Sharks, Steve Parker and Jane Parker. Firefly Books, Buffalo, New York, 2005.

A Laboratory by the Sea: The Mount Desert Island Biological Laboratory 1898–1998, Franklin H. Epstein (editor). River Press, Rhinebeck, New York, 1998.

Mote Marine Laboratory Center for Shark Research, www.mote.org/index.php?src=gendocs&link=SharkResearch&submenu=Research.

NOAA Fisheries: Shark Web Site, www.nmfs.noaa.gov/sfa/hms/sharks.html.

The Shark Almanac: A Fully Illustrated Natural History of Sharks, Skates, and Rays, Thomas B. Allen. Lyons Press, Guilford, Connecticut, 2003.

Sharks, Skates, and Rays: The Biology of Elasmobranch Fishes, William C. Hamlett. Johns Hopkins University Press, Baltimore, Maryland, 1999.

HORSESHOE CRABS

The American Horseshoe Crab, Carl N. Shuster, H. Jane Brockmann, and Robert B. Barlow (editors). Harvard University Press, Cambridge, Massachusetts, 2004.

Crab Wars: A Tale of Horseshoe Crabs, Bioterrorism, and Human Health, William Sargent. University Press of New England, Lebanon, New Hampshire, 2006.

Extraordinary Horseshoe Crabs (Nature Watch), Julie Dunlap. Carolrhoda Books, Minneapolis, Minnesota, 1999.

Horseshoe Crab, www.horseshoecrab.org/.

Smithsonian Marine Station at Fort Pierce, www.sms.si.edu/IRLSpec/Limulu_polyph.htm.

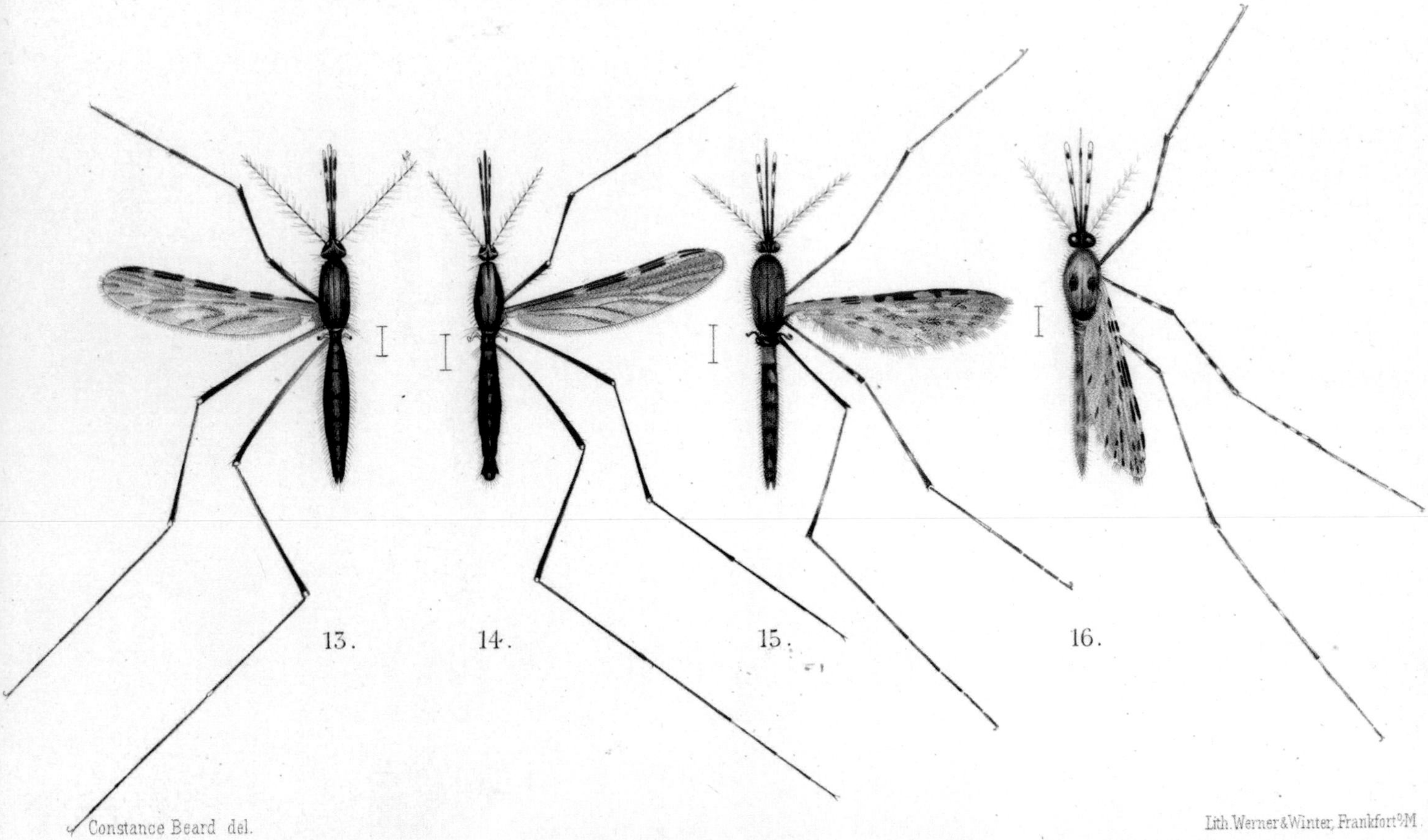

Plate IV.

13.   14.   15.   16.

Constance Beard del.   Lith.Werner&Winter, Frankfort⁰⁄M.

CHAPTER 7

ECOSYSTEM DISTURBANCE, BIODIVERSITY LOSS, AND HUMAN INFECTIOUS DISEASE

David H. Molyneux, Richard S. Ostfeld, Aaron Bernstein, and Eric Chivian

Man is a part of nature, and his war against nature is inevitably a war against himself.

—RACHEL CARSON

When we develop an infectious illness, we tend to believe that we caught it from another person, who in turn caught it from someone else, and that the germ that made us ill had never resided in any species other than our own. But this belief, it turns out, is false more times than not. For most human infectious diseases—some 60 percent— the pathogen has lived and multiplied in other organisms before having been transmitted to people.

Such pathogens are integral parts of ecosystems, complex networks of other organisms that govern their emergence, transmission, and spread. Included in these networks are, for example, the insect vectors that are able to transmit pathogens to humans, the reservoir or host species that serve as sites for pathogens to multiply in and be available for such transmission, and other species that support, or interfere with, the interactions among pathogens, vectors, and hosts. It should come as no surprise, then, that biodiversity loss, which can change the abundance of and relationships among these organisms and their physical and chemical environments, has major implications for the spread of human infectious disease.

About 132 of 175, or around 75 percent, of emerging infectious diseases have reached humans having been first associated with other organisms (see box 7.1). Emerging infectious diseases are diseases that are increasing in incidence or

TYPES OF INFECTIOUS AGENTS AND MODES OF TRANSMISSION

Pathogens, which will also be referred to as *infectious agents* in this chapter, come in a variety of forms. Those that are responsible for human infections can be divided into several groups: viruses, bacteria, fungi, single-celled protozoa such as the parasite that causes malaria, worms (including roundworms, flatworms, and tapeworms), and the recently discovered *prions*, which are infectious proteins. Some 1,415 different pathogens are known to cause disease in humans—217 viruses and prions, 538 bacteria, 307 fungi, 66 protozoa, and 287 worms, though new agents are constantly being identified, as occurred with the sudden acute respiratory syndrome virus, or SARS, outbreak of 2002.[a] This emphasizes that there is likely to be a vast amount of microbiological diversity that remains to be discovered and characterized. Of these 1,415, almost two-thirds occur mainly in nonhuman vertebrate hosts, called *zoonotic reservoirs*. In most of these cases, people who become infected are "accidental" victims.

To attack humans, infectious agents use one or more modes of transmission. They can enter our bodies by penetrating our skin or by being ingested when we drink contaminated water or consume contaminated food. They can also invade us during sexual intercourse or when we inhale them into our lungs. Skin penetration can occur from wounds (e.g., bites or scratches inflicted by a mammal), the bites of *insect vectors* (e.g., mosquitoes, ticks, or other insects that transfer infectious agents from one host to another), or by the burrowing of a parasite (e.g., schistosomes and hookworm larvae). Some infectious agents may have several modes of entry. The bacterium that cause tularemia ("rabbit fever"), for example, can be transmitted by tick bites, inhalation, or eating infected meat.

geographic range and include new types of infections resulting from changes in recognized organisms, old infections spreading to new areas or populations, and previously unrecognized infections (which are typically found in ecologically disturbed areas). The realization that such a high percentage of these human diseases arise from animals, when examined in the context of the present biodiversity crisis, suggests that species loss and the resultant effects on ecosystems may already be dramatically altering the landscape of human infectious diseases.

The diversity of human infectious agents and the diseases they can cause make generalizing about the effects of biodiversity loss and ecosystem disturbance on human health difficult. Nevertheless, patterns do exist, and in this chapter we present examples that illustrate some of the general principles that have been identified.

ECOSYSTEM DISTURBANCES AND THEIR EFFECTS ON INFECTIOUS DISEASES

In this chapter, we use the term *ecosystem disturbance* to refer to a change in an ecosystem's composition or function that occurs as a result of human activities. Although natural processes can also alter ecosystems in fundamental ways, in this chapter we emphasize human-caused disturbances. Some ecosystem disturbances may

Table 7.1. Summary of Infectious Diseases Presented in This Chapter (all statistics are global unless stated otherwise)

Disease	Infectious Agent	Species Involved in Disease Transmission and Mode of Transmission	Comments
Argentine hemorrhagic fever	Junin virus	Contact with feces, urine, or saliva of infected Corn Mice (*Calomys musculinus*), when fecal matter in dust becomes airborne during grain processing or when rodents are accidentally caught in harvesters	The Junin virus is responsible for severe illness in those who go unvaccinated. Bleeding from the gastrointestinal tract and urinary tract can occur, in addition to neurological symptoms such as tremors, seizures, and coma.
Babesiosis	Species of the infectious protozoan *Babesia*, especially *B. microti* (in North America) and *B. divergens* (in Europe)	Bites by species of infected ixodid ticks, especially *Ixodes scapularis* (formerly known as *I. dammini*) and *I. ricinus*	Symptoms develop gradually and include fever, chills, muscle aches, and anemia.
Cholera	*Vibrio cholerae*	Consumption of contaminated food or water	Because the *Vibrio* bacterium causes profuse, watery diarrhea and vomiting, cholera can, if left untreated, cause fatal dehydration in 25–50% of those infected. In 1992, the seventh cholera pandemic began with the appearance of a new strain of *Vibrio cholerae* (O139) that is present across Asia, Africa, and South America. The bacteria are thought to arrive in new coastal areas via bilge water of ships or through the conveyance and dumping of untreated sewage water, a practice that occurs in both the developing and developed world. The present pandemic has hit Africa hardest, where 80% of the approximately 100,000–200,000 yearly cases have occurred since 1995.
Cryptosporidiosis	Primarily *Cryptosporidium parvum* and *C. hominis*	Consumption of food or water contaminated with *Cryptosporidium*	*Cryptosporidium parvum* is a microscopic, protozoan parasite that forms cysts in the host intestine. These cysts are present in massive quantities in the feces of some animals, particularly livestock, and they can contaminate watersheds. Symptoms include diarrhea and abdominal cramping, sometimes associated with fever.
Dengue fever	Dengue fever virus	Mostly bites from infected *Aedes aegypti* mosquitoes	Known as "break-bone" fever, dengue fever produces a debilitating illness with significant fever, body aches, and headache. Subsequent infections with a different strain of the virus can cause a hemorrhagic fever, which, if left untreated, carries a 50% mortality rate. The hemorrhagic fever is particularly dangerous in children. Dengue fever virus is the world's most prevalent mosquito-borne virus, with fifty to one hundred million cases each year worldwide.

Continued

TABLE 7.1. (CONTINUED)

DISEASE	INFECTIOUS AGENT	SPECIES INVOLVED IN DISEASE TRANSMISSION AND MODE OF TRANSMISSION	COMMENTS
Hantavirus pulmonary syndrome	Hantavirus	Contact with urine or feces of infected Deer Mice (*Peromyscus maniculatus*), or other rodents that carry the virus	This disease occurs throughout the Americas. Hundreds of cases are reported annually, and of those infected, about one-third die from infection. The form of the virus present in the Americas infects the lungs and fills them with fluid.
HIV/AIDS	Human immunodeficiency virus	Contact with contaminated human bodily fluids; most frequently transmitted by sexual activity or intravenous drug use with a contaminated needle	The burden of disease from HIV/AIDS is borne overwhelmingly by sub-Saharan Africa, where nearly 70% of the 40 million people infected with HIV worldwide live. Roughly three million people die from HIV infection every year. The virus destroys the host's immune system and can render an infected individual helpless to defend against a variety of opportunistic infections that a normal immune system could easily fight off. When people infected with HIV contract certain kinds of opportunistic infections or have extreme depletion of certain immune cells, they are said to have the acquired immune deficiency syndrome, or AIDS.
Influenza	Influenza virus	Contact with respiratory droplets or contaminated surfaces	Sudden onset of fever, aches, and fatigue herald infection with the influenza virus. The illness can last for 10–14 days. In an average year, three to five million people contract influenza and 250,000–500,000 die from the disease.
Japanese encephalitis	Japanese encephalitis virus	Bites by infected *Culex* mosquitoes	This is a major cause of encephalitis (inflammation of the brain) in Asia with 30,000–50,000 cases per year, associated with pig reservoirs and rice ecosystems.
Kyasanur forest disease	Kyasanur forest disease virus	Bites by one of eight infected tick species, primarily *Haemaphysalis spinigera*	This is a viral hemorrhagic fever only known in and around Karnataka, India.
Leishmaniasis	About 21 species from the single-celled protozoan parasite genus *Leishmania*	Bites from infected sandflies—more than 30 species from two genera, *Lutzomyia* and *Phlebotomus*	Leishmaniasis comes in several forms, predominantly cutaneous (i.e., on the skin) and visceral (i.e., in internal organs). Cutaneous leishmaniasis manifests at the site of infected sandfly bites in the form of ulcerated sores. In visceral leishmaniasis, the parasite can infect the liver, spleen, and bone marrow and is usually fatal if left untreated. The U.S. Centers for Disease Control and Prevention estimates that each year 1.5 million people worldwide contract cutaneous leishmaniasis, and 500,000 others visceral leishmaniasis. The disease occurs in 88 countries, though 90% of cases occur in India, Bangladesh, Nepal, the Sudan, and Brazil.

Disease	Infectious Agent	Species Involved in Disease Transmission and Mode of Transmission	Comments
Leptospirosis	*Leptospira* bacteria (spirochetes)	Direct exposure to the spirochetes via blood or urine, or consumption of infected water or meat	Some 160 mammalian species are infected with leptospires, though in most cases the bacteria do not harm the host. Although several hundred serotypes have been identified, only a small number cause human disease. Leptospirosis has two clinically defined syndromes. The first, known as anicteric leptospirosis (*anicteric* refers to an absence of icterus or jaundice, a yellow discoloration of the skin or eyes that can result from liver disease), is much like influenza, with fever, headache, vomiting, and muscle aches. The second, known as Weil's syndrome, can be deadly, because the parasite compromises liver and kidney functions and can increase the risk of internal bleeding.
Loiasis	*Loa loa* (African Eye Worm)	Bites by infected *Chrysops* species (a genus of deer fly)	Loiasis is a disease of western and central Africa and is caused by adult filarial worms (parasitic roundworms) that survive just under the skin, with larvae that circulate in the blood. Occasionally, the adult worms may migrate to the eye, which can cause severe irritation and inflammation, as well as swellings in the skin marking the paths of migration.
Lyme disease	*Borrelia burgdorferi*	Bites from infected ixodid ticks, especially the Blacklegged Tick (*Ixodes scapularis*) and Sheep Tick (*I. ricinus*)	The most common vector-borne disease in the United States, Lyme disease, caused by a spirochete bacterium, produces nonspecific symptoms in most people, such as fever, headaches, muscle aches, joint aches, and fatigue. In some cases, neurological disease can appear, including meningitis and a facial droop known as Bell's palsy.
Lymphatic filariasis	Mostly caused by the roundworm *Wuchereria bancrofti* (90%), and most of the remainder by *Brugia malayi*	Bites from infected *Culex* species (esp. *Cx. pipiens*) in most urban and semiurban areas, *Anopheles* mosquitoes in the more rural areas of Africa and elsewhere, and *Aedes* and *Mansonia* species in Southeast Asia and Pacific islands	Around the world, 120 million people in 83 countries are infected with lymphatic filarial parasites, and it is estimated that more than one billion (20% of the world's population) are at risk of acquiring infection. The parasites invade the lymphatic system and can obstruct flow, leading to local inflammation and swelling. If left untreated, the disease can cause elephantiasis, a condition where limbs swell to huge proportions, and hydroceles (swelling of the scrotum) can form.
Malaria	*Plasmodium vivax*, *P. ovale*, *P. malariae*, and *P. falciparum*	Bites from infected *Anopheles* mosquitoes	Malaria affects 300–500 million people and kills between one and three million of them, especially young children, each year. The protozoan parasites can cause a high fever, shaking chills, vomiting, and anemia. *P. falciparum* gives rise to the most severe disease.

Continued

TABLE 7.1. (CONTINUED)

DISEASE	INFECTIOUS AGENT	SPECIES INVOLVED IN DISEASE TRANSMISSION AND MODE OF TRANSMISSION	COMMENTS
Nipah encephalitis	Nipah virus	Contact with feces of infected pigs or their oral and nasal mucus secretions that become aerosolized via coughing	The illness presents with fever, headache, and drowsiness and can progress to life-threatening changes in blood pressure and to inflammation of the brain, leading to seizures and coma. Nipah has a high mortality rate, approaching 75% of those infected.
Onchocerciasis or river blindness	*Onchocerca volvulus*, a filarial nematode parasite	Bites from infected *Simulium* flies, known as blackflies, the larval stages of which live near fast-flowing rivers and streams, that bite mostly during the day	The tiny larvae (microfilaria) produced by the adult worm live in the skin and invade the eye, producing severe itching and, potentially, blindness. The disease is still a public health problem in more than 27 countries, most of which are located in Africa and in Central and South America.
Salmonellosis	The bacterium *Salmonella enteriditis*	Contact with food or water contaminated with the bacteria	Salmonella infection causes severe gastrointestinal cramping and diarrhea. The disease afflicts more than one million Americans and causes 500 deaths each year.
SARS (severe acute respiratory syndrome)	SARS virus	Source of human infection is not entirely clear: Some horseshoe bat species (*Rhinolophus* spp.) in China carry the virus; Palm Civets (members of the mongoose family Viverridae), shown to carry SARS-like viruses, may be a reservoir host	The first case of SARS occurred in Guangdong province, China, in November 2002. From there, it spread in only a few months across the world, infecting more than 8,000 people and killing almost 800. SARS starts with fever, headache, and other nonspecific symptoms. After a period of a few days, most patients go on to develop a life-threatening pneumonia.
Schistosomiasis	The parasitic worms *Schistosoma mansoni, S. haematobium, S. intercalatum, S. japonicum,* and *S. mekongi*	Exposure to water contaminated with the worms; snail species from the genera *Bulinus, Oncomelania, Biomphalaria,* and *Neotricula* serve as intermediate hosts	More than 200 million people suffer from schistosomiasis, and millions more are at risk of contracting the disease. Symptoms vary greatly depending upon the particular species. Urinary schistosomiasis from *S. haematobium* entails infection of the bladder that can lead to bloody urine and, in some cases, bladder cancer. Intestinal schistosomiasis, caused by *S. mansoni, S. intercalatum,* and *S. japonicum,* can affect the liver and colon and produce bloody diarrhea and liver failure. The parasites can metastasize to the brain or lungs once established in the body.
Trypanosomiasis	Single-celled protozoan parasites known as trypanosomes; subspecies of *Trypanosoma brucei* cause African sleeping sickness, and *Trypanosoma cruzi* causes Chagas disease	African trypanosomiasis: bites from infected tsetse flies (*Glossina* spp.) Chagas disease: bites from infected insect species from the subfamily Triatominae, known as kissing bugs	African sleeping sickness is aptly named, because the parasite can cause somnolence in the late stage of the disease when the brain is invaded, though this symptom typically follows a painless skin sore, fever, and headaches. The disease is endemic to sub-Saharan Africa. Chagas disease occurs primarily in rural Central and South America and can affect the colon, esophagus, and heart, leading to chronic constipation, difficulty swallowing, and life-threatening heart arrhythmias, respectively. A total of roughly 16 million people are infected in Latin America.

Disease	Infectious Agent	Species Involved in Disease Transmission and Mode of Transmission	Comments
Tularemia	The bacterium *Francisella tularensis*	Inhaling or having direct contact with the infectious bacteria, which can be carried by animals, including rabbits, rodents, and hares, as well as being bitten by an infected tick (e.g., *Dermacentor andersoni*, the Rocky Mountain Wood Tick; *D. variabilis*, the American Dog Tick; *D. occidentalis*, the Pacific Coast Dog Tick; and *Amblyomma americanum*, the Lone Star Tick) or by an infected tabanid fly (e.g., a horse fly)	Infection with *F. tularensis* can bring about several syndromes, the most common of which entails rapid onset of a high fever, headache, chills, and generalized body aches. At the site of the insect bite, an ulcerated lesion develops, eventually capped by a nonhealing scab, that is associated with marked swelling of nearby lymph glands.
West Nile encephalitis	West Nile virus	Bites by infected species of *Culex* mosquitoes, especially *Cx. pipiens* and *Cx. quinquefasciatus*	Although typically a mild illness, infection with West Nile virus can become severe if the virus reaches the central nervous system, where it can produce weakness, severe headache, and confusion. Wild bird populations are important reservoir hosts.
Yellow fever	Yellow fever virus	Bites by infected *Aedes* mosquitoes	Symptoms develop in the week after being bitten by an infected mosquito, including fever, headache, and vomiting. Around 15% of those infected will progress within one day from the onset of symptoms to a "toxic" phase of the disease in which the kidneys and liver may cease to function and from which only about 50% survive. Some 200,000 cases occur each year, and about 30,000 people die from the virus, predominantly in tropical Central America and Africa. Forest monkeys are reservoir hosts. An effective vaccine is available.

occur as the direct and immediate consequence of a specific human activity, such as agricultural development, water resource management, deforestation, or mining. With others, years may pass between the activity (e.g., the burning of fossil fuels) and its ultimate effect (e.g., erosion or forest fires from droughts caused by global warming).

The causes of ecosystem disturbances come in many forms and include changes in local average temperatures or in the degree of their variability; changes in water cycles, for example, in the timing, intensity, and spatial distribution of precipitation; changes in the distribution and availability of surface waters from irrigation or the building of dams; changes resulting from pollution, including pesticides and excessive nutrients such as nitrogen and phosphorus; and the effects of urbanization. However, of all the causes of ecosystem disturbance, habitat destruction and fragmentation resulting from the conversion of natural habitats into fields for growing crops or raising animals, or for human settlements, appear to have had the most impact on human infectious diseases.

To demonstrate how land-use changes can affect the spread of infectious diseases, consider the conversion of the pampas in Argentina (the pampas are the fertile, grassy plains that cover more than 300,000 square miles [more than 777,000 square kilometers] in southern South America) into farms growing monocultures of corn (maize) that occurred in the 1950s. This conversion increased the abundance of Corn Mice (*Calomys musculinus*) next to human settlements and resulted in an outbreak of Argentine hemorrhagic fever (AHF). What was discovered was that the cornfields provided an ideal habitat for the Corn Mice, but not for other native rodents that had competed with, and thereby had helped control, populations of *C. musculinus* in the pampas. Corn Mice, the natural reservoir for the Junin virus that causes AHF, shed millions of virus particles in their urine, feces, and saliva. By causing population surges of this mouse species, this ecosystem disturbance increased human exposure to the virus and resulted in major outbreaks of AHF.[1] AHF has a mortality rate of 15 to 30 percent and can cause severe gastrointestinal bleeding, as well as seizures and coma.

Habitat fragmentation and destruction can also result in an increased risk of human infectious disease by affecting predators more than their prey, the latter of which are more likely to be disease reservoirs or vectors. Predators tend to be more sensitive to habitat disruption, primarily for two reasons. First, they are almost always much less abundant than their prey, and sparse populations are more likely than abundant ones to be lost from their habitat. Second, predator populations usually require large intact areas to meet their dietary needs, so when portions of these are fragmented or destroyed, they are more likely to suffer sharp declines.

Without predator control, some prey species such as rodents, having high reproductive rates, can explode in population size, with significant consequences when they serve as disease reservoirs for the transmission of infectious agents to humans. The same is also likely to be true for prey that are disease vectors, such as for some mosquito species, but the role of predators in controlling such vector populations is less well understood.[2]

In the following sections, we explore the ways in which the major causes of ecosystem disturbance—deforestation, water management practices, agricultural development, and climate change—influence human infectious diseases.

Alterations of Forest Ecosystems

The world's forests are home to many species that are involved in infectious disease transmission. The major insect vector groups—*Anopheles*, *Aedes*, *Culex*, and *Mansonia* mosquitoes; *Simulium* blackflies; the New-World *Lutzomyia* sandfly vectors of *Leishmania* species; the *Chrysops* fly vector of *Loa loa*; and the *Glossina* tsetse fly species that transmit trypanosomes—all contain species that depend on forest ecosystems and on woodland savannas. As described in chapter 1, forests around the world are undergoing unprecedented changes. The implications of such changes for human infectious diseases are many.

Deforestation creates new edges and interfaces that can promote population growth in animal reservoir hosts and vectors and, in addition, often attract people to settle in these potentially risky edge zones. Several diseases—leishmaniasis, yellow fever, trypanosomiasis (both African sleeping sickness and Chagas disease), and Kyasanur forest disease among them—are acquired by insect vectors biting humans at or near the interface between forest and human settlements. In addition, some

Figure 7.1. The Sandfly (*Lutzomyia longipalpis*). The Sandfly is an important vector of visceral leishmaniasis, also known as kala-azar, in Central and South America. (Courtesy of Jose Ribeiro, National Institute of Allergy and Infectious Diseases.)

animal reservoir hosts tend to increase in abundance and become concentrated near forest edges, compounding the risk of human exposure to the pathogens they carry. For example, some studies in North America have shown that the White-Footed Mouse (*Peromyscus leucopus*), a reservoir for the pathogens that cause Lyme disease and babesiosis, increases in abundance near forest–field edges.

An additional risk for exposure to infectious disease and, in particular, vector-borne disease comes from traveling deep within previously undisturbed forest. This risk increases when migration into the forest occurs in the context of some deforestation activities, such as clear-cutting, road building, and mining, that provide new edges and interfaces within the forests themselves, and that also tend to involve people who, unlike those who are indigenous inhabitants of the forest, have little or no immunity to local endemic diseases. Outbreaks of yellow fever, leishmaniasis, and malaria have occurred in workers engaged in such deforestation activities and in settlers at the forest's edge, as the result of such increased contact with vectors.

The destruction of forest habitat can also result in the replacement of the most common vector species with a more effective disease vector, such as one of the *Anopheles* species replacing a more benign native mosquito. Such has been the case following deforestation in some parts of Southeast Asia and Amazonia.[3] The following mechanisms seem to be involved. For one, deforestation and the road building that accompanies it cause forest floor depressions that allow standing pools of water to form. The associated removal of groundcover plants and organic debris on the forest

Figure 7.2. Interface Between a Forest and Human Settlement in Manaus, Brazil. The proximity of housing to rainforest habitat contributes to outbreaks of leishmaniasis and other infectious diseases. (© David H. Molyneux.)

floor, both of which normally serve to drain standing water, increases the likelihood that these depressions will be filled and become ideal breeding sites for *Anopheles* mosquitoes. In addition, removal of the trees tends to reduce the acidity of the pools (some trees acidify standing water as their leaves break down and form organic acids), making it more suitable habitat for some *Anopheles* larvae, which prefer a more alkaline environment. Finally, deforestation increases ambient light and temperature on the forest floor, leading to more photosynthesis by algae in the water, the mosquito larvae's main food source. The penetration of sunlight to the pools, in particular, has been associated with increased breeding among *Anopheles* mosquitoes.

Sometimes the vector species favored by deforestation is less effective at transmitting disease than the original vector(s). However, in general, this is not the case. While deforestation results in an overall decrease in forest mosquito biodiversity, the surviving dominant species are almost always more effective vectors for malaria, for reasons that are not well understood. This has been observed as a consequence of deforestation essentially everywhere that malaria occurs, for example, in East Africa when forests were cleared for rice production, in the Kanchanaburi Province of Thailand when they were cut for cane sugar cultivation, and in Indonesia when they were converted into fish farms. It has also followed settlement-related deforestation in India, and even in the United States from the early eighteenth well into the nineteenth century, especially along parts of the Mississippi River as far north as Illinois.

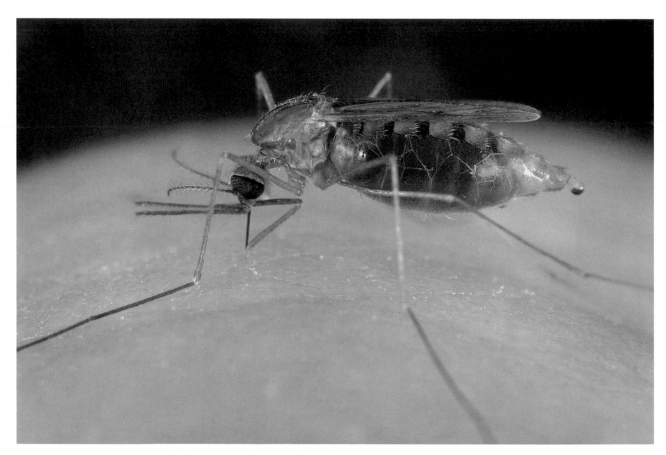

Figure 7.3. *Anopheles freeborni* **Mosquito.** This female *Anopheles freeborni*, known as the Western Malaria Mosquito, is having a blood meal. The malarial parasite must mature in the mosquito's gut for more than a week before the disease can be transmitted to another person via the mosquito's saliva. (Photo by James Gathany, Centers for Disease Control and Prevention.)

In the past few decades, deforestation in the Amazon has led to a proliferation of *Anopheles darlingi*, the mosquito species that is highly effective at transmitting malaria to humans in this region and that has, in some instances, replaced some twenty other *Anopheles* species that were present when the forests were intact. Perhaps *An. darlingi* benefits more than other *Anopheles* species from the alterations in aquatic habitats mentioned above that are associated with deforestation. Similar patterns have been observed in Southeast Asia, where the diversity of mosquito populations is reduced through deforestation, but where species that are effective malaria vectors rapidly adapt to new habitats.

Deforestation can also influence human infectious diseases carried by certain snails. Forests contain streams, rivers, lakes, and ponds that are shaded by overhanging trees, which serve to maintain relatively constant water levels. These conditions, which provide a broad array of habitats, often tend to favor a diverse freshwater snail fauna. Deforestation leads to more sunlight penetrating into forest water bodies, resulting in more vegetation growth, more variable water levels, and at times, even a disappearance of surface waters entirely. These changes inevitably alter snail diversity in the forests. Few of the original snail species can adapt to these new deforested conditions, and the ones that can adapt well to open areas are generally also those better able to serve as intermediate hosts (organisms that support an immature or nonreproductive form of a parasite) for the parasitic flatworms known as schistosomes that cause the disease schistosomiasis (formerly called bilharzia). Once again, deforestation, by altering natural forest biodiversity, increases the risk of human infectious disease.

A compelling example of how deforestation can spread schistosomiasis within a human population comes from Cameroon.[4] There, the forest ecosystem, with its shaded ponds and slow-moving streams, predominantly supported one species of snail, *Bulinus forskalii*, an intermediate host for *Schistosoma intercalatum*, a schistosome that causes little illness in humans. However, following deforestation, exposure of the forest water bodies to the sun favored another snail, *Bulinus truncatus*, which became the dominant snail species. *B. truncatus* is an effective intermediate host for another schistosome, *S. haematobium*, which causes urinary tract disease in people, so with deforestation, urinary schistosomiasis became a significant public health problem in parts of Cameroon (see also discussion below about *B. truncatus* and schistosomiasis in Senegal).

In the Philippines, snails from the genus *Oncomelania* function as the intermediate host for the schistosome species *Schistosoma japonicum*, and these snails are able to live both inside and outside of the forest. Before the start of large-scale deforestation, *S. japonicum* was found primarily in rodent populations and rarely, if ever, infected humans. Following major deforestation in the Philippines, however, the schistosome moved into humans, perhaps in part because of larger human settlement in the formerly forested areas and the increased degree of contact between humans and the *Oncomelania* snails.

Still further consequences for the spread of infectious disease can come from planting new forests. Reforestation may make use of nonnative plants, which may provide better conditions for some vectors that are capable of increasing the transmission of infectious agents. The introduction of immortelle trees (so-named because the trees' flowers maintain their color and shape when dried) in the 1940s to Trinidad from Peru to provide shade for cocoa crops, for example, resulted in a malaria epidemic. The trees

(*Erythrina glauca* and *E. microptery* were used) created habitat for bromeliads (a diverse group of plants that includes pineapples and Spanish moss). These plants, in turn, provided suitable habitats in the spaces between their leaves where water would accumulate and allow for the proliferation of larvae of the mosquito *Anopheles bellator*, a malaria vector. The combination of an explosion in *An. bellator* populations and a local population of cocoa farm workers that were brought into contact with them fueled the epidemic.[5] Leishmaniasis in South America, malaria in Thailand, and trypanosomiasis in Africa all increased in incidence when native plants were replaced by alien species, because in each case, their vectors were better able to adapt to the new conditions.

Water Management

Irrigation, dams, and small impoundments of water such as those formed by microdams (e.g., for fish farms) cause disturbances to ecosystems that often have dramatic effects on infectious diseases, particularly those transmitted by mosquitoes and snails.

Extensive irrigation can lead, for example, to serious problems with malaria. In the 1990s, "irrigation malaria" was endemic and widespread in a population of about 200 million people in rural India. This public health catastrophe was the result, in part, of an inadequate assessment of the health impacts of irrigation and of what measures were necessary to mitigate them. Poorly maintained systems allowed irrigated water to seep into surface waters, which created, along with the poor drainage and the rise in water tables associated with irrigation, conditions suitable for the breeding of the major malaria vector in India, *Anopheles culicifacies*. These systems also led to the development of slow-running streams favored by another major malaria vector in the region, *An. fluviatilis*. Increases in the populations of both vectors led to marked increases in the number of cases of malaria in these irrigated regions in India.[6–8]

For the last fifty years, dam construction has been one of the major causes of environmental degradation and of significant human health consequences in many parts of the world, including both losses of biodiversity and increases in infectious

Figure 7.4. Microdam Forming a Small Lake in Burkina Faso, Africa. Such water impoundments can create new breeding grounds for mosquitoes and snails that carry human disease. (© David H. Molyneux.)

diseases. River ecosystems typically contain a diverse flora and fauna that the river's variable flow patterns and distinctive river bottom habitats support. Damming tends to severely disrupt and, in many cases, destroy these ecosystems. Changes in the river's flow alter habitats both within and around the river, both upstream and downstream of dams, and typically few of the original species that had lived in these ecosystems manage to survive. Some others that had been rare, however, may become more abundant. Some of the species that tend to be better adapted to habitats that develop after dams are created are better able to promote disease transmission to humans, such as some insect vectors, and snails, the intermediate hosts for schistosomiasis. Furthermore, downstream spillways (a spillway is a structure that conveys surplus water over or through a dam) have been known to provide ideal breeding sites for *Simulium* blackflies, the vectors of the worm *Onchocerca volvulus* that causes onchocerciasis, or river blindness. These spillways also create potential larval habitats for mosquito vectors of malaria and other diseases, because they can lead to the formation of adjacent still-water pools.

Construction of the Diama Dam in Senegal created outbreaks of intestinal and urinary schistosomiasis that affected thousands of people upstream, causing serious health problems in a population that had been essentially free of the disease (see box 7.2).

Similarly, the Aswan High Dam on the lower Nile in Egypt and the Blue Nile irrigation project in Sudan have resulted in millions of Nile Delta inhabitants having a high and chronic risk of exposure to schistosomiasis. In the case of the Aswan High Dam, explosive growth of phytoplankton and other vegetation in Lake Nassar (the lake the dam created) has been a boon for fish populations; 80,000 tons of fish are now caught each year from the lake. Fishermen from other regions, some carrying *S. haematobium*, moved into the area seeking to take advantage of this new bounty, and as the *B. truncatus* snail population also increased significantly (because of a dramatic increase in the numbers of water plants in the lake), the stage was set for an outbreak of urinary schistosomiasis. Downstream from the dam, irrigation practices moved from seasonal to perennial, and this change altered habitats in such a way as to favor *B. truncatus* in Upper Egypt, as well, and led to more infections with urinary schistosomes. As a result, the prevalence of *S. haematobium* in Upper and Middle Egypt leapt from about 6 percent before 1960 to nearly 20 percent in the 1980s. In Lower Egypt, the same happened with intestinal schistosomiasis, but to an even greater degree.[9–12] There, the intermediate hosts were *Biomphalaria* snails.

The rising water tables and reduced water flow in small irrigation canals in the Nile Delta that accompanied the building of the High Dam has also fostered the spread of lymphatic filariasis between 1970 and 1990, because these effects increased available *Culex pipiens* (a mosquito vector of filariasis) breeding sites.[13]

Agricultural Development

Livestock and game form a key link in a chain of disease transmission from animal reservoirs to humans. Given that livestock often act as intermediaries between a wild animal reservoir of infection and a human host (described further below), changes in livestock management can have serious consequences for human health by promoting the emergence of new pathogens and the reemergence of old ones.

BOX 7.2

Dams, Irrigation, and Schistosomiasis in Senegal

Constructed in 1985, about 40 kilometers (25 miles) from the mouth of the Senegal River, the Diama Dam was intended to prevent the intrusion of salt water into the river during the dry season. While it did reduce salinity, the dam also dramatically altered the region's ecology. One organism that made its appearance and prospered after the dam was built was the snail *Biomphalaria pfeifferi*, an important intermediate host for *Schistosoma mansoni* (the parasite that causes intestinal schistosomiasis). The main snail species that *B. pfeifferi* replaced in many areas around the river, *Bulinus globosus*, is not a *S. mansoni* host. Unknown to the Diama region prior to 1985, *S. mansoni* quickly took hold in the local human population. *S. mansoni* eggs were first noted in 1988 in stool samples of a single person in Richard-Toll, a town approximately 130 kilometers (about 80 miles) upstream of the dam. By the end of 1989, almost 2,000 people were positive for *S. mansoni*, and by August 1990, fully 60 percent of the town's population of more than 50,000 were infected. During this period, *B. pfeifferi* became the dominant snail species in the area around Richard-Toll, comprising about 70 percent of the total number of snails collected. And almost half of them were infected with *S. mansoni*.

Scanning Electron Micrograph of the Larva of *Schistosoma mansoni*. The larval form of the schistosome, known as a cercaria, burrows through human skin to initiate an infection. *S. mansoni* causes intestinal schistosomiasis. (© Jonathan Emerson Kohler, University of Washington.)

Before the dam, seawater would flow into the Senegal River and make its waters salty, a condition that is known to interfere with the ability to reproduce for both schistosomes and their intermediate pulmonate snail hosts (pulmonate snails are freshwater snails, descended from terrestrial species, that have lungs and breathe air). It was therefore believed that the decreased salinity in the river after the dam was built, along with the river's becoming less acidic and having more stable water levels, favored

Shell of the Snail *Bulinus globosus*, an Important Intermediate Host for *S. haematobium*. (Photo by Thomas K. Kristensen, Mandahl-Barth Research Center for Biodiversity and Health, Denmark.)

pulmonate snails over other native snails and led to a greater incidence of schistosomiasis.

The Diama Dam was also responsible for the emergence of schistosomiasis in the region for other reasons. A substantial rise in irrigation for the cultivation of rice, made possible by the dam, for example, generated additional habitats for the snail *Bulinus senegalensis*, yet another intermediate host for *S. haematobium*, the schistosome that causes urinary schistosomiasis. But it was primarily other conditions that changed as a consequence of the dam that resulted in a major outbreak of urinary schistosomiasis. The most prevalent and important intermediate host for *S. haematobium* in the region was the snail *B. globosus*, but it did not cause significant disease, probably because the local strain of *S. haematobium* it carried had existed in the region for very long periods of time, leading to some degree of immunity among local populations. After the dam was built, urinary schistosomiasis increased greatly, partly because populations of another snail, *B. truncatus*, which did not carry the local *S. haematobium* strain, grew significantly. In addition, large numbers of immigrant farm workers moved into the area, eager to be a part of the greatly expanded irrigated rice production. Some of these farmers, coming from other parts of Africa, including Mali, carried with them another strain of *S. haematobium*, for which *B. truncatus* snails were competent intermediate hosts. Local populations had little to no immunity to this alien *S. haematobium* strain, and as a result, a major outbreak of urinary schistosomiasis ensued.

Both intestinal and urinary forms of schistosomiasis continue to increase in prevalence and distribution in the Senegal River Basin. The history of schistosomiasis along the Senegal River upstream from the Diama Dam provides a cautionary tale about the potential effects of dam construction on the spread of vector-borne diseases and illustrates the complexity of interactions, and the difficulty in predicting them, among pathogens and hosts that follow the disruption of some natural ecosystems.

In the 1980s *Salmonella enteritidis*, found in poultry and their eggs, emerged world-wide as a major human disease-causing bacterium. Research on its spread suggests that *S. enteritidis* filled an ecological niche on large poultry farms that had been vacated by the nonpathogenic—at least for humans—*Salmonella gallinarum*. *S. gallinarum* had long been a significant burden for the poultry industry, because it caused typhoid in fowl. As a result, large numbers of poultry farmers in the United States, and in some other countries such as England, adopted a "test and slaughter" method of disease control, in which any bird that tested positive for the disease was killed. This method, combined with greater antibiotic use, largely eliminated *S. gallinarum* in commercial poultry flocks in the United States and England in the 1970s. By, 1985, however, the incidence of *S. enteritidis* infections in U.S. poultry jumped to five times the levels of 1976. Evidently, *S. gallinarum* had competitively excluded *S. enteritidis* from colonizing fowl, so when *S. gallinarum* was no longer present, *S. enteritidis* infections in poultry increased markedly.[14] Although rates of infection have since fallen, current estimates suggest that more than 200,000 people in the United States still become ill each year with *S. enteritidis* at a cost of more than US$1 billion.[15,16]

Some features of *S. enteritidis* help to explain its ability to displace *S. gallinarum*. For one, *S. gallinarum* has no reservoir other than domestic poultry and waterfowl, whereas *S. enteritidis* is also harbored by rodents, so even when infected animals are killed, bacteria can repeatedly reinfect poultry flocks. Moreover, *S. enteritidis*, in contrast to *S. gallinarum*, often infects poultry without producing symptoms, facilitating the spread of infection, because infected animals are not identified and thus are not treated or killed. Once poultry have become infected with *S. enteritidis*, other aspects of commercial poultry operations contribute to its spread, including concentrated rearing practices, where tens or hundreds of thousands of animals are housed at a single site, enabling rapid disease spread, first within the facility, and then to sites that can be thousands of miles away when poultry are exported.

Intensive rearing of livestock, particularly poultry and pigs together, creates an ideal setting for generating infectious disease outbreaks of other pathogens, as well. These agricultural practices, which are widespread in some Asian countries such as China, can promote the evolution of new, virulent forms of the influenza virus, with serious consequences for human health worldwide (see section on pathogen diversity below).

JAPANESE ENCEPHALITIS

Pigs can also contribute to the spread of Japanese encephalitis (JE), a viral disease associated with rice ecosystems in Asia (encephalitis is an inflammation of the brain, usually caused by a virus). Regionwide, an estimated 50,000 human cases occur each year, with 20 percent of those infected dying from the disease, and an additional 20 percent becoming disabled.[17] The disease exacts a considerable social and economic toll. The primary vector throughout Asia, *Culex tritaeniorhynchus*, breeds abundantly in flooded rice fields, as does another important vector, *Culex vishnui* (in India, Thailand, Sri Lanka, and Taiwan). Several other *Culex* species breed in a variety of habitats, some associated with irrigated rice. The transmission cycle of JE involves an amplifying host (a host in which a pathogen multiplies rapidly, resulting in an amplification of its spread by vectors), usually the domestic pig, although in some instances ardeid birds (the ardeids are the family of long-legged, long-necked wading birds that include herons and egrets) are also involved.

JE is endemic (meaning that it occurs continuously, without external inputs, in a specific region) in irrigated paddy rice field areas of Thailand and China. In recent years there have also been outbreaks in the Terai region of Nepal, and in Sri Lanka, accompanying the introduction of new irrigation systems adjacent to areas where pigs are raised. These conditions favor the emergence of JE and the potential for epidemics, because vectors of the disease and amplifiers of the viral pathogen are brought together.[17,18]

NIPAH VIRUS

In 1998, Nipah virus first emerged in Malaysian pigs as a respiratory and neurological disease and then jumped to humans with lethal consequences. Nipah virus can cause severe encephalitis and vascular disease in people, and in the Malaysian outbreak, those with Nipah virus infections had a mortality of about 40 percent. More than 100 people died. Essentially all of those who became infected had direct contact with pigs, either in pig-rearing facilities or in slaughter houses. The rapid spread of Nipah virus among pig farms throughout peninsular Malaysia and into Singapore was most likely caused by an intense sell-off of infected pigs following the initial outbreak. In order to control the outbreak, millions of pigs were slaughtered, hundreds of farms were closed, and tens of thousands of jobs were lost. Nipah virus cost the Malaysian government more than $350 million and destroyed the livelihoods of countless numbers of people.[19]

The origin of Nipah virus has been traced back to two native species of fruit bats or "flying foxes" from the genus *Pteropus* (and the family Pteropodidae): the Malayan Flying Fox (*Pteropus vampyrus*), which is ubiquitous throughout peninsular Malaysia, Borneo, and Thailand, and the Variable Flying Fox (*P. hypomelanus*). Pteropid bats, which do not develop clinical signs of disease when infected with Nipah virus, are considered the natural reservoir for henipaviruses (the viral genus to which Nipah belongs) and have most likely co-evolved with these viruses over time. Their range includes the Old-World tropical regions, from Madagascar eastward through Southeast Asia, Australia, and the South Pacific islands. The bats serve vital roles in tropical ecosystems, acting as seed dispersers and pollinators for rainforest plants. Hunting and deforestation threaten this ecologically important group of animals throughout their range.[20]

Several factors may have coincided to cause the 1998 Nipah virus outbreak. First, there had been extensive deforestation in the rainforests where fruit bats live and feed, both for timber and for agricultural expansion and intensification, the latter mainly involving the planting of rubber trees and oil palms. As discussed

Figure 7.5. Malayan Flying Fox (*Pteropus vampyrus*). (Photo by Thomas Kunz, Boston University.)

in chapter 3 (page 105), there has been a massive growth in oil palm plantations in the region over the past forty years. Second, and perhaps most important, while pig farms in Malaysia historically were small operations, farm sizes had been increasing markedly in the decades prior to the Nipah outbreak. It is clear that flying foxes have long roosted in the region where Nipah virus first emerged, and it is possible that spillovers from bats to pigs have occurred in the past. But evidence now suggests that without a high density of pigs to support an outbreak, the virus would have likely run its course, with pigs either recovering or succumbing to the disease. A lack of human infections and Nipah's resemblance to other diseases in pigs may have resulted in the infection not being detected in the past. It is now believed that it was the density of pigs that may have been the critical factor in triggering the Nipah outbreak in 1998 (the farm where the outbreak occurred had 30,000 pigs at that time)—that a threshold had been reached that would sustain the infection among the pigs and allow the Nipah virus to spread to farm workers.

But how did the Nipah virus jump from bats to pigs and then to people? While fruit bats had clearly been present in the vicinity of the farm in the past, their numbers may have been greater at the time of the outbreak. With deforestation leading to a reduced availability of natural food resources in Malaysia (they generally feed on figs and other wild fruits), they may have been more likely to seek out fruit from commercial orchards. When these orchards were on pig farms, such as was the case with the farm where Nipah first broke out (mango trees had been planted to provide shade for the pig enclosures), spread of the virus from the bats to pigs became possible. Because Nipah virus is known to be present in fruit bat saliva, urine, and feces, it is likely that infected pteropid bats passed the infection onto the pigs via partially eaten dropped fruit and via their excreta. Once infected, the pigs developed severe coughs, and the large numbers of pigs were able to sustain the Nipah infection and to pass it on to people who came into contact with their infected feces or with their oral and nasal mucus secretions that had become aerosolized via coughing.

The combination of a high density of domestic animals in close proximity to wildlife, the loss of natural food resources for wildlife, and the presence of fruit orchards next to susceptible domestic animals were all conditions, established by humans, that contributed to the emergence of Nipah virus disease in people. Nipah virus has since emerged in Bangladesh, where annual outbreaks in people occurred between 2001 and 2005. The mortality rate has been as high as 75 percent. Domestic animals, however, have not been involved in the Bangladeshi outbreaks, which presumably have occurred through direct transmission from pteropid bats to humans or, in some cases, by consumption of contaminated fruit juices.

Bats have also recently been discovered to be hosts for the viruses of two other major emerging infectious diseases: severe acute respiratory syndrome, or SARS, and Ebola. In the former case, four different species of the insectivorous (insect eating) horseshoe bats of the genus *Rhinolophus* have been found to carry the SARS virus. They may be its natural reservoir host in the wild. After the virus was initially discovered in Palm Civets (*Paguma larvata*) in live animal markets in Guangdong Province, China, where the disease first emerged, more than 10,000 of the animals were destroyed. But, in fact, the civets may have caught the virus from the horseshoe bats, which are commonly sold in the same markets for food and for use as traditional medicines.[21]

With the Ebola virus, no wild reservoir for the virus had been found since the first outbreak in Africa in 1976 until 2005, when it was detected in three fruit bat species, *Hypsignathus monstrosus, Epomops franqueti*, and *Myonycteris torquata*, in Gabon and the Republic of Congo.[22] In both SARS and Ebola, the bushmeat trade has been involved in the transmission of the disease to humans—Palm Civets and horseshoe bats in China in the case of SARS and primates with Ebola in West–Central Africa.

Urbanization

People in all countries of the world are moving to cities. By 2007, for the first time in human history, the population of cities will be greater than that in rural areas. Most of this urban population growth will occur in the developing world, which is already home to seven of the world's ten cities with populations that exceed fifteen million. The repercussions of this transition include habitat destruction, subsequent ecosystem changes, and, as this chapter illustrates in several other contexts, consequences for the spread of infectious disease.

Urban construction will sometimes drain wetlands and other standing water, eliminating potential breeding sites for some mosquito populations, thereby serving to reduce the incidence of some infectious diseases. However, urbanization can also

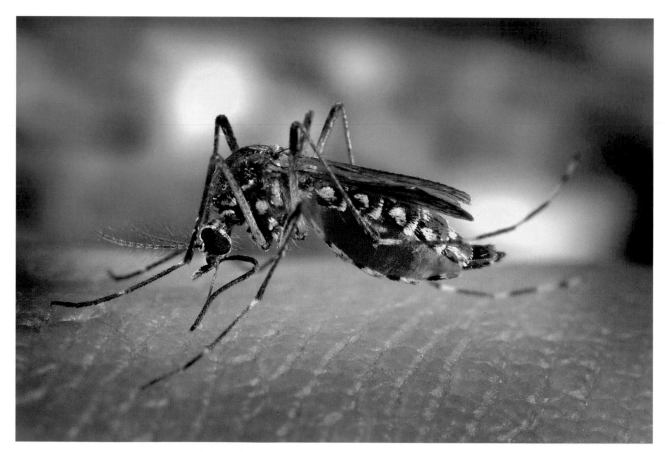

Figure 7.6. *Aedes aegypti* Mosquito. *Ae. aegypti* is the main vector in the world for yellow fever and for dengue fever, the latter caused by the world's most prevalent mosquito-borne virus. *Ae. aegypti* is widely distributed, being found in tropical and semitropical areas on all continents. (Courtesy of the Centers for Disease Control and Prevention.)

Figure 7.7. Blacklegged Tick (*Ixodes scapularis*). Also called the Deer Tick, *I. scapularis* is the vector of Lyme disease in the eastern United States, and also carries the disease babesiosis. (Courtesy of Scott Bauer, U.S. Department of Agriculture.)

Figure 7.8. White-Footed Mouse (*Peromyscus leucopus*), the **Primary Reservoir for Lyme Disease in the Eastern United States**. *P. leucopus* is found throughout eastern and central regions of the United States, extending north into Canada and south to the Yucatan in Mexico. (© Jim Schulz/Chicago Zoological Society.)

increase the incidence of other vector-borne diseases. One example is the urbanization-caused increase in dengue fever in Singapore, Rio de Janeiro, and Jakarta, where rain gutters, old tires and buckets, broken bottles embedded in concrete walls, cans, polystyrene containers, and other urban repositories of standing water have become preferred breeding sites for the mosquito *Aedes aegypti*, the major vector of dengue virus (as well as yellow fever virus in Africa). Impoverished populations on the periphery of cities in some developing countries, who will have greater contact with natural habitats, will be at even greater risk than those in the centers.[23]

Recent studies document how urban and suburban sprawl in the eastern United States has increased the human risk of exposure to Lyme disease, primarily because such development has resulted in the fragmentation of forests and in increased numbers of people living adjacent to these fragments.[24] The Blacklegged Tick (*Ixodes scapularis*) is the vector of Lyme disease, as well as of several other pathogens in the eastern United States, and the primary reservoir for Lyme disease in this region is a common rodent, the White-Footed Mouse (*Peromyscus leucopus*).

Mice live in many different habitats, from pristine old-growth forest to degraded woodlots, garden sheds, and even kitchens. Several studies have demonstrated that populations of White-Footed Mice become very concentrated in small forest fragments, probably due to the absence of other vertebrate species that prey upon, or compete with them (forest fragmentation, as discussed above, affects predators over prey disproportionately). As a consequence, tick populations in small forest fragments have many White-Footed Mice, but few other mammalian hosts, on which to feed, resulting in a high proportion of the ticks being infected and able to infect people. In contrast, in more extensively forested areas, the combination of fewer White-Footed Mice and more abundant, alternative, reservoir-incompetent hosts (hosts that do not pass on the Lyme bacteria to ticks that bite them, or do so poorly) results in a lower proportion of the tick population being infected. It has been determined that forest fragments less than about 3 hectares (7.4 acres) present the greatest risks for acquiring Lyme disease because of these pathogen–vector–host relationships.[24] Further examples of how urbanization affects transmission of infectious diseases are given below in the context of changes to biodiversity.

Vector, Pathogen, and Host Diversity and Human Infectious Disease

Biodiversity, as defined in chapter 1, is the variety of life on Earth at all levels of biological organization, from the genes within local populations of species, to the species themselves composing local communities, to these communities making up the biological constituents of ecosystems. Biodiversity as it relates to infectious diseases can be thought of at any of these levels. The incidence of infection may be influenced by the genetic makeup of pathogens, vectors, and hosts; by the numbers of species in each of these groups; by the diversity of habitats available in an ecosystem; and by variations in human behavior, such as the application of pesticides, that select for certain species over others.

In some cases, greater biodiversity will be associated with an increased incidence of disease. Tropical areas, for example, harbor greater pathogen diversity, and therefore people living there may have more risk of infection than will those in species-poor boreal regions. In other cases, greater biodiversity acts as a buffer to risk, as occurs when a diverse assemblage of predators and competitors controls the abundance of rodent host reservoirs, such as is the case with Lyme disease described above.

Moreover, changes in the various components of biodiversity may be even more influential to disease incidence than those in the absolute amount of biodiversity, either up or down, in a given ecosystem.[25,26] In Uganda, for example, the expansion and movement of cattle populations into areas previously inhabited by native ungulates (a large group of mammals that have hoofs, e.g., antelopes and cows), combined with the invasion of abandoned cropland by the nonnative plant *Lantana camara*, is believed to have contributed to changes in tsetse fly (*Glossina*) distribution that initiated epidemics of African sleeping sickness (ASS) in the 1980s.[27,28] These outbreaks, which began in Busoga (a region of Uganda), are thought to have resulted from coffee and cotton plantations having been abandoned because farmers could no longer make a living with these crops under rules imposed by the Amin regime; the collapse of tsetse control programs and health services; the rapid spread of *Lantana* into areas where the farms had been, creating suitable habitat in its dense vegetation for the proliferation of a tsetse fly vector of ASS, *Glossina fuscipes*; and the introduction of cattle, which provided a highly competent reservoir host for a subspecies of the parasite that causes ASS, *Trypanosoma brucei rhodesiense*. *G. fuscipes* is a generalist vector that will feed on cattle, as it will on any available host.[29] The movement of cattle in Uganda continues to this day to influence the spread of sleeping sickness in that country.[30] In this case, replacement of native biodiversity with nonnative alien and invasive species, rather than changes in the absolute number of species, was responsible for the infectious disease outbreaks.

Figure 7.9. Tsetse Fly (*Glossina pallidipes*), Shimba Hills, Kenya. Of the twenty-three known tsetse fly species, all but three transmit trypanosomes to people. (© 1999 Steven Mihok.)

Diversity of Vectors

The major vector-borne pathogens and the diseases caused by them are concentrated in the tropics, with the majority being found in tropical rainforests and in the woodlands and savannas at the edges of these ecosystems. Within the tropics, undisturbed ecosystems tend to have the greatest diversity of disease vector species. Disturbed ecosystems, on the other hand, tend to have lower species diversity, but the disturbances appear to favor the success of "generalist" vectors that bite a wider variety of animals, have broader geographical distributions, and can thrive in a greater variety of habitats.

In some cases, more disease vectors in an ecosystem may increase the chances that people will acquire a vector-borne disease. For instance, transmission of West Nile viral encephalitis, which first appeared in North America in 1999, might be facilitated when at least two species of mosquito are involved in spreading the virus among hosts. Bird-feeding mosquitoes, such as *Culex tarsalis*, are effective at transmitting West Nile virus among birds, but because these mosquitoes are unlikely to bite people, other mosquitoes are necessary for the disease to be transmitted to us. One such species, *C. pipiens*, bites both birds and people and so is capable of transmitting the West Nile virus from birds to people. Recent studies indicate that *C. pipiens* can also transmit the infection between people. However, since *C. pipiens* mosquitoes bite birds infrequently, a separate vector (e.g., *C. tarsalis*) that maintains the disease in bird populations increases the chances that people will become infected.[31]

A similar relationship between vector species and human disease may exist with the transmission of Lyme disease in California. One species of tick, *Ixodes spinipalpis*, is responsible for maintaining an infection cycle of the Lyme bacterium, *Borrelia burgdorferi*, within rodent reservoirs, but it rarely bites people. Another tick, *I. pacificus*, is able to transmit the infection from rodent hosts to humans. The presence of both species, therefore, increases the risk that people in California will become infected with Lyme. However, in areas of California where the Western Gray Squirrel (*Sciurus griseus*) is abundant, Lyme disease risk can be high even when most of the ticks are *I. pacificus* and few *I. spinipalpis* are present. Western Gray Squirrels are highly competent Lyme disease reservoirs, and *I. pacificus* frequently bites both squirrels and people.

Although a diverse vector community by itself may increase the likelihood of disease transmission to people in some cases, most situations are far more complex. For some vector-borne infections, for example, it is the specific characteristics of the vector community (e.g., susceptibility, feeding habitats, biting behavior) that are of key importance, not the number of different vector species or their abundance. An example is given in box 7.3.

Diversity of Pathogens

As with vectors, greater pathogen diversity is typically associated with a higher risk of pathogen transmission to humans. However, this is not always the case. For example, the presence of the measles virus can generate an immune response that can protect against other infectious diseases such as whooping cough. In this case, higher diversity of pathogens reduces the burden of disease. In addition to existing diversity, the potential to mutate and produce new genetic forms within a pathogen species allows for rapid evolutionary change by natural selection. The ability of the genome of some pathogens to mutate easily and the impacts of such mutations on outbreaks of infectious diseases have demonstrated that epidemics arise not necessarily from the existence of high genetic diversity among pathogens but from their capacity to diversify genetically. For instance, the high genetic mutation rate of the human immunodeficiency virus facilitates its ability to evade our immune systems, because it can change its appearance faster than the time it takes our immune cells to recognize it and develop an effective defensive response.

Another example of the importance of pathogens' ability to undergo rapid genetic change comes from the influenza virus. These viruses infect horses, species of swine and birds, and of course, humans. The seasonal influenza that circulates in the human population each year is a genetically distinct strain of the virus that arises in East Asia and then travels around the globe. Most often, only small changes materialize in the flu virus genome, producing viruses that are similar to those that have circulated in the past, and that are therefore familiar to the immune systems of people who have been previously infected. As a result, they generally cause only mild disease.

Wild and domestic birds serve as repositories for avian influenza viruses, most strains of which do not produce significant illness in them. When ducks and chickens are raised near pigs, as commonly occurs in East Asia, they can transmit their flu strains to the pigs. Both mammalian and avian types of influenza viruses replicate well in pigs and are able to exchange their genetic information, with the result that new, and potentially virulent, flu strains can be formed. These may have major changes in their genomes, such as having one of the eight strands of genetic material that make up the flu viral genome replaced with a new one. Pigs have therefore been called "mixing vessels" for flu viruses. Humans, too, can serve as "mixing vessels," with avian and human influenza viruses exchanging genetic information. Such exchanges can create viruses that no living person's immune system has ever been exposed to, leading to the possibility of a severe global pandemic.

It is also possible for avian viruses from birds to be transmitted directly to humans, with potentially catastrophic consequences. In 1997 in Hong Kong, a strain of influenza virus in poultry was found to infect workers who slaughtered chickens.

IRRIGATION AND CHANGES IN MOSQUITO FAUNA: A CASE STUDY FROM SRI LANKA

In response to growing energy needs in Sri Lanka, the Mahaweli River dam project was conceived, and construction of a cascade of large dams along the river began in 1976. Nearly thirty years later, this project continues to have serious impacts on the spread of infectious diseases in the region. After the dams were built, and forests were subsequently converted to irrigated rice farms, shaded streams and forest pools were replaced by a multitude of exposed habitats, including rice fields, canals, small reservoirs, and temporary rainwater pools. Not surprisingly, these new conditions translated into dramatic shifts in the kind and number of mosquitoes present. In the span of three years, overall mosquito species richness dropped 20 percent, from forty-nine species that had been present in the forests, to thirty-nine species after their conversion to irrigated farmland, and only 60 percent of these once prevalent forest species remained abundant after irrigation development. But, at the same time, ten mosquito species that had not been common in the forests became dominant after the forests disappeared, including several important disease vector species, such as *Anopheles culicifacies* and other malaria-carrying *Anopheles* species; *Aedes albopictus*, a vector for dengue fever; *Mansonia* species that carry filariasis; and *Culex* vectors of Japanese encephalitis. In addition, several other malaria vectors, including *An. subpictus*, increased in numbers and became significant participants in malaria transmission.

Since then, this highly virulent strain, known as H5N1, has spread from Asia to Europe, Africa, and Oceania, mostly via migratory birds. In its wake, more than 200 million domesticated poultry have been slaughtered in an attempt to stem the spread of disease. According to the World Health Organization, as of September 10, 2007, 328 people have been infected globally and 200 have died from the disease (see www.who.int/csr/disease/avian_influenza/country/en/), almost all of whom had direct contact with infected birds. The high mortality rate of the H5N1 virus is likely a product of its innate virulence combined with the fact that human immune systems are not familiar with it. Should the H5N1 virus mutate so that person to person spread becomes possible, an influenza pandemic would almost certainly result, which experts expect may be as devastating as the "Spanish Flu Pandemic" of 1918–1919 that infected some one in five people living at that time and killed at least twenty-five million people worldwide in one year.[32]

Figure 7.10. Pigs and Chickens Being Raised Together in Vietnam. (Courtesy of the U.N. Food and Agriculture Organization/D. Nam.)

A high degree of genetic diversity occurs within local populations of the bacterium that causes Lyme disease, *B. burgdorferi*. In the eastern United States, as many as fifteen different strains of *B. burgdorferi* can coexist in stable populations in some areas. Only four of these strains have been recovered from Lyme disease patients, suggesting that the other eleven are not infectious to humans. The four infectious strains all are acquired by ticks from White-Footed Mice; two of them can also be acquired from Eastern Chipmunks (*Tamias striatus*) or Short-Tailed Shrews (*Blarina brevicauda*). If prior exposure to one of the strains that does not cause illness in humans results in an enhancement of that individual's immune response to those strains that do cause illness, then the presence of high bacterial diversity among *B. burgdorferi* strains could play a protective role. Such a scenario for Lyme disease is plausible but remains speculative.[33,34]

Genetic diversity within bacterial populations also allows for rapid evolution of antibiotic resistance. Antibiotics tend not to kill all the bacteria they are targeted to eradicate. One reason for this is because in any given population of bacteria, some may be resistant to the effects of a specific antibiotic. So the use of antibiotics themselves can select for the survival of bacteria resistant to these antibiotics. Too short a course or too low a dose of antibiotics tends to eliminate only the most susceptible strains, allowing resistant ones to proliferate, whereas too long a course can cause sustained selection for initially rare but more antibiotic-resistant types.

Large-scale rearing of livestock has been a fertile ground for the creation of antibiotic-resistant bacteria. In the United States, livestock receive roughly three times the amount of antibiotics by weight per year as do people—some 20 million pounds (slightly more than 9 million kilograms).[35] Of this amount, the vast majority, some experts estimate more than 90 percent, is used in subtherapeutic doses to prevent infection and to stimulate more rapid growth. About one-fifth of the antibiotics used in livestock in the United States are nearly identical to those used in people, setting the stage for the development of human bacterial infections that are resistant to commonly prescribed antibiotics.[36] Antibiotic resistance in some strains of *E. coli*, *Campylobacter jejuni*, and *Salmonella enteritidis*, the three major bacterial causes of human gastrointestinal illness, have already been tied to the use of antibiotics in livestock. (See section in chapter 8 on antibiotic resistance in *Campylobactor*, page 366.) Antibiotic resistance can emerge in a matter of weeks, as has been documented in poultry, and widespread prevalence of resistant bacteria has been demonstrated to occur within a few years after the introduction of a new antibiotic in human populations.

Diversity of Hosts

High species diversity within communities of vertebrates can reduce the risk of a disease being transmitted to humans, a phenomenon termed the "dilution effect."[37] For many diseases, only a few species are suitable, or "competent," hosts for the pathogen. Others, called "incompetent" hosts, might also be exposed to the pathogen but will not support its proliferation (in some cases, because their immune systems kill it) or its transmission to vectors that feed upon them. The most "competent" hosts are considered disease reservoirs, and most of these tend to be abundant, widespread, and resilient species, able to thrive in heavily disturbed habitats. They also appear to be the favorite sources of blood meals for disease vectors. It may be that natural selection favors pathogens and vectors that are able to thrive in, or on, common, widespread host species, as opposed to pathogens and vectors that are restricted to rarer or more sensitive hosts. Often, though not always, these more common and widespread host animals are rodents.

Because many of the vectors (usually mosquitoes or ticks) that transmit zoonotic pathogens are host generalists (i.e., they tend to feed on a wide variety of vertebrates), a large proportion of the vectors are likely to be feeding on "incompetent hosts" in communities that contain a high diversity of vertebrate species. As a result, in such communities, the pathogens are "diluted" among these hosts, the vectors are less likely to become infected, and the human risk of disease is therefore reduced.

In contrast, in communities with few vertebrate species, vectors have fewer alternatives for their blood meals. However, because highly "competent" reservoir hosts such as rodents tend to thrive in species-poor communities, as well as in species-rich ones, vectors in species-poor communities are more likely to be feeding upon these "competent" reservoirs and to become infected, and as a result, the risk of human infection is greater.

The "dilution effect" has been shown to operate with Lyme disease, West Nile virus disease, and hantavirus disease, and it is likely that it applies to other diseases as well. For this to occur, the following conditions must be met:

1. The disease vector must feed on a wide variety of host species.
2. The hosts must differ in their ability to transmit the pathogen to the vector (which depends on how readily the host can be infected, when it is infected, and how well the pathogen replicates in it).

3. The most competent reservoir host must be abundant and widespread and able to persist in degraded habitats where species diversity is low.

When these conditions occur, vertebrate communities with high species diversity will contain a greater proportion of "incompetent" reservoir hosts that will serve to deflect vector blood meals away from the most competent reservoirs, reducing infection prevalence in the vectors and the risk of transmission to humans. A more diverse community also has more predators and competitors, both of which tend to reduce the abundance of competent reservoirs, such as the White-Footed Mouse in eastern U.S. forests, further reducing the risk of human disease.[26]

A parallel phenomenon to the "dilution effect," known as the "decoy effect," may also occur with snails and schistosomiasis, where the presence of a diversity of snail species, some of which are "incompetent" intermediate hosts for schistosomes, may lower the risk of human exposure to the disease.[38,39] With disturbances such as deforestation or the building of dams, snail diversity may be compromised, and more "competent" hosts may be able to increase their numbers, resulting in outbreaks of schistosomiasis.

BIOLOGICAL CONTROLS

Perhaps one of the most important, although also one of the most poorly understood, controls on the spread of human infectious diseases comes in the form of organisms that prey upon vectors and intermediate hosts, or that neutralize infectious agents. One example may be illustrated by the outbreak of schistosomiasis around Lake Malawi in the early 1990s. In this case, a growing human population in the area led to overfishing that reduced the abundance of some fish that are major snail predators, such as the cichlid fish *Trematocranus placodon*. With reduced numbers of *T. placodon*, snails in the lake, including the competent intermediate host for *Schistosoma haematobium*, the snail *Bulinus nyassanus*, were able to increase their populations. Greater numbers of *B. nyassanus*, along with greater numbers of people living around the lake, appeared to have set the stage for the epidemic of urinary schistosomiasis that ensued.[40]

Biological controls come in many forms. Some bacteria, for example, have proven themselves to be effective at destroying mosquito larvae. Toxins derived from the bacterium *Bacillus thuringiensis israelensis*, known as Bti, and *Bacillus sphaericus*, specifically target mosquito larvae. (*B. thuringiensis* is the same bacterium that is the source for Bt toxin genes that have been inserted into some genetically modified crops [see chapter 9, page 384], but the Bt toxin genes are different from those that code for mosquito larval toxins.) Bti has its greatest potency with *Aedes* mosquitoes and, in some studies, with *Ae. aegypti* (the major vector of yellow fever and dengue).[41] It also kills the larvae of blackflies (*Simulium*). The toxin produced by *B. sphaericus* is particularly effective against *Culex* mosquitoes that can carry West Nile virus and some equine encephalitis viruses. *B. sphaericus* toxins work in pools of standing water that contain high concentrations of organic matter, conditions where Bti is less effective. One major advantage of these toxins over most chemical pesticides is that they are nontoxic to humans, and when used in appropriate amounts, they are more specific to their target organisms, making them less harmful to the environment.[42]

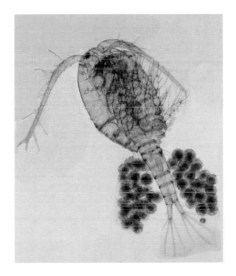

Figure 7.12. *Mesocyclops* species. (© Sonja Carlson, University of New Hampshire Center for Freshwater Biology, cfb.unh.edu/CFBkey/index.html.)

Figure 7.13. *Toxorhynchites splendens* larvae. The larvae of *Toxorhynchites* species may be important mosquito predators in some areas. (© Kosol Charernsom, Kasetsart University, Thailand.)

Dozens of fungal species have been tested for their insecticidal potential, as well. Two fungal species (*Metarhizium anisopliae* and *Beauveria bassiana*), in particular, have shown promise. Although such microbes alone are not likely to control mosquito populations, they can be important component parts of mosquito control strategies.[43]

Bacteria have also been useful in other ways to control infectious diseases. A bacterial cocktail has been shown to prevent colonization of pathogenic strains of *Salmonella* in the intestines of commercially reared poultry. Composed of twenty-nine different species that normally inhabit chicken intestines without causing illness (these bacteria are known as commensals — see box 3.1 on Microbial Ecosystems in chapter 3), the cocktail serves as a substitute for a mother hen, who would normally pass such commensal organisms to her offspring. Since chickens in commercial production facilities are raised without such maternal exposures, they have no natural source for their commensal bacterial flora.[44]

Recent research on the use of copepods (small crustaceans that live in both fresh and salt water) for the control of dengue fever in Vietnam has shown that these creatures can markedly reduce transmission of this disease. Where copepods of the local genus *Mesocyclops* were introduced into breeding areas for *Aedes aegypti*, mosquito populations fell by more than 90 percent over two years. The incidence of dengue plummeted at sites where copepods had been introduced, with the number of infections dropping by an average of more than 75 percent during the first year of study compared to locales where the copepods had not been introduced. No cases of dengue at all occurred in the copepod sites over the subsequent two years.[45]

Larger organisms may also help control some vectors of human diseases. Mosquitoes of the genus *Toxorhynchites*, for example, are known to prey on other mosquito species in many parts of the world, including the United States, consuming as many as 400 mosquito larvae during their larval stage of development. However, what role, if any, they play in controlling populations of mosquitoes that are vectors of human disease is not clear.[46]

Several species of fish, particularly in the context of rice farming, have also been shown to be highly effective at lowering the abundance of mosquito larvae. Their rediscovery as mosquito control agents was spurred by the development of resistance among some mosquitoes to pesticides. One remarkable success has been in India, where the introduction of fish into rice farms has helped to lower the burden of malaria by several hundred thousand cases between 1998 and 2003. In some villages, such as Puram in southern India, the introduction of guppies (*Poecilia reticulata*) into wells and streams has led to the elimination of the disease.[47] By contrast, during the same time period, in certain regions of India where DDT, but not guppies, was used for mosquito control, malaria incidence actually increased.[48]

In many ecosystems, a complex food web makes predicting the effects of losing predators that control vector populations difficult if not impossible. The introduction of exotic species often comes with considerable risk to the well-being of native species (see section on invasive species in chapter 2, page 47). In Australia, for example, the introduction of

Figure 7.14. Red Fox (*Vulpes vulpes*) with a Vole. (© 1992 Steve Kaufman.)

Figure 7.15. Redtail Hawk (*Buteo jamaicensis*) with an Eastern Grey Squirrel (*Sciurus carolinensis*). (© James F. Harrington, Saugus Photos Online.)

Gambusia holbrooki, known as the Mosquitofish because of a belief in the species' ability to control mosquito populations, has become pervasive in several freshwater ecosystems and has caused displacement of many native fish. These displaced fish include the Red Finned Blue Eye (*Scaturiginichthys vermeilipinnis*) and Edgbaston Goby (*Chlamydogobius squamigenus*), which, ironically, may have a greater capability to control mosquitoes than does *G. holbrooki*. *G. holbrooki* and the related species *G. affinis* are native to southern and eastern parts of the United States but have been distributed so widely that they may be among the most common freshwater fish species in the world.

Mammalian predators such as bobcats, weasels, and foxes, together with avian predators such as hawks and owls, are capable of regulating the abundance of small rodents in boreal and temperate zones. When these predators are experimentally excluded, rodent populations often swell. Because rodents are reservoirs for so many human infectious diseases, the expectation is that maintaining, or even enhancing, the populations of rodent predators will reduce the incidence of rodent-borne diseases.[2]

SPECIES EXPLOITATION AND THE CONSUMPTION OF BUSHMEAT

As discussed in chapter 2, an increasing human population in many parts of the developing world has both intensified the demand for food and made accessible, through the construction of roads as part of mining and logging operations, forested areas that were previously unreachable. These conditions have contributed to a rise in the hunting and eating of bushmeat. In the forests of West–Central Africa, the consumption of primate bushmeat has been implicated in the emergence of HIV/AIDS. Researchers who have studied the history of this disease now generally agree that the viral strain that has caused the pandemic HIV-1 originated from a closely related virus, known as a simian immunodeficiency virus, or SIV, that infects a particular subspecies of chimpanzee in West–Central Africa, *Pan troglodytes troglodytes*.[49,50] Genetic studies indicate that sometime between 1910 and 1950 was the last time that an SIV from this chimpanzee subspecies, called SIVcpz, shared a common ancestor with the human virus, HIV-1, providing strong evidence that SIVcpz was transferred from *P. troglodytes troglodytes* to people during this period[51] (see also discussion in chapter 6, page 244). Precisely when the virus entered humans is not known, but the transfer likely occurred when hunters butchered and handled SIVcpz-infected chimpanzee meat.

There are other SIVs besides SIVcpz. One from the Old-World monkey, the Sooty Mangabey (*Cercocebus atys*), SIVsm, has been shown by genetic sequencing to be the source for HIV-2, a strain of HIV that has been largely confined to West Africa.[52] By studying SIVsm in Sooty Mangabeys and HIV-2 disease in humans (and in several

Figure 7.16. Juvenile Sooty Mangabey. (Photo by W. Scott McGraw, Ohio State University.)

species of macaques that can develop fatal AIDS-like syndromes when experimentally infected with SIVs), we may better understand HIV-1 and AIDS.[53]

Recent studies of blood taken from primates kept as pets or killed by hunters in West–Central Africa by Martine Peeters (from the Institut de Recherche pour le Développement in Montpellier, France), Beatrice Hahn, and others have demonstrated that as many as thirty nonhuman primate species in Africa may carry SIVs. Prevalence rates range from 4 percent up to 60 percent in wild communities and vary depending on the virus and the primate species. There may be thirteen or more distinct SIV lineages. Because these primates carry SIVs without signs of disease, some have speculated that current populations of African nonhuman primates may be the survivors of ancient simian pandemics.[54]

Since nonhuman primates carry numerous other viruses in West–Central Africa, and perhaps in other parts of Africa, as well, indigenous populations are likely to be having ongoing exposures to these other viruses, mostly by hunting nonhuman primates, but in some cases by keeping them as pets.[54] For instance, antibodies to simian foamy virus (SFV), which is a retrovirus found in nonhuman primates (a retrovirus is a virus that stores its genes as RNA, but can transcribe them into DNA and integrate them into its host cell genome), have been found by Nathan D. Wolfe and his group at the UCLA School of Public Health in about 1 percent of nearly 1,100 hunters sampled in Central African forests, demonstrating that transmission of these viruses from nonhuman

Figure 7.17. Gorilla Slaughtered for Bushmeat. Exposure to the blood of primates, as is occurring with this man's handling of a butchered gorilla, allows for the transmission of primate viruses to people. (© Karl Ammann, karlammann.com.)

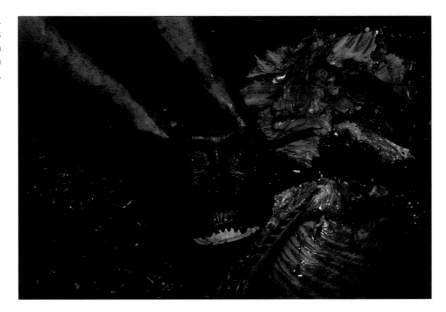

primates to people has occurred. Sequence analysis revealed three geographically independent SFVs from two different monkey species, de Brazza's Monkey (*Cercopithecus neglectus*) and the Mandrill (*Mandrillus sphinx*).[55] Because it is widely accepted that the HIV/AIDS pandemic began with such transmissions, we must ask ourselves whether human populations are now being exposed to nonhuman primate viruses that may cause yet other global pandemics in future decades.

In 2005, two previously unknown retroviruses related to HIVs, called human T-cell leukemia viruses, or HTLVs, were found in primate hunters in Cameroon. These were called HTLV-3 and HTLV-4. HTLV-3 is nearly identical genetically to a simian virus, STLV-3, which has been found in monkeys from very diverse habitats in East, Central, and West Africa, making it likely that the hunters acquired this virus from nonhuman primates.[56] Although HTLV-3 and HTLV-4 are not presently linked to any human diseases, another HTLV, HTLV-1 has been. HTLV-1 can, in about 1 percent of those infected, cause a difficult to treat blood cancer, adult T-cell leukemia, as well as a neurological disease known as HTLV-1–associated myelopathy that causes debilitating weakness. An estimated 10 to 20 million people worldwide have been infected with HTLV-1, and roughly 2 to 5 percent of them are expected to become ill.[57] The incidence of infection is not equal in all parts of the world: In the United States and Europe, the incidence is still very low (0.05 percent), but in endemic areas, including southern Japan and parts of the Caribbean, several percent of the population is infected, with intravenous drug users and people who receive multiple blood transfusions at higher risk. HTLV viruses are not routinely screened for by blood banks in Africa, adding to fears that they could be spread by blood transfusions.[58]

It is imperative that we understand the dynamics of viral infections acquired from nonhuman primates by studying these infections, and that large, healthy wild populations of these primates are maintained so that we may be able to do so. And, of course, it is essential that much greater efforts be made to educate people to stop all hunting of nonhuman primates, so that they are not exposed to their potentially deadly viruses (including Ebola), and so that we can save these magnificent creatures for their own sakes, and for the critically important information they may be able to give us.

CLIMATE CHANGE AND ITS EFFECTS ON INFECTIOUS DISEASES

The evidence that Earth's climate is warming due to the release of greenhouse gases from human activity is now overwhelming, and the impacts of global climate change on biological systems that control the emergence and spread of human vector-borne infectious diseases are becoming clearer. As described in chapters 2 and 3, climate change can disrupt ecosystems in ways that affect the populations of, and relationships among, disease vectors, hosts, and pathogens.

Warming temperatures, for one, will affect these organisms in many ways, altering their ranges, feeding, and reproductive habits, their ability to defend themselves, and other vital behavioral and physiological processes that, when disturbed, may change their capacity to cause human disease. Infectious agents such as protozoa, bacteria, and viruses, and their vectors—mosquitoes, ticks, and other insects—do not possess organ systems that allow them to regulate their body temperatures, and thus they are particularly sensitive to fluctuations in ambient temperatures. Climate warming can also differentially change the timing of various biological cycles of organisms that have co-evolved, decoupling exquisitely timed events that are essential for survival.

Global climate change has also brought about an acceleration of the world's hydrological cycle. This has led to more evaporation of water from the land and the oceans and an overall increase in cloud cover and precipitation, and in the frequency and severity of extreme weather events worldwide—both torrential rains and floods, and heat waves and droughts. Drought- and flood-driven human migrations can influence the spread of infected people (e.g., into and out of regions where malaria occurs) and increase the numbers of people living with poor sanitation in crowded refugee settlements and camps. Severe storms can also cripple public health infrastructure, increasing the likelihood and the consequences of disease outbreaks.

The Effect of Climate Change on Pathogens

Our growing understanding of marine coastal ecosystems has led to a greater appreciation of the role of climate warming on the spread of certain human diseases, notably cholera, caused by the bacterium *Vibrio cholerae*. The seasonality of cholera epidemics in some parts of the world has been linked to that of plankton blooms. Studies in the Bay of Bengal, for example, that examined sea-surface temperatures, the appearance of chlorophyll-containing phytoplankton, and the incidence of cholera support the hypothesis that warmer water promotes phytoplankton growth and can lead to cholera outbreaks. Under some conditions, *V. cholerae* bacteria can assume a dormant state in which they are not capable of causing infections. However, vibrios can revert to an infectious form in the presence of high nutrient concentrations and warmer sea-surface temperatures. Warm, nutrient-rich waters can also foster the growth of algae that serve as food for copepods, to which *Vibrio cholerae* attach and on which they feed (see figure 7.18). Copepods may in turn be consumed by shellfish and attach to finfish. These marine reservoirs for *V. cholerae* facilitate its long-term persistence in certain regions, such as in the estuaries of

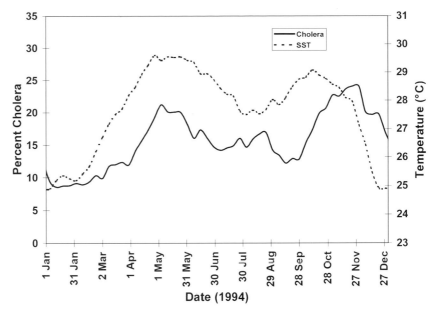

Figure 7.18. Graph Showing the Relationship Between Sea-Surface Temperatures (SST) and Cholera Case Data in Bangladesh, 1994. (Re-printed with permission from R.R. Colwell, Global climate and infectious disease: The cholera paradigm. *Science*, 1996;274:2025–2031. © 1996 American Association for the Advancement of Science.)

Figure 7.19. Microscopic View of a Female Copepod Carrying *Vibrio cholerae* Bacteria. (Courtesy of Drs. Rita Colwell and Anwarul Huq, University of Maryland.)

the Ganges and Bramaputra rivers in Bangladesh, and allow for regular outbreaks to occur.

As ocean temperatures are predicted to rise over the next century from global warming and more intense El Niños, there is concern that cholera outbreaks will increase in number and intensity.[59][61]

Climate-change–driven flooding due to heavy storms may also contribute to cholera outbreaks by increasing nutrient runoff from the land to coastal waters and triggering algal blooms. And it may produce conditions favorable to the spread of other human infectious diseases on land, as well. One example is leptospirosis, a water-borne bacterial disease. Outbreaks of leptospirosis typically follow floods that drive rodents from their burrows and that spread water contaminated by the urine of infected animals. Such was the case in the leptospirosis outbreaks that occurred after severe flooding from Hurricane Mitch inundated Nicaragua, Honduras, and Guatemala in 1998 (cases of malaria, dengue fever, and cholera also surged after the storm). In some urban areas of Brazil, such as in Rio de Janeiro, leptospirosis usually occurs during the summer rainy season when flooding affects low-lying areas that are densely populated by infected rats.[62]

Increased rainfall and flooding can also promote outbreaks of cryptosporidiosis, a diarrheal disease of humans caused by the protozoan parasite *Cryptosporidium parvum* that can live in the intestines of humans, farm animals, wild animals, and pets. Heavy rains can wash the parasites (in resistant forms called oocysts) contained in the waste of farm animals into reservoirs, and, because most public drinking water facilities are not able to filter out or treat water to kill the microscopic oocysts, such rains can result in high oocyte counts in drinking water and epidemics of cryptosporidiosis. These conditions occurred in the 1993 outbreak in Milwaukee, Wisconsin, when more than 400,000 people came down with the disease and more than 100 died from it, all of whom were people with weakened immune systems.[63]

The Effect of Climate Change on Vectors

Climate change can also have profound effects on disease vectors. Mosquito reproduction and survival depend upon temperature, and research on mosquito populations has shown that, in general, they reproduce more, and also bite more often, with increasing temperatures. When certain temperatures are exceeded, however, mosquito populations may drop. These data enter into models that predict that rising temperatures associated with climate warming will allow mosquitoes to expand their ranges into higher elevations and higher latitudes, and to increase population sizes within their traditional ranges. Higher temperatures can also speed up the development of malarial parasites or viruses in mosquitoes and hence increase their transmission, provided that there is not a decline in mosquito longevity as a result of excessive temperatures.

The same situation seems to apply to some tick species. From the mid 1980s until the late 1990s, there was a substantial increase in the incidence of tick-borne encephalitis (TBE) in Stockholm County in central Sweden. This disease is carried by the European Tick (*Ixodes ricinus*), also the main vector in Europe of Lyme disease. It was found that *I. ricinus* populations shifted their range to higher latitudes, related to fewer winter days with minimum temperatures lower than −12 degrees Celsius (around 10 degrees Fahrenheit), and increased their concentrations in their traditional range, in association with milder winters and extended spring and autumn seasons (see figure 7.20). These effects were associated with the increased number of TBE cases.[64,65]

In the case of dengue fever, research in Honduras, Nicaragua, and Thailand has shown that fluctuations in climate are linked to variations in the abundance of *Aedes aegypti* mosquitoes. Warmer weather was associated with more abundant mosquitoes and more cases of dengue fever. With malaria, studies in the highlands of central Ethiopia that covered the period from 1968 to 1993 have shown an association between a warming climate and an increased incidence of malaria, as has also been true for malarial outbreaks in highland towns of Kenya. But in Kenya, other factors, such as the immune status of those who came down with the disease and the effectiveness of public health measures, have also been implicated in the observed increased malaria transmission.[66]

Extreme weather events also have implications for vectors, though the effects of both flooding and drought are highly variable and situation dependent. In some cases, very heavy rains may decrease vector populations, because they may wash away eggs and larvae from breeding sites and thus decrease the likelihood of transmission, at least initially, after a storm. In other cases, flooding leads to more bodies of standing water, providing greater numbers of breeding sites for mosquitoes, and may drive some vertebrate reservoirs of disease into closer proximity to people, both leading potentially to more human disease. Paradoxically, drought may also facilitate population increases of some mosquito species that are important disease vectors. For instance, in urban settings droughts may result in the formation of small pools of highly concentrated, nutrient-rich water in drains, favoring the breeding of *Culex* species, including *C. quinquefasciatus* and *C. pipiens* that carry the West Nile virus.

Because climate change is projected to lead to more frequent and more severe droughts and floods in coming decades, it is highly likely that we will see more frequent and more severe outbreaks of some vector-borne human infectious diseases.

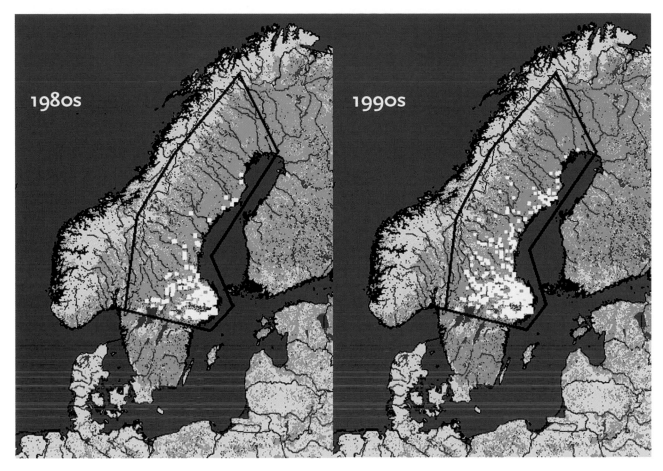

Figure 7.20. Change in the Distribution and in Concentrations of the European Tick (*Ixodes ricinus*) in Central Sweden. White squares illustrate districts in Sweden where *I. ricinus* ticks were reported to be present in the early 1980s and in the mid 1990s. The black line outlines the study region. There was an increase in the abundance of *I. ricinus* and a northern shift in its distribution from the 1980s to the 1990s. (From E. Lindgren, L. Tälleklint, and T. Polfeldt. Impact of climatic change on the northern latitude limit and population density of the disease-transmitting European tick, *Ixodes ricinus*. *Environmental Health Perspectives*, 2000;108(2):119–123.)

Figure 7.21. The Mosquito *Culex quinque-fasciatus*, the Main Vector for West Nile Virus in Southeastern United States. (Courtesy of the Centers for Disease Control and Prevention.)

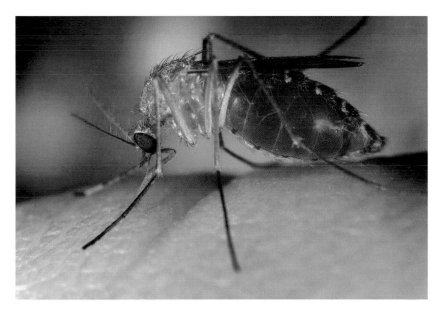

The Effect of Climate Change on Reservoir Hosts

Rodents are key reservoir hosts for many pathogens, including those that cause hantavirus pulmonary syndrome (HPS), Argentine hemorrhagic fever, Lyme disease, leishmaniasis, babesiosis, bubonic plague, and tularemia. For these diseases and others, increased rodent abundance is generally associated with an increased risk of human disease. For pathogens transmitted directly from rodent excreta, such as hantaviruses and leptospirosis bacteria, more rodents mean higher deposition rates of pathogens and the possibility of greater human exposure. For disease agents transmitted from rodents to humans by vectors, such as Lyme disease (ticks) or plague (fleas), higher densities of rodent reservoirs can result in higher rates of contact between vectors and reservoirs, a higher infection rate in vectors, and consequently, a higher risk of transmission to humans. One of the characteristics of climate change is not only an increased incidence of extreme weather events such as floods and droughts, but also greater weather variability, perhaps in part associated with climate-driven, more intense El Niño Southern Oscillation cycles, where floods may follow droughts. Such variability can lead to rapid population growth of some rodents, as was seen in the outbreak of HPS in the Four Corners area of New Mexico in the late spring of 1993 and again in 1998. In the 1993 outbreak, populations of Deer Mouse (*Peromyscus maniculatus*) increased tenfold in association with heavy winter snows and spring rains that followed a prolonged drought.[66] Similar patterns driven by climate change could occur in many other settings, with other reservoir hosts, vectors, and pathogens.

CONCLUSION

Predicting how human activities that disrupt ecosystems and result in losses in biodiversity affect the spread of human infectious disease is a daunting task. It is difficult enough to understand the complex relationships among pathogens, hosts, and vectors in undisturbed natural systems. For some diseases, such as malaria, schistosomiasis, and Lyme disease, our understanding of the impacts of human-caused ecological change is relatively well developed. In most cases, however, these associations are less well understood. There is limited knowledge, for example, about how the loss of some vector predators such as amphibians, which have among the highest known proportion of species at risk of extinction, may influence the emergence of new diseases or the spread of established ones. Nor is much research being done in the Arctic, which may be a sentinel site for observing the effects of climate change on the transmission of some infectious diseases because of its more rapid warming. And it is only very recently that a possible reservoir host for the deadly disease Ebola has been discovered and that the influence of genetic diversity among hosts, pathogens, and vectors has been explored.

Despite many uncertainties, clear examples now exist, and clear patterns are beginning to emerge, that demonstrate increases in vector-borne human infectious diseases as a result of such activities as deforestation, agricultural development, the building of dams, urbanization, and climate warming. Much greater attention to how these and other activities affect human disease is warranted in coming years as human populations increase and as such activities intensify.

◆ ◆ ◆

Suggested Readings

Beasts of the Earth: Animals, Humans, and Disease, E.F. Torrey and R.H. Yolken. Rutgers University Press, New Brunswick, New Jersey, 2005.

Climate Change Futures: Health, Ecological and Economic Dimensions, Paul R. Epstein, 2005; see chge.med.harvard.edu/research/ccf/.

The Coming Plague: New Emerging Diseases in a World Out of Balance, Laurie Garrett. Penguin Books, New York, 1995.

"Control of Human Parasitic Diseases." *Advances in Parasitology*, Volume 61, 2005.

Emerging Infectious Diseases, www.cdc.gov/ncidod/EID/ (this Centers for Disease Control and Prevention journal focuses on emerging diseases and includes many articles that discuss the role of ecosystem change in outbreaks of infectious disease).

The Great Influenza: The Epic Story of the Deadliest Plague in History, John Barry. Penguin Group, New York, 2004.

Human Frontiers, Environments, and Disease: Past Patterns, Uncertain Futures, Tony McMichael. Cambridge University Press, Cambridge, U.K., 2001.

Rats, Lice and History, Hans Zinsser. Bantam Books, New York, 1965.

SARS: A Case Study in Emerging Infections, Angela R. McLean, Robert M. May, John Pattison, and Robin A. Weiss (editors). Oxford University Press, New York, 2005.

Six Modern Plagues and How We Are Causing Them, Mark Jerome Walters. Island Press, Washington, D.C., 2003.

"Unhealthy Landscapes: Policy Recommendations on Land Use Change and Infectious Disease Emergence," Jonathan A. Patz et al. 2004. *Environmental Health Perspectives*, Vol. 112, Issue 10; see www.chponline.org/docs/2004/112-10/toc.html.

World Health Organization, www.who.int.

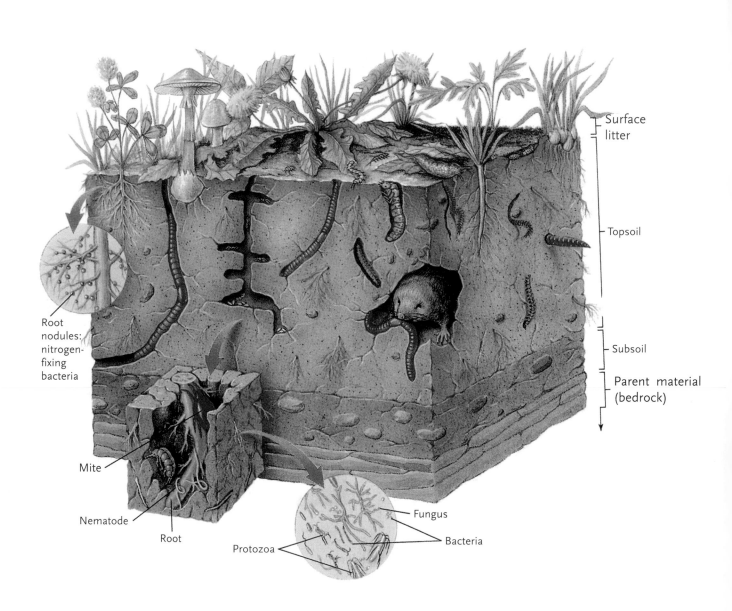

Surface
litter

Topsoil

Subsoil

Parent material
(bedrock)

Root
nodules:
nitrogen-
fixing
bacteria

Mite

Nematode

Root

Protozoa

Fungus

Bacteria

CHAPTER 8

BIODIVERSITY AND FOOD PRODUCTION

Daniel Hillel and Cynthia Rosenzweig

When earth is rich it bids defiance to droughts, yields in abundance, and of the best quality. I suspect that the insects which have harassed you have been encouraged by the feebleness of your plants; and that has been produced by the lean state of the soil.

—THOMAS JEFFERSON, in a letter written to his daughter Martha Randolph in 1793

Human beings, originally children of Nature, representing one species among many that shared the terrestrial environment, have gradually increased in numbers and expanded the extent and range of their activity so that we have now gained dominance over, and drastically modified, entire terrestrial and even marine biomes throughout the world. (Biomes are the world's major biological communities, defined by their predominant vegetation types, e.g., forests, deserts, and grasslands.) As a result, numerous other species have been deprived of their natural habitats and have been endangered or even eradicated. As mentioned in chapter 1, recent calculations suggest that rates of species extinctions are now on the order of 100 to 1,000 times greater than those before humans dominated Earth. For some well-documented groups, extinction rates have been even greater. Unless checked, the continued increase of human population and the intensified manipulation of the environment for short-term advantage are more than likely to result in serious consequences for human health. Having tampered with Nature in hopes of gaining control over it, humans are now more dependent on Nature than ever, especially on the diversity and intrinsic mutuality of all life forms that are its essential features.

A crucial imperative is to ensure the adequate production and supply of food for a human population of more than six billion and increasing, in a world in which terrestrial and aquatic resources already have been seriously degraded or depleted. Despite the lower fertility levels projected and the increased mortality risks to which some populations are being subject, the population of the world is expected to increase from approximately 6.5 billion at present to around 8.9 billion by 2050 (8.9 billion is the "best guess" mid-range estimate, made in 2004, by the U.N. Populations Fund).[1] The yearly addition of some seventy-seven million people on average poses many

(left)

Soil Organisms. An enormous diversity of organisms, as shown in this cross section of temperate soil, make soils fertile and food production on land possible. (From Peter H. Raven and Linda R. Bert (editors), *Environment*, 3rd ed. © 2001, Harcourt, Inc., reprinted with permission from John Wiley & Sons, Inc.)

extremely difficult challenges for human beings, especially in producing adequate supplies of food. The world's current mean population density of fifty people per square kilometer (slightly more than seventeen people per square mile) is projected to rise to seventy people per square kilometer by 2050, and since only about 10 percent of land is arable (i.e., suitable for agriculture), population densities per unit of arable land are, in fact, roughly ten times higher than these figures.[1,2] Given the poverty and famine that prevail in several regions, and the predicted change of Earth's climate (which in its normal state is already inherently unstable), it is an open question whether, and how, humanity can provide for itself while avoiding irreversible damage to natural ecosystems and their biodiversity. Increasing awareness of the issue and the development of new methods for conserving and managing food producing ecosystems, on land and in the oceans, offer hope for some progress in this difficult task. Using the promise inherent in such methods, however, must be constrained by an understanding of the potential problems and hazards they pose.

HISTORICAL BACKGROUND

For the greater part of their history, *Homo sapiens* have roamed over the landscape in small bands, subsisting as hunters and gatherers, and occasionally as scavengers. Being omnivorous, they availed themselves of a variety of food sources, opportunistically and eclectically, gathering edible plants and killing some animals for their meat (as well as for their skin, bones, antlers, and other usable parts). In time, humans learned to manipulate their environment, initially through the creation of fires. Although their lives were physically rigorous, they were venturesome and adaptable enough to spread out from their native African savanna into all the habitable continents. Relying on their ingenuity and tool-making ability, they adapted to widely varying environments—from icy northern Eurasia to arid central Australia.

A dramatic change in human lifestyle began toward the end of the epoch known by geologists as the Pleistocene (which lasted from about 1.8 million to about 11,000 years ago) and the beginning of the geological period that we are currently in, called the Holocene, which immediately followed the Pleistocene. That change evidently took place earliest in the Near East, some 10,000 to 12,000 years ago, during what archaeologists call the Neolithic Age. As the last ice age ended, the warming trend gave rise to a profusion of plant and animal life in that region, which afforded the human populations living there an abundance of food sources and of favorable sites for regular, and eventually permanent, habitation.

As groups of humans shifted from nomadic to sedentary living and began to form settlements, they also learned, after collecting seeds of wild plants such as wheat and barley, to domesticate selected plants. Thus, agriculture began. At first, it was in the form of rain-fed farming in relatively humid areas; later, water was provided to crops by irrigation from rivers in river valley farms. Simultaneously, animal husbandry developed, based on the herding of livestock such as sheep, goats, and cattle, both in conjunction with village-based farming and in the context of an alternative lifestyle called pastoralism, a semi-nomadic, subsistence pattern of living characterized by tending herds of animals.

Of the many plants with edible products, several were found suitable for early domestication. Prominent among these were selected species of the Gramineae family (the cereal grains of wheat, barley, oats, rye, and sorghum), the Leguminosae family (peas, lentils, chickpeas, and several types of beans), vegetables of various genera, and a number of fruit-bearing woody plants or trees (olives, grapes, almonds, pomegranates, figs, and dates). Only a limited number of animals lent themselves conveniently to domestication. Breeding programs, along with natural hybridization, played a pivotal role in shaping the genetic and evolutionary trajectories of domesticated animal species.[3]

Consequently, human societies abandoned their prior lifestyle as roaming hunter-gatherers, and as they became sedentary producers of food, they came to depend on their managed crops and livestock for subsistence. Agriculture created plants and animals (e.g., wheat, rice, maize [corn], cattle, swine, and poultry) that are now some of the most prevalent and widespread organisms on Earth, and thanks to these organisms, humans indeed have become the world's dominant species. A mutual dependency thus developed between humans and the organisms they domesticated. However, in some cases, such as with corn, as described by Michael Pollan and others, it is not always clear which species has been the domesticator and which the domesticated.[4]

The same processes of transition to an agricultural or pastoral economy that first took place in the Near East also appeared independently in several other centers, and then rapidly spread from these places as well.[5] These include, among others, southern and eastern Asia, Central Africa, and Central America, each with its own indigenous selection of plants and animals able to be domesticated. In all those locations, the agricultural transformation improved food security and thereby set in motion a progressive increase of human population density. So productive was the enterprise of agriculture that over time an ever-decreasing number of farm workers were able to feed ever-larger numbers of people. Urban centers then developed in which people engaged in a variety of other occupations (e.g., industry, art, science, medicine, and institutionalized religion), thus creating the basis for complex civilizations.[6]

A less auspicious consequence of these same developments was a narrowing of the variety of foods that served to sustain populations. The domesticated lifestyle provided only a limited number of tended species and strains instead of the wide selection of types and sources of food that humans previously had been able to collect or hunt in the wild. As the variety of foods was reduced, so too was the nutritional balance and quality of the diet. The study of archaeological remains from around the world reveals that the shift from hunting and gathering to increased nutritional focus on domesticated grains that occurred around 10,000 years ago coincided with a decline in health, including increased evidence of dental disease, iron-deficiency anemia, infections, and bone loss.[7] Moreover, reliance on a small number of crops and animals maintained in managed sites also made societies vulnerable to production failures resulting from the vagaries of weather, as well as from pests and diseases of crops and livestock. People living in close communities, and eventually in cities, themselves became more vulnerable to communicable diseases.

So the great advantages of domestication were not without attendant disadvantages. However, the increase in population numbers and densities made possible by the initial successes of agriculture did not allow a return from the domestication of crops and animals to the lifestyle and economy of nomadic hunting and gathering.

Humans also changed biologically, because of the selective pressures of living in built environments and changes in diet associated with being increasingly sedentary.[8] The agricultural transformation thus became effectively irreversible.

As long as human exploitation of the land and its biotic resources was restricted to small enclaves, the surrounding expanses of relatively undisturbed natural ecosystems remained intact, with their biodiversity preserved. But, as the extent and intensity of human exploitation of the land increased, along with an increase in populations, natural habitats were reduced and fragmented. This process of human encroachment has continued and accelerated over hundreds of years, until, today, nearly half of Earth's continental surface is under direct human management, with croplands and pastures making up most—around 80 percent—of this amount.[9] A similar process

Figure 8.1. Severe Soil Erosion on a Wheat Farm in Washington State. The farmer is using a probe to take a soil sample for measuring such things as its nutrient availability and organic matter. (Photo by Jack Dykinga, U.S. Department of Agriculture.)

has occurred in Earth's freshwater and oceanic ecosystems. Even where humans have not intervened directly, the secondary effects of their activity (e.g., the chemical residues of industrial production) have caused indirect deleterious effects.

Within agricultural lands themselves, poor management practices have led to their degradation. Removing the vegetative cover and pulverizing the soil by tillage or by the trampling of livestock or machinery has made the soil vulnerable to wind erosion during dry periods and to water erosion during rainstorms. In extreme cases, fertile topsoil has been completely scoured away, and less fertile subsoil (or even the sterile bedrock) has been exposed. Soil productivity is thus greatly impaired, as is its capacity to support various forms of life.

Quite another process of soil degradation occurs in irrigated lands, particularly in river valleys located in arid regions. There, the traditional practice of flood irrigation with large volumes of water causes much percolation through the soil. This tends to raise the water table, to saturate the soil excessively (a phenomenon called waterlogging), and to accumulate salts at or near the soil surface (a process called soil salinization), all of which destroy soil productivity.

Fortunately, the picture is not entirely bleak. Many of the ills just described can be prevented or alleviated. New trends and opportunities offer hope that further threats to biodiversity can be avoided. For example, human population growth seems to be slowing. Moreover, agriculture has already begun to develop and adopt better production methods coupled with biological control and conservation practices that are aimed at preserving, and even enhancing, the diversity of life in agricultural systems. These approaches are impelled by a growing recognition of the indispensable importance of biodiversity to agriculture.

AGRICULTURE

Dependence of Agriculture on Biodiversity

All the plants whose products are used by humans, either directly or indirectly via plant-consuming animals, were derived originally from wild ancestors. So were all domesticated animals. Domesticates were selected and bred for their desirable traits, although these traits have been the ones most advantageous for farmers and consumers, not necessarily those most desirable for the crop or livestock species themselves. But as environmental circumstances and stresses changed, as the requirements and preferences of humans changed, and as some domesticated organisms became vulnerable to certain diseases and pests, the need arose repeatedly to breed new varieties.

Traditionally, agricultural breeding has been done with close genetic relatives. These are either wild genotypes, or domesticated varieties or strains of the relevant organisms. Genetic diversity in a crop species is often considered a resource for future crop improvement. Different strains may contain different genes, and sometimes may include genes that impart resistance to certain pests or environmental stresses. Recently, new possibilities for genetic manipulation have arisen that transfer desired

TABLE 8.1 THE FIFTEEN MOST IMPORTANT FOOD CROPS IN TERMS OF PRODUCTION

PLANT CROP	TYPE OF CROP	WORLD PRODUCTION (× 1,000 METRIC TONS)
Sugarcane	Sugar plant (stem)	1,290,345
Corn (maize)	Cereal grain	712,334
Rice, paddy	Cereal grain	629,881
Wheat	Cereal grain	625,151
White potato	Ground crop (tuber)	320,978
Soybean	Legume	214,849
Cassava (manioc)	Ground crop (root)	208,559
Barley	Cereal grain	137,553
Sweet potato	Ground crop (root)	122,883
Sorghum	Cereal grain	59,154
Peanuts (ground nuts)	Legume	37,763
Millet	Cereal grain	30,533
Oats	Cereal grain	23,589
Beans, dry	Legume	22,880
Rye	Cereal grain	15,200

Source: FAOSTAT, Core Production Data from the U.N. Food and Agricultural Organization, available from faostat.fao.org/ [cited September 14, 2007].

traits not just between strains of the same species, but even from one species to another. This technology potentially greatly enlarges the range of genetic resources available to agriculture, though the new techniques also present new hazards (see chapter 9). Either way, breeding plants and animals for agricultural purposes was and remains dependent on Nature's rich array of life forms, that is, on its natural biodiversity.

Of all the myriad species of plants or animals whose products can be useful to humans, agriculture directly uses only a few hundred. Among these, just eighty crop plants and fifty animal species provide most of the world's food. According to the U.N.'s Food and Agriculture Organization, a total of only twelve plant species provide approximately 75 percent of our total food supply,[10] and only fifteen mammal and bird species make up more than 90 percent of global domestic livestock production.[11] However, what is not generally appreciated is that those relatively few species depend vitally for their productivity on hundreds of thousands of other species. Among the latter are insects and birds that pollinate crop flowers and feed on crop pests.

Even more numerous and varied are the microbial species that live on and in plants and animals, and that are especially abundant in the soil. They, too, help to protect against pests, as well as to decompose residues (including pathogenic and toxic agents), transmute them into nutrients for the continual regeneration of life, and form and stabilize soil structure. Agricultural productivity and sustainability benefit from microorganisms in many ways, including the conversion by bacteria of elemental nitrogen from the atmosphere into nitrogen compounds, soluble ammonium and nitrates, that serve as essential nutrients for plants. Nitrogen-fixing bacteria may be symbiotic, such as *Rhizobium* bacteria that attach themselves to the roots of legumes, or they may be free-living. Quite another function is fulfilled by mycorrhizal fungi, which live in association with crop roots and facilitate the uptake of phosphorus and other relatively immobile nutrients (described in boxes 8.5 and 8.6 below).

Pollinator species, and biological control agents that prey on insect and other kinds of pests generally live in natural or seminatural ecosystems,[12] underscoring the importance of maintaining undisturbed areas adjacent to agricultural tracts, such as hedgerows between fields. Clearing away such ecosystems in the belief that such action prevents the invasion of pest species into fields and orchards sometimes actually does more harm than good by depriving agriculture of beneficial organisms.[13]

In ways both visible and invisible, agriculture depends on biodiversity. This dependence not only operates in the present but also provides an insurance policy for the future. Genetic diversity in wild populations, for example, can protect crops from future outbreaks of pests and diseases, and from such disturbances as climate change, by serving as a pool for the natural and guided (by hybridization) selection of new, better adapted, and more resistant organisms. Diminution of that diversity endangers agriculture just as it endangers all the other, inherently interdependent processes of life on Earth.[14]

Functions of Biodiversity in Agriculture

Growing conditions differ from place to place due to differences, for example in soil, water availability, temperature, exposure to sun and winds, day length, and the prevalence of diseases and pests. They also differ from season to season due to the variability of climate. Pure stands of genetically similar, or essentially identical, plants, selected because of their ability to grow well under the specific conditions of a particular place, are therefore at greater risk when these conditions change than are genetically diverse stands. Genetically diverse crops can better survive in environments in which conditions fluctuate, because some are vulnerable to certain changes but others are not. Though such diverse crops may not provide yields that are as great during favorable or normal seasons, they are more likely to provide an adequate, and perhaps even greater, yield during unfavorable seasons. Pure stands, lacking genetic variety and hence an adaptability to changing conditions, may be devastated during such seasons, for example, those characterized by extreme weather.[15]

DISEASE CONTROL

Genetic diversity is thus likely to reduce the odds of crop failure and to contribute to greater stability of production, benefits that are also found in the mixed-species and multispecies cropping systems common to subsistence farms. The vulnerability of monocultures to disease illustrates this value of genetic diversity. Pathogens spread more readily, and epidemics tend to be more severe, when the host plants (or animals) are more genetically uniform, numerous, and crowded, like battalions of identical soldiers in close formation, because the pathogens encounter less resistance to spreading their infections than they do in mixed stands. Owing to their high densities and the large areas over which they are grown, both crop plants and livestock are repeatedly threatened by ever-new infestations by pests and diseases. Existing pests and diseases are continually evolving strains that overcome the innate defenses of particular strains or breeds, as well as the chemical treatments applied by farmers.

Many historical examples can be cited to prove that monoculture stands, or concentrations of crops and livestock with uniform genetic traits, though they may be more productive in the short run, entail the risk of succumbing, sooner or later, to changing conditions. Catastrophic outbreaks of disease, invasions of insects, and climatic anomalies have caused many wholesale crop and animal destructions in the past. Such episodes have resulted in famine, especially where, in the absence of sufficient diversity, no varieties or breeds were present that could withstand the destructive outbreaks.

Among the many examples of disastrous outbreaks are the infestation of stem rust on wheat in Roman times; the mass poisoning from ergot-tainted rye during the Middle Ages in Europe; the failure of French vineyards in the late nineteenth century secondary to invasion by an aphid known as the Grape Phylloxera (*Daktulosphaira vitifoliae*), which carried downy mildew disease; and the potato famine that hit Ireland in the 1840s and 1850s. The latter was caused by the fungus *Phytophthora infestans*, which arrived accidentally from North America and attacked the genetically uniform potato stock that had long served as the mainstay of Irish farms. As a result, more than a million people died from starvation or typhus and other famine-related diseases, and 1.5 million people immigrated to North America during the famine years alone.[16]

The massive concentration of agricultural production (and of food consumption) in only three primary crops—wheat, rice, and maize (corn)—that together account

Figure 8.2. Different Varieties of Potatoes. Growing a variety of potato strains in the same field protects against crop failures from disease or extreme weather. (Courtesy of the U.S. Department of Agriculture.)

Figure 8.3. Stem Rust in Wheat. This wheat disease, caused by the fungus *Puccinia graminis*, has undergone a resurgence in recent years. (Courtesy of Jacolyn A. Morrison, Cereal Disease Laboratory, Agricultural Research Service, U.S. Department of Agriculture.)

for more than half of the global totals for nutritional energy derived by people from crop plants is particularly worrisome. In principle, such a concentration creates vulnerability. One example of this vulnerability is the recent outbreak of scab—*Fusarium*, or head blight—on wheat and barley in the states of Minnesota and North and South Dakota. Many farmers in areas where scab has been severe have been forced to abandon farming for lack of alternative crops to grow profitably.[17] Another is the resurgence of stem rust on wheat caused by a new strain of the same fungus that led to epidemics in Roman times, *Puccinia graminis*. *P. graminis* had caused huge wheat losses during the first half of the twentieth century and then reappeared in Uganda in 1999, spreading over the next four years (spores are carried by the wind and the clothing of travelers) to Kenya and Ethiopia.[18] If control strategies are not widely and promptly implemented, it is believed, this new stem rust strain (*P. graminis f. sp. tritici*, also called Ug99) could spread beyond eastern Africa to the Middle East and Asia and significantly affect wheat production in these regions.[19] Ultimately, the best insurance against the future failure of wheat, rice, and maize crops is the enhancement of biodiversity, both to increase the diversity of strains for these crops and to discover appropriate substitutes for them.

Other experiments have shown reductions in the severity of wheat diseases when mixtures of wheat varieties are planted together and, on a commercial scale, in the severity of Barley Powdery Mildew (caused by the fungus *Erysiphe graminis*) when different varieties of Barley (*Hordeum vulgare*) are interplanted.[20,21] And it is widely known that when mixtures of different crop species are planted, such as maize and beans (as was widely practiced by Native Americans), yields are greater, as is resistance to the spread of pests, diseases, and weeds (see below).[22]

INSECT PESTS

Small-scale farmers in the tropics have long used crop diversification as a way of minimizing the risk of crop failure, for example, as a result of pest infestations.[23] Experiments have demonstrated that the differences in pest abundance between diverse and simple agricultural systems can be explained, in part, by the ability of nonhost species to disrupt pests from attacking their main hosts effectively. This phenomenon largely applies to so-called "specialist herbivores," that is, those that have specific host targets.

Several mechanisms seem to be involved in diverse systems that interfere with an insect's host-seeking behavior. These include camouflage—the host plant is guarded from insect pests by the presence of other plants that conceal it; crop background—certain pests prefer certain backgrounds of a particular color and/or texture; masking or dilution of attraction stimuli—the presence of some nonhost plants mask or dilute the attractant stimuli of the host plant, leading to a breakdown or reorientation of feeding and reproduction by the insect pest; and repellent chemical stimuli—aromatic odors given off by certain plants that disrupt the insect's host-finding ability.

BOX 8.1

GENETIC DIVERSITY AND DISEASE CONTROL IN RICE

Rice Plants in Flooded Paddy Fields in Taiwan. Typically such fields are terraced in Yunnan Province in China because of the mountainous terrain. (© Corbis Corporation)

Until about 100 years ago, when farmers planted *monocultures* (one type of plant rather than several mixed types of plants), they did so, for example, by growing different crops such as wheat or maize or rice, each in its own field.

Gradually, farming became more restricted so that a greater reliance was placed on fewer and fewer crops. Many farmers in many parts of the world began to grow just one species of crop, such as in vast areas of the U.S. farm belt where, for example, only maize or only soybeans are grown. In recent decades, monocultures are more likely to be composed of specific varieties within species, or even specific genetic differences within varieties.[a] This progressive reduction in the diversity of crops puts them at greater risk for the spread of infection: If one plant is susceptible to a certain infectious agent, then that infection is capable of spreading to other similar or identical plants in the field. The standard response among most farmers growing monocultures that are vulnerable to such infections, for example, to a particular fungal disease, has been to rely on the development of new resistant varieties

developed by hybridization (or perhaps now by genetic engineering) or on new fungicides.

But other strategies, potentially less damaging to the environment, are beginning to be more widely practiced. Subsistence farmers in Asia and in other parts of the world have known for centuries, and perhaps for millennia, that growing crop mixtures is more productive than growing single varieties (Darwin wrote about this in *The Origin of Species* for growing wheat). Until fairly recently, however, the mechanisms of this better productivity were not well understood. In a seminal experiment involving thousands of farmers and more than 3,300 hectares (about 8,154 acres) in Yunnan Province in China in 1998 and 1999, Youyong Zhu and his co-workers studied genetically diversified rice crops (*Oryza sativa*) to test the effect of such plantings on rice blast disease, caused by the fungus *Magnaphorthe grisea*. Yunnan Province has a cool, wet climate that fosters the development of rice blast. To control it, farmers have traditionally made multiple fungicide applications to rice plant foliage.

When disease-resistant varieties of hybrid rice were planted alongside disease-susceptible varieties

BOX 8.1 (CON'T.)

of glutinous rice (a type of rice used mainly in Chinese cooking for desserts), glutinous rice yields increased by 89 percent, and the severity of their blast infections decreased by 94 percent, when compared to glutinous rice grown in monoculture. Blast severity also decreased, although to a lesser extent, among the hybrid varieties. Blast was controlled so well that by the end of the two-year experiment farmers completely stopped applying fungicides, and the practice of rice variety intercropping expanded to involve more than 40,000 hectares (about 99,000 acres) by the year 2000.[b]

At one survey site in 1999, data collected about microclimates in the rice plant canopies provided one explanation for the dramatic results of the intercropping. The data demonstrated that the height differences between the taller glutinous and shorter hybrid rice varieties created a physical barrier that resulted in temperature, humidity, and light conditions that were less favorable for rice blast disease than those present in the canopy microclimates of monocultures, where the crop heights were uniform.[b] Dilution of the rice blast pathogen was also thought to be a factor in reducing disease severity in glutinous rice, because of the increased distance between susceptible plants in mixed fields as opposed to those grown in monoculture. Another reason for the success of the rice intercropping may be that with mixed varieties, an immunization process is at work. If a particular rice variety, like the hybrid rice in this experiment, is exposed to a particular pathogen strain to which it is resistant, such as a strain of rice blast, it can develop a generalized immune response that may serve to protect it against other, genetically different, pathogen strains that normally would cause infection. As a result, the spread of infection in the field may be inhibited.[a] In mixed plantings, competition may develop between those pathogens that are better adapted to specific planted varieties and those

Rice Neck Blast Infection in a Variety of Rice Known as Wells Rice, Arkansas, 2009. (© Rick Cartwright, University of Arkansas Division of Agriculture.)

that are better adapted to the combinations. By changing the mixtures that are planted in successive years, a farmer might be able to stay ahead of these adaptations and further lower disease incidence.

Some mechanisms interfere with pest populations as a whole, including mechanical barriers, such as companion crops that block herbivores from moving across polycultures. There are also microclimate influences, as shown in the rice blast study described in box 8.1, which may also cause insects to experience difficulty in locating and remaining in suitable microhabitats.

Other field studies have supported the hypothesis that increasing crop diversity will decrease pest abundance. For instance, Lepidoptera (butterflies and moths) larvae, which bore into maize and sorghum plants, constitute one of the major constraints to efficient maize and sorghum production in the developing world. The effect of using an agroforestry system called "alley planting," involving maize interspersed with hedgerows of the tree legume *Leucaena leucocephala*, spaced at 3-meter (~10-foot) intervals, has been studied in western Kenya. By several measures, the maize–*Leucaena* intercrop plots conferred significant protection against borers when compared to plots in which maize was grown on its own. For example, the abundance of adult, larval, and pupal stages of Maize Stem Borers (*Busseola fusca*) was reduced,

Figure 8.4. Alley Planting of Maize in Africa. Maize planted between rows of the nitrogen-fixing legume *Leucaena leucophalia* on a farm at the headquarters of the International Institute for Tropical Agriculture in Ibaddan, Nigeria. Such alley farming is becoming increasingly popular in Africa and in other parts of the world. (© Musa Usman, International Institute for Tropical Agriculture, Ibaddan, Nigeria, www.iita.org.)

there was less damage to maize foliage and to plant stems, the number of borer entry and exit holes were fewer, and, of greatest significance, there was less maize mortality. The reduced numbers of pests in the maize–*Leucaena* plots were associated with greater yields per plant and for the plot as a whole, even though the presence of *L. leucocephala* reduced the number of maize plants by 25 percent.[24] The legume, with nitrogen-fixing bacteria attached to its roots, also served to fertilize the soil for the maize, no doubt also contributing to the greater yields.

During a good part of the twentieth century, farmers throughout the world have relied heavily on chemical pesticides. But often these pesticides kill natural enemies of the pests and provoke resistance in the pests they are intended to kill. The absence of natural enemies may allow even benign insects to increase their populations to such an extent that not only do they become pests in their own right, but they may also be able to acquire resistance to pesticides. This pattern is known as the "pesticide treadmill." In Central America, for instance, a host of predatory and parasitic arthropods was removed from agricultural systems, and their loss resulted in greater problems, to the point that the cotton industries of Guatemala, El Salvador, and Nicaragua were severely damaged.[12]

In the last decades of the twentieth century, an increasing awareness of the limitations and damages associated with chemical pesticides has led to the development of sophisticated techniques referred to as "integrated pest management" (IPM).[25]

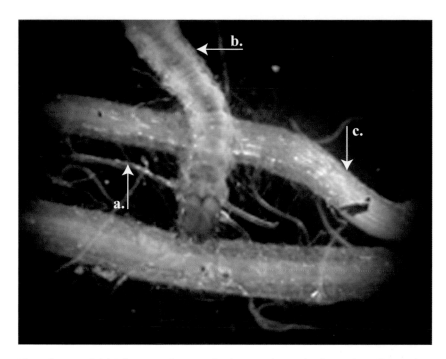

Figure 8.5. Beneficial Soil Nematodes. Beneficial nematodes (a) that live in the soil enter their hosts, such as this Western Corn Rootworm (*Diabrotica virgifera virgifera*) (b) that is eating a corn root tip (c), through natural openings such as the mouth or anus. Once inside, they release their own intestinal endosymbiotic bacteria (see box 3.1 on microbial ecosystems in chapter 3), that digest the host and multiply, supplying food for the nematode, which in turn grows and multiplies. When the host is fully digested, the nematode molts and breaks out into the soil again, looking for new hosts. (Photo by Sergio Rasmann and Matthias Held, University of Neuchâtel, Switzerland.)

Such methods are based on the judicious combination of biological controls, either applied directly or encouraged through the maintenance of undisturbed habitat such as hedgerows and other natural noncrop habitats bordering fields (see chapter 3, page 95), together with sparing applications of chemicals only when absolutely necessary (see chapter 9, page 400).[26]

The biological control component of IPM, in turn, depends on ecosystem biodiversity. For example, spiders are one of the species that show some of the greatest potential as biological control agents.[27] Others include various species of nematodes, wasps, ladybugs, and lacewings. Still others, such as shrews (which, along with moles, belong to the mammalian order Insectivora); frogs, toads, and salamanders; dragonflies and damselflies; praying mantises; bats; and birds, eat a wide variety of insects, slugs, worms, snails, and other organisms that attack crops. They may also help control agricultural pests in some fields and gardens, but because some of these are generalist predators and may also be eating beneficial organisms, their net contribution to pest control may often be difficult to determine. Although our knowledge of natural biological controls is rudimentary, it is clear that they play critically important roles in controlling crop pests.

BIRDS

Investigations into the relationship between agricultural intensification, with its associated widespread use of fertilizers and pesticides, and the collapse of Europe's

farmland bird populations have found significantly greater declines in bird populations and contractions in bird ranges in countries with more intensive agriculture. The effects were so great as to be discernable at a continental level, making them comparable in scale to deforestation and global climate change as major anthropogenic threats to avian biodiversity.[28]

Comparisons of bird populations in agricultural fields with those in native savannah and grasslands in the Serengeti have revealed similar patterns. Substantial, but previously unnoted, declines in bird biodiversity were discovered in the agricultural lands there. The abundance of bird species found in these fields was only 28 percent of that in native savannahs. Insect-eating and grain-eating species were the most affected, particularly ground-feeding and tree species, with as many as 50 percent of the species of both groups not found at all in the agricultural sites.[29] Although there was a concurrent decline in insects in the agricultural regions, it is predicted that the great reduction in insectivorous birds will likely affect the ability of these Kenyan farmers to control future insect-pest outbreaks. Also, the lack of raptors in the agricultural sites, particularly those that consume rodents (e.g., the Black-Shouldered Kite [*Elanus caeruleus*] and the Long-Crested Hawk Eagle [*Spizaetus ayresii*], both of which are abundant in the Serengeti savannah), may be contributing to the frequent explosions in rodent populations, such as those seen for the Natal Multimammate Mouse (*Mastomys natalensis*), in agricultural zones.[30]

POLLINATORS

There are thought to be more than 100,000 different pollinator species on Earth. Declines in their numbers, reaching 70 percent in some places, have been reported in every continent except Antarctica.[31,32] The consequences of such steep declines in pollinators for the world's food supply are potentially enormous. While the majority of the world's staple crops (wheat, rice, maize, potatoes, yams, and cassavas) are either wind or self-pollinated, or are propagated vegetatively (e.g., by stolons [aerial shoots from a plant that produce new root systems and new offshoots] or by rhizomes), many other important agricultural species rely on pollinators.[33] For instance, more than 80 percent of the 264 species grown as crops in the European Union are dependent on insect pollination.[34,35] In addition, the yield of tomatoes, sunflowers, olives, grapes, and soybeans—all major crops—is optimized by regular pollination.[33] Fruit trees and legumes may be particularly hard hit by a loss of pollinators, especially since they are grown intensively (see the opening figure for chapter 3, which illustrates the impacts on apple orchards in a region of Nepal when native bees went extinct).

In late 2006 in the United States, honeybees began dying in great numbers along the East Coast of the United States, as well as in Texas, California, and other states. A total of some twenty-four states were affected, and losses of up to 70 percent of hives were reported. The disease has been named colony collapse disorder (CCD) and is causing great alarm among beekeepers and farmers who grow such crops as alfalfa, almonds, apples, oranges, peaches, blueberries, and cranberries that are all heavily dependent on honeybee pollination. Billions of dollars in agricultural losses are expected from the declines. As of early 2007, the cause of CCD had not been determined,[36,37] and there was concern among some researchers that certain

(left)

Figure 8.6. Other Beneficial Organisms. (a) Brachonid wasp eggs (probably the brachonid *Cotesia congregata*) on a Tomato Hornworm (*Manduca quinquemaculata*). The eggs hatch and digest the worm. (© Jill M. Nicolaus, 2004.) (b) Cross Spider or Cross Orbweaver (*Araneus diadematus*). (Courtesy of Dawn Hudson, Dreamstime.com.) (c) Green Lacewing (*Chrysoperla* sp.) larva eating white fly nymphs. Green Lacewing larvae have been called "aphid lions" and are voracious predators of many agricultural pests. (Courtesy of Jack Dykinga, Agricultural Research Service, U.S. Department of Agriculture.) (d) Fourteen-Spot Ladybugs or Lady Beetles (*Propylea quatuordecimpunctata*) devouring an aphid on a pea plant. Scientists believe Fourteen-Spot Lady Beetles may help control Russian Wheat Aphids (*Diuraphis noxia*) that now infest seventeen Great Plains and western states in the United States. (Courtesy of Scott Bauer, Agricultural Research Service, U.S. Department of Agriculture.) (e) *Diapetimorpha introita* wasp is preparing to lay an egg in the tunnel of a Corn Earworm (*Helicoverpa zea*) pupa. (Courtesy of Scott Bauer, Agricultural Research Service, U.S. Department of Agriculture.) (f) Big-Eyed Bug (*Geocoris* sp.), having glued a whitefly to a leaf, can devour its prey at its leisure. (Courtesy of Jack Dykinga, Agricultural Research Service, U.S. Department of Agriculture.)

BOX 8.2

BENEFICIAL INSECTS AND RICE PRODUCTION IN INDONESIA

Indonesia, the world's most densely populated country, during its very long history of agriculture has concentrated on the production of rice grown in flooded fields called paddies (fields are flooded to keep weeds in check). In the past, paddy rice was interspersed with other types of crops, and such diverse vegetation landscapes were common in many of the country's agricultural areas. In the 1960s, however, high-yielding varieties of rice were introduced, together with the chemical insecticides, herbicides, and fertilizers they required, and multiple cropping systems were replaced by monocultures, including in the northern part of West Java, the country's rice bowl. Concurrently, the use of insecticides increased dramatically.[a,b] For a while, rice production increased, but this came at the expense of exploding pest populations in many areas. In response, even larger insecticide applications were made, but the pest attacks did not subside. And, in 1974, a previously minor pest, the Brown Plant Hopper (BPH; *Nilaparvata lugens*), became a major threat to rice production in Indonesia.[b,c,d]

To counter these attacks, Indonesian scientists and those at the International Rice Research Institute in the Philippines (one of the research centers that is part of the Consultative Group on International Agricultural Research, or CGIAR) developed rice varieties that were resistant to BPH, but farmers also continued spraying insecticides. This was an unfortunate strategy, because BPH is an "insecticide-induced resurgent pest," and within a short period of time, BPH became a major pest again. The heavy use of insecticides, it is believed, actually accelerated the adaptation of pests to resistant crop varieties for the following reasons: It released the pests from control by natural enemies, because beneficial predatory insects and parasitoids are often more sensitive to insecticides than are the pests themselves, and it selected for genetic variants within insect pest populations that were capable of surviving on the resistant crop varieties. As a result, there was a resurgence of BPH, even with the introduction of these BPH-resistant rice varieties.[d,e]

In 1986, the Indonesian government decided that its insecticide policies were not working and began a national program, sponsored by the U.N.'s Food and Agricultural Organization, that trained farmers to engage in farming practices in their paddy rice fields that encouraged the presence of beneficial insects. Fifty-seven insecticides were banned for rice production. What was discovered was that large populations of detritus and plankton-feeding insects began to develop in flooded rice fields in the absence of the insecticides, and that these, in turn, led to a wide variety of generalist predator and parasitoid insects that were able to reproduce rapidly. By the time the rice plants began to develop and herbivores began to attack them, beneficial insects were present in sufficient numbers to be able to control them. With a return to these natural control systems, some rice farmers in Indonesia have been able to keep a variety of rice pests—leaf rollers, stem borers, leaf hoppers, and plant hoppers, among others—in check.[d]

Identifying these beneficial insects and understanding their life cycles have become a major priority for many agricultural scientists and farmers in Indonesia, and great strides have been made. Of the generalist predators, some of the most important are the coccinellid beetles, such as the species *Harmonia octomaculata* (which feed on the eggs, nymphs, and adults of many types of insect pests, including BPH), ground beetles (the *Carabidae*) such as *Ophionea nigrofasciata* (which can consume up to five rice leaf-rollers or BPH larvae a day), and the staphylinids, such as *Paederus fuscipes* (which are very common in rice paddy fields and prey on rice borer moths and eggs, as well as on plant hoppers).[f] Some of the major parasitoids that have been identified are the wasps that lay their eggs on the eggs, larvae, and pupae of various insect pests, such as *Tetrastichus schoenobii* (which attack the eggs and pupae of stem borers), *Telenomus rowani* (which consume the eggs of rice stem borers), and braconid wasps such as *Cotesia angustibasis* (which are commonly found in paddy rice fields that are free of pesticides and that are major parasitoids of leaf roller larvae) (see figure 8.6a).[f]

By not using pesticides (saving significant amounts of money in the process), by encouraging the presence of natural beneficial insect predators and parasitoids, and by a return in some areas (e.g., the western and central parts of Java) to multiple cropping and intercropping systems using different rice varieties, some farmers in Indonesia have been able to match, and even to exceed, the rice yields from conventional farming practices, including those using high-yielding varieties. And they have done so with lower costs and with less harm to the environment.[g] Using biological controls to grow paddy rice has widespread applicability far beyond Indonesia.[h]

BOX 8.3

HARVESTING PEST SPECIES AS FOOD

Insects as Food. Various insects and spiders are sold as food at outdoor markets in Phnom Penh, Cambodia. (Photo by Rhymer Rigby, www.rhymer.net.)

Another alternative to using pesticides in some regions involves harvesting pests for use as human food.[a] Rice-field grasshoppers (primarily the species *Oxya volox*), major rice pests, were formerly a common food ingredient known as *metdugi* in Korea, but their use as food declined as pesticide use increased during the 1960s and 1970s. In 1982, as some insecticide spraying decreased and grasshopper populations increased, *metdugi* began to be sold once again as a foodstuff. For older Koreans, eating *metdugi* was nostalgic, bringing back a taste from the past.[b] This decline in insecticide use, spurred by renewed interest in *metdugi* and by a growing desire among some Koreans to eat pesticide-free rice, led to the development of organic rice farming in various regions of Korea. The transition to organic rice farming was also attractive because yields were the same as in sprayed fields and the organic rice commanded higher prices.

Other countries have also had success with harvesting agricultural pests. Grasshoppers are a favorite food in many parts of the Philippines, and as a result, the fields from which they are harvested are generally not sprayed with chemical pesticides. These grasshoppers are fed to pasture-raised chickens (and to cows and fish), which can then be sold for higher prices. The same is true in Thailand. In 1983, local Thai officials encouraged villagers to collect 10 tons of pest grasshoppers for food, because chemical control efforts had been unsuccessful, and by 1992, a small farmer in Thailand could earn up to US$120 per half-acre of grasshopper harvesting, twice as much as he or she could from maize.[c,d]

Figure 8.7. Various Pollinators. (a) Monarch Butterfly (*Danaus plexippus*) on Red Clover (*Trifolium pratense*). (Courtesy of Caroline Henri/ Dreamstime.com.) (b) Honeybee (*Apis mellifera*). (Courtesy of U.S. National Institutes of Health.) (c) Female Black-Chinned Hummingbird (*Archilochus alexandri*) by a Bottlebrush (*Callistemon ridigus*) flower. (Courtesy of Paul Wolf/Dreamstime.com.) (d) Lesser Long-Nosed Bat (*Leptonycteris curasoae*), which is listed by the International Union for the Conservation of Nature and Natural Resources as Endangered, pollinating a Saguaro (*Canegiea gigantea*) flower. (© Merlin D. Tuttle, Bat Conservation International, www.batcon.org.) (e) Bumblebee (*Bombus pratorum*). (Courtesy of Paul Morley/Dreamstime.com.) (f) A wasp (*Polistes dominula*) on stonecrop (*Sedum* sp.). (Courtesy of Janice Muskopf/ Dreamstime.com.) (g) Common Rose Beetle (*Cetonia aurata* L.) on white cherry flowers. (Courtesy of Steffen Foerster/Dreamstime.com.) (h) Painted Lady Butterfly (*Vanessa cardui*) on Sunflower (*Helianthus annuus*). (Courtesy of Lloyd Clements/Dreamstime.com.)

pesticides, such as the insect neurotoxin imidacloprid, banned in France because of suspected honeybee toxicity, might be involved.[38] A study published in *Science* has since demonstrated a strong correlation between the presence of a virus, called Israeli Acute Paralysis Virus or IAPV, and CCD hives. IAPV was first detected in Israel, where infected bees became paralyzed and died outside of the hives (a characteristic of CCD). The virus may have been carried into the United States by infected bees, perhaps from Australia, and resulted in the epidemic of CCD. But it is still not clear whether IAPV is the cause of CCD, either by itself or in combination with other factors.[39]

When compared to wind-pollinated plants, or plants that are pollinated by a broad range of organisms, plants that have specific animals pollinating them, such as figs that are pollinated only by fig wasps, have the lowest risk of pollen being wasted during transport. These same plants, however, also have the highest risk of pollination failure if their pollinators are lost.[40] For this reason, a decline in biodiversity may have cascading effects on species survival, because it may disrupt these close-knit, highly efficient, co-evolved relationships. Just as a high diversity of pollinators may help increase the diversity of plants, a high diversity of plants supports more pollinators. A recent study in the United Kingdom and the Netherlands, for example, shows a marked parallel decline in bee species in the two countries in recent decades and in the plant species that depend on them to reproduce.[41]

In agricultural regions, crops may be isolated from the habitats that support the pollinators they depend on to be productive. Experiments on isolated "islands" of radish and mustard plants, which were set up in an agricultural landscape at varying distances from a species-rich grassland, showed that increasing isolation resulted in fewer bee visits per hour to the radish and mustard islands, and also in reductions in the diversity of the visitors. In addition, the development of fruits and seeds declined with increasing isolation from the grasslands.[42] In another study, the amount of woody border had a significant positive effect on the overall diversity of insect families (see figure 8.4; see also figure 3.11 in chapter 3) in agricultural fields.[43]

CULTIVATED PLANTS AND WILD RELATIVES

Although clear benefits exist in planting agricultural lands near wild ones, this practice is not without potential downsides. One often-problematic outcome that has gained attention occurs when cultivated crops breed with their wild relatives. The growing of Sugar Beets (*Beta vulgaris L.*) in France provides an example. While some Sugar Beet seed production fields near the Golfe du Lion (that portion of the Mediterranean that stretches from the border of Catalonia in Spain to Toulon in France) are many kilometers from wild Sea Beets (*Beta vulgaris maritima*) growing along the water, and at least 1 kilometer away from those growing in the inland valleys, the Sea Beets nevertheless have been able to pollinate the cultivated Sugar Beets. As a result, by the mid-1970s, these Sugar Beet fields had become pocked with beets that were flowering prematurely, or "bolting," a trait of Sea Beets. Subsequent investigation found that these bolters were the result of cultivated beets having been pollinated by their wild relatives.[44]

The problem occurs in the reverse direction, as well. Just as genes from wild relatives can move into domesticated crops, those from domesticated crops can also move into wild populations. Individual crop plants typically contain less genetic variation than do individual populations of their wild relatives.[45] The evolutionary result of continued and substantial gene flow from a single crop cultivar (cultivars are specific crop varieties that can be reliably cultivated by seed or by grafting) to a wild population would be a decrease in the wild populations' genetic diversity. It is also possible that some wild species might suffer from sterility and have their populations reduced because of assimilation with a crop species, and might even become extinct. A recent literature review found twenty-eight well-documented examples in which hybridization between crops and their wild relatives led to new plant types that became either weeds in agricultural ecosystems or invasives in the wild.[46] And it has been found that spontaneous hybridization between a given crop and at least one wild relative is the rule rather than the exception. For the twenty-five most important food crops, all but three have some evidence for hybridization with one or more wild relatives, causing a wide array of effects.[44]

For instance, natural hybridization with cultivated rice has caused the near extinction of the endemic Taiwanese wild rice *Oryza rufipogon formosana*.[47] Collections of this wild rice over the last century have shown a shift toward characteristics of the cultivated species and a decline in fertility. Throughout Asia, typical specimens of other subspecies of *O. rufipogen*, and of the wild rice *O. nivara*, are rarely found because of such hybridization with cultivated rice crops.[48] Also, hybridization with cultivated maize may have played a role in the extinction of some populations of wild maize that were its original ancestors.[49]

While growing plants near natural areas does have many benefits, the dangers of gene flow need to be considered. Surrounding a field with plants such as hemp, which interfere with the spread of pollen and thereby prevent contamination from plants outside the crop space, is one possible solution. Similarly, "trap crops" or forest border hedgerows for field crops might be beneficial, not only in preventing gene flow but also because they offer other benefits of biodiversity such as pest management.[50]

GENETIC BASES OF AGRICULTURAL CROPS

Genetic diversity within each species of crop, among its wild progenitors or relatives as well as its cultivated varieties and strains, is of obvious and immediate importance to agriculture. Traditional methods of plant breeding, based on the selection and cross-breeding (hybridization) of genetically distinct strains, are still the most commonly used. They have been and continue to be employed, for example, in efforts to improve crop resistance to fungal diseases or insect infestations, as well as to environmental stresses such as heat and dry spells or excess salinity.

The preservation of genetic diversity among wild plants can best be achieved in the natural setting, within native habitats and natural ecosystems, while that of agricultural cultivars can most effectively be accomplished in designated fields and greenhouses. When such methods of living-plant preservation are neither practical nor sufficient, seed stocks of numerous species and varieties must be preserved in specially organized and carefully maintained collections. Such collections can serve as genetic pools, from which plant breeders may draw genes that can impart new varieties with superior tolerance or resistance to pests, diseases, or weather anomalies.

Figure 8.8. Seeds in Kew Gardens Seed Bank. Seeds from close to 12,000 different plant species are kept at Kew Gardens in airtight containers at minus 20 degrees Celsius (minus 4 degrees Fahrenheit) and around 15 percent relative humidity. These conditions are expected to preserve them, in most cases, for many decades. (© Board of Trustees of the Royal Botanic Gardens, Kew.)

The need for improved varieties arises repeatedly, as new pests appear or as old pests themselves acquire immunity to prior modes of control.

Large seed-storage facilities called seed banks have been organized, such as the U.S. National Plant Germplasm System in Colorado, maintained by the U.S. Department of Agriculture, which contains more than a quarter of a million different varieties; the national gene bank in the African country of Malawi, which stores some 8,000 varieties of native crops and fruits; and the International Plant Genetics Resources Institute in Rome, Italy (one of the institutes of the Consultative Group on International Agricultural Research, or CGIAR) that oversees plant germplasm collections worldwide.[51–53] The seed bank at Kew Gardens in London is also a prime repository for plant seeds.

Seed banks hold large collections of indigenous cultivars and wild relatives of crop species, as well as modern crop varieties and special breeding stock. They are intended to preserve seeds essentially indefinitely, by keeping them dry and at subfreezing temperatures. Great progress has been achieved in organizing and maintaining such facilities, yet much more can and should be done to enlarge, improve, and coordinate the various seed banks throughout the world. One plan, for example, funded by the Norwegian government, involves the creation of a "doomsday vault," containing some three million seeds, representing all known varieties of the world's food crops, built in permafrost, deep within a sandstone mountain, on the Norwegian Arctic island of Spitsbergen. The repository is being built in response to the concerns of some seed scientists who fear that existing seed banks will not be able to withstand such potential catastrophic events as nuclear war, rising sea levels, a collapse in electrical power systems, earthquakes, asteroid strikes, or terrorism.[54]

Soil Biodiversity

Despite the fact that we walk on and over them every day of our lives, soils remain among the least known habitats on Earth. It is unfortunately all too easy to take them for granted, yet growing evidence indicates that soils may be one of the most species-rich habitats on the planet.[55,56] Almost every phylum known above ground is represented in soil, and each has a wealth of species diversity. It is estimated that few of these species, however, perhaps fewer than 10 percent, have been identified and described.[57]

Soil organisms contribute to a wide range of essential soil ecosystem services.[58] Life in soils (see the opening figure for this chapter) includes vertebrates (e.g., prairie dogs, gophers, moles, lizards, and pack rats); macrofauna (large invertebrates up to several centimeters [or a few inches] long, e.g., ants, termites, millipedes, spiders, centipedes, earthworms, enchytraeids [small pale worms], isopods [woodlice], and snails); micro- and mesofauna, less than a millimeter to a few millimeters in length, such as the tardigrades (water bears), rotifers (wheel animals), nematodes (round worms), mites, and members of the Order Collembola (springtails); and the microbes—algae, lichens, protozoa, fungi, bacteria, archaea, and viruses.[55,59] Most (90–95 percent) of the species of microflora and microfauna that have been described are thought to be rare.[60] The abundance of soil organisms is truly astounding. A cubic

Figure 8.9. Collembola (Springtail). This springtail, of the hypogastrurid family, is a fungal feeder. There are about 7,500 known species of Collembola worldwide, and they are nearly ubiquitous in terrestrial ecosystems. Collembola are among the oldest known terrestrial animals, with fossils dating from 400 million years ago. (Photo by Mark St. John, Natural Resource Ecology Laboratory, Colorado State University.)

meter (around 35 cubic feet) of grassland soil can harbor many billions of organisms—10 million nematodes, 45,000 earthworms and enchytraeids, 48,000 mites and Collembola, hundreds of thousands of protozoa, algae, and fungi, and billions of bacteria[61]—the majority of which have not been identified.

Baseline information on the distribution, abundance, dynamics, and interactions of individual soil species and their influence on ecosystem functioning is generally lacking on local and global scales.[62] Moreover, information on what key species are necessary for soil ecosystem functioning, both in the short and long term, is also lacking. It is essential that we have better baseline measurements of soil biodiversity if we are to begin to monitor soil disturbances over time.

FUNCTIONS OF SOIL BIOTA

Table 8.2 lists different members of the soil biota and the soil ecosystem services they perform. Species in the soil are directly involved in ecological services that sustain human populations.[58] *Saprophytic organisms* are those that obtain their nutrients from dead and decaying plant or animal matter. *Actinomycetes* are bacteria that possess the ability, like fungi, to form mycelium-like, branching filaments. *Diazotrophic* means nitrogen-fixing. The *rhizosphere* is the region surrounding the roots of plants. The organisms represented in table 8.2 perform the following ecosystem services:[58]

- Maintain soil fertility by decomposing organic matter and recycling nitrogen and carbon and other nutrients

- Modify soil structure and the dynamics of water storage and flow by aggregating or clumping soil particles (which serves to retain moisture)

- Help mix organic matter and microscopic life throughout soils, redistributing nutrients

- Influence carbon storage in soils and the flow of trace gases

- Contribute to air and water purification by degrading pollutants

- Enhance the amount and efficiency of how vegetation acquires nutrients

- Affect plant community diversity and plant fitness through numerous associations

These associations can be *mutualistic*, where both species benefit from each other, or *parasitic*, where one species benefits at the expense of the other. Through these many connections, soil biota have essential and intimate links to ecosystem functioning, not only in the soils themselves, including those of freshwater and marine sediments, but also in aboveground terrestrial and aquatic systems.[58,60,62]

The contribution of a particular species to ecosystem functioning in soils is difficult to isolate given the complexity of species interactions. However, by lumping

TABLE 8.2. SOIL ECOSYSTEM SERVICES PERFORMED BY DIFFERENT MEMBERS OF THE SOIL BIOTA

FUNCTIONS	ORGANISMS INVOLVED
Maintenance of soil structure	Earthworms, arthropods, soil fungi, mycorrhizae, plant roots, and some other microorganisms
Regulation of soil hydrological processes	Mostly invertebrates such as earthworms and arthropods, and plant roots
Gas exchange and carbon sequestration	Mostly microorganisms and plant roots; some carbon protected in large compact biogenic invertebrate aggregates
Soil detoxification	Mostly microorganisms
Decomposition of organic matter	Various saprophytic and litter-feeding invertebrates (detrivores), fungi, bacteria, actinomycetes, and other microorganisms
Suppression of pests, parasites, and diseases	Mycorrhizae and other fungi, nematodes, bacteria, and various other microorganisms, Collembola, earthworms, and various predators
Sources of food and medicines	Plant roots, various insects (crickets, beetle larvae, ants, termites), earthworms, vertebrates, microorganisms, and their byproducts
Symbiotic and asymbiotic relationships with plants and their roots	Rhizobia, mycorrhizae, actinomycetes, diazotrophic bacteria, and various other rhizosphere microorganisms
Plant growth control (positive and negative)	*Direct effects:* plant roots, rhizobia, mycorrhizae, actinomycetes, pathogens, phytoparasitic nematodes, rhizophagous insects, plant growth promoting rhizosphere microorganisms, biocontrol agents. *Indirect effects:* most soil biota.

Source: Brown, G.G., Bennack, D.E., Montanez, A., Braun, A., and Bunning, S. 2001. What is soil biodiversity and what are its functions? U.N. Food and Agriculture Organization Soil Biodiversity Portal. For further information, please visit www.fao.org/ag/AGL/agll/soilbiod/default.htm.

organisms into groups based on their function, as well as on their morphology, physiology, and the source of their food, the role of groups of organisms in soil food webs can begin to be understood. For example, all mite species with similar mouthparts can be considered part of a fungal-feeder functional group, while termites, earthworms, and ants belong to a group that functions to improve aeration and water infiltration by creating tunnels in the soil. Most organisms have more than one function. For example, earthworms are both organic matter transformers, engulfing organic matter and transforming it into defecated pellets, and soil aerators.

Understanding how soil biota sustain soil fertility and contribute to the recycling of nutrients and to plant productivity requires an understanding of such things as the genetic variation within and among populations of soil organisms, soil species richness and composition, and the diversity of functional groups. However, with some exceptions, there is little to no understanding about the global distribution of soil biodiversity across thousands of soil types. The biodiversity within soil food webs is better known for intensively managed ecosystems of high economic value, such as those of agriculture, forestry, and rangeland/pastures, than it is for less managed ecosystems.

HOW A FALLEN LEAF DECOMPOSES

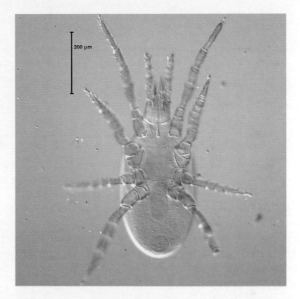

Woodlouse. There are some 3,500 species of woodlice worldwide. They are crustaceans, not insects, and are related to crabs and lobsters. They are among the only members of this subphylum to have invaded land without the need to return to water at any phase of their life cycles, but they are restricted to damp places, for example, in leaf litter or decaying bark, and they still breathe using gills. They are critically important organisms in helping to break down and recycle organic material, including rotting plants, returning essential nutrients to the soil. For more information on woodlice, see en.wikipedia.org/wiki/Woodlouse. (© Woody Thrower, www.snark.com/~woody/wordpress/.)

***Hypoaspis similisetae* Mite.** This mite, a member of the mesostigmatid family, is a predator of other microarthropods and of nematodes. (Courtesy of Mark St. John, Natural Resource Ecology Laboratory, Colorado State University.)

When a leaf falls to the ground, many organisms are involved in its decomposition. Large invertebrates such as woodlice feed on the leaf, ripping, tearing, and shredding it. At the same time, microbes, including bacteria and saprophytic fungi, begin the process of decay, digesting portions of the leaf. They are eaten by small invertebrates such as nematodes, which are in turn ingested by mites, all of which are part of the soil food web. By all of these actions involving countless species that live on and in the soil, the leaf is thus slowly broken down into smaller and smaller fragments and is finally transformed, along with those organisms that have fed on it and that also eventually die and decompose, into carbon, nitrogen, and other chemicals that return to the soil and contribute to its fertility.

SOIL HABITAT DISTURBANCE

Disturbance to the soil habitat in natural ecosystems affects soil biodiversity both directly and indirectly, through cascading effects on other soil properties, such as water permeability, salinity, erosion, and carbon, nitrogen, and oxygen content. All of these, in turn, may be affected by a loss in soil biodiversity.[63] Such disturbances can also affect ecosystem functioning. Land-use changes, invasive species, the deposition of acid rain and nitrogen compounds from the atmosphere, and pollution by sewage, excess fertilizers, and toxic chemicals all can alter soil communities and the plants they sustain, affecting plant fitness and composition.[64]

Land-use change is the major driver that affects soils, such as the conversion of natural systems to agriculture, altering not only the diversity of plant species in these systems, but also that of soil microbes, mycorrhizae, nematodes, termites, beetles, and ants.[65–70] An example is what has happened to some soils in the Amazon following the

BOX 8.5

THE MYCORRHIZAE

The discovery of the fungi known as mycorrhizae was made in the late nineteenth century by a German forestry scientist, Albert Bernhard Frank, who was investigating how to grow the much sought after European food delicacy, truffles. Frank and others since found that truffles and other mycorrhizal fungi live in vast networks in the soil in symbiotic relationships with plants, from which they receive sugars and other compounds and to which they provide essential nutrients and other services. Mycorrhizae are found in all terrestrial habitats that have topsoil—in boreal, temperate, alpine, and tropical regions. Some 90 percent of all vascular plant families contain species associated with mycorrhizae. The more we learn about the mycorrhizae, the clearer it becomes that they play, and have played perhaps since the time that plants first colonized the land some 400 million years ago, vital roles in maintaining the health and survival of plants and plant communities.[a,b]

About 100,000 fungal species have been identified and named, but it is estimated that this number constitutes only about 6 to 7 percent of the total of 1.5 million or more species.[c] Estimating the number of mycorrhizal fungi presents a particular challenge because it is often difficult to determine whether specific fungi are true mycorrhizae or are saprophytes that digest and recycle dead organic matter.

The mycorrhizae are divided into two main groups. The best known, the ectomycorrhizae, produce their fruiting bodies as mushrooms that surface near the trees with which they are symbiotic. These fungi produce a covering or mantle on the root and grow in the outer region of the root between the cells, forming a net. All the mushrooms of the same type in a given area are likely to be part of the same organism, which can extend great distances underground.

The other group of mycorrhizae is known as arbuscular mycorrhizae, which spend their entire lives underground. Both types extend fine threads called *hyphae* (singular *hypha*) from the web network they construct called the *mycelium*, which surrounds and, in the case of arbuscular mycorrhizae, penetrates the cells of the plant root tip. Hyphae are approximately one 1/60th the thickness of plant root tips and can, as a result, penetrate tight spaces in the soil that the roots themselves cannot reach.

Forming vast networks, which in a cubic yard of soil can reach thousands of miles in length if the individual strands are stretched out end to end,[d] mycorrhizae bring many times the amount of phosphorus[e] (some believe hundreds and perhaps even thousands of times more),[f] and possibly also water, nitrogen, and other essential nutrients, to plants than they could obtain on their own. (For more on mycorrhizal nutrient transport and transfer, see www.biology.duke.edu/bio265/jlp13/myco.php?t=nutrient.) It may be more than just their enormous surface areas that allow mycorrhizae to bring large quantities of essential nutrients to plants. Their hyphae are thought, in addition, to play an active role in absorbing phosphorus, even at very low concentrations, so that it becomes available to plant roots.[g,h] In return, the plant provides the fungus with the sugars, starches, proteins, and lipids it needs to survive.

Mycorrhizae perform other essential services as well. Their intricate underground web holds onto soil nutrients and moisture and serves to prevent erosion and protect plants from drought. Some mycorrhizae are able to bind toxic metals such as cadmium and make

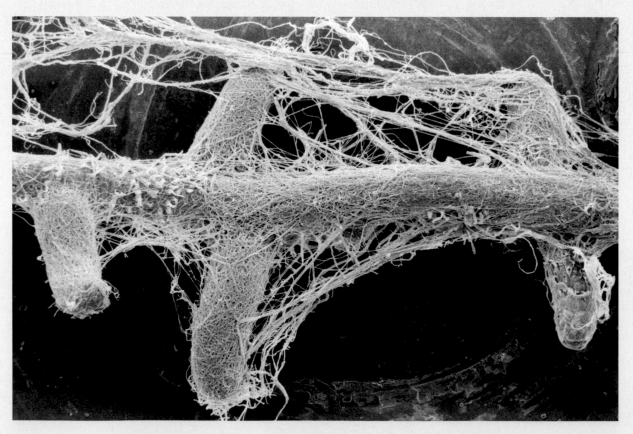

Scanning Electron Micrograph of Yellow Birch (*Betula alleghaniensis*) Root Tips Colonized by the Ectomycorrhizal Fungus *Paxillus involutus*. (Photo provided by R. Larry Peterson, University of Guelph, Canada.)

BOX 8.5 (CON'T.)

them unavailable to plants (perhaps also keeping them from becoming airborne),[d] while others help increase plant uptake, such as that by the Chinese Brake Fern (*Pteris vittata* L.; see also figure 3.10 and discussion in chapter 3, p. 94), of toxic substances such as arsenic.[i] Some are also thought to protect plants from certain diseases. It has been observed, for example, that arbuscular mycorrhizae protect some plants from root pathogens.[j] Their extensive web, moreover, serves to link one plant to another, so that nutrients and sugars may be shared among adjacent plants. Such an availability of sugars in areas where mycorrhizae are prevalent may also make it possible for some newly sprouted seeds to grow under low-light conditions. Because of mycorrhizae, some plant species, such as some types of orchids, have dispensed with photosynthesis altogether, instead tapping into the nutrients supplied by mycorrhizal webs.[k]

Research has documented that greater species richness of arbuscular mycorrhizae in soils, translates into greater plant diversity, higher levels of nutrient capture by the plants, and greater total ecosystem productivity.[l,m] This finding provides yet another example that illustrates how aboveground and belowground organisms are intimately linked in terrestrial plant communities.[n]

There is evidence that human activity, such as the release of nitrogen compounds into soils from fertilizers and from acid rain, and exposing soils to increased concentrations of atmospheric carbon dioxide and to ozone pollution, can affect the diversity, abundance, and functioning of mycorrhizal communities.[o,p,q,r] Given the increasing recognition that mycorrhizae are central to the health, diversity, and productivity of plant communities, it is critically important that we understand better how they function and that we do everything we can to preserve them.

Larch Tree Seedling Mycorrhizae. The white threads are the hyphae of the fungus; the thicker red/brown ones are the Larch roots. (Photo by R. Finlay; © PlantWorks Ltd. U.K., www.plantworkuk.co.uk.)

BOX 8.6

NITROGEN-FIXING BACTERIA

German scientists in the late nineteenth century discovered the phenomenon of nitrogen fixation. In investigating the fertilizer needs of various crops, Hermann Hellriegel and Hermann Wilfarth found that some plants in the legume family—such as beans, peas, alfalfa, lupines, and vetch—grew well in soils deprived of nitrogen. When they sterilized the soil, however, this did not occur, suggesting to them that some organism in the soil, later isolated in 1888 as a bacterium from the genus *Rhizobium* by the Dutch microbiologist Martinus Biejerinck, was providing the nitrogen. It is now known that such legumes live in association with their *rhizobia*, the nitrogen-fixing bacteria that live symbiotically with leguminous plants, residing in small nodules that develop from their root cells. And it is now clear that these organisms, like the mycorrhizae, are among the most important of any on the planet.

Fifty-three species of rhizobia from twelve genera have been identified, and there are thought to be another 100 to 200 nitrogen-fixing bacterial species that are free-living—in soils, in animals such as some termites and a wood-boring mollusk known as a "ship-worm," and in aquatic environments where they are key "primary producers," at the base of freshwater and marine food chains. Many nitrogen-fixing bacteria are members of a family called the Cyanobacteria, which are ancient, self-sufficient organisms that are able to fix both carbon (from carbon dioxide) by photosynthesis,

Soybean Root Nodules Containing Nitrogen-Fixing Bacteria. (Courtesy of Scimat/Photo Researchers, Inc.)

and nitrogen by converting one molecule of nitrogen gas, which cannot be utilized by plants, into two molecules of ammonia, which can be.

Some of the complexities of the relationship between legumes and nitrogen-fixing bacteria are beginning to be understood, for example, the workings of genes and molecular signals in the plant and the bacteria that lead to their symbiotic union, bypassing the antimicrobial defenses the plant erects against all other soil bacteria;[a] the workings of the enzyme nitrogenase, which catalyzes the reaction that converts nitrogen into ammonia; and the ways that bacteria and

---

transformation from rainforest to pasture—an exotic earthworm species (*Pontoscolex corethrurus*) has become the predominant invertebrate in these new pasture lands, now constituting 90 percent of the biomass.[71] Such transformations also increase soil compaction and change its texture, affecting the diversity and abundance of some vertebrates and larger invertebrates that are dependent on specific soil types.[72]

USING BIODIVERSITY TO IMPROVE SOILS

In general, it is easier to sustain soils and prevent their degradation than to restore them. Efforts at soil reclamation across large scales, whether after fires or disturbances caused by intensive chemical use in agriculture or forestry, have focused on supplying a sufficient amount of organic matter to the soils in the form of plant litter or animal wastes in order to increase fertility and improve a soil's water-holding capacity. The addition of earthworms has also led to enhanced productivity.[73] In these situations, the objective is not to recreate the original species diversity present in the natural soils, but to enhance vegetation growth by restoring the functioning of the soil community.

BOX 8.6 (CON'T.)

plants work together to shield nitrogenase from oxygen, which would otherwise destroy it.[b]

Some nitrogen-fixing bacteria are associated only with certain specific legumes, while others show less specificity and are able to colonize many different ones. One strain of a nitrogen-fixing *Rhizobium* species called NGR234, for example, is able to live symbiotically with more than 112 different legumes.

A variety of environmental changes may affect plants and their rhizobia. When the soil is overly acidic from acid rain, for example, root colonization by some nitrogen-fixing bacteria can be inhibited,[c] but the effects of soil acidification are highly variable, with some rhizobial species being acid-tolerant. In addition, the effects of acidification may relate, in some cases, to soil levels of aluminum, mobilized by low pHs, that may be toxic to some nitrogen-fixing bacteria.[d] Some pesticides inhibit nodule formation and plant growth in various legume–rhizobial systems, with the effects varying with the stage of plant growth, the pesticide used, and the plant species involved.[e] Extreme weather with high temperatures and severe drought (which are increasingly likely to occur in some regions as a result of global climate change) may have significant impacts on rhizobial survival. In western Australia, for example, rhizobial mortality has been observed when soil surface temperatures reached 50–60 degrees Celsius (122 to 140 degrees Fahrenheit)[f] As a result, some nitrogen-fixing bacterial strains with greater resistance to extreme weather conditions are being tested in these regions.

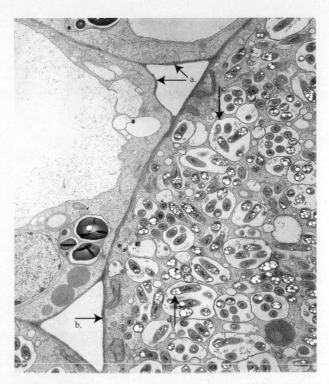

A Transmission Electron Micrograph of a Soybean Root Nodule in Cross Section. (a) Soybean root tip cells. (b) Root nodule. (c) Symbiosome (a membrane-enclosed compartment containing symbiotic organisms, in this case, soybean rhizobial bacteria). (d) The rhizobial bacteria originally Gram-negative bacilli in soil, they undergo rapid multiplication in nodule cells, change shape, and lose their motility, and are referred to as bacteroids. (© E.H. Newcomb & S.R. Tandon/Biological Photo Service.)

"No-till" or "low-till" agriculture, the practice of not (or minimally) plowing or otherwise disturbing the soil during the period from planting to harvesting, leaves greater amounts of plant organic matter in soils. Such practices also maintain mycorrhizal networks, preserving their extensive nutrient absorbing and transport systems. Over time, the food web in such undisturbed soils begins to mimic the functions of that present in natural systems.[66,74] In no-till or low-till fields, in comparison to those that are intensively tilled, earthworms are found in greater densities, generally improving water infiltration and providing channels that facilitate root penetration (although increased nutrient leaching can also occur); in addition, moisture is enhanced, carbon is sequestered in greater amounts, and soil quality and structure are improved.[75,76]

Soil biota are increasingly being used as indicators to assess soil quality, based on the knowledge that changes in species richness and abundance alter the dynamics of the soil food web and affect plant growth.[77,78] Much attention now is also being focused on managing the amount of carbon (in plant organic matter) that is stored in soils, in order to reduce atmospheric carbon dioxide concentrations and benefit global carbon nutrient cycles.[79]

Figure 8.10. Conservation Tillage with Soybeans. This soybean farm in central Iowa uses low-till practices to prevent erosion and to leave greater amounts of nutrients in the soil. (Courtesy of Lynn Betts, Natural Resources Conservation Services, U.S. Department of Agriculture.)

LIVESTOCK PRODUCTION

The Increasing Demand for Foods of Animal Origin

The last thirty or so years have been characterized by a dramatic growth in a middle class in a number of countries, including India, China, Taiwan, South Korea, Hong Kong, Singapore, Malaysia, and Indonesia in Asia, and Chile and Brazil in South America. Today, in the postindustrial age, global economic development continues to increase the buying power of tens of millions of people and, in tandem, to increase the worldwide demand for foods of animal origin.

The overall demand for meat, fish, dairy products, and eggs around the world has increased markedly. From 1977 to 2003, world meat production increased from 117 million metric tons (MMT) to 253 MMT per year, representing a per capita increase from 27.6 kilograms (about 61 pounds) per person to 40.3 kilograms (about 89 pounds) per person per year. Consumption of milk in developing countries is projected to grow at a rate of 3.3 percent a year through 2020 to reach a total consumption of 431 MMT, compared to 185 MMT in 1993. Similarly, the growth rates for beef and pork consumption through 2020 are each projected at 2.8 percent annually.[2,80]

In some countries, notably in Asia, economic development has led to enormous increases in meat consumption. In China, for instance, per capita meat consumption, which was 10.3 kilograms per person in 1977, reached 52.4 kilograms per person by 2002, a 500 percent increase.[2] Because China is a nation of 1.25 billion people, this increase in demand for meat, mainly in the form of pork and poultry products, raises significant questions about how livestock production can be increased in coming years in a sustainable manner to meet demand. To date, the global response has largely been an increase in intensive animal production, primarily in industrialized nations, along with an introduction of these practices into some developing countries.

The intensification in animal production has been associated with an increased use of antibiotics, growth-promoting hormones, and other chemicals in livestock; an increase in the use of nonrenewable resources, especially fossil fuels; increased risks of rapid disease transmission in large groups of confined animals (and concerns about their welfare in these conditions); and adverse environmental impacts when large concentrations of animal wastes contaminate soils and groundwater. Some scientists have also voiced concern about the contribution that livestock make to the accumulation of greenhouse gases associated with global warming.[81]

Furthermore, intensive livestock production is associated with an increased use of cereal grains as livestock feed that could instead have been used for direct human consumption. As was mentioned above, with the area of globally available cropland essentially fixed, meeting the food demands of a growing population and, in particular, maintaining adequate cereal grain supplies are serious challenges. Demand for grains to be used as animal feed in developing countries is expected to double between now and 2020.[80]

While modern, intensive animal production methods have been largely successful at providing increased amounts of wholesome foods of animal origin at reasonable prices, some critics contend that the market price reflects a failure to factor in the environmental and social costs of modern livestock production practices.

Importance of Livestock to the Poor

Despite the documented successes of global economic development, the rewards have not been evenly distributed among the world's citizens. The income of the richest fifth of the world's population is seventy-four times that of the poorest fifth, and their share of the world gross domestic product is 86 percent compared to 1 percent for the bottom fifth.[82] Such disparities in income reflect disparities in access to food. While some newly prosperous nations in East Asia have increased their importation of food in response to economic growth and a demand for more diverse and abundant foodstuffs, many less prosperous countries, for example, in sub-Saharan Africa, are increasing their food importations because of expanding populations and a diminished capacity for domestic food production. In the latter cases, there have been actual decreases in per capita production of basic cereals. It is estimated that between 1990 and 2020, the gap between the production of, and demand for, cereals in sub-Saharan Africa is likely to widen from 1 MMT to 27 MMT.[83]

An estimated 1.2 billion people live in poverty, earning less than $1 per day, and 840 million of them are undernourished.[84] The vast majority of the world's poor and undernourished live in the rural areas of developing countries, especially in Asia and Africa. Many of the rural poor remain partially or wholly dependent on livestock for their livelihood. Pastoralists and landless peasants graze livestock for food and income, while subsistence farmers rely on livestock to plow their fields, draw water from wells for irrigation, provide manure for crop fertilization and cooking fuel, and bring crops to market. In addition, livestock add variety to diets through meat, eggs, or dairy products.

There are large expanses of land, particularly in semiarid regions of North Africa, the Middle East, and South and Central Asia where, due to lack of rainfall, the grazing of livestock on grasslands is the only productive agricultural use of the land. Many nomadic, or pastoralist, cultures are based on this animal grazing activity. While clearly a productive use of the land that contributes greatly to the human food supply, pastoralism can also be associated with environmental problems, notably, desertification (the process in which formerly productive land becomes desert) in semiarid regions, and deforestation when forests are cleared for livestock grazing (these are discussed in more detail below). While traditional livestock grazing methods have often been carried out in relative harmony with local ecosystems, emerging pressures such as population growth, urban expansion, loss of traditional grazing rights, and the failure to develop and promote adequate marketing channels for livestock products have forced traditional peoples into situations where overstocking and overgrazing are becoming more common. Nevertheless, considerable opportunities exist to better integrate livestock production into traditional farming systems so that animals contribute to replenishing and sustaining agricultural ecosystems rather than to destabilizing them.

Livestock Production Systems

Extensive livestock production, or pastoralism, utilizes naturally occurring grasslands and water sources as the inputs for animal production, while the various forms of intensive livestock production are linked in some manner to crop production. Traditionally, the relationship has been complementary, with animals closely integrated into crop production activity. Livestock provide draft power for working the fields. Crop byproducts not directly edible by humans, such as stalks, hulls, and leaves, can be fed to livestock, which are able to convert them into milk, meat, or eggs. Animal manure is returned to the fields as fertilizer, thus promoting the success of the next agricultural crop. Even pastoralist herders, who often do not raise crops of their own, may enter into arrangements with crop farmers so that the herder's livestock are permitted to graze a farmer's stubble field in exchange for the soil-enriching manure that the animals deposit there. Such traditional systems represent a closed resource loop, in which few external inputs are required, and resources available within the system, such as manure and crop byproducts, are recycled. In such systems, cultivated grains are used primarily to feed the farm family, and animals are fed crop residues largely for maintenance rather than production.

Modern intensive livestock-production systems, which have become industrialized, such as commercial beef feedlot or poultry broiler production, often disrupt the traditional links between animal and crop agriculture. Many commercial livestock production systems represent open resource loops that require substantial external inputs to maintain the high levels of productivity for which they are noted. Livestock may be kept in large housing facilities, often at great distances from croplands, with feed trucked in for the animals. In turn, livestock housing may be surrounded by insufficient land to allow use of animal manure as fertilizer, leading to waste disposal problems. In 1997, the U.S. Department of Agriculture estimated that animals in the U.S. meat industry produced a total 1.4 billion tons of waste, which is roughly 130 times the amount of waste produced by the entire population of the United States each year.[85] Extensive use of fossil fuels in the form of fertilizer, diesel fuel, heat, and electricity is also required. For beef produced via the modern feedlot system, the fossil fuel energy input is 35 kilocalories for every kilocalorie of beef produced. (In the United States, fully 17 percent of all fossil fuel use goes toward the production of food.)[85] To support the high level of production that modern, selectively bred livestock are capable of, it is often necessary to feed considerable amounts of cereal grains to livestock, thus making them major contributors to the problem of putting people in direct competition with livestock for the grain produced on the world's limited croplands. As demand for livestock products continues to increase and cropland area remains finite, the challenge for meeting that demand is considerable.

Livestock and the Environment

Much has happened since the Neolithic period to disturb the natural equilibrium that existed among early domesticated animals, the environment that sustained them, and the people who managed them. Over time, as human and animal populations expanded, the carrying capacity of ancient grazing and farming lands was exceeded, and the land itself became degraded.[86] People and their animals sought new lands in new regions, often not as hospitable as their original homelands. Those who migrated northward into Europe, for example, had to deal with a climate that did not allow for year-round grazing and had to provide housing and winter feed, requiring an increased expenditure of both energy and raw materials.

In some places, large herds of domestic animals displaced wild animals, and natural habitat was transformed to support widespread livestock production. For example, as recently as the nineteenth century on the Great Plains of the United States, human settlers replaced wild buffalo and biologically diverse grasslands with domesticated cattle and cereal crops, driving the buffalo and some native prairie grasses to near extinction. A similar process is under way in Latin and South America in our

own time, where deforestation is occurring on a large scale to produce grazing land for cattle (as well as farmland, mainly for the growing of soybeans), generally at the expense of the enormous biodiversity of the forests.[87]

Livestock production has, during its history, become increasingly associated with environmental disturbance. Extensive grazing and ranching have been linked to desertification and deforestation and other forms of land degradation and to a loss of biodiversity. And, in addition to causing excessive natural resource consumption and the contamination of soil and groundwater with chemical pollutants and animal wastes (the U.S. Environmental Protection Agency estimates that farming practices account for 70 percent of the pollution found in the country's rivers and streams),[85] intensive livestock production has been implicated in the eutrophication (the over-enrichment of aquatic systems with nutrients) of lakes, rivers, estuaries, and coastal waters; the formation of marine "dead zones" (see "Livestock Production Affecting Biodiversity," below, and discussion of dead zones in chapter 2, page 52); and the emergence and spread of some diseases, such as cryptosporidiosis and salmonellosis (see discussion of these topics in chapter 7, page 319).

While livestock production generally has negative environmental impacts, livestock can and do exert positive effects on the environment as well. In mixed farming systems, they provide manure in proper amounts for soil enrichment, reducing the need for and use of chemical fertilizers for crops. One ton of cow manure contains about 8 kilograms of nitrogen, 4 kilograms of phosphorus, and 16 kilograms of potassium. In addition to providing these essential nutrients for crop production, manure, unlike commercial chemical fertilizers, helps to replenish organic matter in soils and to increase topsoil biodiversity, both of which are critical to proper aeration, water infiltration and retention, erosion resistance, mineral retention, and nutrient cycling.

Animal manure is also used widely around the world as a source of household energy for cooking or heating fuel, substituting for fossil fuels or firewood. This reduces the use of nonrenewable resources and eases pressure on forests. Manure can be burned in dried cakes, or can be transferred to biodigesters to generate burnable

Figure 8.12. Methane Generators in Mian Yang, Sichuan Province, China. Methane, produced by anaerobic bacteria from human and animal wastes in these belowground concrete containers, can supply as much as 60 percent of a typical family's cooking and heating needs. The leftover effluent can be used as a food supplement for pigs or as fertilizer for crops. (Photo by Paul Henderson.)

methane gas. It is estimated that throughout rural villages in China, there are some five million biogas chambers producing methane for household use in cooking and illumination.[88]

Animals also provide power for agricultural work and transportation, substituting for tractors and other vehicles that burn fossil fuels and pollute the air. For some steep terrain and for delicate soils, the use of animals for cultivation and other field-work, instead of farm equipment, is essential, making farming under such conditions possible and reducing the risk and rate of soil erosion. While the general tendency has been to modernize agriculture by introducing mechanization, there is considerable opportunity to improve the productivity and efficiency of small-scale agriculture around the world through use of animal power. In Uganda, for example, some 90 percent of the power employed in agriculture is still produced by human labor using hand tools, while only about 8 percent is provided by animals (mainly oxen and donkeys) and 2 percent by tractors.[89] While the purchase, maintenance, and running of a tractor is beyond the financial means of most African small farmers, the role of animals as sources of power could be greatly expanded at reasonable cost and with considerable benefit for both agriculture and the environment.

Genetic Base of Livestock Species

The richness of biodiversity is reflected in the enormous range of adaptations that living organisms express in order to successfully exploit specific ecological niches. Domestic animals also reflect such a wealth of genetic diversity and adaptation. There are an estimated 7,600 breeds of domesticated cattle, goats, sheep, buffalo, yaks, pigs, horses, rabbits, chickens, turkeys, ducks, and geese worldwide. These breeds display a remarkable variety of phenotypes, physiologic adaptations, immunological defenses, and behavior patterns that make them well suited to widely differing but highly specific environments, purposes, and production systems. Tragically, the pool of genetic diversity available to much of humanity's livestock is dwindling at a rapid rate, with about 190 breeds having gone extinct in the past 150 years (60 of these in the past five years alone) and a further 1,500 at risk of extinction.[90]

The great extent to which genetic variation can be expressed in domestic animals as different physical and behavioral traits is exemplified in the diversity of dog breeds that have been developed by human societies, which despite ranging from Chihuahuas to Saint Bernards, all represent a single species, *Canis familiaris*. The diversity within livestock species is also evident, for example, with distinct breeds of goats selected for cashmere fiber production (the Chinese Liaoning goat), mohair fiber production (the Angora goat), high milk production (the Toggenburg goat), meat production (the Boer goat), or adaptation to arid climates (the Black Bedouin goat), with all of these breeds being different variants of the one goat species, *Capra hircus*.

Genetic diversity in animals, as in plants, provides the variation within populations that makes it possible for them to adapt to changing natural conditions. With domestication, much of this diversity is lost, with some traits that are most desirable for human purposes selectively bred for (e.g., milk yield, leaner beef, the daily rate of weight gain, or reproductive performance), while others, not of obvious value, are selected against. Under natural conditions, such uniform populations would be less able to survive, but domesticated animals do not have to survive in the wild, being largely dependent on humans for water, food, shelter, and protection against predators and disease. So, as long as conditions can be held constant, this loss of diversity is not a problem. But if these conditions should change, for example, if there were a severe heat wave (which are expected to increase in frequency and intensity with global warming) or if there were an outbreak of a new bacterial, viral, or fungal disease to which a particular, genetically similar or identical group of domesticated animals were vulnerable, then the entire uniform population would be at risk. This is in contrast to a genetically diverse population in which it is likely that there would be some animals that were resistant.

It is this vulnerability that is of increasing concern to those seeking to preserve different varieties of domesticated animals and genetic diversity within those varieties. And it is this vulnerability that causes many to worry that, in a world where the climate is increasingly unstable as a result of human activity, and where there seems to be an ever greater likelihood of newly emerging infectious diseases, the current drive toward increasing monocultures among domestic animals is very unwise.

In the past, farmers had a greater variety of cattle or chickens or other livestock on their small subsistence farms. Today, that diversity is rarely found. For example, in the United States, Holstein Cows now make up more than 90 percent of U.S. dairy cattle, and White Leghorn Chickens produce almost all of the country's white eggs.[91] The same is largely true in other parts of the world.

Indigenous livestock breeds frequently possess traits that are highly desirable for promoting sustainable agriculture in difficult environments. Such traits need to be better characterized, and their genetic transmission better understood. Moreover, populations that possess them should be identified and protected from extinction. Their genetic potential needs to be preserved intact but, at the same time, more fully utilized through selective breeding and cross-breeding to produce new breeds that combine adaptability and hardiness along with increased productivity.

Figure 8.13. Different Breeds of Pigs and Sheep. Farmers used to have a variety of domestic animal breeds on their farms, as in this nineteenth century engraving. (From Solon Robinson (editor), *Facts for Farmers and the Family Circle*. A.J. Johnson, Cleveland, Ohio, 1867.)

There is growing interest, and an increasingly organized effort, in conserving, protecting, and promoting the use of domestic breeds of animals. The U.N. Food and Agriculture Organization (FAO), by establishing a Global Program for the Management of Farm Animal Genetic Resources, has given a high priority to the conservation of livestock breeds. The program supports the use of indigenous breeds in sustainable agriculture, as well as efforts to collect, store, and preserve genetic material for conservation and research. The FAO also maintains a Domestic Animal Diversity Information System (DAD-IS; accessible at www.fao.org/dad-is/) that provides regional inventories of indigenous breeds of all domestic animal species and the status of existing populations.

Threats to Livestock Production from Global Environmental Change

While livestock in some cases contribute to environmental degradation, including to the loss of biological diversity, they themselves suffer its adverse consequences. Global warming, for example, may compromise livestock health and productivity in a number of ways. In arid and semiarid zones, such as the Sahel of Africa, increased temperatures may cause an overall reduction in soil moisture and reduced vegetation cover. Because grass cover and water supply are

Figure 8.14. Different Breeds of Poultry. (From Solon Robinson (editor), *Facts for Farmers and the Family Circle*. A.J. Johnson, Cleveland, Ohio, 1867.)

the limiting factors in cattle grazing, many pastoralists, particularly those who run cattle in mixed herds with smaller stock, may find their opportunities to raise cattle drastically restricted. Camels and goats, better adapted to dry conditions, will become increasingly important to arid zone people dependent on livestock for survival. The cultural implications for traditional cattle herders, and the negative economic impact in the region, will quite likely be severe.

In temperate regions, especially in more northerly latitudes, global warming may present increased opportunities for grazing because warmer temperatures will mean longer growing seasons.[92] Also, increases in atmospheric carbon dioxide concentrations may encourage leaf expansion and the preferential growth of vegetative crops and pastures over cereal crops. In addition, grazing lands at higher altitudes, or upland pastures, may in some cases become more lush.[93]

But there may also be negative impacts from global warming on livestock in temperate zones, including those caused by extreme weather events (both torrential rains and flooding, and prolonged droughts) and by changes in the life cycles of pests and diseases.[94] For example, a longer grazing season, with greater dependency on grass as feed, could lead to unanticipated nutritional disorders, such as magnesium deficiency early in the grazing season, or cobalt and selenium deficiencies later in a grazing season extended by global warming.[95] In addition, patterns of gastrointestinal parasitic diseases could alter significantly because warmer winters would allow for greater overwintering of some nematode ova and larvae in soils and encourage earlier and more rapid development to infective stages in the early spring.[95,96] Such

LIVESTOCK AND POULTRY GENETIC RESOURCES IN CHINA

China has extremely rich livestock and poultry genetic resources. According to 1989 statistics, for example, there were 596 livestock breeds, breed groups, and types across the country. The many variants of yaks, goats, and poultry in China demonstrate this genetic richness.[a]

- Yak: Originally from the Qinghai-Tibet Plateau, which is more than 3,000 meters (or roughly 9,800 feet) above sea level, the yak (all domesticated yaks are from the same species, *Bos grunniens*) produces meat, milk with a high fat content, and high-quality fiber. It can be also be used for transportation. According to the U.N. Food and Agriculture Organization, there are twelve different yak breeds in China. They are the essential livestock of local herdsman, being their means of producing both food and fiber.

- Goat: Famous goats in China include Ningxia's Zhongwei Goat, Liaoning's Cashmere Goat, Jining's Qing Goat, and Chengdu's Ma Goat. The Zhongwei Goat produces good-quality white fur, while Liaoning's produces high yields of long cashmere fibers, and the Qing (a prolific breed averaging more than two births a year), a mixture of black and white wool. Chengdu's Ma Goat can produce milk that has an extremely high fat content (6.5 percent), almost double that of European breeds.

- Poultry: The majority of China's poultry are dual-purpose breeds, producing both meat and eggs. Hetian Chicken (all chickens are from the species *Gallus domesticus*) is used as a meat broiler, with characteristics of thin skin, slender bones, and tender and delicious meat. Pekin Duck (all domesticated ducks, except the South American Muscovy Duck, are variants of the species *Anas domesticus* and originally came from the wild Mallard) is used for preparing the world-famous roasted duck. The Gaoyou Duck is used to prepare salted duck and is renowned for its two-yolk eggs. Xianju Chickens can lay 200 eggs a year, with each egg averaging 40 grams (about 1.4 ounces). Shaoya Ducks can lay 280 to 300 eggs a year, with each being from 60 to 65 grams. The Huo Goose (almost all Chinese domesticated geese, including the Huo Goose, are variants of the species *Anser cygnoides*) can lay only 150 eggs a year, but the eggs average 128 grams (more than a quarter of a pound each!).

Because of a general neglect of local breeds, China's rich livestock and poultry genetic resources are under threat. Surveys made in the 1970s and 1980s showed that ten local breeds had disappeared, nine were in danger of extinction, and twenty were reduced in population size. This trend is continuing with the development of intensive animal husbandry. The Chinese government is attempting to reverse this trend and protect these extremely valuable genetic resources.

seasonal changes in feed availability and parasite infectivity may require alteration of breeding and birthing schedules for various species of grazing livestock to avoid the occurrence of nutritional deficiencies and parasitic infections from pastures altered by climate change. Tick-borne diseases could also become more widespread as the ranges of these arthropod vectors expand in response to warmer temperature and moister conditions. (See "The Effect of Climate Change on Vectors" in chapter 7, page 320.) New vector-borne diseases may be introduced as temperatures become milder and vectors, such as the midge that carries the virus for bluetongue in ruminants, are able to successfully survive the winter in places where they were unable to do so before.[97] Livestock diseases with wildlife reservoirs, such as tuberculosis, may become more difficult to control as warmer temperatures and increased vegetation allow greater winter survival rates for, and expansion of, wildlife populations capable of maintaining the disease.[95]

Livestock Production Affecting Biodiversity

Landscape transformation, involving deforestation, land degradation, and desertification, is the process best known by which livestock contribute to the loss of biodiversity. The clearing of forests for the purpose of cattle grazing generally destroys the habitat for fauna associated with the forest ecosystem. Cattle raising has been an important cause of forest loss on a regional basis in Latin America, especially in Costa Rica and Brazil.[98]

In grassland systems, overstocking, overgrazing, and improper management of grazing herds can result in a loss of plant biodiversity, degradation of soil through compaction and erosion, and disruption of the normal regenerative cycle of grassland flora. When livestock are allowed to graze along streams, severe erosion of the banks can result, with a loss of streamside vegetation and negative impacts on aquatic ecosystems. Such transformations can adversely affect animal species both directly and indirectly. In Australia, for instance, conservationists are concerned that the clearing of coastal lands for cattle pastures is producing large amounts of silt, which is washing into the ocean and contributing to the death of corals in the Great Barrier Reef.[99,100] However, when properly managed and sited, livestock can contribute to the health and diversity of grassland ecosystems through the enrichment of soil with manure or by the dissemination of plant seeds that pass through their digestive tracts or that cling to their hooves and coats.

The goat is often cast as the culprit when land degradation is observed and goats are present. However, it must be recognized that goats, noted for their ability to survive under harsh conditions, may remain in degraded environments often after people and other animals that were primarily responsible for the degradation have abandoned these areas. The casual observer arriving late on the scene may conclude that it was the goats that were responsible. It is certainly true that goats, if not properly managed, can indeed contribute to further land degradation, but this is often not the case. The eastern Mediterranean, or ancient Levant, region is an important example. After centuries of deforestation and continuous cultivation without sufficient reenrichment of the soils, land in the region became so degraded that most types of cropping and cattle grazing became impossible. Rural people turned to goat herding, because goats represented one of the few agricultural enterprises that such a degraded landscape was able to support.

In intensive production systems, eutrophication (see chapter 2, page 52) of lakes, ponds, rivers, and estuaries associated with excessive nutrient deposition derived from animal manure can have adverse affects on freshwater ecosystems by promoting algal blooms and depriving them of oxygen, killing resident organisms. In some marine habitats, animal wastes, such as from large-scale swine and poultry farms, have been implicated in algal blooms that produce toxic substances, such as those involving the dinoflagellate *Pfiesteria piscicida*. This organism caused massive fish kills in the 1990s off the coasts of North Carolina and Maryland and resulted in various neurological disorders in people who were exposed.[101,102] Other harmful algal blooms (HABs), resulting in part from animal wastes finding their way into rivers, estuaries, and coastal marine environments, have contaminated shellfish and caused toxic paralytic and amnesic disorders among those who consumed them.[103,104] Aquaculture activities may also play a role in promoting the growth of harmful algae. In the past, only a few regions of the United States were affected by HABs. Now, virtually every U.S. coastal state has reported serious outbreaks, which may be responsible for more than $1 billion in losses in the last two decades through direct impacts on coastal resources and communities.[105]

Transmission of infectious diseases is another way that livestock can affect biodiversity. The first major infectious disease outbreak in wild animals that was described in some detail was the rinderpest panzootic that occurred in wild African ungulates (hooved mammals, e.g., antelopes and llamas) following the introduction of infected cattle into sub-Saharan Africa in the 1890s. (*Rinderpest* is an acute, contagious viral disease with high rates of mortality, mainly found in cattle and characterized by ulceration of the animal's digestive tract.[106,107] A *panzootic* is the animal version

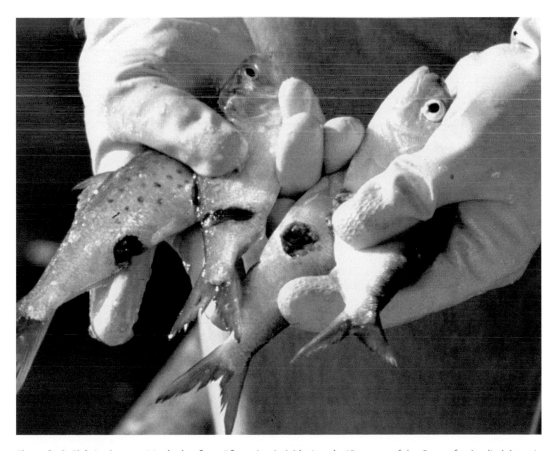

Figure 8.15. Scanning Electron Micrograph of the Dinoflagellate *Pfiesteria piscicida*. (Courtesy of the Center for Applied Aquatic Biology, North Carolina State University.)

Figure 8.16. Fish Lesions on Menhaden from *Pfiesteria piscicida* Attack. (Courtesy of the Center for Applied Aquatic Biology, North Carolina State University.)

of a pandemic, namely, an infectious disease that affects many animal species over a large area.) Pastoralists who depended on cattle for their livelihood during this outbreak suffered immensely from loss of their animals. But the effect on wildlife populations was also devastating. An estimated 90 percent of Cape Buffalo (*Syncerus caffer*) populations in Kenya were lost in the epidemic. In some areas, populations of certain species, such as Roan Antelope (*Hippotragus equinus*), Greater Kudu (*Tragelaphus strepsiceros*), and Bongo (*Tragelaphus eurycerus*), were so reduced that they never recovered. Some evidence suggests that the mortality of wild ruminants was so high that tsetse fly populations in some regions died out for lack of suitable hosts to feed upon.[108] To prevent the spread of rinderpest into southern Africa, extensive game-proof fences were erected, which disrupted the migratory patterns of some wild ungulates that had moved regularly over long distances in search of food and water. Though now largely under control, rinderpest still affects wildlife. In 1994 and 1995, an outbreak of the disease in Tsavo National Park in Kenya eliminated 60 percent of the Cape Buffalo and 60 percent of the Lesser Kudu (*Tragelaphus imberbis*).[109]

The spread of rabies and distemper to wild canids, including to African Wild Dogs (*Lycaon pictus*), the Bat-Eared Fox (*Otocyon megalotis*), jackals (various *Canis* species), and Hyenas (*Crocuta crocuta*), as well as to Lions (*Panthera leo*) and possibly other wild cats in Africa, is linked to closer contacts between wildlife and domestic dogs. Such contact has occurred because of increased human settlement on the peripheries of game parks and reserves, and an increased presence on rangelands of pastoralist livestock grazers that herd their animals with dogs.[110]

In 2001, the occurrence of foot-and-mouth disease in the United Kingdom, Ireland, the Netherlands, and France posed a serious threat to susceptible wild ruminants,

---

BOX 8.8

THE USE OF ANTIBIOTICS IN LIVESTOCK PRODUCTION

Antibiotics are also widely fed to livestock and extensively used in aquaculture, even in the absence of infection (see also discussion of antibiotics in chapter 7, page 310). Such preventive use has become necessary to avoid the spread of infections in overly crowded confines, but antibiotics are mainly used with livestock as growth promoters. While infections are prevented, chronic antibiotic treatment ensures that any bacteria that do survive will likely be resistant to the antibiotic being used and, in some cases, to related antibiotics that share similar chemical structures. Just such a situation occurred when enrofloxacin, an antibiotic used only in livestock and belonging to a class of antibiotics known as fluoroquinolones, was administered to large numbers of chickens in the mid-1990s. Within a few years, strains of *Campylobacter jejuni*, a bacterium that lives in chickens and is the most common cause of food-borne human diarrhea in the United States, appeared that were resistant to treatment with enrofloxacin. The resistant bacteria spread to humans, and in 2001 the U.S. Food and Drug Administration reported that more than 11,000 human cases of infectious diarrhea had been caused by resistant *C. jejuni* that had originally evolved in chickens.[a] This alone would have been cause for worry, but to make matters worse, these bacteria were resistant not only to enrofloxacin, but also to human antibiotics closely related to it, such as the fluoroquinolone ciprofloxacin. Given the appearance of such cross-resistance, infectious disease specialists have voiced concern that continued imprudent antibiotic use in domestic animals may breed still more antibiotic-resistant organisms, which may further diminish the effectiveness of our already strained arsenal of antibiotics. The World Health Organization has recently called for a ban on the routine use of such antibiotics in livestock because of these concerns.[b]

some of them endangered species housed in zoos. The outbreak renewed the policy debates about vaccinating endangered animal populations in zoos, parks, and reserves. During the 2001 outbreak, the European Union passed a resolution allowing endangered species in zoos to be vaccinated against the foot-and-mouth virus, but only when an outbreak of the disease had been confirmed within 25 kilometers of the zoo.[111]

FOOD FROM AQUATIC SYSTEMS

The Marine Food Chain

The production of seafood is one of the many goods and services generated by marine ecosystems. Human societies have been harvesting this bounty from the very earliest times, for example, from the Red Sea as long ago as the Middle Paleolithic period, some 100,000 years ago.[112] Initially, relatively few species were harvested and in small numbers, and the impacts on stocks and on marine bio diversity must have been minimal. However, as populations increased, and as fishing technologies improved, marine species began to be fished at unsustainable levels. Such overexploitation is not just a present-day phenomenon, however. Green turtles in the Cayman Islands, for example, were so overharvested for local consumption and trade during the eighteenth century that by 1800 there were none to be found in this region.[113] During the same century, the Steller's Sea Cow was hunted to extinction.

During the past hundred years, marine resources have been exploited as never before, threatening biodiversity in all the oceans. In this section we look at the role of harvesting seafood in endangering marine species and in disrupting marine ecosystems (See also chapter 2, page 44, for a discussion of overexploitation of seafood.)

Million tonnes

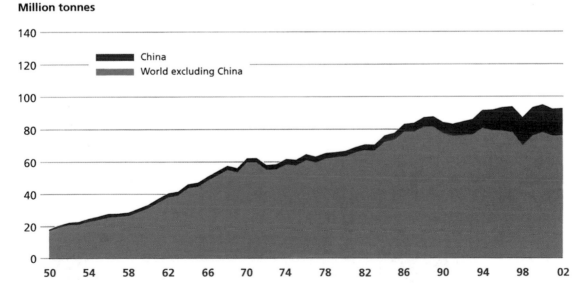

Figure 8.17. Total World Marine Fisheries Catch by Year in Million Metric Tons (MMT; 1 MMT, equals 1.1023 tons). (From *State of World Fisheries and Aquaculture (SOFIA) 2004.* U.N. Food and Agriculture Organization, 2004.)

Biodiversity of Marine Food Species

Fish are the dominant group within the marine food chain in terms of global catches. However, only forty of the world's roughly 20,000 identified fish species are taken in large quantities. Many other species are also taken by some multi-species tropical fisheries. According to the U.N.'s Food and Agriculture Organization, the top ten species caught in marine waters in 2002 were, in order, Anchoveta (*Engraulis ringens*), Alaska Pollock (*Theragra chalcogramma*), Skipjack Tuna (*Katsuwonus pelamis*), Capelin (*Mallotus villosus*), Atlantic Herring (*Clupea harengus*), Japanese Anchovy (*Engraulis japonicus*), Chilean Jack Mackerel (*Trachurus murphyi*), Blue Whiting (*Micromesistius poutassou*), Chub Mackerel (*Scomber japonicus*), and Largehead Hairtail (*Trichiurus lepturus*). Their total catch was about 27 MMT (almost 30 million tons) out of the global marine total of about 84 MMT (almost 93 million tons) that year. In the oceans, as on land, relatively few species make up a significant percentage of the total used as food.[114] But in contrast to the situation on land, where a dozen plant species supply 75 percent of the "global larder,"[115] and just four crops—wheat, rice, corn, and potatoes—account for more food production than all other crops combined (sugar cane is not included here because it is not a food crop per se but is used to produce sucrose, or table sugar, in addition to other products such as molasses), in the oceans the ten most harvested species comprise only about one-third of the total.

The extent to which fish and other harvested seafood depend on the biodiversity of the ecosystems in which they are found is a critical but poorly understood issue. Scientists have a limited grasp of whether different species in marine ecosystems perform the same functional roles, so it is often not possible to determine whether the loss of any one of them interferes with specific forms of ecosystem functioning. Nor is it well understood how the loss (or addition) of "keystone predators" or entire trophic groups would disrupt marine food webs. (*Keystone predators*, e.g., sea otters, play important roles in controlling prey species such as sea urchins, whose populations, if left unchecked, would expand rapidly and erode biodiversity, in the case of sea urchins by overfeeding on kelp, a dominant species and key habitat for many other species. *Trophic groups* are groups of different species that are organized hierarchically by their method of feeding, according to their position in the marine food chain—carnivores are at a higher trophic level than herbivores, which are higher than decomposers.) It is also not known what the loss of some marine ecosystems, for example, those that serve as fish and shellfish breeding grounds and nurseries (e.g., in estuaries, coastal wetlands, mangroves, sea grass beds, and coral reefs), means to seafood production. These and other questions that relate to the complex interrelationships in marine food webs need much more attention.

Marine Fisheries

Fish and other marine organisms make a very important contribution to the human diet, accounting for a significant part of people's protein intake, particularly in some coastal developing countries. Seafood protein is easily digestible and of high quality, containing a good mix of essential amino acids (*amino acids* are the building blocks of proteins, some of which—*essential amino acids*—we cannot synthesize and therefore must be supplied in our diets). For centuries marine food supplies depended on the catch obtained from fisheries, using simple technologies of hooks and lines, harpoons, and beach seines (shallow-water nets, also called draw or sweep nets, hauled in from the shore). In recent decades, advanced technologies such as fleets of efficient trawlers, navigation equipment, echo sounders, and sonar have turned fishing from a chance-driven, food-gathering operation into a highly predictable, harvesting one.

As a result of these developments, the annual catch, which was about 40 MMT per year (about 44 million tons) in the 1940s, in recent years has more than doubled.

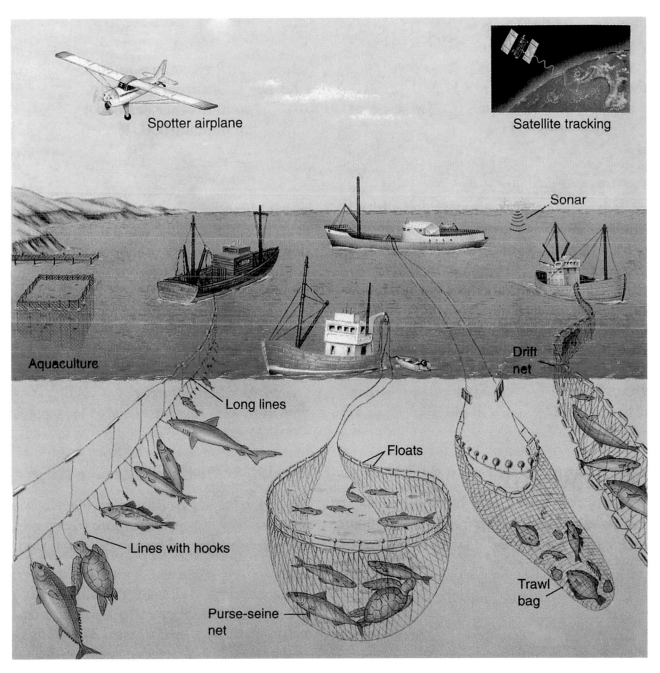

Figure 8.18. Modern Commercial Fishing Methods. (From Peter H. Raven and Linda R. Bert (editors), *Environment*, 3rd ed., © 2001, Harcourt, Inc., reprinted with permission from John Wiley & Sons, Inc.)

This rise has caused the collapse of some fisheries and the endangerment of many others around the world, such as those in the George's Bank (off the coast of New England and Nova Scotia) and in the Bering Sea (although the Bering Sea has seen a decrease in some species groups, e.g., certain marine mammals, and an increase in others).[116] Research on the dynamics of fish populations has led to international treaties that have allocated quotas for the different fishing nations and fishing grounds. But while these treaties have tended to alleviate overfishing in general, they have not had much bearing on individual fish species, which in some cases have become endangered, such as both the Southern Blue Fin Tuna (*Thunnus maccoyii*) and the Atlantic Cod (*Gadus morhua*). In the last fifteen years, the annual landing of the world marine fisheries has been maintained at a stable figure of about 85 MMT.

Only about 75 percent of captured fish are used directly as human food. The remainder produces fishmeal and fish oil, which are widely used for feeding chickens, pigs, and other farm animals, as well as in aquaculture and as both fertilizers and food additives.[117] The 60 MMT or so per year of captured fish used directly as human food, while an enormous amount, seems miniscule in comparison to the anticipated future world fish food demand, which is projected to reach 120 MMT by 2010.[118] With fisheries declining in productivity, there seems little possibility that the global demand for fish, if it is wild caught, can be met.

Direct consumption of fish and shellfish worldwide provides about 6 percent of all protein, and 15 percent of the animal protein, consumed by humans.[119] Of the total world population of more than six billion, an estimated one billion, principally from Africa and Southeast Asia, rely on fish and shellfish as their staple protein. For example, in Bangladesh and Indonesia, about 50 percent of average daily protein intake comes from fish; in Sierra Leone and Ghana, it climbs to more than 60 percent.[120] A further 5 percent of total human protein consumption comes indirectly from livestock fed with fishmeal.

The ocean accounts for more than 75 percent of the annual fish catch, and about 95 percent of this is taken from coastal waters. These areas are by far the most productive parts of the ocean and include, in addition to the estuaries, mangroves, marshes, seagrass beds, and coral reefs, areas of upwelling, nutrient-rich ocean waters such as those off the coast of Chile and Peru. These "food factories" generally lie within 370 km (200 nautical miles) off the coast, that is, within what are designated Exclusive Economic Zones (EEZs) where, under the Law of the Sea, specific countries have rights to all marine resources, including fisheries. Yet it is these same narrow coastal strips that are subject to the greatest stresses in the marine environment.[121]

The U.N. Food and Agriculture Organization has estimated that about 70 percent of commercial marine fisheries are being fished unsustainably and that these practices have reached crisis proportions.[114] What are the impacts on marine biodiversity of such overexploitation? For one, there are reductions in the stock size of fishery species as a result of overharvesting, and many species have become threatened, even those, as recent research has indicated, that are widely distributed.[122] Less obvious are indirect impacts, including increased mortality of nontarget species, such as unwanted fish, dolphins, and sea turtles caught as bycatch; the loss of some species secondary to physical damage to their habitats, such as coral reefs damaged by anchors, or seabeds stripped by trawl nets; and the effects on marine life by what is

called "fishing down marine food webs," where the excessive capture of fish at higher trophic levels (see description above) leads to fish being caught at lower levels, destabilizing the entire food web.[123] Fishery species face particularly acute problems when critical life-cycle phases are disrupted by heavy fishing and other environmental pressures, for example, when there is contamination by sewage, agricultural nutrient discharge, persistent organic pollutants, or heavy metals, particularly in spawning and nursery areas.[124]

In addition, modern fishing practices that may result in even more far-reaching consequences, both biological and social. For example, fishing itself can serve as a mediator of evolution, by changing the age at which sexual maturity is reached in a target species, thus affecting its reproductive status, and by favoring the survival of smaller fish.[125] And it can result in long-lasting economic and sociopolitical changes that are global in extent. The complexity of downstream effects from overfishing on biodiversity and on human health can be seen in the impacts of such fishing practices off the coast of West–Central Africa, mainly by fleets from the European Union, on people in the region. In this case, residents from Cameroon, Ghana, and other neighboring coastal countries, deprived of marine fish for food, have turned instead to increased hunting of bushmeat from the forests, threatening some native species[126] and potentially exposing themselves, for example, when they are killing and eating primates, to various primate viruses that may cause serious human diseases (see chapter 7, page 315).[127,128]

Freshwater Fisheries

Freshwater fisheries produce approximately one-quarter of the world's food fish, more than 31 MMT (about 34 million tons) per year in 2001 (global fish production for 2001 from all sources amounted to about 129 MMT).[114] Freshwater totals include both fish that are captured and those grown by aquaculture. But in contrast to marine fisheries, where most of the catch is landed by industrialized fleets coming from a small number of countries, freshwater fishing is more likely to operate on small, local scales and to be found in rural areas in developing countries, outside the purview of those who gather statistics. As a result, freshwater fishing totals are likely to be significant underestimates.

Increasing degradation of rivers, lakes, and streams, and of their watersheds (e.g., by deforestation) and growing levels of pollution of freshwater systems are endangering freshwater biodiversity and contributing to the growing global shortage of food from aquatic sources. About 20 percent of the world's freshwater fish are threatened (see "Habitat Loss: Fresh Water" in chapter 2).[129] In some parts of the world that have been studied, the situation is even more dire, such as in the Mediterranean region, where more than 50 percent of known endemic freshwater fish have been listed as threatened.[130] In addition, adequate flows no longer reach the deltas of many rivers during average flow years, including the Nile, the Yellow River in China (Huang He), the Amu Darya and Syr Darya in Central Asia, and the Colorado River in the United States and Mexico. This leads to coastal nutrient depletion, loss of habitat for native fisheries, plummeting populations of birds, shoreline erosion, and consequent adverse effects for many local communities.[131,132]

Figure 8.19. Freshwater Basket Fishing by Woman from the Mbukushu Tribe in Botswana. (© 2005 Frans Lanting/www.lanting.com.)

quaculture, the raising of aquatic organisms under controlled conditions, a practice that most likely began in China several thousand years ago, has more in common with livestock production on land than it does with fishing. The organisms must be fed, they need to be confined, their diseases need to be prevented or treated, and their wastes must be disposed of. Aquaculture can involve either freshwater or marine organisms. In the latter case, it is generally called *mariculture*. In response to a growing demand for fish, and a diminishing supply globally, there has been enormous growth in aquaculture during the past twenty years, particularly in China, which now accounts for more than two-thirds of global aquaculture production.[133]

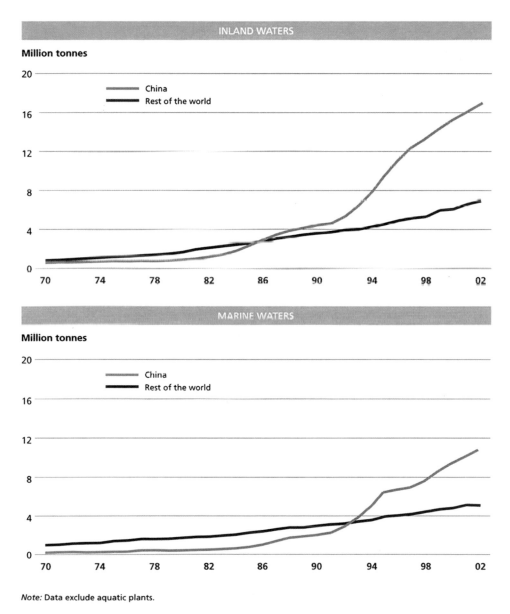

Note: Data exclude aquatic plants.

Figure 8.20. Aquaculture Production in Marine and Inland Waters since 1970. (From *State of World Fisheries and Aquaculture (SOFIA)* 2004. U.N. Food and Agriculture Organization, 2004.)

Many freshwater species are raised in aquaculture facilities. These include Striped Bass (*Morone saxatilis*) and species of trout, catfish, freshwater prawns, and crawfish, among others. But by far the most commonly raised freshwater organisms, especially in China, are the fish tilapia and various species of carp. Catfish, tilapia, and carp are herbivores and are largely fed high-protein vegetable diets derived from corn, wheat, cottonseed, peanuts, or soybeans. The feed of trout and Striped Bass contains high levels of fishmeal and fish oil, both of which are gradually being replaced by plant-based ingredients.

MARICULTURE

Mariculture is the captive production of marine organisms in seawater. It is of particular importance for feeding local populations in some arid or semiarid regions, where agricultural food production may be limited by drought or a shortage of fresh water. As a result of advances in research and development in the last few decades, mariculture is growing at an average yearly rate of 6 to 10 percent.[117] The U.N. Food and Agriculture Organization (FAO) reports an increase in mariculture from roughly 5 million metric tons (MMT; about 5.5 million tons) in 1990 to more than 11 MMT in 1997. In the same period, freshwater aquaculture grew from about 8 MMT per year (about 8.8 million tons) to more than 17 MMT annually. According to the FAO, global aquaculture from fresh and salt water in the year 2003 amounted to about 35.5 MMT, or roughly one-quarter of all fish consumption.[133] More than 8 MMT of farmed seaweeds are also produced each year, with China being the largest producer.

Marine fish production is presently practiced in floating net cages that are suspended from rafts and stocked with fingerlings (small, young fish) produced in land-based hatcheries or obtained from the wild. Aquacultured shellfish, such as clams, oysters, and mussels, grow attached to netting or long lines suspended from rafts or floats. The fish are generally given commercially prepared feed in the form of pellets until they are marketed. Some carnivores such as tuna are fed whole fish. Cage farms require relatively protected areas such as fjords and protected bays, and are usually situated near the shore so that feeding and maintenance can be easily accomplished. In the last decade or so, however, because of increased concerns about the environment and the development of better cage and mooring technologies, farms are being moved farther away from the shoreline, often to relatively open-sea conditions. Such farms generally have large feed-storage capacity, are equipped with feeding machines, and have their own energy supply systems.

AQUACULTURE AND THE ENVIRONMENT

While aquaculture has enormous potential to help feed rapidly growing world populations, it also carries significant risks to aquatic biodiversity, both in freshwater ecosystems and in the oceans. Aquaculture facilities in many parts of the world are acutely aware of these risks, which producers and scientists alike are actively attempting to address. We will review them here.

Antibiotics given to aquacultured fish and shellfish can be harmful to marine wildlife[134,135] and can lead to the development of antibiotic-resistant bacteria that could infect people[136,137] (also see discussion of antibiotic resistance in chapter 7, page 310). Effluents containing feed particles, fish cadavers, and feces can contaminate surrounding areas with high levels of nutrient pollution. This is a particular problem for cages and pens in waters that are shallow or that have little tidal flushing. Such conditions can lead to a loss in water quality, with resultant eutrophication and lowered oxygen levels, endangering local flora and fauna.[138,139] Nutrient pollution can also be a problem in poorly managed freshwater aquaculture facilities.

Disease

Raising large numbers of fish and shellfish in tightly confined pens and cages risks outbreaks of infectious diseases. Aquaculture farmers make great efforts to avoid such conditions, but infections occur nevertheless. White spot and yellowhead virus infections, for example, have broken out in shrimp farms in Asia, reaching at times epidemic proportions,[140] and both pathogens have shown up in wild and farmed shrimp populations in the United States.[141] Infections with

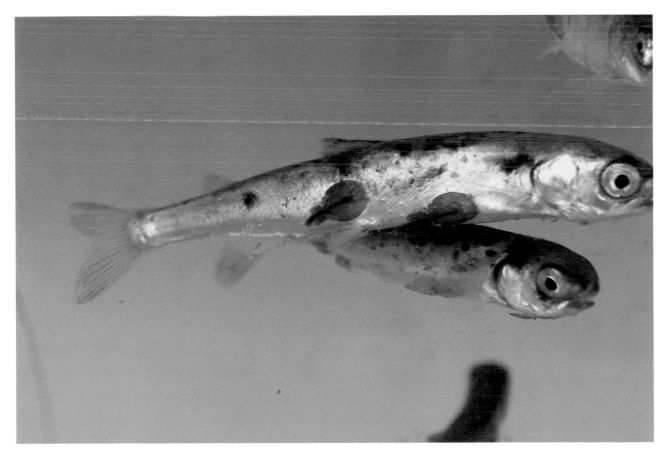

Figure 8.21. Sea Lice (*Lepeophtheirus salmonis*) on Pink Salmon (*Oncorhynchus gorbuscha*) Fingerlings. (Photograph by Alexandra Morton, www.raincoastresearch.org.)

bacteria such as *Escherichia coli*, *Shigella*, and *Vibrio cholera* have also been reported, although these are rare and often associated with poor farm management. Sea Lice (*Lepeophtheirus salmonis*) are also a common problem with farmed finfish such as salmon and can cause significant mortality in wild salmon that become infected.[142,143]

Escape

Despite improved confinement technologies, the problem of escape of farmed organisms remains high. According to the World Wildlife Fund in the United Kingdom, for example, as many as a half million farmed fish escape into Norwegian coastal waters each year, and in Scotland, in January 2005 alone, around 630,000 farmed fish escaped.[144] They can carry with them their diseases and parasites and can infect wild stocks. But the greatest potential danger with escapees is their ability to interbreed with wild populations, diluting genes in their hybrid offspring that had initially evolved for successful life in the wild, and thereby imperiling their survival.[145] This potential for hybridization between farmed and wild fish is of particular concern with Atlantic Salmon and Atlantic Cod, both of which are already endangered in the wild (in some catches as many as 40 percent of the salmon caught in the North Atlantic have been found to come from mariculture facilities).[146] The problem with escape and

Figure 8.22. Coastal Aquaculture Ponds in Malaysia Developed from Mangrove Forests. Malaysia has lost half of its mangrove forests—a million and a half acres—in the past fifty years. (Courtesy of Tim Laman/National Geographic Image Collection.)

hybridization could become even greater if fertile, genetically modified salmon and other fish begin to be farmed.[147]

Destruction of Habitat

Hundreds of thousands of acres of mangroves and coastal wetlands have been destroyed to create ponds for the aquaculture of shrimp and Milkfish (*Chanos chanos*, a fish that normally lives in the open ocean but has been widely cultivated for food). This practice, particularly widespread in Southeast Asia, has caused extensive damage to coastal ecosystems[148] (see figure 8.22). An estimated 65,000 hectares (around 160,000 acres) of mangroves have been made into shrimp ponds in Thailand alone in recent years.[149] Such mangrove habitats are not only extremely biologically rich themselves but are also breeding grounds and nurseries for wild finfish and shellfish, and their destruction can result in large losses in wild populations and have major impacts on coastal marine biodiversity. As many as one-third of fish (excluding bycatch) that are caught each year in Southeast Asia, for example, are mangrove dependent. Moreover, mangroves are essential to the health of coral reefs and seagrass beds, both among the greatest repositories of marine biodiversity.[150] And finally, mangroves and coral reefs form natural buffers that protect coastal lands and the people that live there.[151,152]

Figure 8.23. A Mangrove Forest Off the Coast of Belize. In this underwater photograph of a mangrove forest, looking up toward the above water canopy, one can see the incredible richness of life blanketing the mangrove roots. (Courtesy of Tim Laman/National Geographic Image Collection.)

Depletion of Wild Populations

Mariculture operations in developing countries, for fish such as Milkfish, mullets, groupers, and some members of the bream family, depend heavily on wild fingerlings to stock their facilities, rather than on those raised in land-based hatcheries. This practice runs the risk of depleting wild populations. In addition, the rapidly increasing demand for fishmeal and fish oil to feed carnivorous aquacultured fish in all countries is associated with further depletion of wild stocks. From 1986 to 1997, eight out of the top twenty species fished—Anchoveta (*Engraulis ringens*), Chilean Jack Mackerel (*Trachurus murphyi*), Atlantic Herring (*Clupea harengus*), Chub Mackerel (*Scomber japonicus*), Japanese Anchovy (*Engraulis japonicus*), Round Sardinella (*Sardinella aurita*), Atlantic Mackerel (*Scomber scombrus*), and European Anchovy (*Engraulis encrasicolus*)—were used to make fishmeal and fish oil for aquaculture, livestock production (the poultry and swine industries, it should be pointed out, are the world's largest consumers of fishmeal), and other products such as pet food.[148] Because it generally takes somewhere between two to five kilograms of wild fish to produce a kilogram of farmed fish, this practice, if continued, will contribute to disrupting already stressed marine ecosystems and to endangering the survival of many marine species, some of which are already classified as overexploited or depleted.[153]

Fish such as anchovies, mackerel, herring, and sardines are important protein sources for large numbers of people around the world, particularly for those in some developing countries. Because they are plentiful, reproduce rapidly, and can be harvested sustainably without disrupting food webs; because they are low on the food chain and therefore not likely to be as contaminated by pollutants such as mercury, and perhaps also by PCBs and other organochlorines (in some regions, e.g., the Baltic Sea, herring have been found to be contaminated with PCBs, although levels have been decreasing over the past 25 years); and because they contain high levels of omega-3 fatty acids (which are associated with cardiovascular health, especially for those who are at risk of a heart attack), these small ocean fish should be consumed in greater amounts by people, in both developing and industrialized countries alike, rather than being fed so widely to animals.

Figure 8.24. Salted Anchovies Being Sold at La Boqueria Market, Barcelona, Spain. (Photo by Eric Chivian.)

THE PROMISE OF SUSTAINABLE
AQUACULTURE

If aquaculture is to be sustainable—and given the rapidly growing deficit between global demands for fish and the capacity of capture fisheries to meet these demands, it must be sustainable—its continued growth must take into account the preservation of healthy freshwater and coastal marine ecosystems. Several practices would help ensure such preservation, while also increasing aquaculture yields:

- Encouraging increased consumption of farmed herbivorous fish: Roughly 80 percent of total global aquaculture involves the farming of herbivores—carp, tilapia, Milkfish, mollusks, and catfish. Growth in culturing these organisms needs to be strongly encouraged, perhaps especially in the industrialized world, and the practice of feeding fishmeal to herbivores, which is rapidly expanding for some fish such as tilapia and carp, and which contributes to depleting wild fish stocks, should be discouraged.

- Developing better plant-based foods for carnivorous farmed fish: Plant-based alternatives to fishmeal and fish oil are being used in carnivorous farmed fish with good results.[154] Vegetable oils such as linseed and rapeseed oil, substituted for fish oil during a portion of the feeding cycle in aquacultured Atlantic Salmon (*Salmo salar*), did not affect growth rates or the health of the fish, while significantly reducing PCB content and only marginally reducing omega 3 levels.[155] Plant-based proteins will need to be supplemented with specific amino acids to give satisfactory results in growth, and higher protein diets will need to be developed.

- Developing land-based mariculture facilities to reduce the problems of nutrient pollution, escape of farmed fish, and destruction of coastal habitats like mangroves and wetlands: Some new mariculture facilities on land rely on principles that were developed for recirculated freshwater aquaculture. One approach has been the development of integrated pond technology, where effluents produced by marine food organisms in land-based ponds, such as fish and shrimp, can serve as nutrients for the production of microalgae or macroalgae. These, in turn, serve as food for oysters, clams, abalone, or sea urchins. As a result, three major crops can be grown—fish or shrimp, algae, and shellfish or sea urchins. In one model, a 1-hectare (~ 2.5-acre) land-based, integrated system produced 25 tons of fish, 50 tons of shellfish, and 30 tons of seaweeds a year.[156] The final effluents from these systems are nutrient-poor and can be recycled or returned to the sea. There is growing interest in China and other Asian countries in such technologies, which may eventually be adapted for the mass production of fish and other aquatic crops with minimal environmental impacts.

Other technologies that have shown promise include the following:

- Highly intensive, environmentally friendly, recirculated aquaculture, also known as RAS, where effluents are treated with nitrifying and denitrifying microbial processes.[157] These systems are very efficient in using land and sea-water resources and in fish production. They can also be insulated thermally from the external environment, thereby allowing production to continue independent of season.

- The practice, largely in China, where four types of carp—Silver Carp (*Hypophthalmichthys molitrix*), Grass Carp (*Ctenopharyngodon idellus*), Common Carp (*Ctenopharyngodon idellus*), and Bighead Carp (*Aristichthys nobilis*)—are produced in the same ponds, because their different feeding habits complement each other and make the most efficient use of the pond's biota.[158] There are other models, also practiced in China, for example, where two other fish species, Black Carp (*Mylopharyngodon piceus*) and the Wuchang Fish (*Megalobrama amblycephala*), have been added to the four-carp system, resulting in a new stable system with high fish productivity.[159]

- The method of salmon farming in Chile that makes use of *Gracilaria chilensis*, a red alga that removes large amounts of dissolved nitrogen and phosphorus coming from salmon cages, with the remainder of these effluents serving to help produce a seaweed crop.[160]

- The possibility of integrating aquaculture with growing land crops, using the effluent, for example, to fertilize tomatoes. Such methods would be of special value in semiarid zones where water is scarce.[161]

Aquaculture can make an enormous contribution to helping feed growing human populations, but sustainable technologies that preserve the freshwater and marine ecosystems upon which it depends will be central to fulfilling its long-term potential.

CONCLUSION

Instead of the widely practiced approach, which treats the growing of crops and the raising of livestock without regard to ecological relationships, new agroecosystem methods strive to integrate food production into the larger environmental domain and to recognize and preserve the role of native fauna and flora in their natural habitats. A more holistic approach to the integration of farming and ecology will lead to improved nutrient recycling, biological pest and disease control, pollination, soil quality maintenance, water-use efficiency, and carbon sequestration. It will also help provide greater resistance to the anticipated impacts—droughts, floods, and heat waves—from global climate change.

For marine resources, protected areas and other conservation efforts are to be encouraged and intensified. Wetlands, mangroves, and coral reefs serve as important nurseries to marine biodiversity and must be protected. Awareness of environmental sustainability has become a central issue in aquaculture policy making, and the aquaculture industry has committed itself to adopting more environmentally friendly technologies.

The enhanced greenhouse effect is expected to result in significant global warming during the course of this century. The potential impacts of climate change and climate variability on biodiversity need to be more fully identified and understood, both in natural and in agricultural systems.

Finally, national and international policies are needed to encourage wide-scale adoption of the agroecosystem paradigm, and thus the conservation of biodiversity, in food-producing systems. This will ensure nutritious food for still-growing global populations, minimize exposure to agricultural chemicals, and promote both human and ecosystem health.

◆ ◆ ◆

Suggested Readings

Agri-Culture: Reconnecting People, Land and Nature, Jules Pretty. Earthscan, London, 2002.

Climate Change and the Global Harvest: Potential Impacts of the Greenhouse Effect on Agriculture, Cynthia Rosenzweig and Daniel Hillel. Oxford University Press, New York, 1998.

Climate Variability and the Global Harvest: Impacts of El Niño and Other Oscillations on Agroecosystems, Cynthia Rosenzweig and Daniel Hill. Oxford University Press, New York, 2008.

Consultative Group on International Agricultural Research (CGIAR), www.cgiar.org/.

Food and Agricultural Organization, www.fao.org.

The Forgotten Pollinators, Stephen L. Buchmann and Gary Paul Nabhan. Island Press, Washington, D.C., 1996.

Guns, Germs, and Steel: The Fates of Human Societies, Jared M. Diamond. W.W. Norton & Company, New York, 1999.

Life in the Soil: A Guide for Naturalists and Gardeners, James B. Nardi, University of Chicago Press, 2007.

Managing the Livestock Revolution: Policy and Technology to Address the Negative Impacts of a Fast-Growing Sector. Agriculture and Rural Development Department, World Bank, Washington, D.C., 2005.

The Omnivore's Dilemma: A Natural History of Four Meals, Michael Pollan. Penguin Press, New York, 2006.

Out of the Earth: Civilization and the Life of the Soil, Daniel Hillel. Free Press, New York, 1991.

Seafood Lover's Almanac, 2nd edition, Mercedes Lee, with Suzanne Iudicello and Carl Safina. National Audubon Society and Blue Ocean Institute, Cold Spring Harbor, New York, 2001.

State of the World's Fisheries 2004, U.N. Food and Agricultural Organization, www.fao.org/docrep/007/y5600e/y5600e00.htm.

Tending Animals in the Global Village: A Guide to International Veterinary Medicine, David M. Sherman. Blackwell Publishing, Ames, Iowa, 2002.

Fred Kirschenmann in a Buckwheat Field on His 3,700-Acre Organic Farm in North Dakota in the Summer of 2006. The trees in the foreground are part of a hedgerow that provides natural habitat to attract predators and parasites that feed on crop pests. (Photo by Constance L. Falk.)

CHAPTER 9

GENETICALLY MODIFIED FOODS AND ORGANIC FARMING

Eric Chivian and Aaron Bernstein

A nation that destroys its soils, destroys itself.

—FRANKLIN DELANO ROOSEVELT

No discussion of biodiversity and food would be complete without an examination of the subjects of genetically modified (GM) foods and organic farming. Both practices have enormous implications for biodiversity and for food production, and as a result, for human health. And debates about their risks and benefits will continue to occupy policy makers and the general public worldwide for many years to come as they grapple with trying to feed rapidly expanding numbers of people using fewer and fewer acres of arable land and fisheries that are being increasingly depleted. As detailed in chapter 8, this issue is one of the central challenges of our time. We recognize that GM foods and organic farming are topics that are widely, and at times heatedly, debated, both generally and within the scientific community, with strong proponents on both sides of the issues. Below, we review the evidence for each of these agricultural technologies as completely as space allows, but we encourage those who would like to delve more deeply into their complexities to consult the references provided and form opinions of their own. The decisions we make in coming years about how we grow food, both for the health of world populations and for that of the global environment, must be based not on politics or vested interests, or on widely quoted but often inadequately studied assumptions, but on objective scientific grounds.

GENETICALLY MODIFIED FOODS

As with aquaculture, GM foods (also called genetically engineered or *transgenic* foods) hold great promise that they may provide one of the solutions to help feed growing world populations.[1] But, as is the case with aquaculture (see chapter 8), there are also potentially large, and often not well understood, risks from GM technologies—to the environment in general and to biodiversity and the functioning of ecosystems in particular. While a comprehensive discussion of this critically important topic is beyond the scope of this book, in this section we review some of the potential benefits and risks of GM crops. A thorough understanding of these could not be more important at the present time, given that (1) in 2005, 222 million acres (about 90 million hectares) of approved GM crops were grown by 8.5 million farmers in twenty-one countries, with U.S. crops making up more than half of this amount (around 123 million acres, an area larger than the state of California);[2] (2) the global acreage devoted to these crops is increasing rapidly (growing by double-digit rates in recent years, 11 percent in 2005); and (3) in 2001, 26 percent of the maize and 69 percent of the cotton,[3] and in 2005, more than 80 percent of the soybeans grown in the United States were GM crops.[4] New figures for 2006, reported by the International Service for the Acquisition of Agri-Biotech Applications, an industry group, show an additional 13 percent increase to 252 million acres (102 million hectares).[5]

Among the major successes cited for the genetic modification of crops are the insertion of Bt genes (which produce insect pathogens, derived from strains of the bacterium *Bacillus thuringiensis*) into maize, potatoes, and cotton to make these crops resistant to certain insect pests,[6] and the insertion of herbicide-tolerance genes that allow GM crops to thrive despite being exposed to certain herbicides.[7] Rice has also been modified—in one case to produce beta-carotene (an antioxidant compound found in carrots and other yellow and orange vegetables that our bodies can convert into vitamin A),[8] and in another to reduce the concentrations of glutelin, a rice protein that is undesirable for sake brewing.

However, behind these and other successes of genetic modification lurk unexpected effects and potential pitfalls. The decrease in glutelin levels in rice, for example, was associated with an unintended increase in levels of compounds called prolamines,[9] which can affect the nutritional quality of rice and increase its potential to induce an allergic response.[10] Modified organisms can also escape from greenhouses and fields and aquaculture cages into natural ecosystems and disrupt their biodiversity. We have already seen the potential of this in aquaculture, where the escape of farmed salmon (which were not genetically modified) is threatening wild salmon stocks in the Atlantic Ocean (see discussion below and "Food from Aquatic Systems" in chapter 8).

The application of genetic transformation techniques to crop plants raises a critically important question: Does this technology offer the potential for mitigating the problem of biodiversity loss? Or the opposite: Does it pose a danger of exacerbating it? Proponents of the new technology contend that it can help intensify production on favorable lands, thereby alleviating the pressure on, and preventing the further degradation of, agriculturally marginal lands and their natural ecosystems. They also say that GM crops can reduce the need for tillage and for various chemicals such as pesticides, thereby enhancing biodiversity.

Opponents of the same technology fear that it can damage biodiversity, for example, by promoting greater use of certain pesticides associated with GM crops that are particularly toxic to many species, and by introducing exotic genes and organisms into the environment that may disrupt natural plant communities and other ecosystems. Still others say that food production problems are generally not biological in origin, but instead lie in such areas as lack of market access, the burdens of developing countries' debts, or poorly developed food processing and transportation infrastructures, none of which GM technologies would serve to address.[11] In addition, it is believed we are still far from reaching the full potential of hybrids for most crops, which can be achieved without further genetic modification,[12] and there are concerns that GM technologies would lessen incentives to develop such hybrids. Other objections pertain to the exclusive commercial appropriation and exploitation of the technology, which may indeed hinder the free exchange of information and ideas that has always been the hallmark of science, to the special detriment of the poorer countries.

Below we review these arguments, pro and con.

Background

Even prior to the advent of GM technology, traditional plant breeding methods have resulted in extensive alterations of crop plants. The genomes of plants used as crops have undergone numerous and, in some cases, considerable changes in the course of improving crop traits. For example, crops as different as broccoli, Brussels sprouts, and cabbage have all been derived from a single species of mustard, *Brassica oleracea*. Classical breeding techniques involving hybridizing different crop varieties (which are variants of the same species) are still the most effective approach for dealing with traits that depend on multiple genes distributed over the entire genome (so-called polygenic traits).[13] On the other hand, genetic engineering may be preferable for manipulating traits that depend on one or only a few genes, although there is now considerable research going on to engineer polygenic traits, as well, such as those involving metabolic pathways.

In the process of domestication, crop and livestock species cease to be "natural"— that is, they lose the ability to survive by themselves in an open environment and instead depend on humans for water, fertilizers or food, and protection against pests, diseases, and predators. Indeed, the genetic alterations achieved through domestication have been profound. Recombinant DNA technology —the transfer of genes from one species to another—however, has added an entirely new dimension to these alterations, opening up possibilities such as the ability to insert bacterial genes into maize, which cannot be achieved by traditional breeding methods.

The genetic modification of food entails the deliberate and specific manipulation of an organism's genome in order to modify aspects of its biology, such as its rate of growth, nutrient composition, resistance to pests and herbicides, tolerance of adverse growing conditions, and durability of the edible product. In some respects, this human-made genetic intervention is equivalent both to the process of random mutation that occurs in Nature as plants reproduce themselves, and to the trial-and-error hybridization used in traditional cross-breeding by agriculturalists. All three involve a haphazard element in that, depending on where the genetic change occurs within the genome, the function of other existing genes within the host genome may be altered.[14]

However, GM technology is different in three respects. First, the genetic change is specific, planned, and deliberately sought, and there is an immediate, direct insertion of genes as opposed to multiple generations of breeding. Second, the introduced genes can come from any species of plant, animal or microbe; that is, the procedure can be *transgenic*. It is this aspect that has been a particular source of public concern and scientific uncertainty. Transgenic techniques transcend the natural barriers between related species (e.g., if pollen from an apple tree [*Malus pumila*, also widely known as *Malus domestica*] is carried to the flower of a Red Raspberry plant [*Rubus idaeus* L.], the raspberry plant will not recognize the apple pollen because they are different species, so the apple pollen cannot fertilize the raspberry ova, even though they both belong to the same plant family, Rosaceae) and, indeed, between distant species, such as the bacterium *Bacillus thuringiensis* and maize, which might bring some surprising changes in the functioning of the altered host genome. Third, the technology requires the use of marker genes to confirm the successful insertion of the index gene. Those marker genes may confer properties, such as antibiotic resistance, which could cause significant problems.

The attainment of higher yields and the development of more environmentally sustainable practices have been, and will continue to be, the main challenges of agriculture. By 2025, the world's approximately eight billion people will require an average world cereal yield of about 4 metric tons per hectare (about 1.75 tons per acre). And, if conventional farming methods continue to be relied upon to the extent they are now, an approximate doubling of the current global use of synthetic nitrogen will be required to produce the needed 3 billion tons of grain.[15] With world population growth, available agricultural land has steadily declined in recent decades, from about 0.5 hectares (about 1.2 acres) per person in the 1960s to less than half that amount at the present time, and by 2050, further reductions are projected to perhaps 0.14 hectares (somewhat more than one-third of an acre) per person.[16] On a global scale, therefore, it will be necessary to increase the per-hectare yields of all the major crops, which, by current methods, may not be possible. Another potential way to achieve the objective of feeding additional populations is to further invade and destroy remaining natural habitats (many of which are already marginal for farming), but this is not considered to be an option.

In too many areas, agricultural practices have been extremely harmful to the environment, and to biodiversity in particular. The negative impacts have increased along with the growth of world population. While the production of a ton of food with modern cultivars (which are particular varieties of plant species or hybrids, selected for certain attributes that are retained following propagation) of wheat or maize may require less land, less energy, and smaller amounts of agrochemicals than those under cultivation thirty years ago, the doubling of world population in the same period has more than offset these technological advances.

Yields of the staple grain crops have been steadily increasing over the past decades, but this trend appears to be leveling out. Even if we extrapolate the gains of the past over the next thirty years, the projection suggests the likely occurrence of grain deficiencies in all major regions of the world, with the exception of Europe and North America.[15]

Potential Benefits of GM Crops

REDUCED AND MORE ENVIRONMENTALLY SOUND AGROCHEMICAL USE

Insect-resistant and herbicide-tolerant GM crops should lead to reduced levels of pesticide and herbicide use and, as a result, to fewer environmental impacts. One should not, for example, have to spray for pests that have already been effectively dealt with by the presence of Bt toxins in the crops, and one should theoretically be able to use lower amounts of herbicides, because those employed with some GM crops are so effective. Indeed, for some GM crops, this seems to be the case, with the trend showing an overall reduction (with the exception of glyphosate-tolerant soybeans) in the use of agrochemicals from 1997 to 1998 in association with increasing levels of GM farm acreage.[17] A recent two-year farm-scale evaluation involving eighty-one commercial fields in Arizona showed a similar decrease in insecticide use with Bt cotton when compared to non-GM cotton.[18]

In China, recent studies have also demonstrated reduced pesticide use when GM crops are planted, both for Bt cotton[19] and for two varieties of GM rice engineered to attack rice pests.[20] Of great importance in both of these cases was an observed reduction in the incidence of pesticide-related illnesses among the Chinese farmers who participated. The same was true in another recent study involving nearly 5,000 small farmers growing Bt cotton in the Makhathini region of South Africa, where lower amounts of pesticides were used, and a decline in cases of pesticide poisonings was reported. In this study, higher cotton yields were achieved as well.[21]

There are also claims that the herbicide glyphosate (the active ingredient in Roundup), widely used with GM crops, may be more environmentally friendly than alternative herbicides, because it is reputed to have a shorter half-life[22] and a lower toxicity for mammals, birds, and fish.[23] But glyphosate and its metabolites may be more persistent in some environments than has been recognized, particularly when extremely large amounts of the chemical are used (as is increasingly the case with GM soybean fields in the United States).[22] There are also published reports that glyphosate may contaminate freshwater systems[24] and that it can cause significant mortality in some North American amphibians (and perhaps in amphibians in other parts of the world), many species of which are already endangered[25,26] (see also the discussion of amphibian declines in chapter 6). Moreover, because resistance to glyphosate is beginning to develop in some weeds, such as Rigid Ryegrass (*Lolium rigidum*), larger doses of glyphosate, or the use of more toxic herbicides, may be required.[17]

With regard to pesticides and Bt crops, it has been found that some farmers may use fewer pesticides initially when compared to conventional crops but that, after several years, they may be using as much or more than they did before. This has occurred in China with some farmers growing Bt cotton, because new cotton pests that had not been problems in the past, leaf bugs called mirids, have emerged in Bt cotton fields, requiring new and extra pesticides to control them. What has been hypothesized is that conventional pesticides and cotton bollworms (which were now absent because of Bt toxins) had served to keep mirid populations in check.[27,28] Given that Bt cotton seed can cost two to three times more than conventional seed, the extra pesticide costs these farmers incurred no longer made using Bt cotton economically advantageous.

The development of resistance to Bt toxins among numerous insect species has also been documented,[29] leading to the possibility of growing insect pest resistance to Bt crops and the need for larger amounts of pesticides. Therefore, it may be too soon to determine whether there will be less use of agrochemicals with GM crops over time compared to conventionally grown crops, or whether the chemicals used on GM crops will be more environmentally friendly than those used on conventional crops (see discussion below).

SOIL CONSERVATION

Genetically modified, herbicide-tolerant (GMHT) crops allow farmers to use what are called postemergent herbicides (e.g., glyphosate), applied later in the growing cycle, rather than being mixed in with the soil when crops are planted. These herbicides serve to promote low-till and no-till practices (see figure 8.10 in chapter 8), and such practices have the potential to lead to increased soil organic matter, higher levels of soil carbon sequestration, and decreased soil erosion, nutrient leaching, and water loss, all of which are beneficial for the environment.[30]

INCREASED YIELD

Improved yield has been one of the main justifications for the development of GM crops. GM crops are expected to provide greater yields, because they are designed to do so and because of their engineered ability to withstand attack by pests and to grow well under less than ideal conditions. For example, GM potatoes have been developed that show significant resistance to attack by the fungus *Phytophthora infestans* that causes Potato Late Blight, the most devastating disease of potatoes.[31] And in a world with more extreme weather events and rising seas secondary to global warming, crops engineered to grow well under conditions of drought or increased salinity, such as rice engineered to withstand drought and salt water[32] and tomatoes that can thrive in salty soils,[33] offer a significant advantage over conventional crops vulnerable to these changes. There are some indications in the United States that GM crops have led to greater yields.[18,34] But it is often not clear whether the differences in yield that have been observed between GM and non-GM crops are the result of other extraneous factors.[17] More studies need to be performed to address this critically important question.

OTHER POTENTIAL BENEFITS FROM GM ORGANISMS

GM plants are also being tested for a variety of other purposes. For example, some transgenic plants are being developed that can be used to remove organic compounds, heavy metals, and other contaminants from the environment.[35] Others are being used as factories to develop biopharmaceuticals, including vaccines.[36]

There are many other promising areas of transgenic research that involve food production and human health. One, for example, is the engineering of mice to carry a gene (*fat-1*) from the roundworm *Caenorhabditis elegans* (see discussion of *C. elegans* in chapter 5, page 180) that enables them to convert their abundant supplies of omega-6 fatty acids to omega-3s, the fatty acids, mostly found in fish oils, that have been shown to promote cardiovascular health as well as other potential health benefits[37] (see chapter 4, page 154). This technology could potentially be adapted to cattle and chickens, so that people could obtain their omega-3 fatty acids from such animal products as meat, milk, or eggs (although this may become less necessary, because free-range, grass-fed cattle may accumulate significant levels of omega-3 fatty acids in their meat, in contrast to grain-fed, feedlot cattle).[38] In another recent development,

Thale Cress plants (*Arabidopsis thaliana*) are able to make both omega-3 and omega-6 fatty acids after the insertion of algal and mushroom genes, which may pave the way for some crop species to be engineered to manufacture them, as well.[39] Transgenic animals are also being developed that produce medicines in their milk, such as transgenic goats that can produce human antithrombin, an anticoagulant protein used to treat antithrombin deficiency, a hereditary disorder.[40]

Potential Risks

RISK OF INVASION

There are also significant potential risks with GM foods, for example, the risk that they may invade natural habitats. Twelve of the world's thirteen most important food crops—including wheat, rice, maize, and soybeans—have been found to hybridize with wild relatives, and for seven of these—wheat, rice, soybeans, sorghum, millet, beans, and sunflowers—such hybridization has led to the development of weedy species.[41] So it should not be surprising that scientists are concerned about hybridization between GM crops and their wild relatives,[42] as well as about the possibility of transgenes (a gene that is artificially inserted into the genome of an organism) being inserted into weeds, making them potentially more invasive.[43] Some scientists believe that there are too many barriers standing in the way of such transfers—that, for example, a pollen grain must fly a certain distance, find an appropriate and mature recipient, pollinate it, and yield a viable seed capable of developing into a nonsterile mature plant, and finally, the progeny of this plant must be able to grow and reproduce. Others are reassured that, despite millions of acres of GM crops having already been planted, there are as yet no reports indicating that transgenes have been transferred to native species. In the only rigorously designed ten-year field experiment with GM oilseed rape, potatoes, maize, and sugar beets, for example, there was no evidence that the transgenic crops were more invasive or more persistent in the wild than their conventional counterparts.[44]

However, there are often long lag times between the introduction of an invasive plant and evidence of its spread. For example, the Catclaw Mimosa (*Mimosa pigra*) took about 100 years after it was introduced before it spread widely in Australia, threatening other plants over large areas.[45] Also unsettling is the experience with introducing GM Oilseed Rape (*Brassica napus*, which produces the oil known as "canola," which stands for Canadian oil low acid) to Canada in the 1990s, when "volunteer" Oilseed Rape plants carrying GM herbicide-tolerant genes were found in many conventional Oilseed Rape farm fields in western Canada after only two seasons of commercial GM cultivation, despite GM Oilseed Rape not being grown in these fields.[46] A similar event occurred in 2001 in Mexico, when GM maize spread among native maize fields in the high valleys of the Sierra Norte region surrounding Oaxaca.[47] Mexico is the world's center for maize biodiversity, where maize had its origins from the wild grass "teosinte."[48] As a result, this intermingling of native and GM maize caused great alarm over the possibility that GM maize genes could contaminate the country's extremely valuable native maize genetic resources.

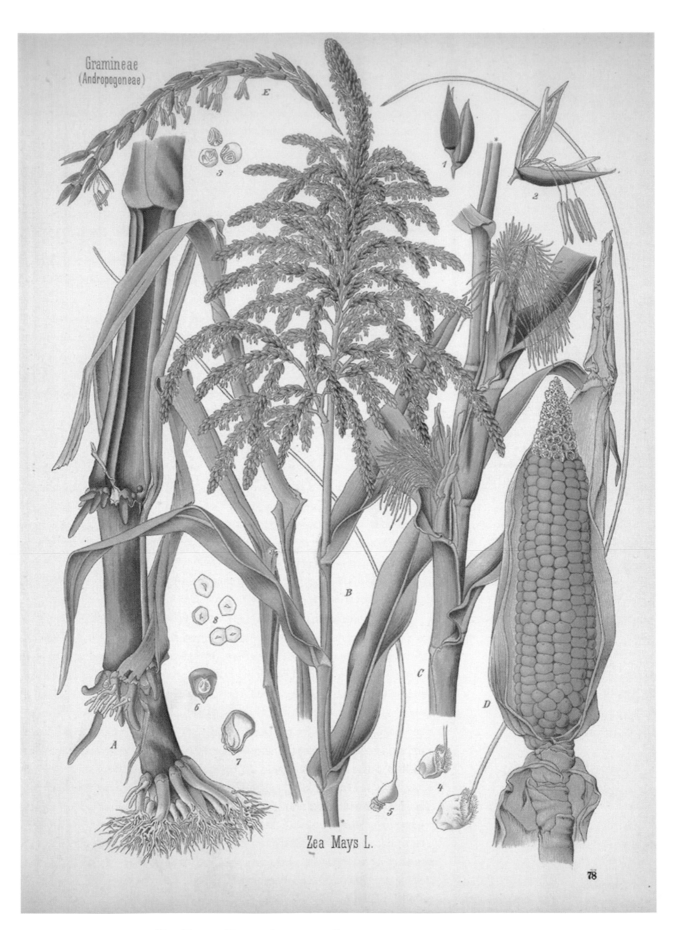

Gramineae
(Andropogoneae)

Zea Mays L.

BOX 9.1

Genetically Modified Salmon

Of the various farmed animals for which transgenic populations currently exist, fish are the ones that may pose the greatest threat for dispersing and surviving in the wild. Transgenic salmon are being engineered to have accelerated growth rates, and because they grow more rapidly, they may prey upon, or displace, their smaller counterparts in the wild. Of particular concern is the so-called "Trojan gene" effect, whereby transgenic salmon, which may not be well adapted for survival in the wild themselves, nevertheless exhibit mating advantages over wild salmon (often a function of body size), potentially driving wild salmon to extinction by successfully breeding with wild individuals, compromising their greater genetic diversity and reducing the fitness of their progeny.[a] Growth-enhanced transgenic Coho Salmon (*Oncorhynchus kisutch*) have, in fact, been shown to display courtship behavior and to be able to spawn with their wild counterparts, producing viable hybrid offspring.[b]

There is also the potential problem of escape of the GM organisms themselves, which could threaten wild populations. This could occur, for example, with GM salmon, compounding the already serious problem that now exists with salmon mariculture, where cultured salmon that are not genetically modified are escaping and endangering wild salmon populations.

EFFECTS ON NONTARGET ORGANISMS

Bt toxin in GM crops may affect nontarget organisms as well as those it specifically targets. Some of these affected organisms may be beneficial insects that are important natural pollinators (e.g., bees, beetles, butterflies, and moths) or important predators of crop pests (e.g., lacewings, ladybird beetles, and parasitic wasps). Others may include soil organisms, such as microbes, and other species that contribute to local biodiversity.[30] There are different types of Bt toxins. For example, one group, formulated from the bacterium *Bacillus thuringiensis* variety *kurstaki*, or Btk, is effective against Lepidoptera—butterflies and moths—and is used to attack the European Corn Borer (*Ostrinia nubilalis*), but it may also be toxic to other Lepidoptera that are crop pollinators. Another group, Bti, is toxic to Diptera, including mosquitoes and blackflies. Still a third group, Btsd (from *Bacillus thuringiensis* variety *san diego*), targets beetles such as the Colorado Potato Beetle (*Leptinotarsa decemlineata*) but may also wipe out ladybird beetles, which are major predators of aphids and other crop pests. Indeed, some studies have shown negative impacts from Bt toxins on beneficial insects such as Green Lacewings (*Chrysoperia carnea*), which are voracious natural predators of soft-bodied crop pests, such as aphids.[49]

Bt toxin can also be released into soils via the roots of GM crops, and although it is generally broken down quickly by microbial activity, it can bind to soil particles,

(left)

Figure 9.1. *Zea mays*. (From F.E. Köhler, *Medizinal-Pflanzen in naturgetreuen Abbildungen mit kurz erläuterndem Texte: Atlas zur Pharmacopoea germanica*, Vol. 3. Gera-Untermhaus, 1883–1914. With permission from Missouri Botanical Garden, 1995–2004; www.illustratedgarden.org/mobot/rarebooks/.)

THE MONARCH BUTTERFLY AND BT CORN

Monarch Butterflies (*Danaus plexippus*), which are frequent visitors to fields across the United States (and can also be found in Australia, New Zealand, Portugal, and Spain), lay their eggs on, and their hatched caterpillars feed on, the leaves of milkweed plants (*Asclepias* spp.—Linnaeus named the milkweed genus after Asclepius, the Greek god of healing, because the plant was widely used in folk medicine) that are often found at the edges of cornfields. In fields planted with Bt corn, Monarchs may be exposed to Bt toxin that is contained in wind-blown corn pollen that coats milkweed leaves adjacent to the fields. The initial finding that Bt corn could lead to mortality among Monarchs raised serious concerns that this was an unintended consequence of planting Bt corn and that there could be other similar nontarget effects.[a] Two years of follow-up studies by many scientists in the United States and in Canada found that the impact of Bt corn pollen on the Monarchs was likely to be negligible.[b] But as an Ecological Society of America 2005 Position Paper[b] makes clear, these studies did not fully address the question about the effects of Bt corn on nontarget species, nor did they assess its effects on agroecosystems:

> Much of the focus of non-target studies has relied on measuring changes in survival and reproduction of a limited number of focal species in laboratory and small-scale field studies, without addressing the potential for community and ecosystem level effects after large-scale introductions.

particularly if the soil has a high clay content and is acidic,[50] and can retain its insecticidal properties for very long periods of time, sometimes for 230 days or more.[51] It is not clear what effects Bt toxins in soils have on soil organisms or soil ecosystem functions.

Bt corn has also been shown to contain a higher content of the compound lignin, a polymer that acts as Nature's cement to create, along with cellulose, strength and flexibility in plant tissues.[52] Lignin resists attack by most microorganisms and is not broken down by anaerobic processes. The environmental implications of increased lignin concentrations, an unintended consequence of genetic engineering, are also not well understood.

INDIRECT EFFECTS

GM crops may also have cascading effects on agroecosystems, for example, threatening species that feed on the pests controlled by Bt toxin, or reducing biodiversity among some species dependent on the weeds controlled by GM herbicides.

An intensive three-year study called the Farm-Scale Evaluations, involving more than 200 sites throughout England and Scotland and carried out by the British government, demonstrates the complexity of such ecosystem effects. The study compared levels of biodiversity in fields of genetically modified, herbicide-tolerant (GMHT) beets, maize, and Oilseed Rape (i.e., they were modified so that they were not harmed by the broad-spectrum chemical herbicides used—in this case, by glyphosate or glufosinate ammonium) with those found in adjacent fields of conventional strains of the same crops. What was discovered, in general, was that GMHT fields had less biodiversity compared to conventional ones, presumably because of reduced weed biomass caused by the herbicides. Specifically, weed biomass decreased more in GMHT Beet and Oilseed Rape fields than in conventional ones, and so did the number of seeds (which farm birds depend on for food), herbivores, pollinators (bees and butterflies), and natural enemies of insect pests. The reverse was true for GMHT maize crops when compared to those grown conventionally, but the researchers noted that the herbicide used for the conventional maize was the potent and persistent chemical atrazine, banned in Europe (see also discussion of atrazine in chapter 6, page 209, and of pollution in chapter 2, page 51), so the comparison in the case of maize may have been flawed.[53,54]

Still other potential indirect effects include the downstream impacts from insects developing resistance to Bt toxins (e.g., the Diamondback Moth [*Plutella xylostella*] has developed resistance to Bt in the field, and several species of other moths, beetles, and flies have done so in the laboratory)[55] and from weeds developing tolerance to glyphosate, as has occurred in Rigid Ryegrass and in three weeds present in soybean fields—Horseweed or Marestail (*Conyza canadensis*—resistant Horseweed is present in more than twelve states), Waterhemp (*Amaranthus rudis* or *A. tuberculatus*), and Ragweed (*Ambrosia artemisiifolia*).[56,57]

HUMAN HEALTH IMPLICATIONS

The potential benefits for human health from GM crops over conventional crops are several, although such comparisons should also be made between GM and other agricultural practices, such as organic farming. If growing GM crops results in the reduced use of toxic chemicals, or a switch to chemicals of lower toxicity, that would be beneficial given the potential role of some pesticides in causing human disease, especially among infants and children.[58] If significantly greater yields were achieved by using GM technologies, particularly in the developing world, where the risk for crop failures because of extreme weather events secondary to climate change will be ever more likely in the coming century, the public health benefits would be enormous. If the nutritional quality of foods could be improved, for example, as has already been done by adding beta-carotene genes to rice to relieve vitamin A deficiency (a condition that afflicts some 400 million people worldwide), great strides in relieving human suffering would be made.[59]

But there are also potential risks. For one, there are the risks that could come from pharmaceutical production in food crop species. The so-called "pharma crops" are grown according to stringent protocols designed to prevent contamination of the food supply. For example, corn that has been genetically engineered to produce drugs such as lactoferrin (an antimicrobial, iron-binding protein, present in high concentrations in human colostrum—the first breast milk secretions) is required by the U.S. Department of Agriculture to be grown at least one mile away from other cornfields. After harvest, such "pharma" corn must be labeled and carefully tracked to avoid mixing it with corn destined for consumption by either humans or livestock. However, scores of recent examples of human error in dealing with GM crops suggest that contamination of food with "pharma crops" is a likely occurrence.[60]

The use of antibiotic resistance genes as markers in GM crops has also raised human health concerns, because such genes could potentially be transferred to bacteria that live in the intestines of cattle and other livestock and in the human gastrointestinal tract and be difficult to treat with antibiotics. Although several scientific reviews have concluded that there is little to no chance of such gene transfer,[61] the editors of this volume believe that using a gene marker that carries a potentially significant human health risk, particularly at a time when we are facing a growing crisis of antibiotic resistance, even if its transfer is extremely unlikely to occur, should be strongly discouraged. This position has been taken by the United Kingdom's Royal Society, which stated "any further increase in the number of antibiotic-resistant microorganisms resulting from transfer of antibiotic-resistance markers from GM foods should be avoided,"[62] and by the Expert Group of the Medical Research Council of the United Kingdom, which has recommended that antibiotic resistance genes be removed from GM foods, even though they considered the possibility of such transfer to be remote.[63]

Another possible human health consideration that has been raised is that potential toxins or allergens could be produced via the transgene itself, or that such compounds might arise inadvertently via other changes in plant chemistry, caused by the action of inserted gene switches and gene promoters or by the accidentally altered functioning of host organism genes.[64,65] This remains a concern, but the potential for GM food allergens has been made less likely by the current practice that prohibits the transfer of genes encoding known allergens or of those from particularly allergenic species.

There is also the possibility that one of the chemicals widely used in GM crops, glyphosate, and perhaps to an even greater extent its commercial preparation Roundup, may act as an endocrine disruptor. A recent study has shown that glyphosate disrupts in human placenta cells the gene expression and activity of the enzyme aromatase, which is responsible for the synthesis of estrogen, at concentrations that are 100 times lower than those recommended for use in GM crops.[66] The addition of surfactants (these are wetting agents that are used to allow easier spreading of a liquid) in Roundup amplified these toxic effects, perhaps because these chemicals facilitated the entry of glyphosate into cells. At higher doses, still below concentrations that are used in agriculture, the toxicity of glyphosate to placental cells could result in human reproduction problems.

Because more than 44 million tons of glyphosate are used in the United States each year (1999 figures)[67] and, by very rough approximation, about double that figure or more globally (given the proportion of U.S. to global GM acreage), exposing millions of agricultural workers, and because glyphosate may persist in soils and contaminate some freshwater ecosystems,[68] thereby entering the water supply and the food chain, we need to have a much better understanding than we do of its potential effects on human health.

SOCIOECONOMIC ASPECTS AND
ETHICAL DIMENSIONS

Another area of concern is that large commercial corporations, under the patent laws and the protection of intellectual property rights, will appropriate the benefits of GM crops to themselves. Many consider it unfair that the culminated work of generations of scientists, researching and publishing openly and cooperatively, should now be claimed as the commercial property of exclusive groups, theirs to grant or withhold according to their profit interests. Apart from the general ethical issues this arrangement raises, there is the specific conflict of interest between the commercial corporations and the people of the developing nations who are most in need of assistance.

Commercial companies obviously aim to profit from their investments in the genetic modification of crops, but they should not be allowed to restrict or deny the benefits of the new technology to developing countries that are unable to afford the associated technology fees. The abuse of commercial power is especially troublesome in cases where the vital genetic resources were extracted initially from the flora (or fauna) of countries too poor to utilize them directly on their own. Patents are the common way to protect research investment. However, the concentration of vital scientific knowledge and its exclusive application to the benefit of a few enterprises should be prevented.

ORGANIC FARMING

Most discussions of farm practices in the scientific literature focus on conventional agriculture, where synthetic fertilizers and various synthetic insecticides and herbicides, all largely derived from petroleum, are widely applied. Up to this point this chapter has been no exception. For example, when we have compared GM to non-GM crops in such areas as the amount of pesticide sprayed, or yields that have been achieved, we have considered only crops that are conventionally grown. The same is true in almost all long-term plans for feeding world populations from the land, which look at conventional and GM food production methods as if these were the only alternatives. But there may also be another option, one that is generally overlooked and that needs to be considered: organic farming (or mixtures of organic and conventional farming known as "integrated farming" or "integrated pest management farming"—see below), which embodies the agroecosystem approach mentioned in chapter 8, and which we now briefly cover.

Organic (also called ecological or bioorganic) farming has been one of the fastest growing segments of agriculture in the United States and in other parts of the world since the early 1990s, increasing by 20 to 25 percent per year.[69] In 2006, there were more than 31 million organically farmed hectares (more than 76 million acres) around the world, with more than a third of this amount found in Australia. In some countries, such as Switzerland and Austria, more than 10 percent of their agricultural land is managed organically. Organic farming is also growing rapidly in China (which now has nearly 3 million hectares that are organically certified), Brazil, Argentina, and some countries in the Middle East, such as the United Arab Emirates. The total market value of organic products worldwide in 2004 was US$27.8 billion.[70] Although the standards for organic production may differ from country to country, and it is these standards that are used to determine organic certification (which carries with it significant financial reward, because organic produce usually sells for a premium), the following three principles are generally adhered to:

- The strict use of organic material—from nitrogen-fixing cover crops and from plant or animal wastes—to fertilize and build up the soil

- The use of biological methods and natural compounds instead of synthetic chemical pesticides to control insect pests and diseases, and mechanical weeding or other methods instead of synthetic chemical herbicides to control weeds

- The conservation of natural ecosystems such as hedgerows within and surrounding organic farms

We consider three main questions in this discussion of organic farming: (1) Is organic food better for human health? (We refer here to direct effects from food, not human health impacts caused by effects on the environment.) (2) Is organic food better for the environment? (3) Can organic farming help answer the world's food problems?

BOX 9.3

VIRGIL, IN PART I OF "THE GEORGICS"

It is of interest that the fertilizing value of legume crops and no-till farming was recognized more than 2,000 years ago by the Roman poet Virgil (70–19 B.C.):

> Sow in the golden grain where previously
>
> You raised a crop of beans that gaily shook
>
> Within their pods, or a tiny brood of vetch,
>
> Or the slender stems and rustling undergrowth
>
> Of bitter lupine . . .
>
> Thus will the land find rest in its change of crop,
>
> And earth left unplowed show you gratitude.

Is Organic Food Better for Human Health?

Although studies comparing the nutritional value of organic with conventional foods have not, in general, shown major differences in the content of such nutrients as carbohydrates, proteins, amino acids, and vitamins,[71] and have been plagued by methodological problems, such as comparing cultivars that had greater differences between them than did the two cultivation systems, some significant differences have emerged. For one, levels of vitamin C and iron seem to be greater in some organically grown crops.[72] For populations in industrialized countries, such increased levels may not be significant, but they may be in some developing countries, where, for example, iron intake may be low. In addition, there may be higher concentrations in organic fruits and vegetables of some secondary metabolites, such as compounds called phenolics, produced by plants to defend themselves against predators, parasites, and diseases, as well as to perform other functions, for example as plant hormones.[73,74] There are 10,000 or more known plant secondary metabolites, and some, such as resveratrol and the flavonoids, are thought to play important roles in helping to prevent cardiovascular disease,[75] and possibly certain types of cancer,[76] because of their antioxidant activity (see also chapter 4, page 154). But other plant secondary metabolites, given that they have evolved for the purposes of plant survival, may have no effect whatsoever on human health, or even a toxic effect, and until careful studies are done to demonstrate that organic crops have higher levels of specific compounds known to be beneficial to human health, it cannot be said that the differences in secondary metabolites that have been reported are significant.

There is also the issue of pesticide use on conventional crops. While most studies have demonstrated that residues of these chemicals on conventional crops are small, and that they fall within ranges that are considered to be safe by various governmental regulatory agencies, there is no question that people eating conventional as opposed to organic food are more likely to be exposed to a wide range of synthetic chemicals, often on a daily basis, and that for many of these compounds, there are few or no data on the human health impacts of chronic, low-level exposures, particularly in combinations.[77] A recent study in Seattle, Washington, for example, comparing two groups of preschool children (ages two through five), where one group ate predominantly organically grown foods and the other predominantly those grown conventionally, showed significantly higher levels (up to six times higher) of organophosphorus concentrations in the urine of the children with the conventional diets, presumably from pesticide residue ingestion.[78] Given that such exposures may be a particular problem for infants and children, whose specific vulnerabilities are generally not factored into most standard exposure regulations,[79] and that there is growing evidence of potential human health problems from chronic exposure to some pesticides,[80–83] this issue alone has lent growing worldwide support to organic methods.

As for eating eggs, milk, cheese, and meat from organically raised chickens, cattle, sheep, goats, and pigs, it is clear that in doing so one can avoid the growth hormones, antibiotics, and some other chemicals that are routinely and widely used in conventional livestock production.

Is Organic Farming Better for the Environment?

Two large surveys—the first a review of the published literature on research in Europe, Canada, New Zealand, and the United States, and the second, a five-year study of farms in England—compared biodiversity in organic versus conventional farms. A clear majority of the seventy-six studies reviewed in the first survey[84] demonstrated that for a broad range of organisms, including birds, mammals, invertebrates, and farmland plants, population abundance and/or richness tended to be higher on organic farms. Importantly, some bird species, such as Skylarks (*Alauda arvensis*) and Lapwings (*Vanellus vanellus*), as well as Greater and Lesser Horseshoe Bats (*Rhinolophus ferrumeouinum* and *Rhinolophus hipposideros*, respectively), Corn Buttercups (*Ranunculus arvensis*), and Red Hemp-Nettles (*Galeopsis ladanum*), all known to have declined in Europe due to agricultural intensification, were shown to be more abundant on the organic farms. The study identified three types of farming practices that were of particular importance to maintaining biodiversity: the prohibition or reduced use of pesticides and inorganic fertilizers; wildlife-friendly management of noncrop habitats and field margins, such as the use of hedgerows; and the juxtaposition of arable fields with those supporting livestock. The study of English farms, the largest survey ever of organic farms, had similar findings. In this case, organic farms on average had 85 percent more plant species, 33 percent more bats, 17 percent more spiders, and 5 percent more birds.[85] Other studies have looked at soil biodiversity in organic versus conventional farms, and one, a 21-year study by the Research Institute of Organic Agriculture in Switzerland, found greater root colonization by mycorrhizae, increased abundance and biomass of earthworms, higher densities of important aboveground arthropod predators such as spiders and carabids (ground beetles), more diverse weed flora, and significant increases in soil microbial diversity.[86] The abundance and diversity of arbuscular mycorrhizal fungi (AMF) in the soil of organic farms have received particular attention, owing to the role that AMF are thought to play in promoting better soil structure, and in improving plant nutrition, resistance to soil-borne pests and disease, resistance to drought, and tolerance of heavy metals (see box 8.5 on the mycorrhizae in chapter 8). And several studies have shown increased AMF richness and abundance, greater crop colonization, and enhanced nutrient uptake in organic than in conventional crops.[87,88]

In addition, energy efficiency, both per unit area and per unit of yield, was greater for organic versus conventional farms, and there was more carbon sequestration in soils, the result of plowing under both crop residues and legume cover crops, all of which contribute to lowering greenhouse gas emissions.[89] One would also expect there to be less runoff of nitrates and phosphates into drinking water and into aquatic ecosystems in organic versus conventional farms, given the very high levels of these fertilizers that are used in conventional systems, but careful studies of these differences remain to be done.

Scientists measured the sustainability of organic, conventional, and "integrated" (a mixture of organic and conventional methods) apple production systems in Washington State over a six-year period. They found that all three systems had comparable apple yields, tree growth, and leaf and fruit nutrient contents. However, the organic and integrated systems had higher soil quality, and when the organic system alone was compared with the conventional and integrated systems, it produced sweeter and less tart apples and had higher profits and greater energy efficiency (7 percent more than the conventional system). The organic system ranked first in overall sustainability, the integrated system second, and the conventional system last.[90]

Can Organic Farming Help Answer the World's Food Problems?

One generally hears that organic farming may be better for the environment, and may even be better for human health, but that it can never replace conventional agriculture and help feed the growing world population. The main reason given is that conventional methods produce consistently higher yields and that organic farms simply cannot compete.[91] The other frequently heard argument is that there is just not enough nonsynthetic nitrogen available to use as fertilizer for organic farms.[92] In this section, we examine these arguments.

The same 21-year study mentioned above also found that organic farms, on average, produced only 80 percent of the yields of the conventional farms.[86] Some other studies have had similar findings. Yet still others show comparable yields. For example, in the longest running organic trial in the world, at the Rothamsted Experimental Station (also known as the Institute of Arable Crops Research) in England, wheat yields on organically manured plots were essentially identical to those from plots fertilized by synthetic fertilizers.[93] And researchers at the Rodale Institute in Pennsylvania, which is dedicated to promoting organic farming, have concluded after more than twenty years of experimentation that organic and conventional maize and soybean plots grown side by side produced roughly the same yields. Of great importance at Rodale, however, was that during drought years, the organic yields were 20–40 percent higher (and in some cases 100 percent higher) than those in conventional plots,[94] a finding of enormous significance given the likelihood of severe droughts in many parts of the world in coming years, particularly in some developing countries, secondary to global warming.[94] University of Essex researchers Jules Pretty and Rachel Hine, looking at more than 200 agricultural projects in the developing world on some nine million farms, with a combined total of almost 30 million hectares (about 74 million acres), found that yields increased by an average of more than 90 percent when these farms converted from conventional to organic approaches. In one of the studies cited, involving 1,000 farmers from the Maikaal District in central India cultivating 3,200 hectares of cotton, wheat, chili, and soy organically, yields were 20 percent higher on average than those on nearby conventionally managed farms.[95] Finally, a University of Michigan study that compared yields of 293 foods (both plant and animal) raised by organic versus nonorganic methods found that worldwide organic yields were on average some 30 percent greater for all plant and animal foods, with the organic yields on average being some 8 percent less than the nonorganic in developed countries, but some 80 percent greater in developing countries.[96]

Critics of organic farming always come back to the issue of there not being enough organic fertilizer to feed the world. The argument talks about lower yields from organic farms (although, as stated above, this does not seem to be the case for many crops), but it mainly relies on looking at the amount of livestock manure that would be needed to raise crops organically, and therefore on the need to create more grazing lands for them, leveling forests and other natural habitats in the process. The same University of Michigan study mentioned above also reviewed seventy-seven studies from temperate and tropical areas and found that greater use of nitrogen-fixing crops in the world's major agricultural regions (a source of nitrogen that is often not emphasized in organic/conventional comparisons, or that is discounted with the argument that farmers cannot afford to plant fields with legumes rather than their own crops) had the potential of yielding for agriculture 58 million metric tons (about 64 million tons) more nitrogen worldwide than is currently being produced synthetically.[96]

The other argument that is made is that organic methods may work well in gardens or in small-scale farming operations, but that they cannot compete on the scales necessary to feed large populations. Two case studies demonstrate that this argument is without merit—the Cuban Organic Farming Experiment and the 3,700-acre (about 1,500-hectare) organic farm in North Dakota owned by Fred Kirschenmann and his family (see boxes 9.4 and 9.5).

INTEGRATED FARMING

Integrated farming is based largely on organic farming, but also makes use of synthetic fertilizers and pesticides.[97] Integrated farming, successfully adopted on a wide scale in Europe, utilizes methods of both conventional and organic production systems in an attempt to optimize both environmental quality and economic profit. Mostly, integrated farming builds the soil with green manure crops and composts, but also adds synthetic fertilizers when needed. Integrated farming tries to keep pests and weeds under control by relying on such practices as crop rotations and intercropping, the use of biological control agents such as predatory insects and soil nematodes, and physical traps for pests that may make use of pheromone lures. (*Pheromones* are chemicals released by organisms, including humans, into the environment to communicate various messages. In agriculture, they are employed to attract various insect pests, with synthetic pheromones being used to mimic species-specific sex attractants.) Synthetic pesticides are used as a last resort and are timed carefully to the pests' life cycles when they can be most effective and when the smallest amounts possible can be used. The least toxic, and generally the most biodegradable, pesticides are chosen. These pest control practices have been referred to as integrated pest management, or IPM, and have been increasingly employed in developing[98] and industrialized countries.[99]

Integrated Crop/Livestock Systems

Organic farms that feed their livestock with the legume crops grown on fallow fields, recycling livestock waste back into the crop fields as fertilizers, are practicing another type of integrated system. These farming systems, where crop cultivation and livestock rearing form integrated components of a single farming system, are called integrated crop/livestock, or mixed farming systems. There

BOX 9.4

The Cuban Organic Farming Experiment

With the collapse of the Soviet Union in 1989–1990, Cuba suddenly found itself with 60 percent lower pesticide and 77 percent lower fertilizer imports, and 50 percent less petroleum available for agriculture. There was also a drop in food imports by more than 50 percent. In response to the looming food shortage crisis, because their agriculture was mostly based on large-scale, capital-intensive, monoculture farming systems that were heavily reliant on synthetic pesticides and fertilizers (derived from petroleum), and on petroleum itself, the Cuban government launched a national effort to convert the nation's modern conventional farming system into an organic one that was low input and essentially self-reliant. Pesticides and herbicides were replaced by pest, pathogen, and weed control methods that made use of biopesticides (e.g., various bacterial and fungal disease agents the Cubans had developed), plant-based pesticides such as Neem (see chapter 4, page 158), natural enemies (e.g., various parasitic and predatory insects), intercropping and crop rotations, and the contributions to weed control made by farm animals. Synthetic fertilizers were replaced by biofertilizers (including *Rhizobium* inoculants for legumes, free-living nitrogen-fixing bacteria, and mycorrhizal fungi), earthworms, compost, animal and green manures, and the integration of grazing animals. There was also a return to animal traction in place of tractors, because of the unavailability of fuel, tires, and spare parts. Woodlands surrounding farm fields were encouraged, and they provided not only forest products (lumber, fuel wood, fruits, nuts, and honey) but also habitat for insect-eating and pollinating birds, insects, and bats. Farms were converted from large, specialized enterprises with one or, at most, only a few products to mixed farming systems producing fruits, vegetables, grains, livestock, and fish. The resulting diversification created a mosaic of land use that served to help buffer against both extreme weather events and infectious diseases in livestock and crops.

In 1993, Cuba greatly reduced the state farm infrastructure, turning farms into cooperatives, a form of worker-owned enterprises, and encouraged the return of urban populations to the countryside. Farmers markets were also reopened. By mid-1995, the food shortage had been overcome, and the 1996–1997 growing season in Cuba recorded its highest ever production levels for ten of the thirteen basic food items in the Cuban diet. Production came primarily from small farms and, in the case of eggs and pork, from booming backyard production. The proliferation of urban farmers who produce fresh produce has been vitally important to the Cuban food supply, with more than 30,000 hectares (around 74,000 acres) in cities devoted to agriculture, producing more than 3 million tons of fresh vegetables each year for some eleven million people. More than 50,000 tons of food were produced annually during the late 1990s in the city of Havana alone.

BOX 9.5

LARGE-SCALE ORGANIC FARMING IN THE UNITED STATES

The Kirschenmann Family Farm, a 3,700-acre farm in south central North Dakota, near the city of Jamestown, has been growing organic crops on a commercial scale since 1976. The farm is still managed, at a distance, by scholar/farmer Fred Kirschenmann, who left the daily operation of the farm in July 2000 to direct the Aldo Leopold Center at Iowa State University. The cash crops on the farm are Hard Red Spring Wheat, Durum Wheat (*Triticum durum*), Winter Rye (*Secale cereale*), Common Flax (*Linum usitatissimum* L.), Sunflowers (*Helianthus annuus*), Millet, Buckwheat (*Fagopyrum esculentum*), and Oats (*Avena sativa*). Alfalfa (*Medicago sativa*) and Sweet Clover (*Melilotus officinalis* or *Melilotus alba*) provide the leguminous cover crops and are used as forage for the livestock, an Angus-cross beef herd of 126 brood cows and 4 bulls, which also graze on organic grass fields.

Central to the farm's success are its crop rotation strategies. The first is an alternation of warm- and cool-season crops to control weeds, which cannot become as well established with the changing environments. The second is a rotation of broad leaf and grassy plants for the purpose of controlling pests and diseases, because such plants are subject to different pest organisms and different diseases. The third principle involves alternating deep-rooted with shallow-rooted plants, because plants with roots at different depths draw their nutrients from different soil levels and can thus prolong the crop rotation cycle. Legumes are used in all the crop rotation cycles because they fix nitrogen and add organic matter to the soil. A typical rotation cycle would be Sweet Clover (legume), Hard Red Spring or Durum Wheat (cool season, grass), Buckwheat (warm season, broad leaf), Rye (cool season, grass), Sunflowers (warm season, deep-rooted, broad leaf), and then back to Sweet Clover. Pest control is achieved through the crop rotations and the maintenance of natural habitats, such as hedgerows separating fields, to encourage natural predators. No biological controls, such as beneficial insects, are added to the system.

Selecting the right crop rotation scheme for a particular farm is a very complex issue. Many factors determine what crops the farmer should select. Different classes of soil have different capabilities for producing various crops. Climate conditions place significant limitations on the types of crops that can be grown in particular landscapes. Available market infrastructures for selling the crops produced may restrict the kinds of crops selected for a crop rotation, as may the type of equipment needed to plant, manage, and harvest the crops. Public policies, which favor the production of some crops over others, may limit a farmer's choices. The demand for a particular crop in the marketplace imposes severe limitations on what a farmer can produce. Farmers can't grow a crop they can't afford to sell, no matter how "beneficial" it may be in a rotation.

Diversified crop/livestock systems enable organic farmers to emulate nature's process of turning all waste in a system into food for another part of the system, producing closed nutrient cycles on the farm. Livestock are fed the legume cover crops of Sweet Clover and Alfalfa. Their manure, mixed with straw, wood chips, and plant wastes, is composted and returned to the fields as fertilizer. Other wastes from cropping systems can also be used. Grain kernels that do not meet quality standards for human foods can be cleaned out of bulk grain supplies and fed to livestock. And turnip and radish greens and other vegetables and wastes can nourish pigs.

Numerous studies have been done on the Kirschenmann Farm, comparing it to conventional farms in the region. Over time, per acre yields were found to be similar to those of conventional farms, with the conventional farms producing slightly greater yields when growing conditions were ideal, but with Kirschenmann farm yields exceeding yields on conventional farms when conditions were poor, for example, in drought years. Although, on average, one year out of four was devoted to legume cover crops and the application of compost on the Kirschenmann farm, with the sacrifice of that year's cash income for each of the fields involved, there were also no costs for pesticides or herbicides or synthetic fertilizers, so that economic returns for the two farm systems were in fact similar. But, of interest, energy use on the Kirschenmann farm when compared to conventional farms in the area was significantly less, as much as 70 percent less.[a] The same was true when other organic farms in each of North Dakota's three ecoregions were compared to conventional farms in the area.[a]

So it would seem that at commercial scales in the United States, an organic farm can compete in yields and in financial returns with conventional farms, at least in the northern Great Plains, when growing such crops as wheat, rye, and oats—and that it can outcompete such farms in terms of lower energy use and when growing conditions are not ideal. In coming years, when the energy costs of running farms are likely to severely cut into profits, not to mention the costs of fertilizers and pesticides derived from oil and natural gas, and when there will be more extreme weather events such as heat waves and droughts from global climate change, the advantages of organic farming, even on large commercial scales, may become even more significant than they are today.

BOX 9.6

TAKAO FURUNO'S RICE-DUCK-LOACH SYSTEM OF AGRICULTURE

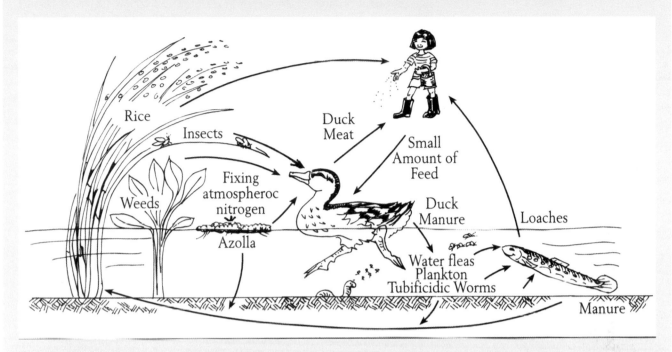

Diagram of Takao Furuno's Integrated Rice-Farming System. (Takao Furuno, courtesy of Tagari Publications, www.tagari.com.) Adapted with permission from Fred Kirschenmann and the Leopold Center for Sustainable Agriculture.

Another approach, practiced with great ingenuity by the Japanese rice farmer Takao Furuno, illustrates how fish aquaculture can be introduced into other food production systems. He adds loaches (bottom-dwelling, omnivorous freshwater fish found throughout Eurasia) to a mixed system of rice and duck farming and, in the process, achieves higher rice yields at lower costs and with less chemical input. While this method may not achieve the high yields hoped for in intensive aquaculture, it nevertheless demonstrates how such mixed farming systems may be productive and sustainable alternatives for subsistence farms in some parts of the developing world.

Takao Furuno, a rice farmer from the village of Keisen in the Fukuoka Prefecture on the island of Kyushu in Japan, decided in 1987 that he would change his typical industrial rice growing practices to incorporate traditional old methods, and at the same time that he would apply the best modern science available. After learning that Japanese farmers in the past used to raise ducks in rice paddies, he decided to introduce into his rice fields the most efficient grazing duck known in his region, the Aigamo, the product of a wild and domestic duck cross. He found that 200 ducks for each hectare (about 2.5 acres) ate the insects on his rice plants so effectively that he no longer needed to use insecticides. Furuno also discovered that as the ducks grew, they would dive and feed on golden snails that were attacking rice roots. He added loaches to this system, because farmers had also done this in the past, and found that the ducks and fish survived well together in his rice fields. The cloudy water kept the ducks from eating the fish. Also, the aquatic fern plant *Azolla* (which has several species), a "weed" that grew on the surface of paddies and choked off rice plants and that had required the application of herbicides for control, was eaten by both the Aigamo ducks and the loaches in quantities sufficient to prevent it from harming the rice. As a result, Takao Furuno no longer needed to apply herbicides. The *Azolla* that were left, in association with nitrogen-fixing cyanobacterium that live in jelly-filled, anaerobic pockets at the tips of *Azolla* leaves, served to fertilize the rice plants, adding to fertilization already provided by droppings from the ducks and the fish.

By adopting these traditional methods, Takao Furuno found that his yields increased by 50 percent and his costs dropped, because he was no longer purchasing insecticides and herbicides. Finally, he planted fig trees along the perimeter of his rice paddies, so instead of being only a rice farmer, he became a producer of rice, duck meat, loaches, and figs, maximizing the use of his 2.4-hectare (6-acre) farm by assembling a highly productive, sustainable, mixed farming system without the addition of chemicals. The principles of this system are described in his book *The Power of Duck*.[a]

are many possibilities, depending on the crops grown and the environmental conditions. One unique approach from Japan, where growing crops is mixed with aquaculture, provides an example of the range of possibilities for such mixed farming systems and illustrates that they may be highly productive and efficient methods for producing a wide variety of foods. These approaches are also widely practiced in China.[100]

CONCLUSION

Genetically modified organisms (GMOs) may soon be developed that produce significantly higher yields of nutritionally improved foods and that withstand some of the environmental stresses agriculture will have to endure increasingly in coming years, such as more intense and frequent extreme weather events, like heat waves and droughts, and soils that contain more salt. They may also be engineered to resist certain diseases that have been catastrophic for farmers around the world, such as rice blast, late potato blight, and wheat rust. And they may be able to do all these things with less harm to the environment and to human health than crop systems managed conventionally. Such improved crops may also be developed by hybridization or by the generation of mutant strains, without resorting to the transfer of genes from one species to another, and it is clear that these techniques have not received the research and development attention they deserve.

There is no question that GMOs have enormous potential in helping to feed the world, perhaps rivaling their already well demonstrated and extraordinary usefulness in biomedical research. But as we point out in this chapter, there is still a great deal we do not know about some of the human health and, particularly, the environmental costs of current GM crop technologies, for example, those that insert Bt and herbicide-tolerant genes. To spread such GM crops around the world, which is now happening at a very rapid rate, before we have more fully understood these costs, especially those of the long term, essentially involves conducting a global environmental experiment. To the editors of this volume, this does not seem wise.

As for organic farming and various integrated farming systems, they have been largely ignored in policy discussions about global food security. Such discussions generally compare only GM with conventional farming. But as more and more studies are demonstrating, organic, and various integrated and mixed farming systems are capable of producing yields that approach, or even exceed, those of conventionally managed systems, particularly during times of drought. And they can do so over large scales and with greater energy efficiency. Organic agriculture is growing very rapidly in industrialized countries as consumers are increasingly interested in buying food free of pesticides and other chemicals. But it may have its most important application in developing countries, particularly as the costs of fossil fuels, and the fertilizers and pesticides derived from them, continue to escalate, and as we enter a world where droughts are increasingly common and where water for irrigation is at a premium. There is also little question that organic and integrated farming methods are less harmful to biodiversity, and most likely also to human health. It is our view that organic agriculture must be included as a major part of any plan that addresses global food security, for as Catherine Badgley and Ivette Perfecto have said in their editorial "Can Organic Agriculture Feed the World?": "A global food system based on agroecological principles is possible and there are urgent reasons to move in this direction."[101]

A healthy debate in the scientific literature is ongoing about GM foods and organic farming, with careful work being done on both the benefits and the risks of these technologies. Scientists of great integrity can be found on both sides of these issues. But all too often, scientists who raise questions about the wisdom of our rapidly growing commitment to GM foods, or those who express support for expanding our use of organic and integrated farming systems, are characterized as uninformed or naive. As was the case with tobacco, there are powerful vested interests involved and enormous financial stakes at play here. And, unfortunately, as was also true for tobacco, at times these attacks and the research behind them have been the work of such interests. But if we are to make decisions about feeding the world in coming decades, decisions that ensure that the greatest number of people will be fed with food that is healthiest both for them and for the environment, we must be sure that the science supporting these decisions has been carefully and objectively obtained, and that it has been fully aired so that people have complete access to all the facts. That is what we have tried to do in this chapter.

◆ ◆ ◆

SUGGESTED READINGS

Biotechnology (GM foods), World Health Organization, www.who.int/foodsafety/biotech/en/.

"Can Organic Farming Feed Us All?" B. Halweil. *World Watch Magazine*, May/June 2006.

The Earthscan Reader in Sustainable Agriculture (Earthscan Readers Series), Jules Pretty (editor). Earthscan, London, 2005

Genetically Engineered Food: Changing the Nature of Nature, Martin Teitel and Kimberly A. Wilson. Park Street Press, Rochester, Vermont, 2001.

Genetically Modified Foods and Organisms, www.ornl.gov/sci/techresources/Human_Genome/elsi/gmfood.shtml (Human Genome Project information).

International Federation of Organic Agricultural Movements (IFOAM), www.ifoam.org/.

International Society of Organic Agriculture Research, www.isofar.org/.

Organic Agriculture at FAO, www.fao.org/ORGANICAG/.

Organic Center, www.organic-center.org/.

Organic Farming, Food Quality and Human Health: A Review of the Evidence, Shane Heaton. Soil Association, Bristol, U.K., 2001; see www.soilassociation.org/.

Organic Farming Research Foundation, ofrf.org/.

"Our Food, Our Future," Donella H. Meadows. *Organic Gardening*, September/October 2000.

Rodale Institute, www.rodaleinstitute.org/.

Rothamsted Archive, www.rothamsted.ac.uk/resources/TheRothamstedArchive.html.

Sustainable Agriculture and Resistance: Transforming Food Production in Cuba, F. Funes, L. García, M. Bourque, N. Pérez, and P. Rosset (editors). Food First Books, Milford, Connecticut, 2002.

CHAPTER 10

WHAT INDIVIDUALS CAN DO TO HELP CONSERVE BIODIVERSITY

Jeffrey A. McNeely, Eleanor Sterling, and Kalemani Jo Mulongoy

Be the change you want to see in the world.

—MOHANDAS K. GANDHI

WHAT ARE WE DOING TO OUR PLANET?

The Millennium Ecosystem Assessment, the most comprehensive inventory of the status of Earth's natural resources, documented early in 2005 that the "ecological footprint" of human beings is becoming ever larger: Our increasing population and consumption of resources are altering and destroying ecosystems at an unprecedented rate.[1] The assessment concluded that human activity has disrupted natural ecosystems more extensively in the past fifty years than in the entire course of human history, as large areas on all continents have been converted to farmland, forests have been felled for timber and to make way for pasture and the growing of crops, and the seas have been plundered for fish and other marine products.

The concept of one's ecological footprint is a useful metaphor that may help people understand the necessity of living their lives sustainably. It seeks to quantify how much biologically productive land area a particular human population, whether it is an individual, a city, a region, or a nation, uses to produce all the resources it consumes and to absorb all its wastes, taking current technologies into account.[2] The Global Footprint Network estimates that humanity's ecological footprint is now more than 20 percent larger than the planet can support at any one time, so we are, in essence, living on 1.2 Earths. Another way to conceptualize our footprint is to consider it in terms of the time necessary for Earth to regenerate what we use. By this framework, it now takes fourteen months for Earth to produce the goods and services that we use up in a single year. Recent estimates have found that North America's ecological footprint is just more than 9 hectares (slightly more than 22 acres) per person; for Western Europe it is about 5 hectares (12.4 acres) per person, and for Asia-Pacific and Africa it is around 1 hectare (almost 2.5 acres) per person. In other words, the average North American has an ecological footprint almost ten times as large as the average African, and more than 4.5 times the footprint that each person on Earth would have to average for human activity to be sustainable.

More than 1.7 billion people are now members of the "consumer class," nearly half of them living in developing countries.[3] Much of their consumption is geared toward goods that are enjoyable but that are not essential for survival. For example, as of 2003, the United States had more private cars than licensed drivers. A comparison of the funding necessary to provide sufficient food, water, and education for the world's poorest people versus what we spend on luxury goods, such as perfume, makeup, cruise vacations, and cosmetic surgeries, is humbling. For example, according to the Society for Aesthetic Plastic Surgery, 1.8 million cosmetic operations were performed in 2003 in the United States alone (and there were an additional 6.4 million nonsurgical cosmetic procedures, e.g., Botox injections), while tens of millions of people in Africa received no health care at all.[4–6]

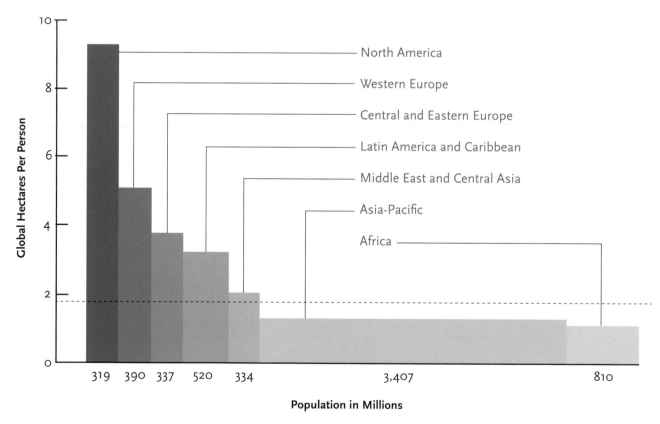

Figure 10.1. "Ecological Footprint," by Region (2001). The dashed line at about 1.8 hectares (almost 4.5 acres) indicates the individual footprint, expressed as the amount of biologically productive land that each person on Earth would have had to average for the total footprint of the world's population to be sustainable. This figure is based on 2001 world population numbers. Today, with greater population, this area would be less. (From Global Footprint Network, National Footprint Accounts, 2004 edition, www.footprintnetwork.org.)

As described in chapter 1, the loss of habitat on land, in lakes and streams, and in the oceans; the release of pollutants into the air, soils, and water; the depletion of the stratospheric ozone layer; invasive species; global climate change; and overfishing, overhunting, and, in general, overexploitation of natural resources all ultimately disrupt the healthy functioning of both natural and domestic ecosystems and threaten the survival of other species. But these alterations to the global environment are the result of human decisions and human behavior, and just as human actions have damaged the global environment, so, too, can they work to preserve and restore it. Many people may feel that they cannot do anything to help solve environmental problems, that the problems are too large, too complicated, and too well entrenched. The authors of this chapter and the editors of this book believe that individuals can make enormous and critically important contributions to protecting the global environment, and that it is never too late to do so. In this chapter we detail some of the things that individuals can do.

WHY DO WE CONSUME SO MUCH?

All of us aspire to achieve a good quality of life for ourselves and for our children. The problem lies in defining what "good" is and in identifying how to get there. What seems the right decision for us individually in the short term may not be the best one for others, including our children, in the long term. We must begin to recognize that almost every action we undertake has some direct or indirect effect on the environment in general, and on biodiversity in particular. This is difficult for many of us to do, given how increasingly removed we have become from the ecosystems that sustain our lives, especially those of us in urban centers in industrialized countries. And, with the global human population estimated by the United Nations to reach nine billion before leveling off by the middle of the twenty-first century,[7] the cumulative effects of human activity on the environment are potentially catastrophic—as the trends highlighted by the Millennium Ecosystem Assessment have so clearly indicated.

Consumption in and of itself is not the problem. Humans need to consume to survive, and in fact, the 2.8 billion people in the world who live on less than $2 per day need to be able to consume more than they do now. Trouble arises when we, individuals and whole populations alike, so overconsume and waste resources, especially those that are nonrenewable such as fossil fuels, that we end up outstripping the ability of Earth to support us. Several factors underlie this drive to consume more than we need. Cultural norms and social influences compel us to dress as our peers do, or drive cars and live in houses that are similar to theirs. Artificially cheap energy and technological advances have made possible an excess of all kinds of goods available to the consumer. Improved transportation brings to consumers in Boston, for example, apples from New Zealand, avocados from Chile, cocoa from Côte d'Ivoire, clothing made in Malaysia, and electronics from China. We find it hard to resist this cornucopia of goods, which seem to arrive effortlessly on our doorsteps, promising luxury and material comfort and which appeal so strongly to our innate desire for pleasure.[3,5]

A large part of the problem is our commitment to a global economic system that takes very little account of the true costs to Earth of our actions. Our ability to radically alter our environment began when our ancestors first tamed fire and fashioned tools for hunting, and it increased markedly with the dawn of agriculture around 10,000 years ago. But the Industrial Revolution, which began around 1760 and accelerated greatly in the middle to late 1800s, marked a quantum leap in our ability to consume the planet's resources and to generate enormous amounts of waste, and has helped to foster a separation between humans and nature.[8] Since then, our national economies and personal lives have been driven by the need to produce, consume, and trade more and more "stuff." We have become convinced that we can take as much as we want to from the environment, and dump as much as we want back into it, as if it were an infinite source and an infinite sink.

We have all become accustomed to paying prices that do not represent the true value of goods. Yet as responsible citizens with an eye toward the world we leave for our children, we need to better understand, and expect to bear, these true costs. Economists measure our well-being with statistics such as gross national product, gross domestic product, the size of foreign currency reserves, or the balance of trade. Corporate executives measure success by the level of consumer demand and the company's bottom line, often focusing only on the short run. This economic system neglects two very important questions: Where does all this "stuff" come from? And what happens to it when we are finished with it? The cost of goods currently does not incorporate the natural resources depleted, or the "ecosystem services" provided by the environment toward their production, disposal, or cleanup of the wastes generated.[9]

The World Bank, some academic economists, and others have begun to develop techniques that attempt to assign more accurate values for ecosystem services.[10] Perhaps we are starting to arrive at a point where we can develop national indicators such as our gross *natural* product, or our Ecosystem and Species Reserves Health Index, or, as in Bhutan, a Gross National Happiness Index (which Bhutan defines by evaluating how well they are doing to promote equitable and sustainable socioeconomic development, preserve and promote cultural values, conserve the natural environment, and establish good governance),[11] but we still have a long way to go. History has shown that we ignore the value of the natural world around us at our peril. Past civilizations have collapsed because they overexploited their natural resources.[8] We must learn from their experiences.

HOW CAN WE CONSERVE BIODIVERSITY?

It often seems to many people, perhaps to most, that they are powerless to influence the forces of environmental destruction, particularly when compared with the large-scale effects of governments or corporations. (Note, however, that some 180 international corporations, including Johnson & Johnson, JPMorgan Chase, Swiss Re, BP, and 3M, are assuming leadership roles in reducing their own environmental footprints and in working to encourage environmentally sustainable practices worldwide, through the World Business Council for Sustainable Development [www.wbcsd.org] and other organizations.) But collectively, smaller scale actions add up, and individuals everywhere are already making a difference in protecting the environment and in serving as models that inspire others to do the same.[4,12] And they are doing so largely without sacrificing their quality of life. How do we lessen our "Ecological Footprints," and improve our health and well-being at the same time, without having to deprive ourselves of all worldly goods, living like monks or hermits? We might consider three main ways:

- We can adopt lifestyles that minimize our "Ecological Footprints."

- We can raise awareness in our homes, workplaces, schools, places of worship, and local communities by discussing how everyday behavior can affect biodiversity in our own back yards, towns, and regions—and how it may have impacts on species and ecosystems thousands of miles away.

- We can support organizations that are working to preserve biodiversity and use our votes to elect environmentally responsible politicians, encourage governments to honor the policies they have put in place, and urge political candidates to put biodiversity concerns at the top of their agendas.

Once a critical mass of the public adopts such behaviors, social pressure makes them part of the culture (e.g., the use of bicycles in Amsterdam).

Lifestyle Choices That Protect Biodiversity

Individuals make choices in their everyday lives that can be good for them as well as for the world's biodiversity. The overarching principle of "reduce, reuse, and recycle" is still very much a valid one, affecting all aspects of our consumption of energy and of products. The Center for a New American Dream (www.newdream.org), the Worldwatch Institute (www.worldwatch.org/features/consumption), the Center for Biodiversity and Conservation at the American Museum of Natural History (research.amnh.org/biodiversity), the United Kingdom's Green Links (www.green-links.co.uk) and Towards Sustainability (www.towards-sustainability.co.uk), India's Centre for Science and Environment (www.cseindia.org), Friends of the Earth Hong Kong (www.foe.org.hk/welcome/geten.asp), the State of the Environment (www.ngo.grida.no/soesa/nsoer/index.htm) in South Africa, the Yonge Nawe Environmental Action Group in Swaziland (www.yongenawe.com/02programmes/iec/iec.html), and many other organizations all have suggestions about how to apply this principle as individuals while still leading active, full, and comfortable lives. Consumer advocates, environmentalists, economists, and policy makers suggest, for example, that we choose goods and services that are publicly provided (e.g., taking public transportation rather than driving where possible, or using the public library instead of buying and discarding books) and buying goods made primarily from recycled materials when available. Above all, do not waste—use only what you need, no more.

The choices we make in three main areas of our lives—the food we eat, the way we live in our homes, and how we transport ourselves—have been identified as having the greatest potential to cause environmental damage and threaten biodiversity. Making better choices in these areas could improve the environment and slow the loss of biodiversity.

THE FOOD WE EAT

Food production has modified vast areas of our planet that were once natural eco-systems, and the rate of this modification and resultant terrestrial habitat loss has accelerated in the past fifty years. In parts of Africa, Asia, and South America, wild animals—sometimes called "bushmeat"—are being harvested at unsustainable rates, affecting everything from song birds to gorillas.[13] But perhaps the greatest direct impact we are having on populations of wild species globally is taking place out of sight, out of mind, below the surface of the oceans.

Food from Aquatic Ecosystems

Global consumption of fish has doubled over the past three decades. Thirty years ago, most people ate almost exclusively wild-caught fish, but since the 1980s wild catches have declined, as most stocks have become fully or overexploited, and consumption of farmed fish (including both marine and freshwater species) has soared. In 1997, the proportion of wild fish in our diet was only about 70 percent of total fish consumption, and today it is even less.

As highlighted in chapters 2 and 8, indiscriminate industrial fishing methods are taking their toll on marine fish stocks worldwide. Particularly vulnerable to over-exploitation are the larger species that are slow to reproduce, such as the Northern Bluefin Tuna (*Thunnus thynnus*), Orange Roughy (*Hoplostethus atlanticus*), the Atlantic Cod (*Gadus morhua*), and many species of sharks. Such fish as tuna and sharks are top predators, and the effects of their removal from ecosystems, while unknown in detail, are likely to be significant, on a scale similar to what has occurred when other, better studied top predators (e.g., wolves and eagles) were removed from their ecosystems.[1] Demand for these species remains high, and as a result their breeding populations are in sharp decline. It is conceivable that some populations of over-exploited fish may never recover, such as the Atlantic Cod found in the Grand Banks off the coast of New England and eastern Canada, perhaps the greatest fishery the world has ever known. Modern fisheries also affect a wide range of marine biodiversity by damaging or destroying sea-bottom habitats. And there is an enormous toll on nontarget species taken as bycatch, such as whales, dolphins, marine turtles, seabirds, and many species of fish that are discarded.

Leading conservation organizations such as the World Wide Fund for Nature (called the World Wildlife Fund in the United States and Canada) have now with-drawn their support for marketing cans of tuna as "dolphin friendly." Although consumer demand for dolphin-friendly tuna did lead to a change in fishing methods, resulting in a reduced toll on dolphins and other cetaceans (the order of marine mam-mals, containing some eighty species, that includes whales, dolphins, and porpoises),

WHOSE RESPONSIBILITY IS SUSTAINABILITY?

The reader will note that in this chapter we overwhelmingly focus on the lifestyles of those who live in industrialized countries; in fact, we concentrate almost exclusively on those who live in the United States. And we devote most of this chapter to people who can afford to make some of the environmentally friendly choices, many of which are more costly, at least in the short run, than other choices that may be more damaging to the environment. It is certainly easier to put together a chapter with such a focus. But there are also good reasons for doing so. For one, although the editors and authors of this book strongly hope that it will reach a very wide audience in developing and industrialized countries alike, most readers, like most of us, are likely to be relatively affluent and to come from the industrialized world. And it is we who ultimately cause the most damage to the global environment, through our excessive burning of fossil fuels, our overconsumption and waste of global resources, our living beyond the capacity of the planet. If we were able to change our ways—above all, if we could learn to live our lives fully conscious of the impacts we have on the environment, using only what we truly need of Nature's bountiful goods and services—then biodiversity and global ecosystems stand a chance.

current methods of catching tuna using long-lines with hooks still kill numerous other species, such as marine turtles and seabirds.

Eating farmed fish may seem a better option, but some fish farming also has a harmful impact on marine biodiversity (see section on aquaculture in chapter 8, page 373). Many farmed fish are fed ground-up fish as fish meal and fish oil, further depleting oceanic stocks of wild species. It takes anywhere from 2 to 5 pounds of wild caught fish to make a pound of carnivorous farmed fish such as salmon. Antibiotics, formerly used only to treat infections, are now routinely used to prevent disease outbreaks in aquaculture farms, and may harm the marine environment. Escapees from fish farms, which are often selectively bred strains of fish species, are altering the genetic balance of their wild cousins, and thereby pose a threat to the viability of the wild stocks. Other problems from aquaculture facilities are the excessive release of nutrients from uneaten food pellets and fish feces that contribute to harmful algal blooms and can threaten marine life (as well as people), and parasites like sea lice and parasitic worms that can infest marine fish farms and infect wild finfish in their vicinity. It must be said that many fish farms, in the United States and in other parts of the world, are becoming more aware of these problems and are trying to address them, but their approach remains the exception rather than the rule for aquaculture.

Freshwater aquaculture is now booming in East Asia and can be much more biodiversity friendly when managed appropriately to ensure that fish cannot escape into the surrounding environment, especially nonnative species that are likely to become invasive. In such closed systems, freshwater aquacultured species, such as tilapia, catfish, and carp that eat plants rather than fish meal, can be raised so that wild fish stocks are not depleted. In China, manure from domesticated animals fertilizes ponds or rice paddies, which then produce algae that can feed as many as four or five species of carp at the same time, because the carps' feeding habits, adapted to different levels of the food chain, serve to complement each other. Such fish "polyculture" is also widely practiced in India.

Similar concerns surround the consumption of some crustaceans. Wild shrimp fisheries have the highest rates of bycatch, harvesting in most cases two pounds, and in some cases more than ten pounds, of accidental victims for every pound of shrimp, including endangered species of marine turtles.[14] And because shrimp are bottom dwellers, the harvesting process tends to be highly destructive of seabeds. Marine shrimp and prawn farms in tropical countries, like carnivorous fish farms, also deplete wild fish stocks and can result in major environmental damage, such as destroying mangrove forests to make way for the farms. Such destruction is reported to have been an important contributing factor in the devastation caused by the December 2004 tsunami in the Indian Ocean[15] (see chapter 3, page 91).

Individually, we can do the following to help sustain marine and freshwater biodiversity:

- Seek out sustainably harvested wild fish such as Alaskan salmon species and Striped Bass (*Morone saxatillis*), and those lower on the marine food chain, such as sardines, anchovies, North Atlantic Mackerel (*Scomber scombrus*), and Atlantic Herring (*Clupea harengus*), which are plentiful (Atlantic Herring may be one of the most abundant fish on Earth) and have the added advantage of containing high levels of omega-3 fatty acids and low levels of such pollutants as mercury. It is not currently possible to say whether sardines, anchovies, North Atlantic Mackerel, and Atlantic Herring also have low levels of PCBs and other organochlorines, because adequate studies on these contaminants have not yet been done. Some herring from the Baltic Sea may have high PCB levels.[16]

- Eat herbivorous farmed fish and shellfish, such as catfish, carp (in Asia), clams, mussels, oysters, and bay scallops.

- Educate ourselves about which species of seafood are under threat and avoid buying them (or eating them in restaurants). Several organizations provide information on what kinds of fish to buy and which ones to avoid. Some of these are listed below by region:

FOR NORTH AMERICA
- Blue Ocean Institute (blueocean.org/seafood)
- Environmental Defense (www.environmentaldefense.org/tool.cfm?tool= seafood)
- Monterey Bay Aquarium (www.mbayaq.org/cr/seafoodwatch.asp)

FOR EUROPE
- World Wildlife Fund (WWF)—Switzerland and International (www. panda.org/downloads/marine/fishguideeng.pdf)
- Marine Conservation Society (www.fishonline.org/information/ MCSPocket_Good_Fish_Guide.pdf)

- Royal Forest and Bird Protection Society (www.forestandbird.org.nz)
- Australian Marine Conservation Society (www.amcs.org.au)
- WWF Hong Kong (www.wwf.org.hk/eng/conservation/wl_trade/reef_fish/online_guide)
- Support organizations that lobby for the reduction or elimination of harmful government fishery subsidies that perpetuate overfishing.

Food from the Land

A quarter of the planet's surface is under cultivation, and when livestock production is taken into account, the proportion of Earth's land surface modified by humans to produce food is even higher.

The most important ecosystem service that agricultural biodiversity provides is, of course, food and food security—but it also provides others, such as nutrient cycling, pest and disease regulation, pollination, maintenance of local wildlife, watershed protection, erosion control, carbon sequestration, and climate regulation. Despite the steady expansion of agricultural ecosystems, agricultural biodiversity is under threat. Worldwide, more than 90 percent of crop varieties have been lost in the past century, and livestock breeds are disappearing at the rate of 5 percent a year. Intensive production methods such as the use of pesticides, the cultivation of monoculture crops, and the loss of field edge habitats are resulting in losses of farmland, wildlife, microbial and invertebrate soil biodiversity, and the diversity of pollinating species and natural predators.[1] Irrigation and livestock wastes from intensive production (e.g., cattle feedlots, and industrial-scale chicken and pig farms) consume water and pollute our waterways, thereby directly affecting wild biodiversity. New technologies, such as genetic engineering, may address at least some of these problems but may also threaten biodiversity in new ways. A precautionary approach is therefore appropriate when applying new agricultural technologies.

Our choice of food can affect our ecological footprint in other ways, as well. Meat production takes disproportionately more resources than do vegetables. It has been estimated that 800 million acres (about 324 million hectares), or 40 percent of U.S. land area, is devoted to raising cattle, with a further 60 million acres (about 24 million hectares) used for growing grain for livestock. Also, the distance our food travels between our farms and our forks—or food miles—is becoming longer each year. Fruit and vegetables on the supermarket shelves often come from halfway around the globe—even when the same fruit or vegetables are in season just down the road. In Britain, the Soil Association (an organization that certifies organic produce) tracked twenty-six ingredients in one basket of food purchased at an organic grocery and found that together they had traveled a distance of 241,000 miles (about 388,000 kilometers)[17]—that's six complete trips around the world. Sometimes, such travel reaches absurd levels, such as the transport of spring water from Fiji to Boston, where tap water is of high quality and where local spring water is widely available. Needless to say, it is fairly pointless to seek out organic food if it has added tons of greenhouse gases to the atmosphere just to reach our plates, contributing to global warming and endangering many species worldwide (see chapter 2, page 63).

As food consumers, we are the principal stakeholders in agricultural ecosystems. Our choices about what we eat can serve to encourage sustainable agricultural production systems that enhance biodiversity.[18]

The following are some of the choices we can make about what we eat that will help preserve biodiversity:

- Seek out farmers markets that offer a range of local fruit and vegetables, as well as dairy products, eggs, meat, and grains, with less packaging (www .localharvest.org/).

- Join a community-supported agriculture initiative (www.csacenter.org) that provides local produce on a regular basis for a preset fee.

- Buy local food in season, and learn how to preserve it throughout the year, so that it does not have to be transported long distances, from other parts of the world, to get to us.

- Buy certified, organically raised meat, dairy products, eggs, cereals, fruit, and vegetables (www.ams.usda.gov/nop/indexIE.htm; www.organicconsumers.org; see also "Organic Farming" in chapter 9).

- Buy shade-grown organic coffee and organic bananas, which help preserve tropical rainforests—look for the Rainforest Alliance Certified label. (The Smithsonian Migratory Bird Center also has a "Bird Friendly" designation for organic coffee, meaning that its cultivation preserves trees and plant diversity.)

- Avoid purchasing foods that list palm oil as an ingredient, such as some margarines and many brands of crackers and cookies. Plantations are proliferating in countries such as Malaysia and Indonesia for the industrial production of oil from Oil Palm Trees (*Elaeis guineensis*), rapidly destroying rainforests in these countries in the process, including habitat for Critically Endangered orangutans and other species. Such widespread deforestation can also increase the risk of some human infectious diseases, such as that caused by the Nipah virus, and vector-borne diseases carried by mosquitoes and snails. Oil palms can be grown sustainably for local populations, as they are in many parts of Africa. In addition, palm oil contains high concentrations of saturated fats and may increase the risk for cardiovascular disease. For our health and for biodiversity in some parts of the world, we should be using olive, soy, and canola oils instead.

- Grow our own fruit and vegetables.

- Eat a diversity of foods, such as different types of grains and potatoes, thus encouraging the production of different varieties.

- Help limit the extent of land under agricultural production by eating less meat and more cereals, fruits, and vegetables, and generally eating lower on the food chain (www.foodalliance.org).

THE WAY WE LIVE IN OUR HOMES

Global climate change, including a rise in global temperatures, changes in rainfall distribution, and increases in extreme weather events, is likely to become one of the greatest threats to biodiversity in the foreseeable future. A recent study has estimated that up to a million species could be at risk worldwide.[19] The signs are already here that global warming is having effects on the biosphere (see chapter 2, page 63). The American Horticultural Society, for example, has had to revise its Heat Zone map of the United States that defines zones by their average number of expected days annually over 86°F. In the Arctic, sea ice is melting so extensively that Polar Bears are beginning to starve because their usual prey, seals, are able to surface at many sites of open water and thus elude capture, instead of at the rare blowholes where Polar Bears once waited for them. Birds, always good indicators of environmental change, are also being affected. In the western Antarctic Peninsula, retreating sea ice is depriving Adélie Penguins (*Pygoscelis adeliae*) of feeding areas, and the number of breeding pairs has declined significantly over the past thirty years. Some coastal seabirds in Britain are failing to breed because changes in water temperature are altering the number and distribution of their principal prey, sand eels. A study of insectivorous birds (e.g., the Blue Tit [*Parus caeruleus*] and the Great Tit [*Parus major*] in the Netherlands) has found that many pairs are failing to produce their usual two clutches of offspring a year, because warmer spring temperatures hasten the development of the insect larvae they depend on for food, out of sequence with the arrival of their hatchlings. Global climate change, driven mainly by the burning of fossil fuels, will also have direct effects on human health and well-being, with increased exposure to heat stress, higher air pollution in our cities, and increasingly severe droughts, floods, and storms affecting communities worldwide. In addition, more and more scientific data point to increases in the occurrence of water, food, and insect-borne diseases.[20]

The energy choices we make in our homes determine our contributions to the burning of fossil fuels and therefore to the emission of greenhouse gases and global climate change. In the United States, total household energy consumption accounted for about 22 percent of national energy consumption in 2001,[21] with heating, hot water, air conditioning, household appliances (ovens, refrigerators, and dryers), and lighting contributing the most toward greenhouse gas emissions. (Some useful figures about regional, national, and global energy use can be found at chemistry.beloit. edu/Warming/pages/emissions.html.)

As individuals, we can have a real influence on reducing the amount of energy we use in our homes and offices, and thereby reducing the extent of climate change. In temperate climates, stopping drafts and optimizing insulation can cut heat losses dramatically, and turning down the thermostat by just 1 degree Fahrenheit (0.6 degrees Celsius) saves about 8 percent in heating costs in an average year. In tropical countries, using traditional architecture rather than modern forms that require air conditioning is often an attractive and energy-saving option. Actions as simple as turning off computers, televisions, and audio equipment when not in use is something we can all do. A 2001 study by the International Energy Agency estimated that the amount of energy used in maintaining such equipment in "standby" mode in several countries ranged from 3 percent of total domestic energy use in Switzerland to a whopping 13 percent in Australia.[21] During the energy crisis of the summer of 2000, with monthly electricity bills rising some 130 percent, and under strong public pressure to conserve energy, the residents of California reduced their average household use of electricity by 12 percent without any obvious reduction in the quality of their lives.[22]

Other choices we make in our homes, such as the clothes, furniture, and building materials we buy, also have an impact on biodiversity.

Here are some things we can do in our homes:

- If buying or building a new house, opt for a smaller, energy-efficient home, with features such as high insulation ratings, storm doors and storm windows in northern regions and light-colored reflective roofing in southern ones (which can reduce air conditioning energy use by 20 percent or more), and alternate energy systems such as photovoltaic panels for electricity, solar hot water heaters, and ground-source heat pumps. (For additional energy-efficient features of homes and communities, see the U.S. Green Building Council website at www.usgbc.org.)

- Choose appliances (refrigerators, ovens, washing machines, etc.) and home heating systems with high energy-efficiency ratings (for energy use of typical home appliances, see www.eere.energy.gov/consumer/ or www.energystar.gov). Many energy-efficient appliances and furnaces will end up paying for themselves over several years with the money saved in energy costs, while also helping to slow the climate change that threatens biodiversity.

- Invest in a low-tech solar clothes drier: Two poles, a length of wire or rope, and some clothes pins should cost about $25 and save hundreds of dollars on electric or gas bills. Your clothes will also smell and feel much nicer.

- Turn off the lights when leaving a room, and turn off or lower the settings of heating or air conditioning when leaving home.

- Replace Edison-era light bulbs with compact fluorescent light bulbs—they use a fraction of the electricity (e.g., a 15-watt compact fluorescent bulb gives off as much light as a 60-watt incandescent bulb) and last up to ten times as long.

- In winter, turn down the thermostat—wear a sweater, and for sleeping, buy warm blankets or down quilts.

- In summer, turn up the thermostat on air conditioners, or turn it off and open some windows.

- Plant trees around our homes, which can reduce energy costs from air conditioning by as much as 25 percent.

- Switch to an energy supplier (electric company) that offers energy from renewable (wind, wave, or solar) power sources. Alternatively, look into purchasing green energy certificates that support alternative energy systems (www.eere.energy.gov).

- Recycle paper, cans, glass bottles, and plastic bottles if our communities participate in these programs.

- Buy cold weather clothing made from plastic that has been recycled and transformed into "fleece"—which has excellent insulating properties.

- Recycle old clothes to charities that accept them for resale.

- Recycle electronic equipment (Cell Phone Recycling and Donation Programs: www.eiae.org/whatsnew/news.cfm?ID=100; Computer Take Back Options www.epa.gov/e-Cycling/donate.htm).

- Buy organic cotton clothes, sheets, and towels—conventionally grown cotton is one of the most pesticide intensive of all crops, and therefore one of the most potentially damaging to birds and other species.

- Buy recycled wood for building or use other materials that are widely available and renewable, such as bamboo and cork.

- Buy recycled fiber carpeting, thereby reducing demand for cotton, wool, or petrochemicals (all of which have impacts on wild biodiversity).

- Avoid buying furniture made from tropical hardwoods (e.g., teak or mahogany), or using such wood for building unless it bears the seal of the Forest Stewardship Council (considered the most rigorous and independent forest certification program) or the SmartWood label from the Rainforest Alliance in the United States, indicating that the wood comes from sustainable forestry operations that minimize the negative impact of logging on biodiversity.

- Do not buy new items made from gold. Gold mining is one of the most environmentally destructive of all industries, destroying rainforests and other habitat, and contaminating surface waters with cyanide and mercury.

- Wherever possible, buy only recycled products for our homes—recycled paper products (towels, toilet paper, and writing bond) save forests and save landfills space.

- Conserve water—the average family in the United States uses 74 gallons (about 280 liters) of water each day, nearly one-third of which goes to flushing toilets.[11] Install low-flow toilets in your homes, take shorter showers, turn off the faucet while brushing your teeth, and plant a native grass lawn. Thirty percent of the water used in New England goes to watering lawns (see below). Conserving water makes more available for lakes, rivers, and streams that support native wild species.

- Reduce junk mail by requesting to be removed from direct mail lists—in the United States, write to the Direct Marketing Association (Mail Preference Service, PO Box 9008, Farmingdale, NY 11735) and ask that mail order companies reduce the number of catalogues they send. In the United States, ninety million trees are cut each year to produce bulk mailings that end up being hauled away by 340,000 garbage trucks, contributing significant greenhouse gases.[23]

HOW WE TRANSPORT OURSELVES

The average U.S. household also emits 3.7 tons of greenhouse gases per year through the use of automobiles, minivans, sport utility vehicles (SUVs), and light trucks. Collectively, transportation accounts for about 28 percent (in 2003) of the total national energy consumption in the United States.

On average, 2.7 people live in each U.S. household, and they travel some 21,000 miles (almost 34,800 kilometers) each year by automobile and an additional 3,150 miles (about 5,070 kilometers) each year by airplane. Newer models of passenger cars are much less gasoline-hungry than older ones and now average about 28.5 miles per gallon (about 12 kilometers per liter). Unfortunately, potential savings in carbon emissions through more fuel-efficient cars has been more than offset by recent trends toward using heavier, four-wheel-drive vehicles, such as minivans, sport utility vehicles, and light trucks, which average only about 20.5 mpg (about 8.7 kilometers per liter) and account for about 26 percent of total energy consumption in the United States for transportation.[24] Is it really necessary to drive to work or the supermarket in a car designed to perform in off-road conditions?

Here are some things we can do to reduce our transportation-related energy consumption:

- Set concrete goals to lessen personal vehicle use; make one trip instead of several a week to the supermarket, and carpool to work.

- Urge local authorities to make our towns more pedestrian and bicycle friendly, and walk or bicycle to work whenever possible (which is already standard practice in the Netherlands, Denmark, China, and many other countries).

- Encourage local authorities to improve public transportation services, and use them (trains, buses).

- Make sure our cars are well tuned, and that their tires are properly inflated; accelerate slowly, and do not drive at excessive speeds—highway driving at around 55 miles per hour (88 kilometers per hour) is generally the most fuel efficient. These practices can make large differences in the amount of fuel we consume.

- Consider buying smaller, more fuel-efficient and less polluting cars, or hybrid gas-electric vehicles.

OTHER THINGS WE CAN DO TO PROTECT BIODIVERSITY

Our Own Back Yards

Managing the green spaces around us for biodiversity is something all of us can do, and even the smallest back yard or apartment balcony can become a haven for wildlife. Here are some tips for use in *our* own back yards:

- Avoid gardening "against the grain": Choose plants suited to our local soil type and climate. Avoid plants that need a lot of fertilizers and water (for suggestions, see www.biodiversityproject.org/5%20Ways%20Campaign/5waysbackyardpress.htm).

- Seek out indigenous trees, shrubs, and flowers (see "Increase Backyard Diversity" [Audubon], www.audubon.org/bird/at_home/wildlife.html, "Rethink Your Lawn" [Audubon], www.audubon.org/bird/at_home/rethink_lawn.html, and "Designing for Wildlife" [Plant Native], plantnative.com/how_wildlife.htm).

- Exercise caution when buying plants: Ensure that they are not invasive and have not been taken from the wild. Rare species such as cycads and tree ferns should come only from certified suppliers (see www.ucsusa.org/invasive_species/what-you-can-do-to-prevent-species-invasion.html, www.invasivespeciesinfo.gov/community/whatyou.shtml).

- Avoid buying flower varieties that have "double" blooms: Insects can't access their nectar because their mouthparts are not adapted to do so.

- For large gardens, leave at least part of it "wild"—a small pile of rotting logs for wood-boring beetles, a dead tree for woodpeckers to nest in, a patch of stinging nettles and other "weeds" for butterfly larvae to feed on.

- Put in plants that attract wildlife, such as the "butterfly bush" (*Buddleja* spp.) for butterflies (for those *Buddleja* spp. that are invasive, make sure seeds are removed before they can be spread by birds).

- If there is space, consider making a small pond: It will soon become a magnet for insects, amphibians, and birds.

- Avoid using pesticides: Choose plants that are naturally resistant to pests (indigenous species usually are), and use biological control agents and natural predators such as nematodes and lady bugs (see "Pesticides: Health and Safety" [EPA], www.epa.gov/pesticides/health/human.htm, "Pesticides: Controlling Pests" [EPA], www.epa.gov/pesticides/controlling/garden.htm, "Beyond Pesticides: Least Toxic Control of Pests in the Home and Garden," www.beyondpesticides.org/alternatives/factsheets).

- Use organic fertilizers and mulches such as barnyard manure and compost made from kitchen waste.

- Establish "natural" lawns that have a variety of plant species: They need less water and less mowing and can be far more attractive to look at than something that looks and feels like the artificial grass of a football field. Or establish a wildflower meadow or mini-prairie. These options support far more biodiversity.

- For small areas of grass, use an old-fashioned, hand-pushed, reel-type mower rather than an electric or fuel-powered one. It is terrific exercise, burning our own calories instead of fossil fuels.

- Provide birds with nest boxes, food in winter, and water all year round, out of reach of cats and dogs.

- Give your kids a small part of our yards to grow their own flowers and vegetables, helping them to learn the values of biodiversity.

- If you play golf, encourage your golf course or club to use natural landscaping and a minimum of pesticides, fertilizers, and water to preserve biodiversity.

- Do not use "bug zappers" because they are enormously destructive to insect diversity, including to beneficial insects, and rarely attract mosquitoes and other biting insects that they are supposed to target. Those that do attract mosquitoes by generating CO_2 from the burning of large amounts of fuel such as propane contribute to global warming.

Conserving Protected Areas

Protected areas such as national parks and nature reserves now cover about 12 percent of Earth's land surface—20 million square kilometers (about 7.7 million square miles). Large protected areas are key to preserving whole ecosystems and their flora and fauna and are especially valuable for conserving populations of species that humans find difficult to live alongside, such as elephants, lions, or wolves. And the establishment of "corridors" between protected areas offers wildlife an ability to move freely throughout their ranges or to shift their ranges in response to global warming. For example, the Yellowstone to Yukon Conservation Initiative extends along 2,000 miles (3,200 kilometers) of the Rocky Mountains in western United States and Canada (the northern Rocky Mountains from Wyoming to the Arctic Circle), and the Mesoamerican Biological Corridor in Central America covers 208,000 square kilometers (about 80,000 square miles), or 27 percent of Mesoamerican territory. (Mesoamerica is the area between central Mexico and the northwest border of Costa Rica that for 3,000 years was home to several pre-Columbian civilizations.) Biological corridors can also reverse the destructive effect of smaller, fragmented protected areas for some species.[25]

Unfortunately, quite a few protected areas, especially in poorer nations, are protected in name only (they are sometimes called "paper parks") because the governments of those countries lack the resources to care for them properly or are unable to manage them effectively because of civil unrest and lawlessness. So the true extent of land under "protection" is not quite as large as it may appear (though other areas without formal designation may serve many of the biodiversity conservation objectives of protected areas). The picture is not rosy for fresh waters or oceans either: Marine protected areas, most of them in shallow coastal waters, cover less than half a percent of the world's seas, despite the solid evidence that such reserves permit the recovery of overharvested and otherwise threatened marine species (see www.world-wildlife.org/oceans/pdfs/fishery_effects.pdf).

Protected areas come in all sizes, and their management goals vary. Some, like many of the wilderness areas in the United States, are managed to preserve the natural condition of their ecosystems. Others, such as the Yellowstone National Park or Tanzania's Serengeti National Park, tolerate limited types of human activity, for example, motorized wildlife viewing or scientific research. For such enterprises to succeed in the long term, local populations must be compensated for any sacrifices they must make to accommodate visitors. Still other protected areas encompass natural ecosystems, largely unaltered by modern development, but allow for varying degrees of consumptive or extractive use (e.g., hunting and fishing by local communities or sustainable timber production).

Many protected areas are lived-in, working landscapes that promote and support traditional livelihoods and cultures in addition to protecting biodiversity. Some examples are the Snowdonia National Park in Wales, Adirondack Park in New York State, Hungary's Hortobágy National Park, and the Champlain-Richelieu Valley that straddles the border between northeastern United States and Canada. Traditional farming methods sympathetic to biodiversity conservation are encouraged in such protected areas, and many of them harbor species that are rare or endangered where modern intensive farming methods prevail. Conservationists believe that landscape models that integrate protected areas within a mosaic of agricultural and urban land uses are the most feasible way to preserve biodiversity in densely populated regions and have great potential in both tropical countries and temperate industrialized countries.[26]

While large protected landscapes are essential to conserve many wide-ranging species and native species communities, the biodiversity value of very small protected areas should not be underestimated. Countless such small reserves exist, sometimes only a few hectares in extent. Many of our neighborhoods have them. They may have been designated to protect a small wetland, create a greenbelt around a town, or preserve a patch of forest or woodland. They can be important local reservoirs for small vertebrates, invertebrates, and plants. One of Britain's smallest nature reserves is a 70-meter (~230-feet) length of hedgerow bank in the county of Suffolk, designated to protect an endangered fungus, the Sandy Stiltball (*Battarrea phalloides*). Florida's 6-acre (~2.4-hectare) Pelican Island is the smallest designated wilderness area in the United States.

Many such local reserves are run by groups of volunteers or local communities. Others are privately owned or managed by conservation organizations. In the United States, more than 5 million acres (about 2 million hectares) are protected by local and regional land trusts through "conservation restrictions" (agreements between landholders and land trusts or government agencies to conserve land by limiting its development). Nongovernmental organizations such as The Nature Conservancy protect a further 7 million acres in the United States. The largest conservation organization in Europe, Britain's Royal Society for the Protection of Birds (www.rsbp.org.uk) has more than a million members and manages 180 nature reserves totaling 320,000 acres (129,000 ha) in the United Kingdom.

To contribute to protected areas near us, we can:

- Help create and perpetuate small protected areas for conserving biodiversity by joining local or national conservation organizations.

- Encourage our city governments to create green spaces and manage them as "natural parks" for local biodiversity. By creating wildlife habitat (e.g., by using indigenous trees and other plants), we will also give our communities opportunities for environmental education as well as for recreation.

- Place a "conservation restriction" on portions of our own property to provide it with ongoing protection.

Raising Awareness

It is a sad fact that a great many of us—especially those of us from inner cities—have very little awareness of the natural world and of our dependence on it. This may be especially true for our children. Few people can name more than a handful of birds, trees, or flowers, if indeed they notice them at all. They don't know where their food and water come from, and they don't know where their waste goes. They do not understand the vast array of ecosystems, for agriculture and watersheds, and for breaking down and absorbing wastes, that need to be intact and functioning properly to keep them healthy and alive. This leads to an inability to recognize their interdependent relationship with other species or to appreciate the role these species play in their health and lives. It can also lead to a poverty of the spirit, because they are deprived of the wonder and the sense of awe and reverence that many experience in recognizing their intimate connection with the natural world.

Some schools are incorporating environmental education in their curricula—but all too often the only exposure our children get to the concept of biodiversity at school is a year or two (at most) of old-fashioned biology. By teaching our children a more holistic view of Nature, starting perhaps with some fundamentally commonsense principles of ecology, we can help them understand our unique role on the planet and, by extension, our responsibility toward it. Only by teaching that we and other species depend on some endangered groups of organisms in all kinds of ways—frogs, for example, for the development of new medicines and as models for medical research, and for their role as key predators of invertebrates (including some insect larvae like mosquitoes, keeping their numbers in check)—and that these organisms have been on Earth for millions and even hundreds of millions of years and have survived the great extinction events, only then can we help young people understand the importance of these and other species in our lives and our interconnectedness with them. Teaching children about their connections with the natural world may be one of the most important things we can do during their school years.

But this education should not be done in an academic setting alone. We must expose our children directly to the biodiversity around them and to ecological principles from an early age.

Some ways we as individuals can help increase awareness are:

- In the home, set an example by reducing our "Ecological Footprints" and explaining why to others, especially to our children.

- Persuade the school board to incorporate environmental awareness into the curriculum.

- As part of this curriculum, encourage the school to develop student gardens, where each child has his or her own plot to grow flowers and even organic vegetables for the school cafeteria. Have the garden plots be sites for biology class projects.

- Work with schools and students to green their campuses (for more information, see www.iisd.org/educate).

- Find ways of reducing our workplace's footprint on biodiversity.

We can tell our families, friends, and neighbors, and those at our workplace, place of worship, and organizations we belong to, what we have done to preserve biodiversity and why it is important. If everyone we know began to follow some of the suggestions mentioned in this chapter, the impact would be considerable.

Supporting Organizations and Political Candidates That Preserve Biodiversity

Joining an environmental conservation organization, and supporting them with our time and money, is another way to have our voices heard. They come in all shapes and sizes, ranging from small ones with a local agenda (greening our city, saving a local pond) to national and international nongovernmental organizations concerned with saving whole ecosystems and species (several are listed in appendix C). Coalitions of such organizations often have huge lobbying power. We can get others to join when we believe strongly in what these organizations are doing.

We can become involved politically in town government if we live in a small municipality, for example, by joining the conservation committee. We can elect environmentally responsible politicians at all levels of government and urge them to place biodiversity concerns at the top of their agendas. We can learn the environmental positions of those who represent us locally, regionally, and nationally, and give them feedback when we support what they stand for and when we do not. We should become informed, be persistent, and organize other constituents of theirs to do the same.

Some cities are already showing leadership in this respect. For example, the International Council for Local Environmental Initiatives (ICLEI; www.iclei.org) is working with 724 cities, towns, counties, and their associations in sixty-eight countries worldwide to develop local strategies to reduce energy consumption and CO_2 emissions, including Atlanta, Chicago, Los Angeles, and many other cities in the United States and hundreds abroad, including Bangkok, Barcelona, Berlin, Cape Town, Kampala, Mexico City, and São Paulo. Work with those leading these initiatives.

Encouraging Governments to Address Biodiversity Loss

The Millennium Ecosystem Assessment was undertaken partly to feed information into four international conventions administered by the United Nations that are concerned with the conservation and sustainable use of biodiversity: the Convention on Biological Diversity, the Ramsar Convention on Wetlands of

International Importance, the U.N. Convention to Combat Desertification, and the Convention on Migratory Species (see appendix B).

These four international agreements are not the only ones focusing on biodiversity—dozens of global and regional agreements are also partly or wholly concerned with saving ecosystems and species. Some of them focus on specific groups of animals (e.g., the Agreement on the Conservation of Albatrosses and Petrels), while others focus on specific habitats or ecosystems (e.g., some of the provisions of the U.N. Convention on the Law of the Sea). Still others, such as the Convention on Trade in Endangered Species of Wild Fauna and Flora (CITES), deal with wildlife trade. The creation of international policy instruments to protect biodiversity dates back almost a hundred years. One of the earliest agreements, the Convention for the Protection of Migratory Birds (a treaty to regulate the hunting of migratory birds found in the United States and Canada), came into force in 1916. Many more followed, such as an agreement on the protection of African wildlife (1933) and the International Convention for the Regulation of Whaling (1946). But most international conventions concerned with protecting whole ecosystems and biodiversity in the broader sense have come into force only during the past fifty years.

Many argue that, as policy instruments, broad-based international conventions are unwieldy, unworkable, or unenforceable and that governments, having signed on to such agreements, sometimes remain unaccountable for their implementation. But it need not be so: As individuals, those of us living in democracies at least should be aware of what commitments our governments have made and should ensure that they live up to them. Actions by governments to reduce their "Ecological Footprints" by signing on to international conventions (and by subsequently enacting legislation at local and national levels) will not have much value unless we as individuals support their efforts.

SOME COLLECTIVE ACTIONS THAT HAVE MADE A DIFFERENCE

- In 2001, the U.S. Forest Service, after receiving 1.6 million comments from concerned citizens, dedicated 58 million acres (about 23.5 million hectares) of wild forests to remain undeveloped for future generations. The challenge is to retain this legacy in the face of changing political conditions.

- In 2004, after receiving more than half a million responses from the public, the British government finally dropped its plans to build a new airport on the Thames estuary, near wetland reserves that were important habitat for waders and other waterfowl.

- The Forest Stewardship Council (FSC) is a stakeholder-owned system that sets international standards for responsible management of the world's forests. Public demand for timber, furniture, and paper from forests that are managed sustainably is soaring. Over the past ten years, more than 50 million hectares (almost 124 million acres) worldwide (including more than 10 million hectares in the United States) have been certified to FSC standards.

INDIVIDUALS WHO HAVE MADE A DIFFERENCE

Wangari Maathai

Professor Wangari Maathai, born in Nyeri, Kenya, founded the Green Belt Movement, which has now worked to conserve the environment and develop communities in Kenya for nearly thirty years, planting nearly thirty million trees across the country and providing cash subsidies to the large number of impoverished women who have done the work. She has been a champion not only for the environment but also for human rights, especially the rights of women. Several other countries have followed the Green Belt Movement's example, and their efforts have been an important component in fighting the deforestation, forest loss, and desertification that plagues Africa and many other regions of the world. Professor Maathai was awarded the 2004 Nobel Peace Prize for her "contribution to sustainable development, democracy, and peace." She was the first African woman to be awarded the Nobel Prize. In addition to her work with the Green Belt Movement, she is Kenya's Deputy Environmental Minister and is the presiding officer of the Economic, Social, and Cultural Council (ECOSOCC) of the African Union and the Goodwill Ambassador for the Congo River Forest Ecosystem.

Pisit Charnsnoh

Pisit Charnsnoh has worked for many years to improve the living conditions in coastal villages in his native Trang Province in southern Thailand. He and his wife founded the Yadfon Association in 1985, which works in thirty communities to protect mangroves and coastal fisheries. The word *yadfon* means "raindrop" in Thai and symbolizes renewal. Mr. Charnsnoh's work has epitomized this renewal in many ways. As a Buddhist, it took him many years to gain the trust of the Muslim coastal villagers, but over time, villagers joined Yadfon and found themselves involved in decision-making roles. They worked to limit the number of commercial shrimp farms in their area, enforce boundaries set for fishing trawlers, restore seagrass beds to provide habitat for the endangered dugong manatee, and encourage better management of the local watershed. Perhaps most important, the villagers banded together to restore a 240-acre (~97-hectare) mangrove forest, which became the first community-managed mangrove forest in the country. Restoration of the mangroves resulted in a 40 percent increase in local fish catch, with the resulting income being used to restore communities. In 2002, Pisit Charnsnoh was awarded the Goldman Prize in Marine Conservation for his work.

Oscar Rivas and Elias Diaz Pena

Oscar Rivas and Elias Diaz Pena have been working since 1991 with communities affected by the Yacyreta Dam project in Paraguay, which flooded the homes of 50,000 people, disrupted fish migration, and altered the region's groundwater system. Mr. Rivas and Mr. Pena formed an organization called Sobrevivencia that mobilized these communities to assess the impacts of another project, the Hidrovia navigation project, that planned to develop a shipping channel using 3,400 kilometers (about 2112 miles) of the Paraguay and Paraná river systems. This project would have had an adverse effect on communities in Paraguay, Argentina, Bolivia, Brazil, and Uruguay and would have endangered the Pantanal, the world's largest wetland area. Sobrevivencia led a coalition of 300 groups of indigenous people, communities that would be affected, and environmentalists. They developed a traveling educational campaign about the impacts of the project and filed a claim with the World Bank and the Inter-American Development Bank asking them to investigate possible violations by Hidrovia's environmental and resettlement policies. This action led to the creation of a new model for evaluating development projects. Oscar Rivas and Elias Diaz Pena won the 2000 Goldman Prize in Rivers and Dams for their work.

TEN THINGS WE ALL CAN DO THAT CAN HELP CONSERVE BIODIVERSITY

1. Take public transportation, bike, walk, or carpool to work at least one day a week, and if you do drive by yourself, drive the most energy-efficient vehicle you can afford.
2. Buy food, preferably organic food—vegetables, fruits, dairy, eggs, and meat—from a farmers market at least one day a week.
3. Eat sustainably harvested seafood and farmed fish that is herbivorous, such as catfish, tilapia, and shellfish. Avoid farmed carnivorous fish such as salmon and shrimp.
4. Install at least one compact fluorescent light bulb in your home—it will save roughly $40 in electricity and replacement bulb costs and reduce carbon emissions by 700 pounds (about 318 kilograms) each year.
5. Turn off lights in empty rooms.
6. Lower the thermostat by at least 1 degree Fahrenheit (about 0.6 degrees Celsius) in winter.
7. Stop using herbicides and pesticides on your lawn.
8. Learn the environmental positions of all those who represent you in government, and support those candidates who have the best records and the best platforms.
9. Tell everyone at home, school, place of worship, and work about what you are doing to conserve biodiversity and ask them to join you.
10. Above all, do not waste—reduce your consumption, buy only what you really need, and reuse and recycle whatever and whenever you can.

NEVER DOUBT THAT A SMALL GROUP OF THOUGHTFUL, COMMITTED CITIZENS CAN CHANGE THE WORLD; INDEED, IT'S THE ONLY THING THAT EVER HAS.

—Margaret Mead

◆ ◆ ◆

Suggested Readings

Collapse: How Societies Choose to Fail or Succeed, Jared Diamond. Penguin Group, New York, 2005.

The Consumer's Guide to Effective Environmental Choices: Practical Advice from the Union of Concerned Scientists, Michael Brower and Warren Leon. Three Rivers Press, New York, 1999.

Contested Terrain: A New History of Nature and People in the Adirondacks, P.G. Terrie. Syracuse University Press, Syracuse, New York, 1997.

Ecosystems and Human Well-being: Synthesis, Millennium Ecosystem Assessment. Island Press, Washington, D.C., 2005.

The End of Poverty: Economic Possibilities for Our Time, Jeffrey D. Sachs. Penguin Press, New York, 2005.

The Last Child in the Woods: Saving Our Children from Nature-Deficit Disorder, Richard Louv. Algonquin Books, Chapel Hill, North Carolina, 2005.

The New Consumers: The Influence of Affluence on the Environment, Norman Myers and Jennifer Kent. Island Press, Washington, D.C., and Covelo Press, London, 2004.

One with Nineveh: Politics, Consumption and the Human Future, Paul and Anne Ehrlich. Island Press, Washington D.C., 2004.

Plan B: Rescuing a Planet under Stress and a Civilization in Trouble, Lester Brown. W.W. Norton & Company, New York, 2003.

Red Sky at Morning: America and the Crisis of the Global Environment, James Gustave Speth. Yale University Press, New Haven, Connecticut, 2004.

State of the World 2004: Special Focus: The Consumer Society, Worldwatch Institute. W.W. Norton & Company, New York, 2004.

APPENDIX A

CO-SPONSORS OF *Sustaining Life: How Human Health Depends on Biodiversity*

Four of the world's leading bodies that work to conserve biodiversity, three that are part of the United Nations—the Secretariat of the Convention on Biological Diversity, U.N. Environment Programme, and the U.N. Development Programme—and one that is a consortium of states, government agencies, and nongovernmental organizations—the International Union for Conservation of Nature and Natural Resources—have co-sponsored this book. They are described below.

THE SECRETARIAT OF THE CONVENTION ON BIOLOGICAL DIVERSITY (WWW.BIODIV.ORG/PROGRAMMES/DEFAULT.SHTML)

The U.N. Convention on Biological Diversity (CBD) is an agreement that has been ratified by 187 countries and the European Union (the United States has not ratified the CBD) to conserve biological diversity, use its components sustainably, and ensure the fair and equitable sharing of the benefits derived from using genetic resources. Parties to the CBD have taken steps to translate the convention into practical action, including the initiation of national biodiversity strategies and action plans in more than 100 countries, raising awareness about biodiversity, and adopting the Cartagena Protocol on Biosafety, an international regulatory framework for the safe transfer, handling, and use of living, genetically modified organisms resulting from modern biotechnology.

The Conference of the Parties (COP) has initiated work on seven program areas that address the biodiversity of marine and coastal ecosystems, agricultural systems, forests, islands, inland waters, dry and subhumid lands, and mountains. Each program establishes its own vision and the basic principles that guide its work, identifies the major issues it will consider, and sets goals that it hopes to achieve and a timetable for achieving them. The COP's agenda also includes a number of other key cross-cutting issues that are relevant to all seven thematic areas, such as ecosystem approach, biosafety, access to genetic resources, traditional knowledge, intellectual property rights, biodiversity indicators, species taxonomy, alien species, public education and awareness, and incentives for meeting program goals.

THE U.N. DEVELOPMENT PROGRAMME (WWW.UNDP.ORG/BIODIVERSITY/PROGRAMMES.HTML)

The U.N. Development Programme (UNDP), the United Nations' global development network that is on the ground in 166 countries, works to help people around the world build better lives and find their own solutions to their national development challenges through access to the knowledge, experience, and resources of UNDP staff and their wide range of partners.

UNDP has made biodiversity for development a prime focus of its Energy and Environment Practice and is now helping more than 140 countries maintain their biodiversity and use it sustainably. Other closely integrated UNDP activities, such as its Biodiversity Global Programme, the Equator Initiative, the Global Environment Facility (GEF), and the GEF's Small Grants Programme leverage change at the local, national, regional, and global levels.

The Biodiversity Global Programme assists developing countries and communities in influencing national and global policies, in benefiting from knowledge about biodiversity, and in advancing their sustainable development and poverty reduction goals. UNDP works to help integrate biodiversity, ecosystem services, protected areas, and other CBD priority areas into national policies and programs, involving such key sectors as agriculture, forestry, fisheries, and energy. These efforts address social, economic and policy frameworks such as the Millennium Development Goals (see www.un.org/millenniumgoals/) and National Sustainable Development Strategies. Activities that specifically involve the relationship between biodiversity and health include empowering local communities and indigenous peoples to protect their traditional knowledge, ensuring the equitable access and the sharing of benefits from biodiversity, and achieving synergies with other multilateral environmental agreements related to biodiversity and ecosystem services.

THE U.N. ENVIRONMENT PROGRAMME (WWW.UNEP.ORG/THEMES/BIODIVERSITY/)

The U.N. Environment Programme (UNEP) is the designated U.N. authority on environmental issues at the regional and global level. Its mandate is to coordinate the development of consensus on environmental policy issues by closely monitoring the global environment and by bringing emerging issues to the attention of governments and the international community for possible action.

Among others, UNEP oversees and supports the following biodiversity conventions and activities: the U.N. Convention on Biological Diversity (see above), the Convention on International Trade in Endangered Species, the Convention on Migratory Species, the Great Apes Survival Project, the Global Programme of Action for the Protection of the Marine Environment from Land-Based Activities, the International Coral Reef Action Network, the International Centre for Integrated Mountain Development, the U.N. Environment Programme's Global Environment Facility Project on Development of National Biosafety Frameworks, the World Conservation Monitoring Center, and Earthwatch.

THE INTERNATIONAL UNION FOR CONSERVATION OF NATURE AND NATURAL RESOURCES (WWW.IUCN.ORG/THEMES/PBIA/THEMES/BIODIVERSITY/WHATWEDO.HTM)

Founded in 1948, the International Union for the Conservation of Nature and Natural Resources (IUCN) brings together states, government agencies, and a diverse range of nongovernmental organizations in a unique world partnership—more than 1,000 members in all—spread across some 150 countries.

The IUCN seeks to influence, encourage, and assist societies throughout the world to conserve the integrity and diversity of nature and to ensure that any use of natural resources is equitable and ecologically sustainable. A central Secretariat coordinates IUCN programs and serves the IUCN membership, representing their views on the world stage and providing them with the strategies, services, scientific knowledge, and technical support they need to achieve their goals. Through its six Commissions, the IUCN draws together more than 10,000 expert volunteers in project teams and action groups, focusing in particular on species and biodiversity conservation and on the management of habitats and natural resources. The IUCN has helped many countries prepare National Conservation Strategies and has demonstrated the application of its knowledge through the field projects it supervises. Operations are increasingly decentralized and are carried forward by an expanding network of regional and country offices, located principally in developing countries.

The IUCN builds on the strength of its members, networks, and partners to enhance their capacity and to support global alliances to safeguard natural resources at local, regional, and global levels.

APPENDIX B

TREATIES, CONVENTIONS, AND INTERGOVERNMENTAL ORGANIZATIONS FOR THE CONSERVATION OF BIODIVERSITY

This appendix contains some of the most prominent international agreements and intergovernmental organizations dealing with biodiversity. For each of them, considerable additional information is available on their respective websites. Numerous other examples could have been included, including regional agreements that are especially important in terms of bringing together countries that share common interests and biological resources.

UNITED NATIONS CONVENTIONS

Convention on the Conservation of Migratory Species of Wild Animals (www.cms.int/about/index.htm)

One of the early environmental conventions, the Convention on the Conservation of Migratory Species of Wild Animals entered into force in 1983 to conserve those species of wild animals that migrate across or outside national boundaries. It calls for the parties to develop and implement cooperative agreements, prohibit taking of endangered species, conserve habitats, and control other adverse factors.

Convention on International Trade in Endangered Species of Wild Fauna and Flora (www.cites.org)

The Convention on International Trade in Endangered Species of Wild Fauna and Flora (CITES), another "first-generation" convention, entered into force in 1975 to ensure, through international cooperation, that the international trade in species of wild fauna and flora does not threaten survival in the wild of the species concerned. It is designed to protect endangered species from overexploitation by means of a system of import–export permits issued by a management authority under the control of a scientific authority in each of the 169 parties.

International Plant Protection Convention (www.ippc.int)

This international treaty facilitates international cooperation to outline actions preventing the spread of invasive pest species of plants and related plant products. The treaty also promotes measures that control invasive pest species. Governed by the Interim Commission on Phytosanitary Measures, this convention has been signed by 144 member nations.

Ramsar Convention on Wetlands (www.ramsar.org)

Also known as the Convention on Wetlands of International Importance especially as Waterfowl Habitat, this convention supports the conservation and wise use of wetlands by national action and international cooperation, as a means to achieving sustainable development throughout the world. With 147 parties, the convention started with a focus on waterfowl but has greatly expanded its mandate and now has listed well more than 1,500 wetlands of international importance, totaling near 130 million hectares.

U.N. Convention to Combat Desertification (www.unccd.int)

This convention entered into force in 1996 and now has about 175 parties. It was established to combat desertification and mitigate the effects of drought in countries experiencing serious drought and/or desertification, particularly in Africa, through effective actions at all levels and supported by international cooperation and partnership arrangements. It encourages parties to prepare national action programs and calls for the developed countries to support their implementation.

U.N. Convention on Biological Diversity—see appendix A

United Nations Framework Convention on Climate Change (unfccc.int/)

This convention entered into force in 1994 and now has some 190 parties. It was established to stabilize greenhouse gas concentrations in the atmosphere that would prevent dangerous anthropogenic interference with the climate system, within a time frame sufficient to allow ecosystems to adapt naturally to climate change. It also intends to ensure that food production is not threatened and to enable economic development to proceed in a sustainable manner. The Kyoto Protocol, which calls for specific measures to reduce greenhouse gasses on the part of the developed countries, entered into force in 2004.

World Heritage Convention (whc.unesco.org/)

This convention also entered into force in 1975 to establish an effective system of collective protection of the cultural and natural heritage of outstanding universal value. It is intended to provide both emergency and long-term protection to natural features and habitats of plants and animals judged to be of "outstanding universal value." It establishes a World Heritage list, which currently has 160 natural sites and 24 mixed cultural and natural sites. It has 180 parties.

UNITED NATIONS SPECIALIZED AGENCIES

U.N. Food and Agriculture Organization (www.fao.org/)

The U.N. Food and Agriculture Organization (FAO) was founded in 1945 to raise the levels of nutrition and standards of living of the populations of the member countries. It seeks to improve the efficiency of production and distribution of all food and agricultural products including forests and fisheries. It is responsible for numerous conventions relating to fisheries, plant genetic resources, and other environmental problems relating to food production.

Global Environment Facility (www.gefweb.org/)

Founded in 1991, the Global Environment Facility (GEF) provides a mechanism for international cooperation through providing new and additional grant and concessional funding to meet the agreed global environmental benefits in several focal areas, including biological diversity, climate change, international waters, and ozone layer depletion. It has provided substantial funding to support the implementation of the U.N. Convention on Biological Diversity, amounting to well more than US$100 million per year, through its implementing agencies (the World Bank, U.N. Environment Programme, and U.N. Development Programme).

U.N. Development Programme—see appendix A

U.N. Educational, Scientific, and Cultural Organization (www.unesco.org or portal.unesco.org/en/)

Founded in 1946, the U.N. Educational, Scientific, and Cultural Organization (UNESCO) contributes to peace and security in the world by promoting collaboration among nations, education, science, culture, and communication. It expects such cooperation to lead to universal respect for justice, the rule of law, and the human rights and fundamental freedoms proclaimed by the Charter of the United Nations for the peoples of the world, without distinction of race, gender, language, or religion. Its environmental activities are especially through its scientific support to governments and through its hosting of the World Heritage and Ramsar Conventions. Its Man and Biosphere Programme has established a worldwide network of 482 biosphere reserves that are managed by the 102 countries in which they are located.

U.N. Environment Programme—see appendix A
World Bank (www.worldbank.org/)

The World Bank, known formally as the International Bank for Reconstruction and Development, was established in 1945 to help raise standards of living in developing countries by channeling financial resources to them from the industrialized countries. It has a large environment department, which helps to ensure that any potentially adverse environmental affects from World Bank–financed projects are addressed and to support the efforts of developing countries to implement sound environmental stewardship. It also is host to the Global Environment Facility.

U.N. World Tourism Organization (www.world-tourism.org/)

The U.N. World Tourism Organization provides a global forum for issues related to tourism policy and tourism logistics. In this role, the organization seeks to develop tourism in a responsible and sustainable manner. The 145 nations and seven territories that are members of the organization use this agency to increase the positive effects of tourism economically, socially, and culturally and to decrease the negative social and environmental effects of tourism.

INTERGOVERNMENTAL ORGANIZATIONS
Consultative Group on International Agricultural Research (www.cgiar.org/)

The Consultative Group on International Agricultural Research (CGIAR) is a strategic alliance of fifteen research centers covering various aspects of agriculture, including fisheries and forestry. Created in 1971, more than 8,500 CGIAR scientists and staff conduct research in more than 100 countries on crops, pro-environment farming techniques, fisheries, livestock and agricultural research services. Thirteen of the individual centers are headquartered in developing countries. Together, the international centers seek to achieve sustainable food security and reduce poverty in developing countries through scientific research and research-related activities. The centers generate global public goods that are freely available to all.

Global Biodiversity Information Facility (www.gbif.org/)

The Global Biodiversity Information Facility is an international organization that is working to make the world's biodiversity data freely accessible anywhere in the world. Its members include countries and international organizations that have agreed to share biodiversity data and to contribute to the development of increasingly effective mechanisms for making those data available via the Internet.

International Centre for International Integrated Mountain Development (www.icimod.org/)

Based in Kathmandu, Nepal, the International Centre for International Integrated Mountain Development promotes the development of economically and environmentally sound mountain ecosystems and improves the living standards of mountain populations, especially in the Hindu Kush-Himalaya region. It works at the interface between research and development and helps generate new mountain-specific knowledge that is relevant to mountain development. The information it produces is freely available through the Internet.

International Centre of Insect Physiology and Ecology (www.icipe.org/)

With headquarters in Nairobi, Kenya, the International Centre of Insect Physiology and Ecology focuses on improving the understanding of insects. It seeks to alleviate poverty, ensure food security, and improve the overall health status of peoples of the tropics by developing and extending management tools and strategies for both harmful and useful insects, while preserving the natural resource base through research and capacity building.

International Tropical Timber Organization (www.itto.or.jp/)

Based in Yokohama, Japan, the International Tropical Timber Organization (ITTO) was established in 1986 to promote the conservation and sustainable management, use and trade of tropical forest resources. Its fifty-nine members represent about 80 percent of the world's tropical forests and 90 percent of the global tropical timber trade. ITTO has also developed a strong forest conservation program that seeks to ensure the conservation of tropical forest resources.

International Union for Conservation of Nature and Natural Resources—see appendix A

Pan-European Biological and Landscape Diversity Strategy (www.strategyguide.org/)

In 1994, the Council of Europe agreed to develop the Pan-European Biological and Landscape Diversity Strategy as a European response to support implementation of the Convention on Biological Diversity. It provides a coordinating and unifying framework for strengthening and building on existing initiatives in Europe to integrate ecological considerations into all relevant socioeconomic sectors. It seeks to increase public participation in conservation activities, thereby building awareness and acceptance of the actions required to conserve biodiversity throughout Europe. The strategy is implemented through a series of five-year action plans that address the most urgent issues as agreed among the European governments.

South Pacific Regional Environment Programme (www.sprep.org/)

Based in Apia, Samoa, the South Pacific Regional Environment Programme promotes cooperation in the Pacific Islands region and provides assistance in order to protect and improve the environment and to ensure sustainable development for present and future generations. It seeks to enable the people of the Pacific Islands to plan, protect, and use their environment for sustainable development. With twenty-one Pacific Island member countries, it seeks to sustain the integrity of the ecosystems of the Pacific Islands region to support life and livelihoods.

Other Relevant Intergovernmental Organizations

- Center for International Forestry Research (www.cifor.cgiar.org/)

- Centro de Investigación Agrícola Tropical (www.ciatbo.org/)

- Conservation of Arctic Flora and Fauna Program (www.caff.is/)

- The Great Apes Survival Project (www.unep.org/grasp/)

- International Council for Game and Wildlife Conservation (www.cic-wildlife .org/)

- Biodiversity International (www.biodiversityinternational.org)

- International Union for the Protection of New Varieties of Plants (www.upov.int/)

- World Agroforestry Centre (www.worldagroforestry.org/)

- World Organization for Animal Health (www.oie.int/)

Appendix C

Nongovernmental Organizations Working to Conserve Biodiversity

Throughout the world, citizens have organized themselves into nongovernmental organizations (NGOs) to support the kinds of measures that have been discussed in this book. In this appendix, we briefly describe the major international NGOs that support conservation, sustainable use, and reduced consumption, as well as a selection of NGOs from various developing countries. In virtually all parts of the world, local NGOs are open for membership to those who would like to support such efforts. This is by no means an exhaustive list, but rather an indicative one that shows the depth and range of civil society organizations that are interested in improving both the quality of the environment and the quality of human life.

International NGOs

BirdLife International (www.birdlife.org/)

BirdLife International is a global partnership of conservation organizations that strives to conserve birds, their habitats, and global biodiversity, working with people toward sustainability in the use of natural resources. BirdLife Partners operate in more than 100 countries and territories worldwide and collaborate on regional work programs in every continent.

Botanical Gardens Conservation International (www.bgci.org.uk/)

The U.K.-based Botanical Gardens Conservation International (BGCI) brings together the 2,000 botanic gardens worldwide to form a community that works in partnership to conserve threatened indigenous species. Through the allocation of grants and the provision of knowledge and support, BGCI works with its member gardens to educate the more than two million park visitors each year and also influence governments on the importance of plant conservation. In addition to Europe, it has regional presence in North and South America, Russia, India, Japan, China, and Africa.

Conservation International (www.conservation.org/)

Conservation International (CI) is a U.S.-based, international nonprofit organization. CI applies innovations in science, economics, policy, and community participation to protect Earth's richest regions of plant and animal diversity in the biodiversity hotspots, high-biodiversity wilderness areas, and important marine regions around the globe. With headquarters in Washington, D.C., CI works in more than forty countries on four continents.

Earthwatch Institute (www.earthwatch.org/)

Earthwatch Institute is an international nonprofit organization, with 50,000 individual members worldwide. They act as a liaison between the scientific community, conservation and environmental organizations, policy makers, business, and the general public. Through their expeditions, Earthwatch Institute enables the general public to assist scientists in the field with research projects, thereby promoting the understanding and action necessary for a sustainable environment.

Fauna and Flora International (www.fauna-flora.org/)

Founded more than 100 years ago, the U.K.-based Fauna and Flora International (FFI) is the longest established international conservation body. FFI acts to conserve threatened species and ecosystems worldwide, choosing solutions that are sustainable, are based on sound science, and take account of human needs.

Friends of the Earth International (www.foei.org/)

Friends of the Earth International is a federation of autonomous environmental organizations from all over the world. Their 1.5 million members and supporters in seventy countries campaign on the most urgent environmental and social issues of the day, while simultaneously catalyzing a shift toward sustainable societies.

Global Footprint Network (www.footprintnetwork.org/)

A small but international, nonprofit start-up, the U.S.-based Global Footprint Network aims to advance the scientific rigor and practical application of the "Ecological Footprint," a tool that quantifies human demand on Nature and Nature's capacity to meet these demands. The network currently has forty-nine partner organizations spanning six continents and twenty-two countries that have adopted the standard as a tool for promoting ecological, social, and economic sustainability.

Greenpeace International (www.greenpeace.org/international/)

Greenpeace is a nonprofit organization with a presence in forty countries. As an independent organization, it relies on contributions from its 2.8 million supporters worldwide as well as foundation grants. Greenpeace focuses on promoting quality public debate on critical global issues threatening biodiversity and the environment, including climate change, forests, oceans, nuclear and toxic chemicals, and genetic engineering.

International Institute for Environment and Development (www.iied.org/)

International Institute for Environment and Development (IIED) is an international policy research institute that provides expertise to support more sustainable and equitable global development. Based in London, it has partnerships with many organizations in the developing world at local, national, and international levels. IIED helps vulnerable groups voice their interests in decision-making processes concerning policy and practice. Environmental sustainability is a core concern but not at the expense of people's livelihoods.

Nature Conservancy (www.nature.org/)

The Nature Conservancy is a large nonprofit organization with more than a million members and supporters. It is dedicated to preserving biodiversity and works to protect land and water areas of high importance. Though its work is primarily based in the United States, the Conservancy also has activities extending to twenty-seven countries worldwide.

TRAFFIC (www.traffic.org/)

TRAFFIC is a wildlife trade monitoring network created as a joint-program by the World Wildlife Fund and the International Union for Conservation of Nature and Natural Resources in the 1970s. TRAFFIC works to ensure that trade in wild plants and animals does not affect the conservation of nature, thereby supporting the Convention on International Trade in Endangered Species (CITES), which is supported by more than 150 countries. TRAFFIC has thirty national and regional offices worldwide that investigate, research, and conduct follow-up actions.

Wetlands International (www.wetlands.org/)

Wetlands International is a nonprofit organization dedicated to wetland conservation and sustainable management. Well-established networks of experts and close partnerships with key organizations have enabled Wetlands International to conduct activities in more than 120 countries worldwide. They also maintain the list of Ramsar sites—wetlands of international importance designated under the Ramsar Convention.

World Resources Institute (www.wri.org/)

The World Resources Institute (WRI) is an environmental think tank that provides governments, international organizations, and private businesses with objective and accurate information on resources and populations. WRI goes beyond research to offer proposals for policy and institutional change that will foster environmentally sound and socially equitable development. The institute's work is carried out by an interdisciplinary staff of scientists and experts assisted by a network of partners in fifty countries.

World Wide Fund for Nature International (www.panda.org/)

Established in 1961, the World Wide Fund for Nature International funds around 2,000 conservation projects in more than 100 countries. The organization has local to global presence with more than fifty country or regional offices (and so is discussed here rather than at the national level). Its aim is to stop the degradation of the planet's natural environment and to build a future in which humans live in harmony with nature by conserving the world's biological diversity, ensuring that the use of renewable natural resources is sustainable and promoting the reduction of pollution and wasteful consumption.

NATIONAL NGOS: AFRICA

Conservation Through Public Health (Uganda; www.ctph.org/)

Conservation Through Public Health is a grassroots nonprofit organization set up in 2002 by Ugandans dedicated to helping their country and region by promoting an integrated approach to wildlife conservation and human public health. They believe that the most efficient and cost effective way of controlling diseases that spread between wildlife, people, and their animals is through merging wildlife conservation with public health programs, particularly around protected areas.

Nigerian Conservation Foundation (www.africanconservation.org/ncftemp/)

The Nigerian Conservation Foundation is the country's foremost nongovernmental organization dedicated to the conservation of biodiversity. Formed in 1980, it aims to stop and eventually reverse the degradation of Nigeria's natural environment. Its three mission themes are biodiversity conservation, sustainable natural resources use, and reduction of pollution and wasteful consumption.

South African National Biodiversity Institute (www.nbi.ac.za)

The South African National Biodiversity Institute was established by an act of parlia ment but operates as a quasi-governmental organization. It carries out programs in conservation, research, education, and visitor services. It is responsible for running the Kirstenbosch National Botanical Garden and also works to conserve the Cape Floral Kingdom with its highly endemic flora.

Wildlife Clubs of Kenya (www.wildlifeclubsofkenya.org)

The Wildlife Clubs of Kenya was formed in 1968 by a group of Kenyan students. It has now greatly expanded and has an environmental education centre at Lake Victoria and at Lake Nakuru, with regional offices in various parts of the country. It also has guest houses, hostels, and camping facilities as a means of ensuring that Kenyan students have a chance to experience the biodiversity of the country. It also publishes a magazine distributed free to students; it is distributed to more than 2,000 school clubs in Kenya.

NATIONAL NGOS: MESO AND SOUTH AMERICA

ARCAS/Association for the Rescue and Conservation of Wildlife (Guatemala; www.arcasguatemala.com)

ARCAS is a nonprofit Guatemalan NGO formed in 1989 by a group of Guatemalan citizens. They initially built a rescue center to care for and rehabilitate wild animals that were being confiscated from the black market by the Guatemalan Government. However, it has since grown into one of the largest and most complex rescue centers in the world. ARCAS also takes on volunteers and leads ecotourism and educational excursions for local children.

CONABIO/National Commission for the Knowledge and Use of Biodiversity (Mexico; www.conabio.gob.mx/)

CONABIO is an interministerial commission created by the president of Mexico in 1992 in order to address problems relating to knowledge of biodiversity and its conservation. CONABIO develops, maintains, and updates the National System of Biodiversity Information as well as supports projects and studies focused on the knowledge and sustainable use of biodiversity. It shares knowledge on biological diversity with governmental institutions and other sectors, follows up on international agreements related to biodiversity, and provides services to the public.

FVA/Vitoria Amazonica Foundation (Brazil; www.fva.org.br)

FVA is a nongovernmental organization founded in 1990. Their activities extend throughout the Amazon region and include scientific, education, social, and economic alternative programs, with the objective of marrying environmental conservation with improvements in quality of life, particularly of the inhabitants of the Rio Negro Basin. FVA believes that an appropriate conservation model for Amazonia will be achieved only by sustainable natural resources use, rooted in respect for the cultural and ethnic diversity of the region.

FARN/Environment and Natural Resources Foundation (Argentina; www.farn.org.ar)

FARN, created in 1985, was established to promote sustainable development through policy, law, and institutional organization. It encourages citizen participation in conserving biodiversity. It has worked extensively on the issue of invasive alien species.

Fundación Pro-Sierra Nevada de Santa Marta/Foundation for the Sierra Nevada of Santa Marta (Colombia; www.prosierra.org)

This NGO has been established especially to conserve the people and the biodiversity of one of the biologically richest parts of Colombia, the Sierra Nevada of Santa Marta. It has a research station, an outreach program to the indigenous peoples, and various projects supporting biodiversity and sustainable development.

Instituto Nacional de Biodiversidad/National Institute for Biodiversity (Costa Rica; www.inbio.ac.cr)

The *Instituto Nacional de Biodiversidad* is a private research and biodiversity management centre, established in 1989 to support efforts to gather knowledge on the country's biological diversity and promote its sustainable use. The institute is based on the premise that the best way to conserve biodiversity is to utilize the opportunities it offers to improve the quality of life of human beings. It has carried out extensive work in inventory and monitoring, conservation, biodiversity informatics, bioprospecting, communications, and education.

ProNaturaleza—Fundación Peruana para la Conservación de la Naturaleza/Peruvian Foundation for the Conservation of Nature (www.pronaturaleza.org/english/index.htm)

Created in 1984, ProNaturaleza is a nonprofit organization dedicated to the conservation of the environment in Peru. They promote and execute conservation projects, develop management schemes for the sustainable use of natural resources, and support the creation of environmental awareness. They have also actively participated in the design of important environmental policies and contributed to the management of thirteen protected areas.

National NGOs: North America and the Caribbean

Canadian Wildlife Federation (www.cwf-fcf.org/)

The Canadian Wildlife Federation (CWF) is one of Canada's largest nonprofit, nongovernmental conservation organizations, with more than 300,000 members and supporters. Since 1962, CWF has advocated the protection of Canada's wild species and spaces through extensive education and information programs. CWF encourages a future in which Canadians may live in harmony with the natural order.

Environmental Foundation of Jamaica (www.efj.org.jm/)

The foundation was established under the program of the U.S. Enterprise of the America's initiative to provide civic organizations with assistance in projects relating to conservation and child development and welfare in Jamaica.

National Wildlife Federation (United States; www.nwf.org/)

Since 1936, the National Wildlife Federation has worked to protect American wildlife through education programs and conservation projects. They benefit from the support of over 4 million members and many affiliated wildlife organizations.

Sierra Club (United States and Canada; www.sierraclub.org/)

The Sierra Club is one of the oldest environmental organizations in North America and is nonprofit and member supported. The club promotes outdoor recreation, protection of communities, and conservation of the natural environment by influencing public policy decisions through the efforts of its 750,000 members.

National NGOs: Asia/Middle East

Bombay Natural History Society (India; www.bnhs.org/)

The Bombay Natural History Society is the largest NGO in the Indian subcontinent engaged in nature conservation research. In existence for over 120 years, the society promotes the conservation of India's natural wealth, protection of the environment, and sustainable use of natural resources for a balanced and healthy development through action based on research, education, and public awareness.

Centre for Biodiversity and Indigenous Knowledge (China; cbik.org/)

Established in 1995 as a membership nonprofit organization, the Centre for Biodiversity and Indigenous Knowledge is dedicated to biodiversity conservation and community livelihood development. As a participatory learning organization, they also produce documents on indigenous knowledge and technical innovations related to resource governance at community and watershed levels, which assists government works.

Haribon Foundation for the Conservation of Natural Resources (Philippines; www.haribon.org.ph)

Haribon started out as a bird-watching society in 1972. However, by 1983 it had evolved to become a membership organization dedicated to the conservation of Philippine biodiversity. It runs projects dedicated to working with people to save habitats, sites, and species throughout the Philippines.

King Mahendra Trust for Nature Conservation (Nepal; www.kmtnc.org.np)

Named after the late King Mahendra who created most of the protected areas in Nepal, the King Mahendra Trust for Nature Conservation was established as a nonprofit NGO with a mandate to work in nature conservation in Nepal. They have successfully undertaken more than eighty small and large projects on nature conservation, biodiversity protection, natural resource management, and sustainable rural development and are supported by a network of partners throughout the world.

Royal Society for the Conservation of Nature (Jordan; www.rscn.org.jo)

The Royal Society for the Conservation of Nature is an independent voluntary organization established in 1966 with the mission of protecting and managing the wildlife and wild places of Jordan. It is one of the few voluntary organizations in the Middle East with such a public service mandate. They manage the natural resources through the setting up of protected areas to safeguard the best wildlife and scenic areas as well as breeding endangered species to save them from extinction. In addition to enforcing governmental laws to protect wildlife and control illegal hunting, they also raise awareness for environmental issues through educational programs in more than 1,000 schools.

Sungi Development Foundation (Pakistan; www.sungi.org/)

Sungi Development Foundation was established in 1989 as a nonprofit and nongovernmental public interest organization by a group of socially and politically active individuals in Pakistan. They aim to bring about policy and institutional changes by mobilizing deprived and marginalized communities with a view to creating an environment in which communities at the local level may be able to transform their lives through the equitable and sustainable use of resources.

NATIONAL NGOs: OCEANIA

Australian Conservation Foundation (www.acfonline.org.au/)

The Australian Conservation Foundation (ACF) is committed to inspiring people to achieve a healthy environment for all Australians. Since the 1960s, ACF has promoted solutions for the environment through research, consultation, education, and partnerships. They work with the community, business, and government to protect, restore, and sustain the Australian environment.

Royal Forest and Bird Protection Society of New Zealand (www.forestandbird.org.nz/)

The Royal Forest and Bird Protection Society is New Zealand's largest national conservation organization with more than 40,000 members in fifty-four branches throughout the country. Its mission is to preserve, and protect the native plants, animals and natural features of New Zealand, and it is active on a wide range of conservation and environmental issues. Much of the on-the-ground conservation work of the society is done by volunteer branch members who run local campaigns and comprehensive conservation programs in their regions.

NATIONAL NGOs: EUROPE

Ecologistas en Acción/Ecologists in Action (Spain; www.ecologistasenaccion.org—in Spanish)

Ecologistas en Acción is a large environmental confederation bringing together more than 300 local ecological groups from all over Spain. It carries out awareness campaigns and also launches public or legal indictments against organizations that damage the environment. It also works to establish eco-friendly alternatives in affected areas.

Naturschutzbund Deutschland (Germany; www.nabu.de)

Established in 1899, Naturschutzbund Deutschland (NABU) is one of the oldest nature conservation NGOs and one of Germany's largest, with more than 385,000 members. Their core work revolves around conservation, but they also address other issues, including renewable energy, climate change, and transport.

Royal Society for Protection of Birds (United Kingdom; www.rspb.org.uk/)

The Royal Society for Protection of Birds (RSPB) was founded in 1889 and has since grown into Europe's largest wildlife conservation charity with over a million members; 150,000 of which are youth members. This U.K.-based charity works to secure a healthy environment for birds and wildlife through the establishing of 182 nature reserves. With a national network of 175 local groups and 13,000 volunteers, the RSPB also tackles related issues from conservation policy to education, from climate change down to damaging local developments.

Stichting Natuur en Milieu/The Netherlands Society for Nature and Environment (www.snm.nl/)

Stichting Natuur en Milieu is an independent organization consisting of eighty professionals all committed to securing a vigorous and healthy natural environment and working for a sustainable society. They publish research and conduct publicity campaigns, thereby stimulating discussion and debate and mobilizing public opinion, putting pressure on key policymakers in the field of nature and the environment.

The Wildlife Foundation of Khabarovsk (Russia; www.wf.ru/)

The Wildlife Foundation is a nongovernmental, nonprofit organization and was founded in 1993 by a small group of Russian ecologists and environmentalists. Their main objective is the preservation of the unique forests and biodiversity of the Russian Far East, including rare and endangered species such as the Amur tiger (Siberian tiger). The foundation also works with indigenous peoples and government officials to protect traditional land use in forested regions.

OTHER IMPORTANT NONGOVERNMENTAL ORGANIZATIONS

- BioNET International (www.bionet-intl.org)

- CAB International (www.cabi.org)

- The Center for International Environmental Law (www.ciel.org)

- Climate, Community and Biodiversity Alliance (www.celb.org/)

- Community Biodiversity Development and Conservation Programme (www.cbdcprogram.org)

- David Suzuki Foundation (www.davidsuzuki.org)

- Defenders of Wildlife (www.defenders.org)

- Edmonds Institute (www.edmonds-institute.org)

- Environment Liaison Centre International (www.elci.org)

- European Centre for Nature Conservation (www.ecnc.nl)

- Foundation for International Environmental Law and Development (www.field.org.uk)

- Global Invasive Species Programme (www.gisp.org)

- Indigenous Peoples' Secretariat on the U.N. Convention on Biological Diversity (Canada; www.cbin.ec.gc.ca/index.cfm?lang=e)

- International Centre for Trade and Sustainable Development (www.ictsd.org)

- International Coral Reef Action Network (www.icran.org)

- International Indigenous Forum on Biodiversity (www.iifb.net)

- International Institute for Environment and Development (www.iied.org)

- International Institute for Sustainable Development (www.iisd.org)

- International Scientific Council for Islands Development (www.insula.org)

- International Seed Trade Federation/International Association of Plant Breeders (www.worldseed.org)

- International Service for the Acquisition of Agri-biotech Applications (www.isaaa.org)

- SWAN International (www.swan.org.tw/eng/index.htm)

- Syzygy (www.syzygy.nl/)

- Tebtebba Foundation (www.tebtebba.org)

- Theme on Indigenous and Local Communities, Equity, and Protected Areas (World Commission on Protected Areas and Commission on Environmental, Economic, and Social Policy of the International Union for Conservation of Nature and Natural Resources; www.tilcepa.org)

- Wildlife Conservation Society (www.wcs.org)

- World Fish Center (www.worldfishcenter.org)

REFERENCES

CHAPTER 1

1. Whittaker, R.H., New concepts of kingdoms or organisms. Evolutionary relations are better represented by new classifications than by the traditional two kingdoms. *Science*, 1969;163(863):150–160.

2. Horner-Devine, M.C., K.M. Carney, and B.J.M. Bohannan, An ecological perspective on bacterial biodiversity. *Proceedings of the Royal Society of London Series B—Biological Sciences*, 2004;271(1535):113–122.

3. Kashefi, K., and D. Lovely, Extending the upper temperature limit for life. *Science*, 2003;301(5635):904.

4. Pimm, S.L., *The World According to Pimm: A Scientist Audits the Earth*. McGraw Hill, New York, 2001.

5. Pimm, S., et al., Human impacts on the rates of recent, present, and future bird extinctions. *Proceedings of the National Academy of Sciences of the USA*, 2006;103(29):10941–10946.

6. Millenium Ecosystem Assessment, *Ecosystems and Human Well-being: Synthesis Report*. Island Press, Washington, DC, 2005.

7. May, R.M., Biological diversity—differences between land and sea. *Philosophical Transactions of the Royal Society of London Series B—Biological Sciences*, 1994;343(1303):6.

8. Winston, J., Systematics and marine conservation, in *Systematics, Ecology and the Biodiversity Crisis*, N. Eldredge (editor). Columbia University Press, New York, 1992, 144–168.

9. Sogin, M.L., et al., Microbial diversity in the deep sea and the underexplored "rare biosphere." *Proceedings of the National Academy of Sciences of the USA*, 2006;103(32):12115–12120.

10. Venter, J.C., et al., Environmental genome shotgun sequencing of the Sargasso Sea. *Science*, 2004;304(5667):66–74.

11. Reaka-Kudla, M., The global biodiversity of coral reefs: A comparison with rain forests, in *Biodiversity II. Understanding and Protecting Our Biological Resources*, M. Reaka-Kudla, D. Wilson, and E. Wilson (editors). Joseph Henry Press, Washington DC, 1997, 83–108.

12. Snelgrove, P.V.R., Getting to the bottom of marine biodiversity: Sedimentary habitats—ocean bottoms are the most widespread habitat on Earth and support high biodiversity and key ecosystem services. *Bioscience*, 1999;49(2):9.

13. Grassle, J.F., and N.J. Maciolek, Deep-sea species richness—regional and local diversity estimates from quantitative bottom samples. *American Naturalist*, 1992;139(2):313–341.

14. van Dam, J., et al., Long-period astronomical forcing of mammal turnover. *Nature*, 2006;443:4.

15. Pimm, S.L., and P. Raven, Biodiversity—extinction by numbers. *Nature*, 2000;403(6772):843–845.

16. Thomas, C.D., et al., Extinction risk from climate change. *Nature*, 2004;427(6970):145–148.

17. Hughes, J.B., G.C. Daily, and P.R. Ehrlich, Population diversity: Its extent and extinction. *Science*, 1997;278(5338):689–692.

18. Hughes, J.B., G.C. Daily, and P.R. Ehrlich, The loss of population diversity and why it matters, in *Nature and Human Society*, P.H. Raven (editor). National Academies Press, Washington, DC, 1998, 71–83.

19. Olson, D.M., and E. Dinerstein, The global 200: A representation approach to conserving the earth's most biologically valuable ecoregions. *Conservation Biology*, 1998;12(3):502–515.

20. Loya, Y., et al., Coral bleaching: The winners and the losers. *Ecology Letters*, 2001;4(2):122–131.

21. Peacock, L., and S. Herrick, Responses of the willow beetle Phratora vulgatissima to genetically and spatially diverse Salix spp. plantations. *Journal of Applied Ecology*, 2000;37(5):10.

22. Hilborn, R., et al., Biocomplexity and fisheries sustainability. *Proceedings of the National Academy of Sciences of the USA*, 2003;100(11):4.

23. Jones, J.C., et al., Honey bee nest thermoregulation: Diversity promotes stability. *Science*, 2004;305(5682):402–404.

24. Nystrom, M., Redundancy and response diversity of functional groups: Implications for the resilience of coral reefs. *Ambio*, 2006;35(1):5.

Box 1.3

a. Häring, M., et al., Independent virus development outside a host. *Nature*, 2005;436:1101–1102.

Additional References

Brook, B.W., N.S. Sodhi, and P.K.L. Ng, Catastrophic extinctions follow deforestation in Singapore. *Nature*, 2003;424(6947):420–423.

Eckburg, P.B., et al., Diversity of the human intestinal microbial flora. *Science*, 2005;308(5728):1635–1638.

Hebert, P.D.N., et al., Ten species in one: DNA barcoding reveals cryptic species in the neotropical skipper butterfly Astraptes fulgerator. *Proceedings of the National Academy of Sciences of the USA*, 2004;101(41):14812–14817.

Levin-Zaidman, S., et al., Ringlike structure of the deinococcus radiodurans genome: A key to radioresistance? *Science*, 2003;299(5604):254–256.

Norse, E., *Global Marine Biological Diversity: A Strategy for Building Conservation into Decision Making*. Island Press, Washington DC, 1993.

Norse, E.A., and J.T. Carlton, World Wide Web buzz about biodiversity. *Conservation Biology*, 2003;17(6):1475–1476.

Pimm, S.L., et al., Bird extinctions in the central Pacific. *Philosophical Transactions of the Royal Society of London Series B—Biological Sciences*, 1994;344(1307):27–33.

Pimm, S.L., et al., The future of biodiversity. *Science*, 1995;269(5222):347–350.

Raven, P.H., and T. Williams (editors), *Nature and Human Society: The Quest for a Sustainable World*. National Academy Press, Washington, DC, 1997.

Roberts, F.A., and R.P. Darveau, Beneficial bacteria of the periodontium. *Periodontology 2000*, 2002;30:40–50.

Rothschild, L.J., and R.L. Mancinelli, Life in extreme environments. *Nature*, 2001;409(6823):1092–1101.

Whitman, W.B., D.C. Coleman, and W.J. Wiebe, Prokaryotes: The unseen majority. *Proceedings of the National Academy of Sciences of the USA*, 1998;95(12):6578–6583.

Woese, C.R., O. Kandler, and M.L. Wheelis, Towards a natural system of organisms—proposal for the domains Archaea, Bacteria, and Eucarya. *Proceedings of the National Academy of Sciences of the USA*, 1990;87(12):4576–4579.

Wu, W.M., et al., Pilot-scale in situ bioremedation of uranium in a highly contaminated aquifer. 2. Reduction of U(VI) and geochemical control of U(VI) bioavailability. *Environmental Science and Technology*, 2006;40(12):3986–3995.

CHAPTER 2

1. Pounds, J.A., et al., Widespread amphibian extinctions from epidemic disease driven by global warming. *Nature*, 2006;439(7073):161–167.

2. Schindler, D.W., et al., Consequences of climate warming and lake acidification for UV-B penetration in North American boreal lakes. *Nature*, 1996;379(6567):705–708.

3. Schindler, D.W., The cumulative effects of climate warming and other human stresses on Canadian freshwaters in the new millennium. *Canadian Journal of Fisheries and Aquatic Sciences*, 2001;58(1):18–29.

4. Kohler, J., et al., Effects of UV on carbon assimilation of phytoplankton in a mixed water column. *Aquatic Sciences*, 2001;63(3):294–309.

5. Tank, S.E., D.W. Schindler, and M.T. Arts, Direct and indirect effects of UV radiation on benthic communities: Epilithic food quality and invertebrate growth in four montane lakes. *Oikos*, 2003;103(3):651–667.

6. U.N. Environment Programme, *Global Environmental Outlook—3*. Earthscan, London, 2002.

7. Pimm, S.L., *The World According to Pimm: A Scientist Audits the Earth*. McGraw Hill, New York, 2001.

8. Heinrich, B., *The Trees in My Forest*. HarperCollins, New York, 1997.

9. UN Population Division, *World Population Prospects: The 2005 Revision*. United Nations, New York, 2005.

10. Pimm, S.L., et al., The future of biodiversity. *Science*, 1995;269(5222):347–350.

11. Pimm, S.L., and P. Raven, Biodiversity—extinction by numbers. *Nature*, 2000;403(6772):843 845.

12. Small, C., and R.J. Nicholls, A global analysis of human settlement in coastal zones. *Journal of Coastal Research*, 2003;19(3):584–599.

13. Pauly, D., and V. Christensen, Primary production required to sustain global fisheries. *Nature*, 1995;374(6519):255–257.

14. Ray, G., et al., Effects of global warming on the biodiversity of coastal marine zones, in *Global Warming and Biological Diversity*, R. Peters and T. Lovevoy (editors). Yale University, New Haven, CT, 1992, 91–104.

15. Roberts, C.M., et al., Marine biodiversity hotspots and conservation priorities for tropical reefs. *Science*, 2002;295(5558):1280–1284.

16. Birkeland, C., *Life and Death of Coral Reefs*. Springer, New York, 1997, 536.

17. Wilkinson, C. (editor), *Status of Coral Reefs of the World: 2004*, Vol. 1. Australian Institute of Marine Science, Townsville, Queensland, Australia, 2004.

18. Watling, L., and E.A. Norse, Disturbance of the seabed by mobile fishing gear: A comparison to forest clearcutting. *Conservation Biology*, 1998;12(6):1180–1197.

19. Roberts, C.M., Deep impact: The rising toll of fishing in the deep sea. *Trends in Ecology and Evolution*, 2002;17(5):242–245.

20. Roberts, J.M., A.J. Wheeler, and A. Freiwald, Reefs of the deep: The biology and geology of cold-water coral ecosystems. *Science*, 2006;312(5773):543–547.

21. Stein, B.A., L.S. Kutner, and J.S. Adams (editors), *Precious Heritage: The Status of Biodiversity in the United States*. Oxford University Press, New York, 2000.

22. Loh, J., et al., The Living Planet Index: Using species population time series to track trends in biodiversity. *Philosophical Transactions of the Royal Society of London Series B—Biological Sciences*, 2005;360(1454):289–295.

23. Stiassny, M.L., The medium is the message: Freshwater biodiversity in peril, in *The Living Planet in Crisis: Biodiversity Science and Policy*, J. Cracraft and F.T. Grifo (editors). Columbia University Press, New York, 1999, 53–71.

24. Chao, B.F., Anthropogenic impact on global geodynamics due to reservoir water impoundment. *Geophysical Research Letters*, 1995;22(24):3529–3532.

25. Office of Surface Mining, *Report on October 2000 Breakthrough at the Big Branch Slurry Impoundment*, Department of the Interior, 2000.

26. Mitchell, J.G., When mountains move. *National Geographic*, 2006;209(3):104–123.

27. Purcell, R.W., *Swift as a Shadow: Extinct and Endangered Animals*. New York: Houghton Mifflin, 2001.

28. Stewart, K.M., The African cherry (Prunus africana): Can lessons be learned from an over-exploited medicinal tree? *Journal of Ethnopharmacology*, 2003;89(1):3–13.

29. Reuters, Global Illegal Wildlife Trade Worth $10 Billion. 2006; available from www.reuters.com [cited August 2, 2006].

30. Eves, H.E., et al., *BCTF Factsheet: The Bushmeat Crisis in West and Central Africa*. Bushmeat Crisis Task Force, Washington, DC, 2002, 2.

31. Christian, M.D., et al., Severe acute respiratory syndrome. *Clinical Infectious Diseases*, 2004;38(10):1420–1427.

32. Lau, S.K.P., et al., Severe acute respiratory syndrome coronavirus-like virus in Chinese horseshoe bats. *Proceedings of the National Academy of Sciences of the USA*, 2005;102(39):14040–14045.

33. Burke, L., L. Selig, and M. Spalding, *Reefs at Risk in South-East Asia*. U.N. Environment Programme–World Conservation Monitoring Center, Cambridge, UK, 2002.

34. Morris, A.V., C.M. Roberts, and J.P. Hawkins, The threatened status of groupers (*Epinephelinae*). *Biodiversity and Conservation*, 2000;9(7):919–942.

35. Jackson, J.B.C., et al., Historical overfishing and the recent collapse of coastal ecosystems. *Science*, 2001;293(5530):629–638.

36. Roberts, C.M., Our shifting perspectives on the oceans. *Oryx*, 2003;37(2):166–177.

37. U.N. Environment Programme, Dugong: Status Report and Action Plans for Countries and Territories. 1999; available from www.unep.org/dewa/reports/dugongreport.asp [cited August 20, 2006].

38. Canadian Department of Fisheries and Oceans, Statistical Services: Commercial Landings. 2006; available from www.dfo-mpo.gc.ca/communic/Statistics/commercial/landings/index_e.htm [cited July 25, 2006].

39. Olsen, E.M., et al., Maturation trends indicative of rapid evolution preceded the collapse of northern cod. *Nature*, 2004;428(6986):932–935.

40. Lewison, R.L., S.A. Freeman, and L.B. Crowder, Quantifying the effects of fisheries on threatened species: The impact of pelagic longlines on loggerhead and leatherback sea turtles. *Ecology Letters*, 2004;7(3):221–231.

41. Myers, R.A., and B. Worm, Rapid worldwide depletion of predatory fish communities. *Nature*, 2003;423(6937):280–283.

42. Pauly, D., et al., Fishing down marine food webs. *Science*, 1998;279(5352):860–863.

43. UN Food and Agriculture Organization, *The State of World Fisheries and Agriculture*. FAO, Rome, 2004.

44. Hobbs, R.J., and L.F. Huenneke, Disturbance, diversity, and invasion—implications for conservations. *Conservation Biology*, 1992;6(3):324–337.

45. Mooney, H., et al. (editors), *Invasive Alien Species: A New Synthesis*. Island Press, Washington, DC, 2005.

46. Suarez, A.V., D.T. Bolger, and T.J. Case, Effects of fragmentation and invasion on native ant communities in coastal southern California. *Ecology*, 1998;79(6):2041–2056.

47. Isard, S.A., et al., Principles of the atmospheric pathway for invasive species applied to soybean rust. *Bioscience*, 2005;55(10):851–861.

48. Garrison, V.H., et al., African and Asian dust: From desert soils to coral reefs. *Bioscience*, 2003;53(5):469–480.

49. Emanuel, K., Increasing destructiveness of tropical cyclones over the past 30 years. *Nature*, 2005;436(7051):686–688.

50. Coote, T., and E. Loeve, From 61 species to five: Endemic tree snails of the Society Islands fall prey to an ill-judged biological control programme. *Oryx*, 2003;37(1):91–96.

51. Forseth, I.N., and A.F. Innis, Kudzu (Pueraria montana): History, physiology, and ecology combine to make a major ecosystem threat. *Critical Reviews In Plant Sciences*, 2004;23(5):401–413.

52. Wiles, G.J., et al., Impacts of the brown tree snake: Patterns of decline and species persistence in Guam's avifauna. *Conservation Biology*, 2003;17(5):1350–1360.

53. Roberts, L., Zebra mussel invasion threatens United-States waters. *Science*, 1990;249(4975):1370–1372.

54. Ludyanskiy, M.L., D. McDonald, and D. Macneill, Impact of the zebra mussel, a bivalve invader—Dreissena-polymorpha is rapidly colonizing hard surfaces throughout waterways of the United-States and Canada. *Bioscience*, 1993;43(8):533–544.

55. Stone, R., Science in Iran—attack of the killer jellies. *Science*, 2005;309(5742):1805–1806.

56. Global Ballast Water Management Programme, The Problem. 2006; available from globallast.imo.org [cited 24 July, 2006].

57. Barel, C.D.N., et al., Destruction of fisheries in Africa's Lakes. *Nature*, 1985;315(6014):19–20.

58. Albright, T.P., T.G. Moorhouse,, and J. McNabb, The rise and fall of water hyacinth in Lake Victoria and the Kagera River Basin, 1989–2001. *Journal of Aquatic Plant Management*, 2004;42:73.

59. Finley, J., S. Camazine, and M. Frazier, The epidemic of honey bee colony losses during the 1995–1996 season. *American Bee Journal*, 1996;136(11):805.

60. Anagnostakis, S.L., Chestnut blight—the classical problem of an introduced pathogen. *Mycologia*, 1987;79(1):23.

61. Kennedy, S., Morbillivirus infections in aquatic mammals. *Journal of Comparative Pathology*, 1998;119(3):201.

62. Harvell, D., et al., The rising tide of ocean diseases: Unsolved problems and research priorities. *Frontiers in Ecology and the Environment*, 2004;2(7):375–382.

63. U.N. Environment Programme, *GEO Yearbook 2003*. UNEP, Nairobi, 2003.

64. Bushaw-Newton, K.L., and K.G. Sellner, Harmful algal blooms, in *NOAA's State of the Coast Report*. National Ocean and Atmospheric Administration, Silver Spring, MD, 1999.

65. Sharpley, A.N., et al., *Agricultural Phosphorus and Eutrophication*, 2nd ed. U.S. Department of Agriculture, Agricultural Research Division, Washington, DC, 2003, 44.

66. Skaare, J.U., et al., Organochlorines in top predators at Svalbard—occurrence, levels and effects. *Toxicology Letters*, 2000;112:103.

67. Hamilton, G., Beluga corpses may hold ugly secrets of the river. *The Gazette* [Montreal], June 13, 1994, A1.

68. Guillette, L.J., et al., Developmental abnormalities of the gonad and abnormal sex-hormone concentrations in juvenile alligators from contaminated and control lakes in Florida. *Environmental Health Perspectives*, 1994;102(8):680.

69. Guillette, L.J., et al., Serum concentrations of various environmental contaminants and their relationship to sex steroid concentrations and phallus size in

juvenile American alligators. *Archives of Environmental Contamination and Toxicology*, 1999;36(4):447.

70. Oaks, J.L., et al., Diclofenac residues as the cause of vulture population decline in Pakistan. *Nature*, 2004;427(6975):630.

71. Green, R.E., et al., Diclofenac poisoning as a cause of vulture population declines across the Indian subcontinent. *Journal of Applied Ecology*, 2004;41(5):793.

72. Swan, G., et al., Removing the threat of diclofenac to critically endangered Asian vultures. *PLoS Biology*, 2006;4(3):395.

73. National Research Council, ed. *The Use of Drugs in Food Animals: Benefits and Risks*. National Research Council, Washington, DC, 1999, 253.

74. Kolpin, D.W., et al., Pharmaceuticals, hormones, and other organic wastewater contaminants in US streams, 1999–2000: A national reconnaissance. *Environmental Science and Technology*, 2002;36(6):1202.

75. Schultz, I.R., et al., Short-term exposure to 17 alpha-ethynylestradiol decreases the fertility of sexually maturing male rainbow trout (Oncorhynchus mykiss). *Environmental Toxicology and Chemistry*, 2003;22(6):1272.

76. Vajda, A.M., et al., Reproductive disruption and intersex in white suckers (Catostomus commersoni) downstream of an estrogen-containing municipal wastewater effluent. *Journal of Experimental Zoology Part A—Comparative Experimental Biology*, 2006;305A(2):188.

77. Fahrenthold, D.A., Male bass in Potomac producing eggs: Pollution suspected cause of anomaly in river's south branch. *Washington Post*, October 15, 2004, A01.

78. Henry, T.B., et al., Acute and chronic toxicity of five selective serotonin reuptake inhibitors in Ceriodaphnia dubia. *Environmental Toxicology and Chemistry*, 2004;23(9):2229–2233.

79. Caldeira, K., and M.E. Wickett, Anthropogenic carbon and ocean pH. *Nature*, 2003;425(6956):365.

80. Driscoll, C.T., et al., Acidic deposition in the northeastern United States: Sources and inputs, ecosystem effects, and management strategies. *Bioscience*, 2001;51(3):180.

81. European Commission: Working Group on Mercury, Ambient Air Pollution by Mercury (Hg) Position Paper. Office for Official Publications of the European Communities, Luxembourg, 2001.

82. Kiely, T., D. Donaldson, and A. Grube, *Pesticides Industry Sales and Usage: 2000 and 2001 Market Estimates*. U.S. Environmental Protection Agency, Washington, DC, 2004.

83. Goldstein, M.I., et al., Monitoring and assessment of Swainson's hawks in Argentina following restrictions on monocrotophos use, 1996–97. *Ecotoxicology*, 1999;8(3):215.

84. Hayes, T., et al., Herbicides: Feminization of male frogs in the wild. *Nature*, 2002;419(6910):895.

85. American Plastic Council, 2005 Sales and Production Data. 2005; available from www.americanplasticscouncil.org/s_apc/sec.asp?CID=296&DID=895 [cited July 30, 2006].

86. Derraik, J.G.B., The pollution of the marine environment by plastic debris: A review. *Marine Pollution Bulletin*, 2002;44(9):842.

87. U.S. Commission on Ocean Policy, An Ocean Blueprint for the 21st Century. 2004; available from oceancommission.gov [cited 30 July, 2006].

88. U.N. Environment Programme, *2004 World Environment Day Global Activity Report*. UNEP, Geneva, 2004.

89. Auman, H.J., et al., Plastic ingestion by laysan albatross chicks on Sand Island, Midway Atoll, in 1994 and 1995, in *Albatross Biology and Conservation*, Graham Robertson and R. Gales (editors). Surrey, Beatty & Sons, Chipping Norton, 1998, 239–244.

90. Thompson, R.C., et al., Lost at sea: Where is all the plastic? *Science*, 2004;304(5672):838.

91. Rex, M., et al., Arctic ozone loss and climate change. *Geophysical Research Letters*, 2004;31(4).

92. Whittle, C.A., and M.O. Johnston, Male-biased transmission of deleterious mutations to the progeny in Arabidopsis thaliana. *Proceedings of the National Academy of Sciences of the USA*, 2003;100(7):4055.

93. de Gruijl, F.R., et al., Health effects from stratospheric ozone depletion and interactions with climate change. *Photochemical and Photobiological Sciences*, 2003;2(1):16.

94. Dudley, J.P., et al., Effects of war and civil strife on wildlife and wildlife habitats. *Conservation Biology*, 2002;16(2):319–329.

95. Westing, A.H., Explosive remnants of war in the human environment. *Environmental Conservation*, 1996;23(4):283–285.

96. Westing, A., Herbicides in war: Past and present, in *Herbicides in War: The Long-Term Ecological and Human Consequences*, A. Westing (editor). Taylor & Francis, London, 1984, 1–24.

97. Dang, H., et al., Long-term changes in the mammalian fauna following herbicidal attack, in *Herbicides in War: The Long-Term Ecological and Human Consequences*, A. Westing (editor). Taylor & Francis, London, 1984, 49–51.

98. Brown, V.J., Battle scars—global conflicts and environmental health. *Environmental Health Perspectives*, 2004;112(17):A994–A1003.

99. Yablokova, O., Oil fires threaten migrating birds. *Moscow Times*, March 28, 2003, 1.

100. Richardson, C.J., and N.A. Hussain, Restoring the Garden of Eden: An ecological assessment of the marshes of Iraq. *Bioscience*, 2006;56(6):477–489.

101. Evans, M., The ecosystem, in *The Iraqi Marshlands. A Human and Environmental Study*, E. Nicholson and P. Clark (editors). Politico's, London, 2002, 201–219.

102. Plumptre, A., The impact of civil war on the conservation of protected areas in Rwanda. 2001; available from www.bsponline.org [cited September 2, 2006]

103. Hart, T., and R. Mwinjihali, Armed conflict and biodiversity in Sub-Saharan Africa: The case of the Democratic Republic of Congo. 2001; available from www.bpsonline.org [cited September 2, 2006].

104. Plumptre, A., et al., Support for Congolese conservationists. *Science*, 2000;288(5466):617.

105. Shambaugh, J., J. Oglethorpe, and R. Ham, Trampled Grass: Mitigating the Impacts of Armed Conflict on the Environment. 2001; available from www.bpsonline.com [cited September 2, 2006].

106. Baldwin, P., *The Endangered Species Act (ESA), Migratory Bird Treaty Act (MBTA), and Department of Defense (DOD) Readiness Activities: Background and Current Law*. Congressional Research Center, Washington DC, 2004.

107. U.S. Congress, HR. 1588 National Defense Authorization Act. 2004.

108. Intergovernmental Panel on Climate Change, Climate Change 2007: Synthesis Report. 2007; available from www.ipcc.ch/ [cited June 30, 2007].

109. Trenberth, K.E., and T.J. Hoar, El Nino and climate change. *Geophysical Research Letters*, 1997;24(23):3057–3060.

110. Trenberth, K.E., et al., Evolution of El Nino-Southern Oscillation and global atmospheric surface temperatures. *Journal of Geophysical Research Atmospheres*, 2002;107(D7–8).

111. Jones, P.D., and M.E. Mann, Climate over past millennia. *Reviews of Geophysics*, 2004;42(2).

112. Thomas, C.D., et al., Extinction risk from climate change. *Nature*, 2004;427(6970):145.

113. Parmesan, C., and G. Yohe, A globally coherent fingerprint of climate change impacts across natural systems. *Nature*, 2003;421(6918):37–42.

114. Root, T.L., et al., Human-modified temperatures induce species changes: Joint attribution. *Proceedings of the National Academy of Sciences of the USA*, 2005;102(21):7465–7469.

115. Parmesan, C., Climate and species' range. *Nature*, 1996;382(6594):765–766.

116. Grabherr, G., M. Gottfried, and H. Pauli, Climate effects on mountain plants. *Nature*, 1994;369(6480):448.

117. Perry, A.L., et al., Climate change and distribution shifts in marine fishes. *Science*, 2005;308(5730):1912–1915.

118. Huisman, J., et al., Reduced mixing generates oscillations and chaos in the oceanic deep chlorophyll maximum. *Nature*, 2006;439(7074):322–325.

119. Behrenfeld, M.J., et al., Climate-driven trends in contemporary ocean productivity. *Nature*, 2006;444(7120):752–755.

120. Doney, S.C., Oceanography: Plankton in a warmer world. *Nature*, 2006;444(7120):695–696.

121. Atkinson, A., et al., Long-term decline in krill stock and increase in salps within the Southern Ocean. *Nature*, 2004;432(7013):100–103.

122. Gross, L., As the Antarctic ice pack recedes, a fragile ecosystem hangs in the balance. *PLoS Biology*, 2005;3(6):1147–1147.

123. Parish, J.K., Dead birds don't lie, but what are they really indicating? Paper presented at the Pacific Seabird Group annual meeting, Girdwood, AK, 2006.

124. Tarling, G.A., and M.L. Johnson, Satiation gives krill that sinking feeling. *Current Biology*, 2006;16(3):R83–R84.

125. Ferguson, S.H., I. Stirling, and P. McLoughlin, Climate change and ringed seal (Phoca hispida) recruitment in western Hudson Bay. *Marine Mammal Science*, 2005;21(1):121–135.

126. Barbraud, C., and H. Weimerskirch, Emperor penguins and climate change. *Nature*, 2001;411(6834):183–186.

127. Sabine, C., et al., Current status and past trends of the global carbon cycle, in *The Global Carbon Cycle: Integrating Humans, Climate, and the Natural World*, C. Field and M. Raupach (editors). Island Press, Washington, DC, 2004, 17–44.

128. Doney, S., The dangers of ocean acidification. *Scientific American*, 2006;294(3):58–65.

129. Visser, M.E., L.J.M. Holleman, and P. Gienapp, Shifts in caterpillar biomass phenology due to climate change and its impact on the breeding biology of an insectivorous bird. *Oecologia*, 2006;147(1):164–172.

130. Bradshaw, W.E., and C.M. Holzapfel, Genetic shift in photoperiodic response correlated with global warming. *Proceedings of the National Academy of Sciences of the USA*, 2001;98(25):14509–14511.

131. Reale, D., et al., Genetic and plastic responses of a northern mammal to climate change. *Proceedings of the Royal Society of London Series B—Biological Sciences*, 2003;270(1515):591–596.

132. Berteaux, D., et al., Keeping pace with fast climate change: Can arctic life count on evolution? *Integrative and Comparative Biology*, 2004;44(2):140–151.

133. Blaustein, A.R., and A. Dobson, Extinctions: A message from the frogs. *Nature*, 2006;439(7073):143–144.

134. Berg, E.E., et al., Spruce beetle outbreaks on the Kenai Peninsula, Alaska, and Kluane National Park and Reserve, Yukon Territory: Relationship to summer temperatures and regional differences in disturbance regimes. *Forest Ecology and Management*, 2006;227(3):219–232.

135. Silliman, B.R., et al., Drought, snails, and large-scale die-off of southern US salt marshes. *Science*, 2005;310(5755):1803–1806.

Additional References

Brashares, J.S., et al., Bushmeat hunting, wildlife declines, and fish supply in West Africa. *Science*, 2004;306(5699):1180–1183.

David D. Doniger, testimony for the Hearing on the Status of Methyl Bromide Under the Clean Air Act and the Montreal Protocol, in *Subcommittee on Energy and Air Quality Committee on Energy and Commerce House of Representatives*. Washington, DC, 2003.

Dudgeon, D., et al., Freshwater biodiversity: Importance, threats, status and conservation challenges. *Biological Reviews*, 2006;81(2):163–182.

Freiwald, A., et al., *Cold Water Coral Reefs: Out of Sight—No Longer Out of Mind*, Cambridge, UK, U.N. Environment Programme–World Conservation Monitoring Center, 2004.

Geiser, D.M., et al., Cause of sea fan death in the West Indies. *Nature*, 1998;394(6689):137–138.

Grossman, D., Spring forward. *Scientific American*, 2004;290(1):84–91.

Gunderson, M.P., G.A. LeBlanc, and L.J. Guillette, Alterations in sexually dimorphic biotransformation of testosterone in juvenile American alligators (*Alligator mississippiensis*) from contaminated lakes. *Environmental Health Perspectives*, 2001;109(12):1257.

Marshall Jones, deputy director, U.S. Fish and Wildlife Service, testimony on the importation of exotic species and the impact on public health and safety, in *The Senate Committee on Environment and Public Works*. Washington, DC, 2003.

Kanyamibwa, S., Impact of war on conservation: Rwandan environment and wildlife in agony. *Biodiversity and Conservation*, 1998;7(11):1399–1406.

Martineau, D., et al., Levels of organochlorine chemicals in tissues of beluga whales (*Delphinapterus-Leucas*) from the St-Lawrence Estuary, Quebec, Canada. *Archives of Environmental Contamination and Toxicology*, 1987;16(2):137.

Martineau, D., et al., Pathology and toxicology of beluga whales from the St-Lawrence Estuary, Quebec, Canada—past, present and future. *Science of the Total Environment*, 1994;154(2–3):201.

Newman, D.J., G.M. Cragg, and K.M. Snader, The influence of natural products upon drug discovery. *Natural Product Reports*, 2000;17(3):215–234.

Orr, J.C., et al., Anthropogenic ocean acidification over the twenty-first century and its impact on calcifying organisms. *Nature*, 2005;437(7059):681.

Pauly, D., et al., Towards sustainability in world fisheries. *Nature*, 2002;418(6898):689–695.

Postel, S.L., G.C. Daily, and P.R. Ehrlich, Human appropriation of renewable fresh water. *Science*, 1996;271(5250):785–788.

Roberts, C.M., and J.P. Hawkins, Extinction risk in the sea. *Trends in Ecology and Evolution*, 1999;14(6):241–246.

Robinson, J.G., K.H. Redford, and E.L. Bennett, Conservation—wildlife harvest in logged tropical forests. *Science*, 1999;284(5414):595–596.

Sammataro, D., U. Gerson, and G. Needham, Parasitic mites of honey bees: Life history, implications, and impact. *Annual Review of Entomology*, 2000;45:519.

Schneider, S.H., and T.L. Root, Ecological implications of climate change will include surprises. *Biodiversity and Conservation*, 1996;5(9):1109–1119.

Siccama, T.G., M. Bliss, and H.W. Vogelmann, Decline of red spruce in the green mountains of Vermont. *Bulletin of the Torrey Botanical Club*, 1982;109(2):162.

Smith, V.H., G.D. Tilman, and J.C. Nekola, Eutrophication: Impacts of excess nutrient inputs on freshwater, marine, and terrestrial ecosystems. *Environmental Pollution*, 1999;100(1–3):179.

Socioeconomic Applications and Data Center, Gridded map of the world. 2005; available from sedac.ciesin.columbia.edu/gpw/index.jsp [cited July 20, 2006].

Van Loveren, H., et al., Contaminant-induced immunosuppression and mass mortalities among harbor seals. *Toxicology Letters*, 2000;112:319.

Visser, M.E., C. Both, and M.M. Lambrechts, Global climate change leads to mistimed avian reproduction, in *Birds and Climate Change*, Moller, M.P., W. Fiedler, and P. Berthold (editors). San Diego, Academic Press, 2004, 89–110.

Vorosmarty, C.J., et al., The storage and aging of continental runoff in large reservoir systems of the world. *Ambio*, 1997;26(4):210–219.

CHAPTER 3

1. Hassan, R., *Millennium Ecosystem Assessment*, Vol. 1, *Ecosystems and Human Well-being: Current State and Trends*, Washington, DC: Island Press, 2005.

2. U.N. Food and Agriculture Organization, The State of World Fisheries and Aquaculture (SOFIA) 2002. FAO, 2002; available from www.fao.org/docrep/005/y7300e/y7300e00.HTM [cited September 20, 2006].

3. U.N. Oceans, UN Atlas of the Oceans. USES: Fisheries and Aquaculture: Fisheries Statistics and Information: Trends: Consumption. 2006; available from www.oceansatlas.org [cited October 2, 2006].

4. Environmental Systems Research Institute, *World Countries 1995*. ESRI, Redlands, CA, 1996.

5. International Energy Agency, *Energy Statistics and Balances of Non-OECD Countries, 1994–95*. IEA, Paris, 1996.

6. Beckett, K.P., P.H. Freer-Smith, and G. Taylor, Urban woodlands: Their role in reducing the effects of particulate pollution. *Environmental Pollution*, 1998;99(3):347–360.

7. Wellburn, A.R., Atmospheric nitrogenous compounds and ozone—is NOx fixation by plants a possible solution? *New Phytologist*, 1998;139(1):5–9.

8. Pawlowska, M., and W. Stepniewski, Biochemical reduction of methane emission from landfills. *Environmental Engineering Science*, 2006;23(4):666–672.

9. Howarth, R., and D. Rielinger, Nitrogen from the atmosphere: Understanding and reducing a major cause of degradation of our coastal waters, in *Science and Policy Bulletin*, No. 8. Waquoit Bay National Estuarine Research Reserve, National Oceanic and Atmospheric Administration, Waquoit, Massachusetts, 2003.

10. Richardson, C., and N. Hussain, Restoring the Garden of Eden: An ecological assessment of the marshes of Iraq. *Bioscience*, 2006;56(6):477–489.

11. Kadlec, R., and D. Hey, Constructed wetlands for river water-quality improvement. *Water Science and Technology*, 1994;29(4):158–167.

12. Yoon, J., D. Oliver, and J. Shanks, Plant transformation pathways of energetic materials (RDX, TNT, DNTs), in *NABC Report 17: Agricultural Biotechnology: Beyond Food and Energy to Health and the Environment*. National Agricultural Biotechnology Council, Ithaca, NY, 2005.

13. Newell, R., Ecological changes in Chesapeake Bay: Are they the result of overharvesting the eastern oyster (Crassostrea virginica)? in *Understanding the Estuary: Advances*

in *Chesapeake Bay Research*, M. Lyncy and E. Krome (editors). Chesapeake Research Consortium, Edgewater, MD, 1988, 536–546.

14. Kemp, W., W.R. Boynton, J.E. Adolf, et al., Eutrophication of Chesapeake Bay: Historical trends and ecological interactions. *Marine Ecology Progress Series*, 2005;303:28.

15. Abramovitz, J., *Imperiled Waters, Impoverished Future: The Decline of Freshwater Ecosystems*. Worldwatch Institute, Washington, DC, 1996, 21.

16. Intergovernmental Panel on Climate Change, Climate Change 2007: Synthesis Report. 2007; available from www.ipcc.ch/ [cited June 30, 2007].

17. Theiling, C., The flood of 1993, in *Ecological Status and Trends of the Upper Mississippi River System*, R. Delaney and K. Lubinski (editors). U.S. Geological Survey Upper Midwest Environmental Sciences Center, La Crosse, WI, 1999.

18. Hey, D.L., and N.S. Philippi, Flood reduction through wetland restoration—the upper Mississippi River basin as a case-history. *Restoration Ecology*, 1995;3(1):4–17.

19. U.S. Geological Survey, National water summary on wetland resources, in *United States Geological Survey Water Supply Paper 2425*. USGS, Reston, VA, 1999.

20. Stokstad, E., After Katrina: Louisiana's wetlands struggle for survival. *Science*, 2005;310(5752):1264–1266.

21. Harrison, P., and F. Pearce, Part II: Ecosystems: Mountains, in *AAAS Atlas of Population and Environment*. American Association for the Advancement of Science, Washington, DC, 2001.

22. National Climatic Data Center, Mitch: The Deadliest Atlantic Hurricane since 1780. 2006; available from www.ncdc.noaa.gov/oa/reports/mitch/mitch.html#INFO [cited October 4, 2006].

23. Thompson, G., Guatemalan Village Overwhelmed by Task of Digging Out Hundreds of Dead from Mud. *New York Times*, October 10, 2005, A13.

24. Raven, P., and J. McNeely, Biological extinction: Its scope and meaning for us, in *Protection of Global Biodiversity: Converging Strategies*, L. Guruswarny and J. McNeely (editors). Duke University Press, Durham, NC, 1998.

25. Alongi, D.M., Present state and future of the world's mangrove forests. *Environmental Conservation*, 2002;29(3):331–349.

26. Harrison, P., and F. Pearce, Part II: Ecosystems: Mangroves, in *AAAS Atlas of Population and Environment*. American Association for the Advancement of Science, Washington, DC, 2001.

27. Arthur, E.L., et al., Phytoremediation—an overview. *Critical Reviews in Plant Sciences*, 2005;24(2):109–122.

28. Singh, O.V., et al., Phytoremediation: An overview of metallic ion decontamination from soil. *Applied Microbiology and Biotechnology*, 2003;61(5–6):405–412.

29. Soudek, P., R. Tykva, and T. Vanek, Laboratory analyses of Cs-137 uptake by sunflower, reed and poplar. *Chemosphere*, 2004;55(7):1081–1087.

30. Connell, S., and S. Al-Hamdani, Selected physiological responses of kudzu to different chromium concentrations. *Canadian Journal of Plant Science*, 2001;81(1):53–58.

31. Purvis, O., et al., Uranium biosorption by the lichen Trapelia involuta at a uranium mine. *Geomicrobiology Journal*, 2004;21(3):159–167.

32. Jellison, J., et al., The role of cations in the biodegradation of wood by the brown rot fungi. *International Biodeterioration and Biodegradation*, 1997;39(2–3):165–179.

33. Paszczynski, A., and R.L. Crawford, Potential for bioremediation of xenobiotic compounds by the white-rot fungus phanerochaete-chrysosporium. *Biotechnology Progress*, 1995;11(4):368–379.

34. Al Rmalli, S.W., et al., A biomaterial based approach for arsenic removal from water. *Journal of Environmental Monitoring*, 2005;7(4):279–282.

35. Rahman, M.M., et al., Chronic arsenic toxicity in Bangladesh and West Bengal, India—a review and commentary. *Journal of Toxicology—Clinical Toxicology*, 2001;39(7):683–700.

36. Adler, T., Botanical cleanup crews: Using plants to tackle polluted water and soil. *Science News*, 1996;150(3):42.

37. Revkin, A., New pollution tool: Toxic avengers with leaves. *New York Times*, March 6, 2001, F1.

38. de Lorenzo, V., Blueprint of an oil-eating bacterium. *Nature Biotechnology*, 2006;24(8):952–954.

39. He, J.Z., et al., Detoxification of vinyl chloride to ethene coupled to growth of an anaerobic bacterium. *Nature*, 2003;424(6944):62–65.

40. Haugland, R.A., et al., Degradation of the chlorinated phenoxyacetate herbicides 2,4-dichlorophenoxyacetic acid and 2,4,5-trichlorophenoxyacetic acid by pure and mixed bacterial cultures. *Applied and Environmental Microbiology*, 1990;56(5):1357–1362.

41. Mitra, J., et al., Bioremediation of DDT in soil by genetically improved strains of soil fungus Fusarium solani. *Biodegradation*, 2001;12(4):235–245.

42. Ralebitso, T.K., E. Senior, and H.W. van Verseveld, Microbial aspects of atrazine degradation in natural environments. *Biodegradation*, 2002;13(1):11–19.

43. Zhang, J.L., et al., Bioremediation of organophosphorus pesticides by surface-expressed carboxylesterase from mosquito on Escherichia coli. *Biotechnology Progress*, 2004;20(5):1567–1571.

44. Lloyd, J.R., and J.C. Renshaw, Bioremediation of radioactive waste: Radionuclide-microbe interactions in laboratory and field-scale studies. *Current Opinion in Biotechnology*, 2005;16(3):254–260.

45. Pimentel, D., Climate changes and food supply. *Forum for Applied Research and Public Policy*, 1993;8(4):54–60.

46. Baskin, Y., *The Work of Nature: How the Diversity of Life Sustains Us*. Island Press, Washington, DC, 1997.

47. Weeden, C.R., A. M. Shelton, and M. P. Hoffman, *Rodolia cardinalis* (Coleoptera: Coccinellidae) Vedalia Beetle, in *Biological Control: A Guide to Natural Enemies in North America*. Cornell University, Ithaca, NY, 2006; available from www.nysaes.cornell.edu/ent/biocontrol/predators/rodolia_cardinalis.html [cited September 12, 2007].

48. Nobre, C.A., P.J. Sellers, and J. Shukla, Amazonian deforestation and regional climate change. *Journal of Climate*, 1991;4(10):957–988.

49. Webb, T.J., et al., Forest cover-rainfall relationships in a biodiversity hotspot: The Atlantic forest of Brazil. *Ecological Applications*, 2005;15(6):1968–1983.

50. Werth, D., and R. Avissar, The regional evapotranspiration of the Amazon. *Journal of Hydrometeorology*, 2004;5(1):100–109.

51. Lavelle, P., et al., Nutrient cycling, in *Ecosystems and Human Well-being*. Island Press, Washington, DC, 2006.

52. Korner, C., Biosphere responses to CO_2 enrichment. *Ecological Applications*, 2000;10(6):1590–1619.

53. Cordell, H.K., et al., Outdoor Recreation Participation Trends, in *Outdoor Recreation in American Life: A National Assessment of Demand and Supply Trends*, K. Cordell (editor). Sagamore, Champaign, IL, 1999, 219–321.

54. Driver, B., Management of public outdoor recreation and related amenity resources for the benefits they provide, in *Outdoor Recreation in American Life: A National Assessment of Demand and Supply Trends*, K. Cordell (editor). Sagamore, Champaign, IL, 1999, 2–15.

55. Wilson, E., and S. Kellert (editors), *The Biophilia Hypothesis*. Island Press, Washington, DC, 1993.

56. Vitousek, P.M., et al., Human appropriation of the products of photosynthesis. *Bioscience*, 1986;36(6):368–373.

57. Imhoff, M.L., et al., Global patterns in human consumption of net primary production. *Nature*, 2004;429(6994):870–873.

58. Vitousek, P.M., et al., Human domination of Earth's ecosystems. *Science*, 1997;277(5325):494–499.

59. Petanidou, T., Sugars in Mediterranean floral nectars: An ecological and evolutionary approach. *Journal of Chemical Ecology*, 2005;31(5):23.

60. Raven, P., S.E. Eichhorn, and R.F. Every, *The Biology of Plants*. W.H. Freeman & Co., New York, 1998.

61. Janzen, D.H., and P.S. Martin, Neotropical anachronisms—the fruits the gomphotheres ate. *Science*, 1982;215(4528):19–27.

62. Daily, G.C., *Nature's Services*. Island Press, Washington, DC, 1997.

63. Pires, M., Watershed protection for a world city: The case of New York. *Land Use Policy*, 2004;21(2):161–175.

64. Postel, S., and B. Thompson, Watershed protection: Capturing the benefits of nature's water supply services. *Natural Resources Forum*, 2005;29(2):98–108.

65. Ricketts, T.H., et al., Economic value of tropical forest to coffee production. *Proceedings of the National Academy of Sciences of the USA*, 2004;101(34):12579–12582.

66. Ricketts, T.H., Tropical forest fragments enhance pollinator activity in nearby coffee crops. *Conservation Biology*, 2004;18(5):1262–1271.

67. Greathead, D.J. The multi-million dollar weevil that pollinates oil palm. *Antenna*, 1983;7:105–107.

68. Intergovernmental Panel on Climate Change, Climate Change 2007: The Physical Science Basis. Summary for Policymakers. 2007; available from www.ipcc.ch/SPM2feb07. pdf [cited March 20, 2007].

69. Cowling, S.A., et al., Contrasting simulated past and future responses of the Amazonian forest to atmospheric change. *Philosophical Transactions of the Royal Society of London Series B—Biological Sciences*, 2004;359(1443):539–547.

70. Wilkinson, C., ed. *Status of Coral Reefs of the World: 2004*, Vol. 1. Australian Institute of Marine Science, Townsville, Queensland, Australia, 2004.

71. Likens, G.E., et al., Recovery of a deforested ecosystem. *Science*, 1978;199(4328):492–496.

72. U.N. Food and Agriculture Organization, Global Forest Resources Assessment 2000; Main Report (Forestry Paper 140). FAO, Rome, 2001.

73. Dregne, H., Land degradation in the drylands. *Arid Land Research and Management*, 2002;16(2):92–125.

74. Erdelen, W., and M.H. Falougi, Preface, in *Combating Desertification: Freshwater Resources and the Rehabilitation of Degraded Areas in the Drylands*. U.N. Educational, Scientific and Cultural Organization (UNESCO), N'Djamena, Chad, 2000.

75. Tucker, C.J., and J.R.G. Townshend, Strategies for monitoring tropical deforestation using satellite data. *International Journal of Remote Sensing*, 2000;21(6–7):1461–1471.

76. *World Urbanization Prospects: The 2001 Revision*. U.N. Population Division, New York, 2001.

77. Lee, S., et al., Impact of urbanization on coastal wetland structure and function. *Austral Ecology*, 2006;31(2):149–163.

78. McNeill, J., *Something New under the sun: An Environmental History of the Twentieth Century*. Penguin, London, 2000, 421.

79. Dahl, T., *Wetlands—Losses in the United States, 1780's to 1980's*. U.S. Fish and Wildlife Service, Washington, DC, 1990, 13.

80. Dahl, T., and G. Allord, Technical Aspects of Wetlands: History of Wetlands in the Conterminous United States. 1997; available from water.usgs.gov/nwsum/WSP2425/history.html [cited October 15, 2006].

81. Aunan, K., T.K. Berntsen, and H.M. Seip, Surface ozone in China and its possible impact on agricultural crop yields. *Ambio*, 2000;29(6):294–301.

82. Paerl, H., Coastal eutrophication in relation to atmospheric nitrogen deposition—current perspectives. *Ophelia*, 1995;41:237–259.

83. Nriagu, J.O., et al., Saturation of ecosystems with toxic metals in Sudbury basin, Ontario, Canada. *Science of the Total Environment*, 1998;223(2–3):99–117.

84. Schoups, G., et al., Sustainability of irrigated agriculture in the San Joaquin Valley, California. *Proceedings of the National Academy of Sciences of the USA*, 2005;102(43):15352–15356.

85. Ghassemi, F., A.J. Jackman, and H.A. Nix, *Salinization of Land and Water Resources*. CAB International, Wallingford, UK, 1995.

86. Kolar, C., and D. Lodge, Freshwater non-indigenous species: Interactions with other global changes, in *Invasive Species in a Changing World*, H.A. Mooney and R. Hobbs (editors). Island Press, Washington, DC, 2000, 3–30.

Box 3.1

a. Horner-Devine, M.C., K.M. Carney, and B.J.M. Bohannan, An ecological perspective on bacterial biodiversity. *Proceedings of the Royal Society of London Series B—Biological Sciences*, 2004;271(1535):113–122.

b. Woese, C.R., Endosymbionts and mitochondrial origins. *Journal of Molecular Evolution*, 1977;10(2):93–96.

c. Zablen, L.B., et al., Phylogenetic origin of chloroplast and prokaryotic nature of its ribosomal-RNA. *Proceedings of the National Academy of Sciences of the USA*, 1975. 72(6):2418–2422.

d. Sagan, L., On the origin of mitosing cells. *Journal of Theoretical Biology*, 1967;14(3):255–274.

e. Alberts, B., et al., *Molecular Biology of the Cell*, 4th ed. Garland Science, Oxford, UK, 2002.

f. Kuroiwa, T., et al., Structure, function and evolution of the mitochondrial division apparatus. *Biochimica et Biophysica Acta—Molecular Cell Research*, 2006;1763(5–6):510–521.

g. Behar, D., et al., The matrilineal ancestry of Ashkenazi Jewry: Portrait of a recent founder event. *American Journal of Human Genetics*, 2006;78(3):487–497.

h. Hooper, L.V., and J.I. Gordon, Commensal host-bacterial relationships in the gut. *Science*, 2001;292(5519):1115–1118.

i. Nakajima, H., et al., Spatial distribution of bacterial phylotypes in the gut of the termite Reticulitermes speratus and the bacterial community colonizing the gut epithelium. *FEMS Microbiology Ecology*, 2005;54(2):247–255.

j. Furla, P., et al., The symbiotic anthozoan: A physiological chimera between alga and animal. *Integrative and Comparative Biology*, 2005;45(4):595–604.

k. Arnold, A., et al., Fungal endophytes limit pathogen damage in a tropical tree. *Proceedings of the National Academy of Sciences of the USA*, 2003;100(26):15649–15654.

l. Fredricks, D.N., Microbial ecology of human skin in health and disease. *Journal of Investigative Dermatology Symposium Proceedings*, 2001;6(3):167–169.

m. Gao, Z., et al., Molecular analysis of human forearm superficial skin bacterial biota. *Proceedings of the National Academy of Sciences of the USA*, 2007;104:2927–2932.

n. Roth, R.R., and W.D. James, Microbial ecology of the skin. *Annual Review of Microbiology*, 1988;42:441–464.

o. Aas, J.A., et al., Defining the normal bacterial flora of the oral cavity. *Journal of Clinical Microbiology*, 2005;43(11):5721–5732.

p. Centers for Disease Control and Prevention, *Third National Health and Nutrition Examination Survey (NHANES III)*. CDC, Atlanta, GA, 1994.

q. Behle, J., and P. Papapanou, Periodontal infections and atherosclerotic vascular disease: An update. *International Dental Journal*, 2006;56(4):256–262.

r. Pangsomboon, K., et al., Antibacterial activity of a bacteriocin from Lactobacillus paracasei HL32 against Porphyromonas gingivalis. *Archives of Oral Biology*, 2006;51(9):784–793.

s. Balakrishnan, M., R. Simmonds, and J. Tagg, Diverse activity spectra of bacteriocin-like inhibitory substances having activity against mutans streptococci. *Caries Research*, 2001;35(1):75–80.

t. Krisanaprakornkit, S., et al., Inducible expression of human b-defensin 2 by Fusobacterium nucleatum in oral epithelial cells: Multiple signaling pathways and role of commensal bacteria in innate immunity and the epithelial barrier. *Infection and Immunity*, 2000;68(5):2907–2915.

u. Kumar, P.S., et al., New bacterial species associated with chronic periodontitis. *Journal of Dental Research*, 2003;82(5):338–344.

v. Lepp, P.W., et al., Methanogenic archaea and human periodontal disease. *Proceedings of the National Academy of Sciences of the USA*, 2004;101(16):6176–6181.

w. Mager, D.L., et al., The salivary microbiota as a diagnostic indicator of oral cancer: A descriptive, non-randomized study of cancer-free and oral squamous cell carcinoma subjects. *Journal of Translational Medicine*, 2005;3:27; available from www.translational-medicine.com/content/3/1/27

x. Backhed, F., et al., Host-bacterial mutualism in the human intestine. *Science*, 2005;307(5717):1915–1920.

y. Breitbart, M., et al., Metagenomic analyses of an uncultured viral community from human feces. *Journal of Bacteriology*, 2003;185(20):6220–6223.

z. Eckburg, P.B., et al., Diversity of the human intestinal microbial flora. *Science*, 2005;308(5728):1635–1638.

aa. Hill, M., Intestinal flora and endogenous vitamin synthesis. *European Journal of Cancer Prevention*, 1997;6(Supplement 1):S43–S45.

bb. Travis, J., Gut check. *Science News*, 2003;163:344.

cc. Comstock, L., and M. Coyne, Bacteroides thetaiotaomicron: A dynamic, niche-adapted human symbiont. *Bioessays*, 2003;25(10):926–929.

dd. Xu, J., and J.I. Gordon, Honor thy symbionts. *Proceedings of the National Academy of Sciences of the USA*, 2003;100(18):10452–10459.

ee. Howell, S., et al., Antimicrobial polypeptides of the human colonic epithelium. *Peptides*, 2003;24:1763–1766.

ff. Stappenbeck, T., L. Hooper, and J. Gordon, Developmental regulation of intestinal angiogenesis by indigenous microbes via Paneth cells. *Proceedings of the National Academy of Sciences of the USA*, 2002;99(24):15451–15455.

Box 3.2

a. U.N. Environment Programme, After the Tsunami: Rapid Environmental Assessment. Protection of People and Property by Healthy Coastal Ecosystems Following the Earthquake and Tsunami of December 26, 2004. 2004; available from www.unep.org/tsunami/ [cited September 10, 2006].

b. Marris, E., Tsunami damage was enhanced by coral theft. *Nature*, 2005;436(7054):1071.

c. Papadopoulos, G.A., et al., The large tsunami of 26 December 2004: Field observations and eyewitnesses accounts from Sri Lanka, Maldives Is. and Thailand. *Earth Planets and Space*, 2006;58(2):233–241.

d. Kunkel, C.M., R.W. Hallberg, and M. Oppenheimer, Coral reefs reduce tsunami impact in model simulations. *Geophysical Research Letters*, 2006;33(23).

e. Wells, S., and V. Kapos, Coral reefs and mangroves: Implications from the tsunami one year on. *Oryx*, 2006;40(2):123–124.

Box 3.4

a. Pantzaris, T., Palm oil uses. *Oleagineux*, 1989;44(6):303–310.

b. Dennis, R., and C. Colfer, Impacts of land use and fire on the loss and degradation of lowland forest in 1983–2000 in East Kutai District, East Kalimantan, Indonesia. *Singapore Journal of Tropical Geography*, 2006;27(1):30–48.

c. Brown, E., and M. Jacobson, *Cruel Oil: How Palm Oil Harms Health, Rainforest and Wildlife*. Center for Science in the Public Interest, Washington, DC, 2005.

d. Rosenthal, E., Once a dream fuel, palm oil may be an eco-nightmare. *New York Times*, January 31, 2007, C1.

e. Hooijer, A., et al., Peat-CO_2, Assessment of CO_2 Emissions from Drained Peatlands in SE Asia. Report Q3943, WL/Delft Hydraulics, Delft, The Netherlands, 2006.

f. Grande, F., J.T. Anderson, and A. Keys, Comparison of effects of palmitic and stearic acids in diet on serum cholesterol in man. *American Journal of Clinical Nutrition*, 1970;23(9):1184–1193.

g. Clarke, R., et al., Dietary lipids and blood cholesterol: Quantitative meta-analysis of metabolic ward studies. *British Medical Journal*, 1997;314(7074):112–117.

h. Shang, J., and H. Kesteloot, Differences in all-cause, cardiovascular and cancer mortality between Hong Kong and Singapore: Role of Nutrition. *European Journal of Epidemiology*, 2001;17(5):469–477.

i. Uusitalo, U., et al., Fall in total cholesterol concentration over five years in association with changes in fatty acid composition of cooking oil in Mauritius: Cross sectional survey. *British Medical Journal*, 1996;313(7064):1044–1046.

j. Vega-Lopez, S., et al., Palm and partially hydrogenated soybean oils adversely alter lipoprotein profiles compared with soybean and canola oils in moderately hyperlipidemic subjects. *American Journal of Clinical Nutrition*, 2006;84(1):54–62.

Box 3.5

a. Costanza, R., et al., The value of the world's ecosystem services and natural capital. *Nature*, 1997;387(6630):253–260.

b. Daily, G.C., et al., Ecology—the value of nature and the nature of value. *Science*, 2000;289(5478):395–396.

c. Gatto, M., and G.A. De Leo, Pricing biodiversity and ecosystem services: The never-ending story. *Bioscience*, 2000;50(4):347–355.

d. Ludwig, D., Limitations of economic valuation of ecosystems. *Ecosystems*, 2000;3(1):31–35.

e. Cohen, J.E., and D. Tilman, Ecology—biosphere 2 and biodiversity: The lessons so far. *Science*, 1996;274(5290):1150–1151.

Additional References

Chigbo, F.E., R.W. Smith, and F.L. Shore, Uptake of arsenic, cadmium, lead and mercury from polluted waters by the water hyacinth Eichornia-Crassipes. *Environmental Pollution Series A—Ecological and Biological*, 1982;27(1):31–36.

Nriagu, J.O., H.K.T. Wong, and R.D. Coker, Deposition and chemistry of pollutant metals in lakes around the smelters at Sudbury, Ontario. *Environmental Science and Technology*, 1982;16(9):551–560.

Pimentel, D., et al., Conserving biological diversity in agricultural forestry systems—most biological diversity exists in human-managed ecosystems. *Bioscience*, 1992;42(5):354–362.

Rhodes, J., A. Kandiah, and A. Mashall, The Use of Saline Waters for Crop Production (FAO Irrigation and Drainage Paper 48). U.N. Food and Agriculture Organization, Rome, 1992.

U.S. Geological Survey, National Water Summary on Wetland Resources (USGS Water Supply Paper 2425). USGS, Reston, VA, 1999.

Swaminathan, M.S., Bio-diversity: An effective safety net against environmental pollution. *Environmental Pollution*, 2003;126(3):287–291.

U.N. Oceans, UN Atlas of the Oceans. USES: Fisheries and Aquaculture: Fisheries Statistics and Information: Trends: Consumption. 2006; available from www.oceansatlas.org [cited October 2, 2006].

World Heritage Committee, IUCN Evaluation of Nominations of Natural and Mixed Properties to the World Heritage List: Report to the World Heritage Committee. IUCN, Cairns, Australia, 2000.

CHAPTER 4

1. Grifo, F., et al., The origin of prescription drugs, in *Biodiversity and Human Health*, F. Grifo and J. Rosenthal (editors). Island Press, Washington, DC, 1997.

2. Newman, D.J., G.M. Cragg, and K.M. Snader, Natural products as sources of new drugs over the period 1981–2002. *Journal of Natural Products*, 2003;66(7):1022–1037.

3. Newman, D.J., G.M. Cragg, and K.M. Snader, The influence of natural products upon drug discovery. *Natural Product Reports*, 2000;17(3):215–234.

4. Efferth, T., Molecular pharmacology and pharmacogenomics of artemisinin and its derivatives in cancer cells. *Current Drug Targets*, 2006;7(4):407 421.

5. Lai, H., T. Sasaki, and N.P. Singh, Targeted treatment of cancer with artemisinin and artemisinin-tagged iron-carrying compounds. *Expert Opinion on Therapeutic Targets*, 2005;9(5):995–1007.

6. Martin, V.J., et al., Engineering a mevalonate pathway in Escherichia coli for production of terpenoids. *Nature Biotechnology*, 2003;21(7):796–802.

7. Ro, D.K., et al., Production of the antimalarial drug precursor artemisinic acid in engineered yeast. *Nature*, 2006;440(7086):940–943.

8. Farnsworth, N., Screening plants for new medicines, in *Biodiversity*, E. Wilson (editor). National Academy Press, Washington, DC, 1988.

9. Farnsworth, N.R., et al., Medicinal plants in therapy. *Bulletin of the World Health Organization*, 1985;63(6):965–981.

10. Jack, D.B., One hundred years of aspirin. *Lancet*, 1997;350(9075):437–439.

11. Huerta-Reyes, M., et al., HIV-1 inhibitory compounds from Calophyllum brasiliense leaves. *Biological and Pharmaceutical Bulletin*, 2004;27(9):1471–1475.

12. Link, K.P., Discovery of dicumarol and its sequels. *Circulation*, 1959;19(1):97–107.

13. Markwardt, F., Hirudin as alternative anticoagulant—a historical review. *Seminars in Thrombosis and Hemostasis*, 2002;28(5):405–413.

14. Whitaker, I.S., et al., Historical article: Hirudo medicinalis: Ancient origins of, and trends in the use of medicinal, leeches throughout history. *British Journal of Oral and Maxillofacial Surgery*, 2004;42(2):133–137.

15. Lee, A.Y.Y., and G.P. Vlasuk, Recombinant nematode anticoagulant protein c2 and other inhibitors targeting blood coagulation factor VIIa/tissue factor. *Journal of Internal Medicine*, 2003;254(4):313–321.

16. Geisbert, T.W., et al., Treatment of Ebola virus infection with a recombinant inhibitor of factor VIIa/tissue factor: A study in rhesus monkeys. *Lancet*, 2003;362(9400):1953–1958.

17. Hayashi, M.A.F., and A.C.M. Camargo, The bradykinin-potentiating peptides from venom gland and brain of Bothrops jararaca contain highly site specific inhibitors of the somatic angiotensin-converting enzyme. *Toxicon*, 2005;45(8):1163–1170.

18. Ondetti, M.A., From peptides to peptidases—a chronicle of drug discovery. *Annual Review of Pharmacology and Toxicology*, 1994;34:1–16.

19. Whitman, W.B., D.C. Coleman, and W.J. Wiebe, Prokaryotes: The unseen majority. *Proceedings of the National Academy of Sciences of the USA*, 1998;95(12):6578–6583.

20. Sogin, M.L., et al., Microbial diversity in the deep sea and the underexplored "rare biosphere." *Proceedings of the National Academy of Sciences of the USA*, 2006;103(32):12115–12120.

21. Bo, G., Giuseppe Brotzu and the discovery of cephalosporins. *Clinical Microbiology and Infection*, 2000;6:6–9.

22. Daniel, T.M., Selman Abraham Waksman and the discovery of streptomycin. *International Journal of Tuberculosis and Lung Disease*, 2005;9(2):120–122.

23. Boothe, J.H., et al., Tetracycline. *Journal of the American Chemical Society*, 1953;75(18):4621–4621.

24. Dimarco, A., M. Gaetani, and B. Scarpinato, Adriamycin (Nsc-123127)—a new antibiotic with antitumor activity. *Cancer Chemotherapy Reports Part 1*, 1969;53(1):33–37.

25. Brown, M.S., and J.L. Goldstein, A tribute to Akira Endo, discoverer of a "penicillin" for cholesterol. *Atherosclerosis Supplements*, 2004;5(3):13–16.

26. Endo, A., The discovery and development of HMG-CoA reductase inhibitors. *Journal of Lipid Research*, 1992;33(11):1569–1582.

27. Tobert, J.A., Lovastatin and beyond: The history of the HMG-CoA reductase inhibitors. *Nature Reviews Drug Discovery*, 2003;2(7):517–526.

28. Vaughan, C.J., Prevention of stroke and dementia with statins: Effects beyond lipid lowering. *American Journal of Cardiology*, 2003;91(4A):23B–29B.

29. Webster, A.C., et al., Target of rapamycin inhibitors (sirolimus and everolimus) for primary immunosuppression of kidney transplant recipients: A systematic review and meta-analysis of randomized trials. *Transplantation*, 2006;81(9):1234–1248.

30. Faivre, S., G. Kroemer, and E. Raymond, Current development of mTOR inhibitors as anticancer agents. *Nature Reviews Drug Discovery*, 2006;5(8):671–688.

31. Lebbe, C., et al., Sirolimus conversion for patients with posttransplant Kaposi's sarcoma. *American Journal of Transplantation*, 2006;6(9):2164–2168.

32. Sehgal, S.N., Sirolimus: Its discovery, biological properties, and mechanism of action. *Transplantation Proceedings*, 2003;35(3A):7S–14S.

33. Kastrati, A., et al., Sirolimus-eluting stents vs paclitaxel-eluting stents in patients with coronary artery disease—meta-analysis of randomized trials. *JAMA*, 2005;294(7):819–825.

34. Morice, M.C., et al., Sirolimus- vs paclitaxel-eluting stents in de novo coronary artery lesions: The REALITY trial: A randomized controlled trial. *JAMA*, 2006;295(8):895–904.

35. Windecker, S., et al., Sirolimus-eluting and paclitaxel-eluting stents for coronary revascularization. *New England Journal of Medicine*, 2005;353(7):653–662.

36. Iakovou, I., T. Schmidt, et al., Predictors, and outcome of thrombosis after successful implantation of drug-eluting stents. *JAMA*, 2005;293(17):2126–2130.

37. Challis, G.L., and D.A. Hopwood, Synergy and contingency as driving forces for the evolution of multiple secondary metabolite production by Streptomyces species. *Proceedings of the National Academy of Sciences of the USA*, 2003;100:14555–14561.

38. Engel, S., et al., Antimicrobial activities of extracts from tropical Atlantic marine plants against marine pathogens and saprophytes. *Marine Biology*, 2006;149(5):991–1002.

39. Hamann, M.T., et al., Kahalalides: Bioactive peptide from a marine mollusk Elysia rufescens and its algal diet Bryopsis sp. *Journal of Organic Chemistry*, 1996;61(19):6594–6600.

40. Smit, A.J., Medicinal and pharmaceutical uses of seaweed natural products: A review. *Journal of Applied Phycology*, 2004;16(4):245–262.

41. Population Council, *Carraguard: A Microbicide in Development*. Population Council, New York, 2004.

42. Kuznetsova, T.A., et al., Anticoagulant activity of fucoidan from brown algae Fucus evanescens of the Okhotsk Sea. *Bulletin of Experimental Biology and Medicine*, 2003;136(5):471–473.

43. Haneji, K., et al., Fucoidan extracted from Cladosiphon okamuranus tokida induces apoptosis of human T-cell leukemia virus type 1-infected T-cell lines and primary adult T-cell leukemia cells. *Nutrition and Cancer*, 2005;52(2):189–201.

44. Schaeffer, D.J., and V.S. Krylov, Anti-HIV activity of extracts and compounds from algae and cyanobacteria. *Ecotoxicology and Environmental Safety*, 2000;45(3):208–227.

45. Zeitlin, L., et al., Tests of vaginal microbicides in the mouse genital herpes model. *Contraception*, 1997;56(5):329–335.

46. Newman, D.J., and G.M. Cragg, Marine natural products and related compounds in clinical and advanced preclinical trials. *Journal of Natural Products*, 2004;67(8):1216–1238.

47. Kortmansky, J., and G.K. Schwartz, Bryostatin-1: A novel PKC inhibitor in clinical development. *Cancer Investigation*, 2003;21(6):924–936.

48. Fayette, J., et al., ET-743: A novel agent with activity in soft tissue sarcomas. *Oncologist*, 2005;10(10):827–832.

49. van Kesteren, C., et al., Yondelis (R) (trabectedin, ET-743): The development of an anticancer agent of marine origin. *Anti-cancer Drugs*, 2003;14(7):487–502.

50. Zelek, L., et al., A phase II study of Yondelis (R) (trabectedin, ET-743) as a 24-h continuous intravenous infusion in pretreated advanced breast cancer. *British Journal of Cancer*, 2006;94(11):1610–1614.

51. Honore, S., et al., Suppression of microtubule dynamics by discodermolide by a novel mechanism is associated with mitotic arrest and inhibition of tumor cell proliferation. *Molecular Cancer Therapeutics*, 2003;2(12):1303–1311.

52. Soriente, A., et al., Manoalide. *Current Medicinal Chemistry*, 1999;6(5):415–431.

53. Paulick, L.M., et al., Pseudopterosins: A potent natural anti-inflammatory agent extracted from Pseudopterogorgia elisabethae a soft coral. *Abstracts of Papers of the American Chemical Society*, 2000;219:U54–U54.

54. Lindquist, N., et al., Isolation and structure determination of diazonamide-A and diazonamide-B, unusual cytotoxic metabolites from the marine ascidian Diazona-chinensis. *Journal of the American Chemical Society*, 1991;113(6):2303–2304.

55. Gerwick, W.H., et al., Structure of curacin-A, a novel antimitotic, antiproliferative, and brine shrimp toxic natural product from the marine cyanobacterium Lyngbya-majuscula. *Journal of Organic Chemistry*, 1994;59(6):1243–1245.

56. Cragg, G.M., and D.J. Newman, Biodiversity: A continuing source of novel drug leads. *Pure and Applied Chemistry*, 2005;77(1):7–24.

57. Ireland, C., et al., Biomedical potential of marine natural products, in *Marine Biotechnology*, D. Attaway and O. Zaborsky (editors). Plenum Press, New York, 1993, 77–99.

58. Fenical, W., et al., New anticancer drugs from cultured and collected marine organisms. *Pharmaceutical Biology*, 2003;41:6–14.

59. Rusch, D.B., et al., The Sorcerer II global ocean sampling expedition: Northwest Atlantic through eastern tropical Pacific. *PLoS Biology*, 2007;5(3):e77.

60. Feling, R.H., et al., Salinosporamide A: A highly cytotoxic proteasome inhibitor from a novel microbial source, a marine bacterium of the new genus Salinospora. *Angewandte Chemie—International Edition*, 2003;42(3):355–357.

61. Piel, J., Bacterial symbionts: Prospects for the sustainable production of invertebrate-derived pharmaceuticals. *Current Medicinal Chemistry*, 2006;13(1):39–50.

62. Barrett, B., Medicinal properties of Echinacea: A critical review. *Phytomedicine*, 2003;10(1):66–86.

63. Flannery, M.A., From Rudbeckia to Echinacea: The emergence of the purple cone flower in modern therapeutics. *Pharmacy in History*, 1999;41(2):52–59.

64. Turner, R.B., et al., An evaluation of Echinacea angustifolia in experimental rhinovirus infections. *New England Journal of Medicine*, 2005;353(4):341–348.

65. Lawvere, S., and M.C. Mahoney, St. John's wort. *American Family Physician*, 2005;72(11):2249–2254.

66. Upton, R. *St. Johns Wort: Hypericum perforatum*. Botannical Booklet Series. 2005; available from www.herbalgram.org/default.asp?c=st_johns_wort [cited September 4, 2006].

67. Marks, L.S., and V.E. Tyler, Saw palmetto extract: Newest (and oldest) treatment alternative for men with symptomatic benign prostatic hyperplasia. *Urology*, 1999;53(3):457–461.

68. Gordon, A.E., and A.E. Shaughnessy, Saw palmetto for prostate disorders. *American Family Physician*, 2003;67(6):1281–1283.

69. Jang, M.S., et al., Cancer chemopreventive activity of resveratrol, a natural product derived from grapes. *Science*, 1997;275(5297):218–220.

70. Kris-Etherton, P.M., et al., Bioactive compounds in foods: Their role in the prevention of cardiovascular disease and cancer. *American Journal of Medicine*, 2002;113(Suppl 9B):71S–88S.

71. Howe, P., et al., Dietary intake of long-chain omega-3 polyunsaturated fatty acids: Contribution of meat sources. *Nutrition*, 2006;22(1):47–53.

72. Ponnampalam, E.N., N.J. Mann, and A.J. Sinclair, Effect of feeding systems on omega-3 fatty acids, conjugated linoleic acid and trans fatty acids in Australian beef cuts: Potential impact on human health. *Asia Pacific Journal of Clinical Nutrition*, 2006;15(1):21–29.

73. Breslow, J.L., n-3 fatty acids and cardiovascular disease. *American Journal of Clinical Nutrition*, 2006;83(6 Suppl):1477S–1482S.

74. Johnson, E.J., and E.J. Schaefer, Potential role of dietary n-3 fatty acids in the prevention of dementia and macular degeneration. *American Journal of Clinical Nutrition*, 2006;83(6 Suppl):1494S–1498S.

75. Furmidge, C., G. Brooks, and D. Gammon, *The Pyrethroid Insecticides: A Scientific Advance for Human Welfare?* Elsevier Press, New York, 1989.

76. Ware, G., and D. Whitacare, *An Introduction to Insecticides*, E. Radcliffe and W. Hutchinson (editors). University of Minnesota, St. Paul, 2004.

77. Bartlett, D.W., et al., The strobilurin fungicides. *Pest Management Science*, 2002;58(7):647–662.

Additional References

American Water Works Association, *Stats on Tap*. 2006; available from www.awwa.org/Advocacy/pressroom/STATS.cfm [cited August 24, 2006].

Buchmann, S., and B. Nabhan, *The Forgotten Pollinators*. Washington, DC, Island Press, 1997.

Center for a New American Dream, *Just the Facts: Junk Mail Facts and Figures*. 2003; available from www.newdream.org/junkmail/facts.php [cited August 24, 2006].

Cragg, G.M., and D.J. Newman, International collaboration in drug discovery and development from natural sources. *Pure and Applied Chemistry*, 2005;77(11):1923–1942.

Cragg, G., and D. Newman, Nature's bounty. *Chemistry in Britain*, 2001;37(1):22–26.

Cragg, G.M., and D.J. Newman, Plants as a source of anti-cancer agents. *Journal of Ethnopharmacology*, 2005;100(1–2):72–79.

Cragg, G.M., and D.J. Newman, Plants as a source of anti-cancer and anti-HIV agents. *Annals of Applied Biology*, 2003;143(2):127–133.

Dernain, A.L., From natural products discovery to commercialization: A success story. *Journal of Industrial Microbiology and Biotechnology*, 2006;33(7):486–495.

Energy Information Administration, *Annual Energy Review: Energy Overview, 1949–2005*. 2006; available from www.eia.doe.gov/emeu/aer/overview.html [cited August 23, 2006].

Goombridge, B., and M. Jenkins (editors), *Ecoagriculture: Strategies for Feeding the World and Conserving Wild Biodiversity*. World Conservation Monitoring Center, Cambridge, UK, 2000.

Hampton, T., Collaboration hopes microbe factories can supply key antimalaria drug. *JAMA*, 2005;293(7):785–787.

International Energy Agency, *Things That Go Blip in the Night. Standby Power and How to Limit It*. International Energy Agency, Paris, 2001.

International Union for the Conservation of Nature and Natural Resources, *Guidelines for Protected Area Management Categories*. IUCN, Gland, Switzerland, 1994.

International Union for the Conservation of Nature and Natural Resources, *Vision for Water and Nature: A World Strategy for Conservation and Sustainable Management of Water Resources in the 21st century*. IUCN, Gland, Switzerland, 2000.

Jensen, P.R., et al., Marine actinomycete diversity and natural product discovery *Antonie Van Leeuwenhoek International Journal of General and Molecular Microbiology*, 2005;87(1):43–48.

Kashman, Y., et al., HIV inhibitory natural-products. 7. The calanolides, a novel HIV-inhibitory class of coumarin derivatives from the tropical rain-forest tree, Calophyllum-lanigerum. *Journal of Medicinal Chemistry*, 1992;35(15):2735–2743.

Kinsley-Scott, T.R., and S.A. Norton, Useful plants of dermatology. VII: Cinchona and antimalarials. *Journal of the American Academy of Dermatology*, 2003;49(3):499–502.

Mueller, R.L., and S. Scheidt, History of drugs for thrombotic disease—discovery, development, and directions for the future. *Circulation*, 1994;89(1):432–449.

Murphy, D., Challenges to biological diversity in urban areas, in *Biodiversity*, E. Wilson (editor). National Academy of Sciences, Washington, DC, 71–76.

Newman, D.J., and R.T. Hill, New drugs from marine microbes: The tide is turning. *Journal of Industrial Microbiology and Biotechnology*, 2006;33(7):539–544.

Prescott-Allen, R., *The Well-being of Nations*. Island Press, Washington, DC, 2001.

Schwartsmann, G., et al., Anticancer drug discovery and development throughout the world. *Journal of Clinical Oncology*, 2002;20(18):47S–59S.

Tran, T.H., et al., A controlled trial of artemether or quinine in Vietnamese adults with severe falciparum malaria. *New England Journal of Medicine*, 1996;335(2):76–83.

Tran, T.H., et al., Dihydroartemisinin-piperaquine against multidrug-resistant Plasmodium falciparum malaria in Vietnam: Randomised clinical trial. *Lancet*, 2004;363(9402):18–22.

U.N. Environment Programme, *Global Environmental Outlook—3*. Earthscan, London, 2002.

Wyler, D.J., The ascent and decline of chloroquine. *JAMA*, 1984;251(18):2420–2422.

Yang, S.S., et al., Natural product-based anti-HIV drug discovery and development facilitated by the NCI developmental therapeutics program. *Journal of Natural Products*, 2001;64(2):265–277.

CHAPTER 5

1. Alberts, B., et al., *Molecular Biology of the Cell*, 4th ed. Garland Science, London, 2002.

2. Maehle, A., and U. Trohler, Animal experimentation from antiquity to the end of the eighteenth century, in *Vivisection in Historical Perspective*, N. Rupke (editor). Routledge, New York, 1990.

3. Nomura, K., et al., A bacterial virulence protein suppresses host innate immunity to cause plant disease. *Science*, 2006;313(5784):220–223.

4. Lolle, S.J., et al., Genome-wide non-mendelian inheritance of extra-genomic information in Arabidopsis. *Nature*, 2005;434(7032):505–509.

5. Piperno, D., and K. Flannery, The earliest archaeological maize (Zea mays L.) from highland Mexico: New accelerator mass spectrometry dates and their implications. *Proceedings of the National Academy of Sciences of the USA*, 2001;98(4):2101–2103.

6. Lorentz, C.P., et al., Primer on medical genomics part I: History of genetics and sequencing of the human genome. *Mayo Clinic Proceedings*, 2002;77(8):773–782.

7. Howard Hughes Medical Institute, *The Genes We Share*. 2006; available from www.hhmi.org/genesweshare [cited September 20, 2006].

8. Waterston, R.H., et al., Initial sequencing and comparative analysis of the mouse genome. *Nature*, 2002;420(6915):520–562.

9. Paigen, K., One hundred years of mouse genetics: An intellectual history. I. The classical period (1902–1980). *Genetics*, 2003;163(1):1–7.

10. Paigen, K., One hundred years of mouse genetics: An intellectual history. II. The molecular revolution (1981–2002) (reprinted from *New Yorker*, 2003). *Genetics*, 2003;163(4):1227–1235.

11. Macario, A.J.L., Heat-shock proteins and molecular chaperones—implications for pathogenesis, diagnostics, and therapeutics. *International Journal of Clinical and Laboratory Research*, 1995;25(2):59–70.

12. Bucciantini, M., et al., Inherent toxicity of aggregates implies a common mechanism for protein misfolding diseases. *Nature*, 2002;416(6880):507–511.

13. van Brabant, A.J., R. Stan, and N.A. Ellis, DNA helicases, genomic instability, and human genetic disease. *Annual Review of Genomics and Human Genetics*, 2000;1:409–459.

14. Brock, T.D., The value of basic research: Discovery of Thermus aquaticus and other extreme thermophiles. *Genetics*, 1997;146(4):1207–10.

15. Karow, J., The "other" genomes. *Scientific American*, 2000;283(1):53.

16. Wood, V., et al., The genome sequence of Schizosaccharomyces pombe. *Nature*, 2002;415(6874):871–880.

17. Holley, R.W., et al., Structure of a ribonucleic acid. *Science*, 1965;147(3664):1462–1465.

18. Hartwell, L.H., Yeast and cancer. *Bioscience Reports*, 2004;24(4–5):523–544.

19. Ankeny, R.A., The natural history of Caenorhabditis elegans research. *Nature Reviews: Genetics*, 2001;2(6):474–479.

20. Kimura, K.D., et al., Daf-2, an insulin receptor-like gene that regulates longevity and diapause in Caenorhabditis elegans. *Science*, 1997;277(5328):942–946.

21. Taub, J., et al., A cytosolic catalase is needed to extend adult lifespan in C-elegans daf-C and clk-1 mutants. *Nature*, 1999;399(6732):162–166.

22. Ingram, D.K., et al., Calorie restriction mimetics: An emerging research field. *Aging Cell*, 2006;5(2):97–108.

23. Dykxhoorn, D.M., and J. Lieberman, The silent revolution: RNA interference as basic biology, research tool, and therapeutic. *Annual Review of Medicine*, 2005;56:401–423.

24. Check, E., A crucial test. *Nature Medicine*, 2005;11(3):243–244.

25. Adams, M.D., et al., The genome sequence of *Drosophila melanogaster*. *Science*, 2000;287(5461):2185–2195.

26. Rubin, G.M., and E.B. Lewis, A brief history of *Drosophila*'s contributions to genome research. *Science*, 2000;287(5461):2216–2218.

27. Janeway, C.A., et al., *Immunobiology: The Immune System in Health and Disease*, 6th ed. Garland Science, London, 2004.

28. Kornberg, T.B., and M.A. Krasnow, The *Drosophila* genome sequence: Implications for biology and medicine. *Science*, 2000;287(5461):2218–2220.

29. Song, Y.H., *Drosophila* melanogaster: A model for the study of DNA damage checkpoint response. *Molecules and Cells*, 2005;19(2):167–179.

30. Driever, W., and M.C. Fishman, The zebrafish: Heritable disorders in transparent embryos. *Journal of Clinical Investigation*, 1996;97(8):1788–1794.

31. Raya, A., et al., The zebrafish as a model of heart regeneration. *Cloning Stem Cells*, 2004;6(4):345–351.

32. Fujisawa, T., Hydra regeneration and epitheliopeptides. *Developmental Dynamics*, 2003;226(2):182–189.

33. Holstein, T.W., E. Hobmayer, and U. Technau, Cnidarians: An evolutionarily conserved model system for regeneration? *Developmental Dynamics*, 2003;226(2):257–267.

34. Newmark, P.A., and A.S. Alvarado, Not your father's planarian: A classic model enters the era of functional genomics. *Nature Reviews Genetics*, 2002;3(3):210–219.

35. Orii, H., et al., The planarian HOM HOX homeobox genes (Plox) expressed along the anteroposterior axis. *Developmental Biology*, 1999;210(2):456–468.

36. Slack, J.M., Regeneration research today. *Developmental Dynamics*, 2003;226(2):162–166.

37. Poss, K.D., M.T. Keating, and A. Nechiporuk, Tales of regeneration in zebrafish. *Developmental Dynamics*, 2003;226(2):202–210.

38. Akimenko, M.A., et al., Old questions, new tools, and some answers to the mystery of fin regeneration. *Developmental Dynamics*, 2003;226(2):190–201.

39. Whitehead, G., S. Makino, C.-L. Lien, and M.T. Keating, Fgf20 is essential for initiating zebrafish fin regeneration. *Science*, 2005(310):1957–1960.

40. Poss, K.D., L.G. Wilson, and M.T. Keating, Heart regeneration in zebrafish. *Science*, 2002;298(5601):2188–2190.

41. Goss, R., *Deer Antlers. Regeneration, Function and Evolution*. Academic Press, New York, 1983.

42. Borgens, R.B., Mice regrow the tips of their foretoes. *Science*, 1982;217(4561):747–750.

43. Illingworth, C., Trapped fingers and amputated finger tips in children. *Journal of Pediatric Surgery*, 1974;9(6):853–858.

44. Heber-Katz, E., et al., The scarless heart and the MRL mouse. *Philosophical Transactions of the Royal Society of London Series B—Biological Sciences*, 2004;359(1445):785–793.

45. Harty, M., et al., Regeneration or scarring: An immunologic perspective. *Developmental Dynamics*, 2003;226(2):268–279.

46. Li, L.H., and T. Xie, Stem cell niche: Structure and function. *Annual Review of Cell and Developmental Biology*, 2005;21:605–631.

47. Parkinson's Disease Foundation, Ten Frequently-Asked Questions about Parkinson's Disease. 2005; available from www.pdf.org/Publications/factsheets/PDF_Fact_Sheet_1.0_Final.pdf [cited September 9, 2006].

48. Sayles, M., M. Jain, and R.A. Barker, The cellular repair of the brain in Parkinson's disease—past, present and future. *Transplant Immunology*, 2004;12(3–4):321–342.

49. Correia, A.S., et al., Stem cell-based therapy for Parkinson's disease. *Annals of Medicine*, 2005;37(7):487–498.

50. Bjorklund, A., et al., Neural transplantation for the treatment of Parkinson's disease. *Lancet Neurology*, 2003;2(7):437–45.

51. Steindler, D., Neural stem cells, scaffolds, and chaperones. *Nature Biotechnology*, 2002;20:1093–1095.

52. Ourednik, J., et al., Neural stem cells display an inherent mechanism for rescuing dysfunctional neurons. *Nature Biotechnology*, 2002;20(11):1103–1110.

53. Teng, Y.D., et al., Functional recovery following traumatic spinal cord injury mediated by a unique polymer scaffold seeded with neural stem cells. *Proceedings of the National Academy of Sciences of the USA*, 2002;99(5):3024–3029.

54. Kim, J.H., et al., Dopamine neurons derived from embryonic stem cells function in an animal model of Parkinson's disease. *Nature*, 2002;418(6893):50–56.

55. Takagi, Y., et al., Dopaminergic neurons generated from monkey embryonic stem cells function in a Parkinson primate model. *Journal of Clinical Investigation*, 2005;115(1):102–109.

56. Dufayet de la Tour, D., et al., {beta}-Cell differentiation from a human pancreatic cell line in vitro and in vivo. *Molecular Endocrinology*, 2001;15(3):476–483.

57. Assady, S., et al., Insulin production by human embryonic stem cells. *Diabetes*, 2001;50(8):1691–1697.

58. Zalzman, M., et al., Reversal of hyperglycemia in mice by using human expandable insulin-producing cells differentiated from fetal liver progenitor cells. *Proceedings of the National Academy of Sciences of the USA*, 2003;100(12):7253–7258.

59. International Diabetes Federation, *Diabetes Atlas*, 2nd ed. International Diabetes Federation, Brussels, 2003.

60. Nottebohm, F., Neuronal replacement in adult brain. *Brain Research Bulletin*, 2002;57(6):737–749.

61. Nottebohm, F., The road we travelled: Discovery, choreography, and significance of brain replaceable neurons. *Annals of the New York Academy of Sciences*, 2004;1016:628–658.

62. Gould, E., et al., Neurogenesis in the neocortex of adult primates. *Science*, 1999;286(5439):548–552.

63. Gould, E., et al., Adult-generated hippocampal and neocortical neurons in macaques have a transient existence. *Proceedings of the National Academy of Sciences of the USA*, 2001;98(19):10910–10917.

64. Eriksson, P.S., et al., Neurogenesis in the adult human hippocampus. *Nature Medicine*, 1998;4(11):1313–1317.

65. Gould, E., Stress, deprivation and adult neurogenesis, in *The Cognitive Neurosciences III*, M. Gazzaniga (editor). MIT Press, Cambridge, MA, 2004.

66. Hoffmann, J.A., et al., Phylogenetic perspectives in innate immunity. *Science*, 1999;284(5418):1313–1318.

67. Beutler, B., Inferences, questions and possibilities in toll-like receptor signalling. *Nature*, 2004;430(6996):257–263.

68. Dempsey, P.W., S.A. Vaidya, and G. Cheng, The art of war: Innate and adaptive immune responses. *Cellular and Molecular Life Sciences*, 2003;60(12):2604–2621.

69. Gura, T., Innate immunity: Ancient system gets new respect. *Science*, 2001;291(5511):2068–2071.

70. Steiner, H., et al., Sequence and specificity of two antibacterial proteins involved in insect immunity. *Nature*, 1981;292(5820):246–248.

71. Raj, P.A., and A.R. Dentino, Current status of defensins and their role in innate and adaptive immunity. *FEMS Microbiology Letters*, 2002;206(1):9–18.

72. Bulet, P., R. Stocklin, and L. Menin, Anti-microbial peptides: From invertebrates to vertebrates. *Immunological Reviews*, 2004;198:169–184.

73. Hancock, R.E.W., and M.G. Scott, The role of antimicrobial peptides in animal defenses. *Proceedings of the National Academy of Sciences of the USA*, 2000;97(16):8856–8861.

74. Zasloff, M., Antimicrobial peptides of multicellular organisms. *Nature*, 2002;415(6870):389–395.

75. Uzzell, T., et al., Hagfish intestinal antimicrobial peptides are ancient cathelicidins. *Peptides*, 2003;24(11):1655–1667.

76. Fudge, D.S., and J.M. Gosline, Molecular design of the alpha-keratin composite: Insights from a matrix-free model, hagfish slime threads. *Proceedings of the Royal Society of London Series B—Biological Sciences*, 2004;271(1536):291 299.

77. Powell, M., S. Kavanaugh, and S. Sower, Current knowledge of hagfish reproduction: Implications for fisheries management. *Integrative and Comparative Biology*, 2005;45(1):158–165.

78. Jorgensen, J.M., et al., *The Biology of Hagfishes*, 1st ed. Chapman & Hall, New York, 1998.

79. Pancer, Z., et al., Somatic diversification of variable lymphocyte receptors in the agnathan sea lamprey. *Nature*, 2004;430(6996):174–180.

Box 5.1

a. International Institute of Islamic Medicine, History of Islamic Medicine. 1998; available from www.iiim.org/iiimim.html [cited September 20, 2006].

b. Karolinska Institutet, Classical Islamic Medicine. 2006; available from www.mic.ki.se/Arab.html [cited September 20, 2006].

c. Margotta, R., *The Story of Medicine*. Golden Press, New York, 1968.

d. Menocal, M.R., *The Ornament of the World*. Little Brown & Company, Boston, 2002.

e. Nuland, S., *Maimonides*. Schocken Books, New York, 2005.

Box 5.3

a. Barnard, N.D., and S.R. Kaufman, Animal research is wasteful and misleading. *Scientific American*, 1997;276(2):80–82.

b. Robbins, F., The use of animals in biomedical research, in *Biomedical Research Involving Animals: Proposed International Guiding Principles*, Z. Bankowski and N. Howard-Jones (editors). Council for International Organizations of Medical Sciences, Geneva, 1984.

Additional References

Akira, S., S. Uematsu, and O. Takeuchi, Pathogen recognition and innate immunity. *Cell*, 2006;124(4):783–801.

Brockes, J.P., and A. Kumar, Appendage regeneration in adult vertebrates and implications for regenerative medicine. *Science*, 2005;310(5756):1919–1923.

Marchalonis, J.J., et al., Natural recognition repertoire and the evolutionary emergence of the combinatorial immune system. *FASEB J*, 2002;16(8):842–848.

Nielsen, C., *Animal Evolution: Interrelationships of the Living Phyla*. Oxford University Press, New York, 2001.

Sanchez Alvarado, A., Regeneration in the metazoans: Why does it happen? *Bioessays*, 2000;22(6):578–590.

Steindler, D.A., and D.W. Pincus, Stem cells and neuropoiesis in the adult human brain. *Lancet*, 2002;359(9311):1047–1054.

Yu, B.P., and H.Y. Chung, Adaptive mechanisms to oxidative stress during aging. *Mechanisms of Ageing and Development*, 2006;127(5):436–443.

CHAPTER 6

1. International Union for Conservation of Nature and Natural Resources, 2006 IUCN Red List of Threatened Species. 2006; available from www.redlist.org [cited August 1, 2006].

2. Mendelson, J.R., et al., Biodiversity—confronting amphibian declines and extinctions. *Science*, 2006;313(5783):48.

3. Stuart, S.N., et al., Status and trends of amphibian declines and extinctions worldwide. *Science*, 2004;306(5702):1783–1786.

4. Kiesecker, J., and A. Blaustein, Influences of egg laying behavior on pathogenic infection of amphibian eggs. *Conservation Biology*, 1997;11(1):6.

5. Señaris, J.C., C. DoNascimiento, and O. Villarreal, A new species of the genus Oreophrynella (Anura; Bufonidae) from the Guiana highlands. *Papéis Avulsos de Zoologia (São Paulo)*, 2005;45(6):61–67.

6. Dupuis, L.A., J.N.M. Smith, and F. Bunnell, Relation of terrestrial-breeding amphibian abundance to tree-stand age. *Conservation Biology*, 1995;9(3):645–653.

7. Petranka, J.W., M.E. Eldridge, and K.E. Haley, Effects of timber harvesting on southern Appalachian salamanders. *Conservation Biology*, 1993;7(2):363–377.

8. Bazilescu, I., Frog trade. TED Case Studies: An Online Journal, 1996; available from www.american.edu/TED/frogs/htm [cited October 7, 2006].

9. Knapp, R.A., and K.R. Matthews, Non-native fish introductions and the decline of the mountain yellow-legged frog from within protected areas. *Conservation Biology*, 2000;14(2):428–438.

10. Vredenburg, V.T., Reversing introduced species effects: Experimental removal of introduced fish leads to rapid recovery of a declining frog. *Proceedings of the National Academy of Sciences of the USA*, 2004;101(20):7646–7650.

11. Beebee, T.J.C., Amphibian breeding and climate. *Nature*, 1995;374(6519):219–220.

12. Blaustein, A.R., et al., Ultraviolet radiation, toxic chemicals and amphibian population declines. *Diversity and Distributions*, 2003;9(2):123–140.

13. Kiesecker, J.M., A.R. Blaustein, and L.K. Belden, Complex causes of amphibian population declines. *Nature*, 2001;410(6829):681–684.

14. Kiesecker, J., and A. Blaustein, Synergism between UV-B radiation and a pathogen magnifies amphibian embryo mortality in nature. *Proceedings of the National Academy of Sciences of the USA*, 1995;92:3.

15. Blaustein, A.R., et al., UV repair and resistance to solar UV-B in amphibian eggs—a link to population declines. *Proceedings of the National Academy of Sciences of the USA*, 1994;91(5):1791–1795.

16. Pierce, B., The effects of acid precipitation on amphibians. *Ecotoxicology*, 1993;2:65–77.

17. Bank, M.S., C.S. Loftin, and R.E. Jung, Mercury bioaccumulation in northern two-lined salamanders from streams in the northeastern United States. *Ecotoxicology*, 2005;14(1–2):181–191.

18. Reeder, A.L., et al., Intersexuality and the cricket frog decline: Historic and geographic trends. *Environmental Health Perspectives*, 2005;113(3):261–265.

19. Johnson, P.T.J., et al., Parasite (Ribeiroia ondatrae) infection linked to amphibian malformations in the western United States. *Ecological Monographs*, 2002;72(2):151–168.

20. Kiely, T., D. Donaldson, and A. Grube, *Pesticides Industry Sales and Usage: 2000 and 2001 Market Estimates*. U.S. Environmental Protection Agency, Washington, DC, 2004.

21. Relyea, R.A., The lethal impact of roundup on aquatic and terrestrial amphibians. *Ecological Applications*, 2005;15(4):1118–1124.

22. Hayes, T.B., et al., Hermaphroditic, demasculinized frogs after exposure to the herbicide atrazine at low ecologically relevant doses. *Proceedings of the National Academy of Sciences of the USA*, 2002;99(8):5476–5480.

23. Hayes, T., et al., Atrazine-induced hermaphroditism at 0.1 ppb in American leopard frogs (Rana pipiens): Laboratory and field evidence. *Environmental Health Perspectives*, 2003;111(4):568–575.

24. U.S. EPA, *Edition of Drinking Water Standards and Health Advisories* (EPA 822-R-02-038). U.S. EPA, Washington, DC, 2002.

25. Rohr, J.R., et al., Exposure, postexposure, and density-mediated effects of atrazine on amphibians: Breaking down net effects into their parts. *Environmental Health Perspectives*, 2006;114(1):46–50.

26. Hayes, T.B., et al., Pesticide mixtures, endocrine disruption, and amphibian declines: Are we underestimating the impact? *Environmental Health Perspectives*, 2006;114(Suppl 1):40–50.

27. Agency for Toxic Substances and Disease Registry, ToxFAQs for Atrazine. 2003; available from www.atsdr.cdc.gov/tfacts153.html [cited October 7, 2006].

28. Trenberth, K.E., et al., The changing character of precipitation. *Bulletin of the American Meteorological Society*, 2003;84(9):1205–1217.

29. Watson, R., and the Core Writing Team (editors), *Climate Change 2001: Synthesis Report*. Intergovernmental Panel on Climate Change, Geneva, 2001, 184.

30. Pounds, J.A., Climate and amphibian declines. *Nature*, 2001;410(6829):639–640.

31. Pounds, J.A., et al., Widespread amphibian extinctions from epidemic disease driven by global warming. *Nature*, 2006;439(7073):161–167.

32. Pounds, J.A., M.P.L. Fogden, and J.H. Campbell, Biological response to climate change on a tropical mountain. *Nature*, 1999;398(6728):611–615.

33. Barrio-Ameros, C., Atelopus mucubajiensis still survives in the Andes of Venezuela. *Froglog*, 2004;66:2–3.

34. Garcia-Perez, J., Survival of an undescribed Atelopus from the Venezuelan Andes. *Froglog*, 2005;68:2–3.

35. La Marca, E., et al., Catastrophic population declines and extinctions in neotropical harlequin frogs (Bufonidae: Atelopus). *Biotropica*, 2005;37(2):190–201.

36. Berger, L., et al., Chytridiomycosis causes amphibian mortality associated with population declines in the rain forests of Australia and Central America. *Proceedings of the National Academy of Sciences of the USA*, 1998;95(15):9031–9036.

37. Daszak, P., A.A. Cunningham, and A.D. Hyatt, Infectious disease and amphibian population declines. *Diversity and Distributions*, 2003;9(2):141–150.

38. Lips, K.R., et al., Emerging infectious disease and the loss of biodiversity in a neotropical amphibian community. *Proceedings of the National Academy of Sciences of the USA*, 2006;103(9):3165–3170.

39. Weldon, C., et al., Origin of the amphibian chytrid fungus. *Emerging Infectious Diseases,* 2004;10(12):2100–2105.

40. Garner, T., et al., The emerging amphibian pathogen Batrachochytrium dendrobatidis globally infects introduced populations of the North American bullfrog, Rana catesbeiana. *Biology Letters,* 2006;2:455–459.

41. Rollins-Smith, L.A., et al., Antimicrobial peptide defenses against pathogens associated with global amphibian declines. *Developmental and Comparative Immunology,* 2002;26(1):63–72.

42. Retallick, R.W.R., H. McCallum, and R. Speare, Endemic infection of the amphibian chytrid fungus in a frog community post-decline. *PLoS Biology,* 2004;2(11):1965–1971.

43. Ouellet, M., et al., Historical evidence of widespread chytrid infection in North American amphibian populations. *Conservation Biology,* 2005;19(5):1431–1440.

44. Harpole, D.N., and C.A. Haas, Effects of seven silvicultural treatments on terrestrial salamanders. *Forest Ecology and Management,* 1999;114(2–3):349–356.

45. Demaynadier, P.G., and M.L. Hunter, Effects of silvicultural edges on the distribution and abundance of amphibians in Maine. *Conservation Biology,* 1998;12(2):340–352.

46. Renda, T.G., Vittorio Erspamer: A true pioneer in the field of bioactive peptides. *Peptides,* 2000;21(11):1585–1586.

47. Daly, J.W., T.F. Spande, and H.M. Garraffo, Alkaloids from amphibian skin: A tabulation of over eight-hundred compounds. *Journal of Natural Products,* 2005;68(10):1556–1575.

48. Saporito, R.A., et al., Formicine ants: An arthropod source for the pumiliotoxin alkaloids of dendrobatid poison frogs. *Proceedings of the National Academy of Sciences of the USA,* 2004;101(21):8045–8050.

49. Dumbacher, J.P., et al., Melyrid beetles (Choresine): A putative source for the batrachotoxin alkaloids found in poison-dart frogs and toxic passerine birds. *Proceedings of the National Academy of Sciences of the USA,* 2004;101(45):15857–15860.

50. Daly, J.W., et al., Evidence for an enantioselective pumiliotoxin 7-hydroxylase in dendrobatid poison frogs of the genus Dendrobates. *Proceedings of the National Academy of Sciences of the USA,* 2003;100(19):11092–11097.

51. Yotsu-Yamashita, M., et al., The structure of zetekitoxin AB, a saxitoxin analog from the Panamanian golden frog Atelopus zeteki: A potent sodium-channel blocker. *Proceedings of the National Academy of Sciences of the USA,* 2004;101(13):4346–4351.

52. Daly, J., et al., A new class of cardiotonic agents: Structure-activity correlations for natural and synthetic analogues of the alkaloid pumiliotoxin B (8-hydroxy-8-methyl-6-alkylidene-1-azabicyclo [4.3.0] nonanes). *Journal of Medical Chemistry,* 1985;28:482–486.

53. Daly, J.W., Thirty years of discovering arthropod alkaloids in amphibian skin. *Journal of Natural Products,* 1998;61(1):162–172.

54. Fitch, R.W., et al., Bioassay-guided isolation of epiquinamide, a novel quinolizidine alkaloid and nicotinic agonist from an Ecuadoran poison frog, Epipedobates tricolor. *Journal of Natural Products,* 2003;66(10):1345–1350.

55. Conlon, J.M., The therapeutic potential of antimicrobial peptides from frog skin. *Reviews in Medical Microbiology,* 2004;15(1):17–25.

56. Jacob, L., and M. Zasloff, Potential therapeutic applications of magainins and other antimicrobial agents of animal origin. *Ciba Foundation Symposium,* 1994;186:197–216.

57. Nelson, E.A., et al., Systematic review of antimicrobial treatments for diabetic foot ulcers. *Diabetic Medicine,* 2006;23(4):348–359.

58. Giacometti, A., et al., In vitro activity of MSI-78 alone and in combination with antibiotics against bacteria responsible for bloodstream infections in neutropenic patients. *International Journal of Antimicrobial Agents,* 2005;26(3):235–240.

59. Erspamer, V., et al., Phyllomedusa skin—a huge factory and store-house of a variety of active peptides. *Peptides*, 1985;6:7–12.

60. Chen, T.B., L.J. Tang, and C. Shaw, Identification of three novel Phyllomedusa sauvagei dermaseptins (sVI-sVIII) by cloning from a skin secretion-derived cDNA library. *Regulatory Peptides*, 2003;116(1–3):139–146.

61. Mor, A., K. Hani, and P. Nicolas, The vertebrate peptide antibiotics dermaseptins have overlapping structural features but target specific microorganisms. *Journal of Biological Chemistry*, 1994;269(50):31635–31641.

62. Amiche, M., et al., Isolation of dermatoxin from frog skin, an antibacterial peptide encoded by a novel member of the dermaseptin genes family. *European Journal of Biochemistry*, 2000;267(14):4583–4592.

63. Pierre, T.N., et al., Phylloxin, a novel peptide antibiotic of the dermaseptin family of antimicrobial/opioid peptide precursors. *European Journal of Biochemistry*, 2000;267(2):370–378.

64. Altman, L., Doctors warn of powerful and resistant tuberculosis strain. *New York Times*, August 8, 2006, A4.

65. Gandhi, N.R., et al., Extensively drug-resistant tuberculosis as a cause of death in patients co-infected with tuberculosis and HIV in a rural area of South Africa. *Lancet*, 2006;368(9547):1575–1580.

66. Zasloff, M., Antimicrobial peptides of multicellular organisms. *Nature*, 2002;415(6870):389–395.

67. Chen, T.B., et al., Dermatoxin and phylloxin from the waxy monkey frog, Phyllomedusa sauvagei: Cloning of precursor cDNAs and structural characterization from lyophilized skin secretion. *Regulatory Peptides*, 2005;129(1–3):103–108.

68. Lazarus, L.H., and M. Attila, The toad, ugly and venomous, wears yet a precious jewel in his skin. *Progress in Neurobiology*, 1993;41(4):473–507.

69. Chen, T.B., et al., Bradykinins and their precursor cDNAs from the skin of the fire-bellied toad (Bombina orientalis). *Peptides*, 2002;23(9):1547–1555.

70. Graham, L.D., et al., Characterization of a protein-based adhesive elastomer secreted by the Australian frog Notaden bennetti. *Biomacromolecules*, 2005;6(6):3300–3312.

71. Nowak, R., Frog glue repairs damaged cartilage. New Scientist, 2004; available from www.newscientist.com/article/dn6492.html [cited August 1, 2006].

72. Myers, C., J. Daly, and B. Malkin, A dangerously toxic new frog (Phyllobates) used by Emberá Indians of western Colombia, with discussion of blowgun fabrication and dart poisoning. *Bulletin of the American Museum of Natural History*, 1978;161(2):307–366.

73. Albuquerque, E., J.W. Daly, and B. Witkop, Batrachotoxin—chemistry and pharmacology. *Science*, 1971;172(3987):995–1002.

74. Wang, S.Y., and G.K. Wang, Voltage-gated sodium channels as primary targets of diverse lipid-soluble neurotoxins. *Cellular Signalling*, 2003;15(2):151–159.

75. Anger, T., et al., Medicinal chemistry of neuronal voltage-gated sodium channel blockers. *Journal of Medicinal Chemistry*, 2001;44(2):115–137.

76. Thouveny, Y., and R. Tassava, Regeneration through phylogenesis, in *Cellular and Molecular Basis of Regeneration: From Invertebrates to Humans*, P. Ferretti and J. Geraudie (editors). John Wiley & Sons, Chichester, UK, 1997, 9–44.

77. Brockes, J.P., A. Kumar, and C.P. Velloso, Regeneration as an Evolutionary Variable. *Journal of Anatomy*, 2001;199(Pt 1–2):3–11.

78. Davis, B.M., et al., Time course of salamander spinal-cord regeneration and recovery of swimming—HRP retrograde pathway tracing and kinematic analysis. *Experimental Neurology*, 1990;108(3):198–213.

79. Whitehead, G.G., et al., Fgf20 is essential for initiating zebrafish fin regeneration. *Science*, 2005;310(5756):1957–1960.

80. Brockes, J.R., and A. Kumar, Plasticity and reprogramming of differentiated cells in amphibian regeneration. *Nature Reviews Molecular Cell Biology*, 2002;3(8):566–574.

81. Nye, H.L.D., et al., Regeneration of the urodele limb: A review. *Developmental Dynamics*, 2003;226(2):280–294.

82. Land, M., and M. Sheets, Heading in a new direction: Implications of the revised fate map for understanding Xenopus laevis development. *Developmental Biology*, 2006;296(1):16.

83. Layne, J.R., and M.C. First, Resumption of physiological functions in the wood frog (Rana-sylvatica) after freezing. *American Journal of Physiology*, 1991;261(1):R134–R137.

84. Future retreat of Arctic Sea ice will lower polar bear populations and limit their distribution. U.S. Geological Survey, Reston, VA, 2007; available from www.usgs.gov/newsroom/special/polar%5Fbears [cited September 12, 2007].

85. Wu, T.-L., D. DiLuciano, and B. Walsh, Bear parts trade. *TED Case Studies*, No. 5, 1997; available from www.american.edu/TED/bear.htm [cited October 7, 2006].

86. Derocher, A.E., N.J. Lunn, and I. Stirling, Polar bears in a warming climate. *Integrative and Comparative Biology*, 2004;44(2):163–176.

87. Krauss, C., Debate on global warming has polar bear hunting in its sights *New York Times*, May 27, 2006, 1.

88. Braune, B.M., et al., Persistent organic pollutants and mercury in marine biota of the Canadian Arctic: An overview of spatial and temporal trends. *Science of the Total Environment*, 2005;351:4–56.

89. Blais, J.M., et al., Arctic seabirds transport marine-derived contaminants. *Science*, 2005;309(5733):445.

90. Muir, D.C.G., et al., Brominated flame retardants in polar bears (Ursus maritimus) from Alaska, the Canadian Arctic, East Greenland, and Svalbard. *Environmental Science and Technology*, 2006;40(2):449–455.

91. Willerroider, M., Roaming polar bears reveal Arctic role of pollutants. *Nature*, 2003;426(6962):5.

92. Lie, E., et al., Does high organochlorine (OC) exposure impair the resistance to infection in polar bears (Ursus maritimus)? Part 1: Effect of OCs on the humoral immunity. *Journal of Toxicology and Environmental Health Part A—Current Issues*, 2004;67(7):555–582.

93. Sonne, C., et al., Is bone mineral composition disrupted by organochlorines in East Greenland polar bears (Ursus maritimus)? *Environmental Health Perspectives*, 2004;112(17):1711–1716.

94. Haave, M., et al., Polychlorinated biphenyls and reproductive hormones in female polar bears at Svalbard. *Environmental Health Perspectives*, 2003;111(4):431–436.

95. Oskam, I.C., et al., Organochlorines affect the major androgenic hormone, testosterone, in male polar bears (Ursus maritimus) at Svalbard. *Journal of Toxicology and Environmental Health Part A—Current Issues*, 2003;66(22):2119–2139.

96. Oskam, I.C., et al., Organochlorines affect the steroid hormone cortisol in free-ranging polar bears (Ursus maritimus) at Svalbard, Norway. *Journal of Toxicology and Environmental Health Part A—Current Issues*, 2004;67(12):959–977.

97. *Arctic Climate Impact Assessment: Impacts of a Warming Arctic.* Cambridge University Press, Cambridge, UK, 2004.

98. Regehr, E., et al., Population decline of polar bears in western Hudson Bay in relation to climatic warming, in *16th Biennial Conference on the Biology of Marine Mammals.* Society for Marine Mammology, San Diego, 2005.

99. Monnett, C., and J.S. Gleason, Observations of mortality associated with extended open-water swimming by polar bears in the Alaskan Beaufort Sea. *Polar Biology*, 2006;29(8):681–687.

100. Ferguson, S.H., I. Stirling, and P. McLoughlin, Climate change and ringed seal (Phoca hispida) recruitment in western Hudson Bay. *Marine Mammal Science*, 2005;21(1):121–135.

101. Beuers, U., Drug insight: Mechanisms and sites of action of ursodeoxycholic acid in cholestasis. *Nature Clinical Practice Gastroenterology and Hepatology*, 2006;3(6):318–328.

102. Shi, J., et al., Long-term effects of mid-dose ursodeoxycholic acid in primary biliary cirrhosis: A meta-analysis of randomized controlled trials. *American Journal of Gastroenterology*, 2006;101(7):1529–1538.

103. Nelson, R.A., Black bears and polar bears—still metabolic marvels. *Mayo Clinic Proceedings*, 1987;62(9):850–853.

104. Floyd, T., R.A. Nelson, and G.F. Wynne, Calcium and bone metabolic homeostasis in active and denning black bears (Ursus-americanus). *Clinical Orthopaedics and Related Research*, 1990(255):301–309.

105. Donahue, S.W., et al., Bone formation is not impaired by hibernation (disuse) in black bears Ursus americanus. *Journal of Experimental Biology*, 2003;206(23):4233–4239.

106. National Osteoporosis Foundation, Fast facts. 2006; available from www.nof.org/osteoporosis/diseasefacts.htm [cited October 8, 2006].

107. Sambrook, P., and C. Cooper, Osteoporosis. *Lancet*, 2006;367(9527):2010–2018.

108. Dennison, E., Z. Cole, and C. Cooper, Diagnosis and epidemiology of osteoporosis. *Current Opinion in Rheumatology*, 2005;17(4):456–461.

109. Johnell, O., and J.A. Kanis, An estimate of the worldwide prevalence, mortality and disability associated with hip fracture. *Osteoporosis International*, 2004;15(11):897–902.

110. Donahue, S.W., et al., Hibernating bears as a model for preventing disuse osteoporosis. *Journal of Biomechanics*, 2006;39(8):1480–1488.

111. National Kidney and Urologic Diseases Information Clearinghouse, Kidney and Urologic Diseases Statistics for the U.S. 2006; available from kidney.niddk.nih.gov/kudiseases/pubs/kustats/index.htm, [cited October 8, 2006].

112. Moeller, S., S. Gioberge, and G. Brown, ESRD patients in 2001: Overview of patients, treatment modalities and development trends. *Nephrology, Dialysis, Transplantation*, 2002;17(12):2071–2076.

113. Meichelboeck, W. ESRD 2005—a worldwide overview. Facts, figures, and trends, in *4th International Congress of the Vascular Access Society*. Karger, Berlin, 2005.

114. Nelson, R.A., et al., Nitrogen-metabolism in bears—urea metabolism in summer starvation and in winter sleep and role of urinary-bladder in water and nitrogen conservation. *Mayo Clinic Proceedings*, 1975;50(3):141–146.

115. Nelson, R.A., et al., Metabolism of bears before, during, and after winter sleep. *American Journal of Physiology*, 1973;224(2):491–496.

116. Nelson, R.A., Urea metabolism in hibernating black bear. *Kidney International*, 1978:S177–S179.

117. Palumbo, P., D.L. Wellik, N.A. Bagley, and R.A. Nelson, Insulin and glucagon responses in the hibernating black bear. *International Conference on Bear Research and Management*, 1983;5:291–296.

118. Ahlquist, D.A., et al., Glycerol metabolism in the hibernating black bear. *Journal of Comparative Physiology B—Biochemical Systemic and Environmental Physiology*, 1984;155(1):75–79.

119. Unterman, T., et al., Insulin-like growth factor-I (Igf-I) and binding-proteins (Igfbps) in the denning black bear (Ursus-americanus). *Clinical Research*, 1992;40(3):A712–A712.

120. Cattet, M., *Biochemical and Physiological Aspects of Obesity, High Fat Diet, and Prolonged Fasting in Free-Ranging Polar Bears*. University of Saskatchewan, Saskatoon, 2000.

121. Ogden, C.L., et al., Prevalence of overweight and obesity in the United States, 1999–2004. *JAMA*, 2006;295(13):1549–1555.

122. Allgot, B., et al., *Diabetes Atlas*. International Diabetes Federation, Brussels, 2003.

123. Groves, C., *Primate Taxonomy*. Smithsonian Institute Press, Washington, DC, 2001.

124. Pontes, A.R.M., A. Malta, and P.H. Asfora, A new species of capuchin monkey, genus Cebus Erxleben (Cebidae, Primates): Found at the very brink of extinction in the Pernambuco Endemism Centre. *Zootaxa*, 2006;(1200):1–12.

125. Thalmann, U. and T. Geissmann, New species of woolly lemur Avahi (Primates: Lemuriformes) in Bemaraha (central western Madagascar). *American Journal of Primatology*, 2005;67(3):371–376.

126. Sinha, A., et al., Macaca munzala: A new species from western Arunachal Pradesh, northeastern India. *International Journal of Primatology*, 2005;26(4):977–989.

127. Ehardt, C.L., T.M. Butynski, and T.R.B. Davenport, New species of monkey discovered in Tanzania: The critically endangered highland mangabey Lophocebus kipunji. *Oryx*, 2005;39(4):370–371.

128. Geissmann, T., Fact Sheet: What Are the gibbons? 2006; available from www.gibbons.de/main2/08teachtext/factgibbons/gibbonfact.html [cited October 8, 2006].

129. MacKinnon, K., Conservation status of Indonesian primates. *Primate Eye*, 1986;29:30–35.

130. U.N. Environment Programme, The Great Apes—The Road Ahead. GLOBIO (Global Methodology for Mapping Human Impacts on the Biosphere), 2002; available from www.globio.info/region/asia/ [cited October 10, 2006].

131. Rijksen, H.D., and E. Meijaard, *Our Vanishing Relative: The Status of Wild Orangutans at the Close of the Twentieth Century*. Kluwer Academic Publishers, Dordrecht, 1999.

132. Butynski, T.M., Africa's great apes, in *Great Apes and Humans: The Ethics of Coexistence*, B. Beck, et al. (editors). Smithsonian Institute Press, Washington, DC, 2001, 3–56.

133. Walsh, P.D., et al., Catastrophic ape decline in western equatorial Africa. *Nature*, 2003;422(6932):611–614.

134. Kalpers, J., et al., Gorillas in the crossfire: Population dynamics of the Virunga mountain gorillas over the past three decades. *Oryx*, 2003;37(3):326–337.

135. Brashares, J.S., et al., Bushmeat hunting, wildlife declines, and fish supply in West Africa. *Science*, 2004;306(5699):1180–1183.

136. Plumptre, A.J., et al., The effects of the Rwandan civil war on poaching of ungulates in the Parc National des Volcans. *Oryx*, 1997;31(4):265–273.

137. Inogwabini, B.I., et al., Status of large mammals in the mountain sector of Kahuzi-Biega National Park, Democratic Republic of Congo, in 1996. *African Journal of Ecology*, 2000;38(4):269–276.

138. Vogel, G., Conservation: Conflict in Congo threatens bonobos and rare gorillas. *Science*, 2000;287(5462):2386–2387.

139. Leroy, E.M., et al., Fruit bats as reservoirs of Ebola virus. *Nature*, 2005;438(7068):575–576.

140. Pourrut, X., et al., The natural history of Ebola virus in Africa. *Microbes and Infection*, 2005;7(7–8):1005–1014.

141. Leroy, E.M., et al., Multiple Ebola virus transmission events and rapid decline of central African wildlife. *Science*, 2004;303(5656):387–390.

142. Leendertz, F.H., et al., Anthrax kills wild chimpanzees in a tropical rainforest. *Nature*, 2004;430(6998):451–452.

143. Koff, R.S., Hepatitis vaccines: Recent advances. *International Journal for Parasitology*, 2003;33(5–6):517–523.

144. Cohen, J., The scientific challenge of hepatitis C. *Science*, 1999;285(5424):26–30.

145. Bukh, J., et al., Studies of hepatitis C virus in chimpanzees and their importance for vaccine development. *Intervirology*, 2001;44(2–3):132–142.

146. Abrignani, S., M. Houghton, and H.H. Hsu, Perspectives for a vaccine against hepatitis C virus. *Journal of Hepatology*, 1999;31:259–263.

147. Esumi, M., et al., Experimental vaccine activities of recombinant E1 and E2 glycoproteins and hypervariable region 1 peptides of hepatitis C virus in chimpanzees. *Archives of Virology*, 1999;144(5):973–980.

148. Sibal, L.R., and K.J. Samson, Nonhuman primates: A critical role in current disease research. *Ilar Journal*, 2001;42(2):74–84.

149. Aggarwal, R., and P. Ranjan, Preventing and treating hepatitis B infection. *BMJ*, 2004;329(7474):1080–1086.

150. Bertoni, R., et al., Human class I supertypes and CTL repertoires extend to chimpanzees. *Journal of Immunology*, 1998;161(8):4447–4455.

151. Guidotti, L.G., et al., Viral clearance without destruction of infected cells during acute HBV infection. *Science*, 1999;284(5415):825–829.

152. Alonso, P.L., et al., Efficacy of the RTS,S/AS02A vaccine against Plasmodium falciparum infection and disease in young African children: Randomised controlled trial. *Lancet*, 2004;364(9443):1411–1420.

153. Heppner, D.G., et al., Towards an RTS,S-based, multi-stage, multi-antigen falciparum malaria. Progress at the Walter Reed Army Institute of Research. *Vaccine*, 2005;23(17–18):2243–2250.

154. Walsh, D.S., et al., Heterologous prime-boost immunization in rhesus macaques by two, optimally spaced particle-mediated epidermal deliveries of Plasmodium falciparum circumsporozoite protein-encoding DNA, followed by intramuscular RTS,S/AS02A. *Vaccine*, 2006;24(19):4167–4178.

155. World Health Organization, Marburg haemorrhagic fever in Angola—update 25. Epidemic and Alert and Response 2005; available from www.who.int/csr/don/2005_08_24/en/index.html [cited October 10, 2006].

156. World Health Organization, Ebola haemorrhagic fever. Fact sheet no. 103. 2004; available from www.who.int/mediacentre/factsheets/fs103/en/index.html [cited August 28, 2007].

157. Formenty, P., et al., Ebola virus outbreak among wild chimpanzees living in a rain forest of Cote d'Ivoire. *Journal of Infectious Diseases*, 1999;179:S120–S126.

158. Formenty, P., et al., Human infection due to Ebola virus, subtype Cote d'Ivoire: Clinical and biologic presentation. *Journal of Infectious Diseases*, 1999;179:S48–S53.

159. Rouquet, P., et al., Wild animal mortality monitoring and human Ebola outbreaks, Gabon and Republic of Congo, 2001–2003. *Emerging Infectious Diseases*, 2005;11(2):283–290.

160. Jaax, N.K., et al., Lethal experimental infection of rhesus monkeys with Ebola-Zaire (Mayinga) virus by the oral and conjunctival route of exposure. *Archives of Pathology and Laboratory Medicine*, 1996;120(2):140–155.

161. Johnson, E., et al., Lethal experimental infections of rhesus monkeys by aerosolized Ebola virus. *International Journal of Experimental Pathology*, 1995;76(4):227–236.

162. Jones, S.M., et al., Live attenuated recombinant vaccine protects nonhuman primates against Ebola and Marburg viruses. *Nature Medicine*, 2005;11(7):786–790.

163. Hopes and fears for rotavirus vaccines. *Lancet*, 2004;365(9455):190.

164. Glass, R.I., et al., Rotavirus vaccines: Current prospects and future challenges. *Lancet*, 2006;368(9532):323–332.

165. Kapikian, A., and R. Canock, Rotaviruses, in *Fields Virology*, B. Fields, D. Knipe, and P. Howley (editors). Lippincott-Raven, Philadelphia, 1996, 1657–1708.

166. Jiang, B.M., et al., Prevalence of rotavirus and norovirus antibodies in non-human primates. *Journal of Medical Primatology*, 2004;33(1):30–33.

167. Dennehy, P.H., Rotavirus vaccines: An update. *Pediatric Infectious Disease Journal*, 2006;25(9):839–840.

168. Joint U.N. Programme on HIV/AIDS, *2006 Report on the Global AIDS Epidemic: May 2006*. UNAIDS, Geneva, Switzerland, 2006.

169. Masupu, K., et al (editors), *Botswana 2003 Second Generation HIV/AIDS Surveillance*. National AIDS Coordinating Agency, Gaborone, Botswana, 2003.

170. Rambaut, A., et al., Human immunodeficiency virus—phylogeny and the origin of HIV-1. *Nature*, 2001;410(6832):1047–1048.

171. Keele, B.F., et al., Chimpanzee reservoirs of pandemic and nonpandemic HIV-1. *Science*, 2006;313(5786):523–526.

172. Stremlau, M., et al., The cytoplasmic body component TRIM5 alpha restricts HIV-1 infection in old world monkeys. *Nature*, 2004;427(6977):848–853.

173. Connor, E.M., et al., Reduction of maternal-infant transmission of human-immunodeficiency-virus type-1 with zidovudine treatment. *New England Journal of Medicine*, 1994;331(18):1173–1180.

174. Van Rompay, K.K.A., Antiretroviral drug studies in nonhuman primates: A valid animal model for innovative drug efficacy and pathogenesis experiments. *AIDS Reviews*, 2005;7(2):67–83.

175. McMichael, A.J., HIV vaccines. *Annual Review of Immunology*, 2006;24:227–255.

176. King, F.A., et al., Primates. *Science*, 1988;240(4858):1475–1482.

177. Burns, R.S., et al., A primate model of parkinsonism—selective destruction of dopaminergic-neurons in the pars compacta of the substantia nigra by N-methyl-4-phenyl-1,2,3,6-tetrahydropyridine. *Proceedings of the National Academy of Sciences of the USA*, 1983;80(14):4546–4550.

178. Alzheimer's Association, Statistics about Alzheimer's Disease. 2006; available from search.alz.org/AboutAD/statistics.asp [cited August 5, 2006].

179. Price, D.L., and S.S. Sisodia, Cellular and molecular-biology of Alzheimers-disease and animal-models. *Annual Review of Medicine*, 1994;45:435–446.

180. Buccafusco, J.J., et al., Differential improvement in memory-related task performance with nicotine by aged male and female rhesus monkeys. *Behavioural Pharmacology*, 1999;10(6–7):681–690.

181. Gandy, S., et al., Alzheimer's A beta vaccination of rhesus monkeys (Macaca mulatta). *Mechanisms of Ageing and Development*, 2004;125(2):149–151.

182. Lemere, C.A., et al., Alzheimer's disease A beta vaccine reduces central nervous system A beta levels in a non-human primate, the Caribbean vervet. *American Journal of Pathology*, 2004;165(1):283–297.

183. Harlow, H.F., The nature of love. *American Psychologist*, 1958;13(12):673–685.

184. Washburn, S.L., and I. Devore, Social life of baboons—a study of troops of baboons in their natural environment in East Africa has revealed patterns of interdependence that may shed light on early evolution of human species. *Scientific American*, 1961;204(6):62–72.

185. Hausfater, G., J. Altmann, and S. Altmann, Long-term consistency of dominance relations among female baboons (Papio-cynocephalus). *Science*, 1982;217(4561):752–755.

186. Fossey, D., *Gorillas in the Mist*. Houghton Mifflin & Company, Boston, 1983.

187. Galdikas, B., *Orangutan Odyssey*. Harry N. Abrams, New York, 1999.

188. van Schaik, C.P., et al., Orangutan cultures and the evolution of material culture. *Science*, 2003;299(5603):102–105.

189. Goodall, J., *My Life with the Chimpanzees*. Simon & Schuster, New York, 1988.

190. Zhou, Z., and S. Zheng, Palaeobiology: The missing link in Ginkgo evolution—the modern maidenhair tree has barely changed since the days of the dinosaurs. *Nature*, 2003;423(6942):3.

191. National Assessment Synthesis Team, U.S. Global Change Research Program. Climate Change Impacts on the United States. The Potential Consequences of Climate Variability and Change. Overview: Alaska. U.S. National Assessment of the Potential Consequences of Climate Variability and Change, 2000; available from www.usgcrp.gov/usgcrp/Library/nationalassessment/overviewalaska.htm [cited October 10, 2006].

192. Malcolm, J., et al., Migration of vegetation types in a greenhouse world, in *Climate Change and Biodiversity*, T. Lovejoy and L. Hannah (editors). Yale University Press, New Haven, CT, 2005.

193. Holsten, E., et al., The Spruce Beetle. 1999; available from www.na.fs.fed.us/spfo/pubs/fidls/sprucebeetle/sprucebeetle.htm [cited August 4, 2006].

194. Egan, T., On hot trail of tiny killer in Alaska. *New York Times*, June 25, 2002, F1.

195. Western Regional Climate Center, Alaska Climate Summaries. 2006; available from www.wrcc.dri.edu/summary/climsmak.htrnl [cited October 10, 2006].

196. Berg, E.E., et al., Spruce beetle outbreaks on the Kenai Peninsula, Alaska, and Kluane National Park and Reserve, Yukon Territory: Relationship to summer temperatures and regional differences in disturbance regimes. *Forest Ecology and Management*, 2006;227(3):219–232.

197. McDonald, G., and R. Hoff, Blister rust: An introduced plague, in *Whitebark Pine Communities: Ecology and Restoration*, D. Tomback, S. Amo, and R. Keane (editors). Island Press, Washington, DC, 2001, 193–220.

198. Kendall, K., and R. Keane, Whitebark pine decline: Infection, mortality, and population trends, in *Whitebark Pine Communities: Ecology and Restoration*, D. Tomback, S. Amo, and R. Keane (editors). Island Press, Washington, DC, 2001, 221–242.

199. Keane, R.E., P. Morgan, and J.P. Menakis, Landscape assessment of the decline of whitebark-pine (Pinus-albicaulis) in the Bob Marshall Wilderness Complex, Montana, USA. *Northwest Science*, 1994;68(3):213–229.

200. Gibson, K., Mountain pine beetle conditions in whitebark pine stands in the greater Yellowstone ecosystem, in *Forest Health Protection*. USDA Forest Service, Missoula, MN, 2006.

201. Powell, J., and J. Logan, Ghost Forests, global warming and the mountain pine beetle. *American Entomologist*, 2001;47(3):160–172.

202. Tomback, D., and K. Kendall, Biodiversity loses: The downward spiral, in *Whitebark Pine Communities: Ecology and Restoration*, D. Tomback, S. Amo, and R. Keane (editors). Island Press, Washington, DC, 2001, 243–262.

203. Petit, C., In the Rockies, pines die and bears feel it. *New York Times*, January 30, 2007, F1.

204. Kizlinski, M.L., et al., Direct and indirect ecosystem consequences of an invasive pest on forests dominated by eastern hemlock. *Journal of Biogeography*, 2002;29(10–11):1489–1503.

205. Shiels, K., and C. Cheah, Winter mortality in Adelges tsugae populations in 2003 and 2004, in *Third Symposium on Hemlock Woolly Adelgid in the Eastern United States*. Forest Health Technology Enterprise Team, USDA, Asheville, NC, 2005.

206. Stevens, W., Ladybugs coming to the rescue of threatened hemlocks. *New York Times*, February 17, 1998, F3.

207. Snyder, C.D., et al., Influence of eastern hemlock (Tsuga canadensis) forests on aquatic invertebrate assemblages in headwater streams. *Canadian Journal of Fisheries and Aquatic Sciences*, 2002;59(2):262–275.

208. Newman, D.J., G.M. Cragg, and K.M. Snader, The influence of natural products upon drug discovery. *Natural Product Reports*, 2000;17(3):215–234.

209. Del Tredici, P., The evolution, ecology, and cultivation of Ginkgo biloba, in *Ginkgo Biloba*, T.A. van Beek (editor). Harwood Academic Publishers, Amsterdam, 2000, 7–23.

210. Hori, T., *Ginkgo biloba, a Global Treasure: From Biology to Medicine*. Springer, New York, 1997, xvii.

211. Sierpina, V.S., B. Wollschlaeger, and M. Blumenthal, Ginkgo biloba. *American Family Physician*, 2003;68(5):923–926.

212. LeBars, P.L., et al., A placebo-controlled, double-blind, randomized trial of an extract of Ginkgo biloba for dementia. *JAMA*, 1997;278(16):1327–1332.

213. Ahn, Y.J., et al., Potent insecticidal activity of Ginkgo biloba derived trilactone terpenes against Nilaparvata lugens, in *Phytochemicals for Pest Control*, P.A. Hedin (editor). American Chemical Society, Washington, DC, 1997, 90–105.

214. Goodman, J., and V. Walsh, *The Story of Taxol: Nature and Politics in the Pursuit of an Anti-cancer Drug*. Cambridge University Press, New York, 2001, xiii.

215. Oberlies, N.H., and D.J. Kroll, Camptothecin and taxol: Historic achievements in natural products research. *Journal of Natural Products*, 2004;67(2):129–135.

216. McGuire, W.P., et al., Taxol—a unique antineoplastic agent with significant activity in advanced ovarian epithelial neoplasms. *Annals of Internal Medicine*, 1989;111(4):273–279.

217. Crown, J., and M. O'Leary, The taxanes: An update. *Lancet*, 2000;355(9210):1176–1178.

218. Kavallaris, M., Discovering novel strategies for antimicrotubule cytotoxic therapy. *EJC Supplements*, 2006;4(7):3–9.

219. Stone, G.W., et al., A polymer-based, paclitaxel-eluting stent in patients with coronary artery disease. *New England Journal of Medicine*, 2004;350(3):221–231.

220. Stone, G.W., et al., Safety and efficacy of sirolimus- and paclitaxel-eluting coronary stents. *New England Journal of Medicine*, 2007;356(10):998–1008.

221. Stromgaard, K., and K. Nakanishi, Chemistry and biology of terpene trilactones from Ginkgo biloba. *Angewandte Chemie-International Edition*, 2004;43(13):1640–1658.

222. Amri, H., K. Drieu, and V. Papadopoulos, Transcriptional suppression of the adrenal cortical peripheral-type benzodiazepine receptor gene and inhibition of steroid synthesis by ginkgolide B. *Biochemical Pharmacology*, 2003;65(5):717–729.

223. Papadopoulos, V., et al., Drug-induced inhibition of the peripheral-type benzodiazepine receptor expression and cell proliferation in human breast cancer cells. *Anticancer Research*, 2000;20(5A):2835–2847.

224. Livett, B.G., What's New in 2004. 2004; available from grimwade.biochem.unimelb.edu.au/cone/new2004.html [cited October 11, 2006].

225. Roberts, C.M., et al., Marine biodiversity hotspots and conservation priorities for tropical reefs. *Science*, 2002;295(5558):1280–1284.

226. Wilkinson, C., ed. *Status of Coral Reefs of the World: 2004*, Vol. 1. Australian Institute of Marine Science, Townsville, Queensland, Australia, 2004.

227. Spalding, M., F. Blasco, and C. Field, *World Mangrove Atlas*. International Society for Mangrove Ecosystems, Okinawa, Japan, 1997.

228. Burke, L., L. Selig, and M. Spalding, *Reefs at Risk in South-East Asia*. U.N. Environment Programme World Conservation Monitoring Centre, Cambridge, UK, 2002.

229. Chivian, E., C.M. Roberts, and A.S. Bernstein, The threat to cone snails. *Science*, 2003;302(5644):391.

230. Weil, E., G. Smith, and D.L. Gil-Agudelo, Status and progress in coral reef disease research. *Diseases of Aquatic Organisms*, 2006;69(1):1–7.

231. Olivera, B.M., Conus peptides: Biodiversity-based discovery and exogenomics. *Journal of Biological Chemistry*, 2006;281(42):31173–31177.

232. Olivera, B.M., et al., Diversity of Conus Neuropeptides. *Science*, 1990;249(4966):257–263.

233. Olivera, B., and L. Cruz, Conotoxins, in retrospect. *Toxicon*, 2001;39:7.

234. McIntosh, J.M., and R.M. Jones, Cone venom—from accidental stings to deliberate injection. *Toxicon*, 2001;39(10):1447–1451.

235. Staats, P.S., et al., Intrathecal ziconotide in the treatment of refractory pain in patients with cancer or AIDS—a randomized controlled trial. *JAMA*, 2004;291(1):63–70.

236. Bowersox, S.S., et al., Selective N-type neuronal voltage-sensitive calcium channel blocker, SNX-111, produces spinal antinociception in rat models of acute, persistent and neuropathic pain. *Journal of Pharmacology and Experimental Therapeutics*, 1996;279(3):1243–1249.

237. Mari, F., and G.B. Fields, Conopeptides: Unique pharmacological agents that challenge current peptide methodologies. *Chimica Oggi—Chemistry Today*, 2003;21(6):43–48.

238. Williams, A.J., et al., Neuroprotective efficacy and therapeutic window of the high-affinity N-methyl-D-aspartate antagonist conantokin-G: In vitro (primary cerebellar neurons) and in vivo (rat model of transient focal brain ischemia) studies. *Journal of Pharmacology and Experimental Therapeutics*, 2000;294(1):378–386.

239. Rajendra, W., A. Armugam, and K. Jeyaseelan, Neuroprotection and peptide toxins. *Brain Research Reviews*, 2004;45(2):125–141.

240. Jimenez, E.C., et al., Conantokin-L, a new NMDA receptor antagonist: Determinants for anticonvulsant potency. *Epilepsy Research*, 2002;51(1–2):73–80.

241. Pinto, A., et al., The action of Lambert-Eaton myasthenic syndrome immunoglobulin G on cloned human voltage-gated calcium channels. *Muscle and Nerve*, 2002;25(5):715–724.

242. Watkins, M., D.R. Hillyard, and B.M. Olivera, Genes expressed in a turrid venom duct: Divergence and similarity to conotoxins. *Journal of Molecular Evolution*, 2006;62(3):247–256.

243. Azam, L., et al., Alpha-conotoxin BuIA, a novel peptide from Conus bullatus, distinguishes among neuronal nicotinic acetylcholine receptors. *Journal of Biological Chemistry*, 2005;280(1):80–87.

244. Nicke, A., S. Wonnacott, and R.J. Lewis, Alpha-conotoxins as tools for the elucidation of structure and function of neuronal nicotinic acetylcholine receptor subtypes. *European Journal of Biochemistry*, 2004;271(12):2305–2319.

245. Janes, R.W., Alpha-conotoxins as selective probes for nicotinic acetylcholine receptor subclasses. *Current Opinion In Pharmacology*, 2005;5(3):280–292.

246. McIntosh, J.M., A.D. Santos, and B.M. Olivera, Conus peptides targeted to specific nicotinic acetylcholine receptor subtypes. *Annual Review of Biochemistry*, 1999;68:59–88.

247. Quik, M., and J.M. McIntosh, Striatal alpha 6* nicotinic acetylcholine receptors: Potential targets for Parkinson's disease therapy. *Journal of Pharmacology and Experimental Therapeutics*, 2006;316(2):481–489.

248. Terlau, H., and B.M. Olivera, Conus venoms: A rich source of novel ion channel-targeted peptides. *Physiological Reviews*, 2004;84(1):41–68.

249. Clarke, S.C., et al., Global estimates of shark catches using trade records from commercial markets. *Ecology Letters*, 2006;9(10):1115–1126.

250. Baum, J.K., et al., Collapse and conservation of shark populations in the northwest Atlantic. *Science*, 2003;299(5605):389–392.

251. Baum, J.K., and R.A. Myers, Shifting baselines and the decline of pelagic sharks in the Gulf of Mexico. *Ecology Letters*, 2004;7(2):135–145.

252. Musick, J.A., et al., Management of sharks and their relatives (Elasmobranchii). *Fisheries*, 2000;25(3):9–13.

253. Ritter, E., Fact sheet: Spiny dogfish, in *Shark Info Research News and Background Information on the Protection, Ecology, Biology and Behavior of Sharks.* 1999; available from www.sharkinfo.ch/SI2_99e/sacanthias.html [cited August 30, 2007].

254. Raloff, J., Clipping the fin. *Science News*, 2002;162(15):232.

255. Burgess, G., ISAF 2004 Worldwide Shark Attack Summary. 2004; available from www.flmnh.ufl.edu/FISH/Sharks/statistics/2004attacksummary.htm [cited October 11, 2006].

256. Lyon, W., Bee and Wasp Stings. 2000; available from ohioline.osu.edu/hyg-fact/2000/2076.html [cited October 11, 2006].

257. Roach, J., Key to lightning deaths: Location, location, location. *National Geographic News*, June 22, 2004; available from news.nationalgeographic.com/news/2003/05/0522_030522_lightning.html [cited August 30, 2007].

258. Environmental News Service, Great white shark protected. *Newswire*, 2004; available from www.ens-newswire.com/ens/oct2004/2004-10-12-03.asp [cited August 30, 2007].

259. Cunningham-Day, R., *Sharks in Danger: Global Shark Conservation Status with Reference to Management Plans and Legislation.* Universal Publishers, Parkland, Florida, 2001.

260. Forero, J., Hidden cost of shark fin soup: Its source may vanish. *New York Times*, January 5, 2006, A4.

261. National Oceanic and Atmospheric Administration, International Commission Adopts U.S. Proposal for Shark Finning Ban. 2004; available from www.publicaffairs.noaa.gov/releases2004/nov04/noaa04-115.html [cited October 11, 2006].

262. Essington, T., et al., Alternative fisheries and the predation rate of yellowfin tuna in the eastern Pacific Ocean. *Ecological Applications*, 2002;12(3):10.

263. Myers, R.A., et al., Cascading effects of the loss of apex predatory sharks from a coastal ocean. *Science*, 2007;315(5820):1846–1850.

264. Lee, A., and R. Langer, Shark cartilage contains inhibitors of tumor angiogenesis. *Science*, 1983;221(4616):1185–1187.

265. Gugliotta, G., FTC Tells Firms to End Shark Cartilage Anti-cancer Claims. *Washington Post*, June 30, 2000, A19.

266. Ostrander, G.K., et al., Shark cartilage, cancer and the growing threat of pseudoscience. *Cancer Research*, 2004;64(23):8485–8491.

267. Loprinzi, C.L., et al., Evaluation of shark cartilage in patients with advanced cancer—a north central cancer treatment group trial. *Cancer*, 2005;104(1):176–182.

268. Batist, G., et al., Neovastat (AE-941) in refractory renal cell carcinoma patients: Report of a phase II trial with two dose levels. *Annals of Oncology*, 2002;13(8):1259–1263.

269. Harvard Medical School, Consumer Health Information. Complementary and Alternative Medicine. Shark Cartilage. 2005; available from www.intelihealth.com/IH/ihtIH/WSIHW000/8513/31402/346293.html?d=dmtContent [cited October 11, 2006].

270. Moore, K.S., et al., Squalamine—an aminosterol antibiotic from the shark. *Proceedings of the National Academy of Sciences of the USA*, 1993;90(4):1354–1358.

271. Kikuchi, K., et al., Antimicrobial activities of squalamine mimics. *Antimicrobial Agents and Chemotherapy*, 1997;41(7):1433–1438.

272. Sills, A.K., et al., Squalamine inhibits angiogenesis and solid tumor growth in vivo and perturbs embryonic vasculature. *Cancer Research*, 1998;58(13):2784–2792.

273. Chopdar, A., U. Chakravarthy, and D. Verma, Age related macular degeneration. *British Medical Journal*, 2003;326(7387):485–488.

274. Ciulla, T.A., et al., Squalamine lactate reduces choroidal neovascularization in a laser-injury model in the rat. *Retina—the Journal of Retinal and Vitreous Diseases*, 2003;23(6):808–814.

275. Genaidy, M., et al., Effect of squalamine on iris neovascularization in monkeys. *Retina—the Journal of Retinal and Vitreous Diseases*, 2002;22(6):772–778.

276. Garcia, C.A., et al., A phase 2 multi-dose pharmacokinetic study of MSI-1256F (squalamine lactate) for the treatment of subfoveal choroidal neovascularization associated with age-related macular degeneration (AMD). *Investigative Ophthalmology and Visual Science*, 2005;46(Supplement S).

277. Herbst, R.S., et al., A phase I/IIA trial of continuous five-day infusion of squalamine lactate (MSI-1256F) plus carboplatin and paclitaxel in patients with advanced non-small cell lung cancer. *Clinical Cancer Research*, 2003;9(11):4108–4115.

278. Zasloff, M., et al., A spermine-coupled cholesterol metabolite from the shark with potent appetite suppressant and antidiabetic properties. *International Journal of Obesity*, 2001;25(5):689–697.

279. Ahima, R.S., et al., Appetite suppression and weight reduction by a centrally active aminosterol. *Diabetes*, 2002;51(7):2099–2104.

280. MacCallum, A., The paleochemistry of the body fluids and tissues. *Physiological Reviews*, 1926;6:316–357.

281. Epstein, F.H., The sea within us. *Journal of Experimental Zoology*, 1999;284(1):50–54.

282. Epstein, F.H., The salt gland of the shark, in *A Laboratory by the Sea: The Mount Desert Island Biological Laboratory 1989–1998*, F.H. Epstein (editor). River Press, Rhinebeck, New York, 1998.

283. Silva, P., R.J. Solomon, and F.H. Epstein, The rectal gland of Squalus acanthias: A model for the transport of chloride. *Kidney International*, 1996;49(6):1552–1556.

284. Goldman, M., et al., Human b-defensin-1 is a salt-sensitive antibiotic in lung that is inactivated in cystic fibrosis. *Cell*, 1997;88:7.

285. Smith, J., et al., Cystic fibrosis airway epithelia fail to kill bacteria because of abnormal airway surface fluid. *Cell*, 1996;85:7.

286. Aller, S.G., et al., Cloning, characterization, and functional expression of a CNP receptor regulating CFTR in the shark rectal gland. *American Journal of Physiology—Cell Physiology*, 1999;276(2):C442–C449.

287. Badani, K.K., A.K. Hemal, and M. Menon, Autosomal dominant polycystic kidney disease and pain—a review of the disease from aetiology, evaluation, past surgical treatment options to current practice. *Journal of Postgraduate Medicine*, 2004;50(3):222–226.

288. Silva, P., et al., Mode of action of somatostatin to inhibit secretion by shark rectal gland. *American Journal of Physiology*, 1985;249(3):R329–R334.

289. Reubi, J.C., et al., Human kidney as target for somatostatin—high-affinity receptors in tubules and vasa-recta. *Journal of Clinical Endocrinology and Metabolism*, 1993;77(5):1323–1328.

290. Ruggenenti, P., et al., Safety and efficacy of long-acting somatostatin treatment in autosomal-dominant polycystic kidney disease. *Kidney International*, 2005;68(1):206–216.

291. Schluter, S.F., and J.J. Marchalonis, Cloning of shark RAG2 and characterization of the RAG1/RAG2 gene locus. *FASEB Journal*, 2003;17(1)470–472.

292. Kasahara, M., et al., The evolutionary origin of the major histocompatibility complex—polymorphism of class-II alpha-chain genes in the cartilaginous fish. *European Journal of Immunology*, 1993;23(9):2160–2165.

293. Flajnik, M.F., Comparative analyses of immunoglobulin genes: Surprises and portents. *Nature Reviews Immunology*, 2002;2(9):688–698.

294. Marchalonis, J.J., et al., Natural recognition repertoire and the evolutionary emergence of the combinatorial immune system. *FASEB Journal*, 2002;16(8)842–848.

295. Tanacredi, J.T., *Limulus in the Limelight: A Species 350 Million Years in the Making and in Peril?* Kluwer Academic/Plenum Publishers, New York, 2001, xvi.

296. Morrison, R.I.G., R.K. Ross, and L.J. Niles, Declines in wintering populations of red knots in southern South America. *Condor*, 2004;106(1):60–70.

297. Osaki, T., et al., Horseshoe crab hemocyte-derived antimicrobial polypeptides, tachystatins, with sequence similarity to spider neurotoxins. *Journal of Biological Chemistry*, 1999;274(37):26172–26178.

298. Ozaki, A., S. Ariki, and S. Kawabata, An antimicrobial peptide tachyplesin acts as a secondary secretagogue and amplifies lipopolysaccharide-induced hemocyte exocytosis. *FEBS Journal*, 2005;272(15):3863–3871.

299. Powers, J.P.S., et al., The antimicrobial peptide polyphemusin localizes to the cytoplasm of Escherichia coli following treatment. *Antimicrobial Agents and Chemotherapy*, 2006;50(4):1522–1524.

300. Tamamura, H., et al., A low-molecular-weight inhibitor against the chemokine receptor CXCR4: A strong anti-HIV peptide T140. *Biochemical and Biophysical Research Communications*, 1998;253(3):877–882.

301. Tamamura, H., et al., T140 analogs as CXCR4 antagonists identified as anti-metastatic agents in the treatment of breast cancer. *FEBS Letters*, 2003;550(1–3):79–83.

302. Tamamura, H., et al., Identification of a CXCR4 antagonist, a T140 analog, as an anti-rheumatoid arthritis agent. *FEBS Letters*, 2004;569(1–3):99–104.

303. Levin, J., P.A. Tomasulo, and R.S. Oser, Detection of Endotoxin in Human Blood and Demonstration of an Inhibitor. *Journal of Laboratory and Clinical Medicine*, 1970;75(6):903–911.

304. Schmid, M.F., et al., Structure of the acrosomal bundle. *Nature*, 2004;431(7004):104–107.

305. Barlow, R.B., J.M. Hitt, and F.A. Dodge, Limulus vision in the marine environment. *Biological Bulletin*, 2001;200(2):169–176.

306. Zhu, Y., et al., The ancient origin of the complement system. *EMBO Journal*, 2005;24(2):382–394.

307. Graham, N., F. Ratliff, and H.K. Hartline, Facilitation of inhibition in compound lateral eye of Limulus. *Proceedings of the National Academy of Sciences of the USA*, 1973;70(3):894–898.

Additional References

Andersen, M., et al., Geographic variation of PCB congeners in polar bears (Ursus maritimus) from Svalbard east to the Chukchi Sea. *Polar Biology*, 2001;24(4):231–238.

Beck, B.B., Disney Institute, and American Zoo and Aquarium Association, *Great Apes and Humans: The Ethics of Coexistence*. Zoo and Aquarium Biology and Conservation Series. Smithsonian Institution Press, Washington, DC, 2001, xxiv.

Beebee, T.J.C., and R.A. Griffiths, The amphibian decline crisis: A watershed for conservation biology? *Biological Conservation*, 2005;125(3):271–285.

Beetschen, J.C., Amphibian gastrulation: History and evolution of a 125 year-old concept. *International Journal of Developmental Biology*, 2001;45(7):771–795.

Blaustein, A.R., and L.K. Belden, Amphibian defenses against ultraviolet-B radiation. *Evolution and Development*, 2003;5(1):89–97.

Blaustein, A.R., et al., Amphibian breeding and climate change. *Conservation Biology*, 2001;15(6):1804–1809.

Brunel, J.M., et al., Squalamine: A polyvalent drug of the future? *Current Cancer Drug Targets*, 2005;5(4):267–272.

Burger, M., et al., Small peptide inhibitors of the CXCR4 chemokine receptor (CD184) antagonize the activation, migration, and antiapoptotic responses of CXCL12 in chronic lymphocytic leukemia B cells. *Blood*, 2005;106(5):1824–1830.

Cragg, G.M., and D.J. Newman, Biodiversity: A continuing source of novel drug leads. *Pure and Applied Chemistry*, 2005;77(1):7–24.

Crown, J., M. O'Leary, and W.S. Ooi, Docetaxel and paclitaxel in the treatment of breast cancer: A review of clinical experience. *Oncologist*, 2004;9:24–32.

Derocher, A., et al., Effects of fasting and feeding on serum urea and serum creatinine levels in polar bears. *Marine Mammal Science*, 1990;6(3):196–203.

Foster, S., and the North American Botanical Council, *Ginkgo: Ginkgo biloba*, rev. ed. American Botanical Council, Austin, TX, 1991.

Gardner, M.B., Simian AIDS: An historical perspective. *Journal of Medical Primatology*, 2003;32(4–5):180–186.

Gorman, J., Gorillas and chimps in peril, report says. *New York Times*, April 7, 2003, 8.

Ha, J.C., et al., Fetal toxicity of zidovudine (azidothymidine) in Macaca nemestrina—preliminary observations. *Journal of Acquired Immune Deficiency Syndromes and Human Retrovirology*, 1994;7(2):154–157.

Hagey, L.R., et al., Ursodeoxycholic acid in the Ursidae—biliary bile-acids of bears, pandas, and related carnivores. *Journal of Lipid Research*, 1993;34(11):1911–1917.

Johnson, P.T.J., Amphibian diversity: Decimation by disease. *Proceedings of the National Academy of Sciences of the USA*, 2006;103(9):3011–3012.

Kalish, M.L., et al., Central African hunters exposed to simian immunodeficiency virus. *Emerging Infectious Diseases*, 2005;11(12):1928–1930.

Laird, D.J., et al., 50 million years of chordate evolution: Seeking the origins of adaptive immunity. *Proceedings of the National Academy of Sciences of the USA*, 2000;97(13):6924–6926.

Lameire, N., et al., Chronic kidney disease: A European perspective. *Kidney International*, 2005;68(S99):S30.

Langer, R., et al., Isolation of a cartilage factor that inhibits tumor neovascularization. *Science*, 1976;193(4247):70–72.

Lehrich, R.W., et al., Vasoactive intestinal peptide, forskolin, and genistein increase apical CFTR trafficking in the rectal gland of the spiny dogfish, Squalus acanthias—acute regulation of CFTR trafficking in an intact epithelium. *Journal of Clinical Investigation*, 1998;101(4):737–745.

Lie, E., et al., Does high organochlorine (OC) exposure impair the resistance to infection in polar bears (Ursus maritimus)? Part II: Possible effect of OCs on mitogen- and antigen-induced lymphocyte proliferation. *Journal of Toxicology and Environmental Health, Part A—Current Issues*, 2005;68(6):457–484.

Lie, E., et al., Geographical distribution of organochlorine pesticides (OCPs) in polar bears (Ursus maritimus) in the Norwegian and Russian Arctic. *Science of the Total Environment*, 2003;306(1–3):159–170.

Livett, B.G., K.R. Gayler, and Z. Khalil, Drugs from the sea: Conopeptides as potential therapeutics. *Current Medicinal Chemistry*, 2004;11(13):1715–1723.

Lundgren, B., et al., Antiviral effects of 3'-fluorothymidine and 3'-azidothymidine in cynomolgus monkeys infected with simian immunodeficiency virus. *Journal of Acquired Immune Deficiency Syndromes and Human Retrovirology*, 1991;4(5):489–498.

Malacinski, G.M., T. Ariizumi, and M. Asashima, Work in progress: The renaissance in amphibian embryology. *Comparative Biochemistry and Physiology B—Biochemistry and Molecular Biology*, 2000;126(2):179–187.

McGuire, W.P., and M. Markman, Primary ovarian cancer chemotherapy: Current standards of care. *British Journal of Cancer*, 2003;89:S3–S8.

Mittermeier, R.A., et al., *Primates in Peril: The World's 25 Most Endangered Primates 2004–2006*. Species Survival Commission, International Primatological Society, Conservation International, 2006.

Mor, A., et al., Isolation, amino-acid-sequence, and synthesis of dermaseptin, a novel antimicrobial peptide of amphibian skin. *Biochemistry*, 1991;30(36):8824–8830.

Nakamura, T., et al., Tachyplesin, a class of antimicrobial peptide from the hemocytes of the horseshoe-crab (Tachypleus-tridentatus)—isolation and chemical-structure. *Journal of Biological Chemistry*, 1988;263(32):16709–16713.

Norstrom, R.J., et al., Chlorinated hydrocarbon contaminants in polar bears from eastern Russia, North America, Greenland, and Svalbard: Biomonitoring of Arctic pollution. *Archives of Environmental Contamination and Toxicology*, 1998;35(2):354–367.

Papadopoulos, V., et al., Peripheral benzodiazepine receptor in cholesterol transport and steroidogenesis. *Steroids*, 1997;62(1):21–28.

Passaglia, C., et al., Deciphering a neural code for vision. *Proceedings of the National Academy of Sciences of the USA*, 1997;94(23):12649–12654.

Piccolino, M., Animal electricity and the birth of electrophysiology: The legacy of Luigi Galvani. *Brain Research Bulletin*, 1998;46(5):381–407.

Pough, F.H., Acid precipitation and embryonic mortality of spotted salamanders, Ambystoma-maculatum. *Science*, 1976;192(4234):68–70.

Reginster, J.Y., and N. Burlet, Osteoporosis: A still increasing prevalence. *Bone*, 2006;38(2, Supplement 1):S4–S9.

Riley, S.P.D., et al., Hybridization between a rare, native tiger salamander (Ambystoma californiense) and its introduced congener. *Ecological Applications*, 2003;13(5):1263–1275.

Rollins-Smith, L.A., and J.M. Conlon, Antimicrobial peptide defenses against chytridiomycosis, an emerging infectious disease of amphibian populations. *Developmental and Comparative Immunology*, 2005;29(7):589–598.

Santiago, M.L., et al., SIVcpz in wild chimpanzees. *Science*, 2002;295(5554):465–465.

Schulze, W., and D. Djuniadl, Introduction of integrated pest management in rice cultivation in Indonesia. *Pflanzenschutz-Nachrichten Bayer*, 1998;51(1):97–104.

Shiels, K., and C. Cheah, Winter mortality in Adelges tsugae populations in 2003 and 2004, in *Third Symposium on Hemlock Woolly Adelgid in the Eastern United States*. U.S. Department of Agriculture and U.S. Forest Service, Asheville, NC, 2005.

Sullivan, N.J., et al., Accelerated vaccination for Ebola virus haemorrhagic fever in non-human primates. *Nature*, 2003;424(6949):681–684.

Tarasick, D.W., et al., Climatology and trends of surface UV radiation. *Atmosphere-Ocean*, 2003;41(2):121–138.

Tavera-Mendoza, L., et al., Response of the amphibian tadpole (Xenopus laevis) to atrazine during sexual differentiation of the testis. *Environmental Toxicology and Chemistry*, 2002;21(3):527–531.

Tilney, L.G., J.G. Clain, and M.S. Tilney, Membrane events in the acrosomal reaction of limulus sperm—membrane-fusion, filament-membrane particle attachment, and the source and formation of new membrane-surface. *Journal of Cell Biology*, 1979;81(1):229–253.

Turtle, S.L., Embryonic survivorship of the spotted salamander (Ambystoma maculatum) in roadside and woodland vernal pools in southeastern New Hampshire. *Journal of Herpetology*, 2000;34(1):60–67.

United States Renal Data System, Annual Data Report. 2005; available from www.usrds. org/atlas.htm [cited August 2, 2006].

Vanrompay, K.K.A., et al., Simian immunodeficiency virus (SIV) infection of infant rhesus macaques as a model to test antiretroviral drug prophylaxis and therapy—oral 3'-azido-3'-deoxythymidine prevents SIV infection. *Antimicrobial Agents and Chemotherapy*, 1992;36(11):2381–2386.

Wolbarsht, M., and S.S. Yeandle, Visual processes in limulus eye. *Annual Review of Physiology*, 1967;29:513–542.

Wolfe, R.R., et al., Urea nitrogen reutilization in hibernating bears. *Federation Proceedings*, 1982;41(5):1623.

World Health Organization, Ebola Hemorrhagic Fever (Fact Sheet No. 103). 2004; available from www.who.int/mediacentre/factsheets/fs103/en/ [cited October 10, 2006].

Xue, J.L., et al., Forecast of the number of patients with end-stage renal disease in the United States to the year 2010. *Journal of the American Society of Nephrology*, 2001;12(12):2753–2758.

Zasloff, M., Magainins, A class of antimicrobial peptides from Xenopus skin—isolation, characterization of 2 active forms, and partial cDNA sequence of a precursor. *Proceedings of the National Academy of Sciences of the USA*, 1987;84(15):5449–5453.

CHAPTER 7

1. Mills, J.N., and J.E. Childs, Ecologic studies of rodent reservoirs: Their relevance for human health. *Emerging Infectious Diseases*, 1998;4(4):529–537.

2. Ostfeld, R.S., and R.D. Holt, Are predators good for your health? Evaluating evidence for top-down regulation of zoonotic disease reservoirs. *Frontiers in Ecology and the Environment*, 2004;2(1):13–20.

3. Walsh, J.F., D.H. Molyneux, and M.H. Birley, Deforestation—effects on vector-borne disease. *Parasitology*, 1993;106:S55–S75.

4. Southgate, V., H. Wijk, and C. Wright, Schistosomiasis in Loum, Cameroun: Schistosoma haematobium, S. intercalatum, and their natural hybrid. *Zeitschrift für Parasitenkund*, 1976;49:149–159.

5. Downs, W.G., and C.S. Pittendrigh, Bromeliad Malaria in Trinidad, British West Indies. *American Journal of Tropical Medicine*, 1946;26(1):47–66.

6. Keiser, J., et al., Effect of irrigation and large dams on the burden of malaria on a global and regional scale. *American Journal of Tropical Medicine and Hygiene*, 2005;72(4):392–406.

7. Tyagi, B.K., and R.C. Chaudhary, Outbreak of falciparum malaria in the Thar Desert (India), with particular emphasis on physiographic changes brought about by extensive canalization and their impact on vector density and dissemination. *Journal of Arid Environments*, 1997;36(3):541–555.

8. Singh, N., R.K. Mehra, and V.P. Sharma, Malaria and the Narmada-river development in India: A case study of the Bargi dam. *Annals of Tropical Medicine and Parasitology*, 1999;93(5):477–488.

9. Abdelwahab, M.F., et al., Changing pattern of schistosomiasis in Egypt 1935–79. *Lancet*, 1979;2(8136):242–244.

10. Cline, B.L., et al., 1983 Nile Delta schistosomiasis survey—48 years after Scott. *American Journal of Tropical Medicine and Hygiene*, 1989;41(1):56–62.

11. Malek, E.A., Effect of Aswan High Dam on prevalence of schistosomiasis in Egypt. *Tropical and Geographical Medicine*, 1975;27(4):359–364.

12. Michelson, M.K., et al., Recent trends in the prevalence and distribution of schistosomiasis in the Nile Delta region. *American Journal of Tropical Medicine and Hygiene*, 1993;49(1):76–87.

13. Harb, M., et al., The resurgence of lymphatic filariasis in the Nile delta. *Bulletin of the World Health Organization*, 1993;71(1):49–54.

14. Rabsch, W., et al., Competitive exclusion of Salmonella enteritidis by Salmonella gallinarum in poultry. *Emerging Infectious Diseases*, 2000;6(5):443–448.

15. Schroeder, C.M., et al., Estimate of illnesses from Salmonella enteritidis in eggs, United States, 2000. *Emerging Infectious Diseases*, 2005;11(1):113–115.

16. Economic Research Service, USDA, Economics of foodborne disease: Salmonella. 2003; available from www.ers.usda.gov/data/foodborneillness/salm_intro.asp [cited August 12, 2006].

17. Hoke, C., and J. Gingrich, Japanese encephalitis, in *Handbook of Zoonoses*, G. Beran (editor). CRC Press, Boca Raton, FL, 1994, 59–70.

18. Peiris, J.S.M., et al., Japanese encephalitis in Sri-Lanka—comparison of vector and virus ecology in different agroclimatic areas. *Transactions of the Royal Society of Tropical Medicine and Hygiene*, 1993;87(5):541–548.

19. Chua, K.B., Nipah virus outbreak in Malaysia. *Journal of Clinical Virology*, 2003;26(3):265–275.

20. Epstein, J.H., et al., Nipah virus: Impact, origins, and causes of emergence. *Current Infectious Disease Reports*, 2006;8(1):59–65.

21. Li, W.D., et al., Bats are natural reservoirs of SARS-like coronaviruses. *Science*, 2005;310(5748):676–679.

22. Leroy, E.M., et al., Fruit bats as reservoirs of Ebola virus. *Nature*, 2005;438(7068):575–576.

23. Gubler, D.J., Epidemic dengue/dengue hemorrhagic fever as a public health, social and economic problem in the 21st century. *Trends in Microbiology*, 2002;10(2):100–103.

24. Allan, B.F., F. Keesing, and R.S. Ostfeld, Effect of forest fragmentation on Lyme disease risk. *Conservation Biology*, 2003;17(1):267–272.

25. Keesing, F., R.D. Holt, and R.S. Ostfeld, Effects of species diversity on disease risk. *Ecology Letters*, 2006;9(4):485–498.

26. Dobson, A., et al., Sacred cows and sympathetic squirrels: The importance of biological diversity to human health. *PLoS Medicine*, 2006;3(6):e231.

27. Abaru, D.E., Sleeping sickness in Busoga, Uganda, 1976–1983. *Tropical Medicine and Parasitology*, 1985;36(2):72–76.

28. Leak, S., *Tsetse Biology and Ecology: Their Role in the Epidemiology and Control of Trypanosomosis.* CABI, Nairobi, 1998.

29. Fevre, E.M., et al., The origins of a new Trypanosoma brucei rhodesiense sleeping sickness outbreak in eastern Uganda. *Lancet*, 2001;358(9282):625–628.

30. Fevre, E.M., et al., A burgeoning epidemic of sleeping sickness in Uganda. *Lancet*, 2005;366(9487):745–747.

31. Kilpatrick, A.M., et al., West Nile virus epidemics in North America are driven by shifts in mosquito feeding behavior. *PLoS Biology*, 2006;4(4):606–610.

32. WHO, Avian Influenza. 2006; available from www.who.int/csr/disease/avian_influenza/en/ [cited August 16, 2006].

33. Brisson, D., and D.E. Dykhuizen, ospC diversity in Borrelia burgdorferi: Different hosts are different niches. *Genetics*, 2004;168(2):713–722.

34. Seinost, G., et al., Four clones of Borrelia burgdorferi sensu stricto cause invasive infection in humans. *Infection and Immunity*, 1999;67(7):3518–3524.

35. Animal Health Institute, Antibiotic Use in Animals Rises in 2004. 2004; available from www.ahi.org/mediaCenter/documents/Antibioticuse2004.pdf [cited September 4, 2007].

36. Mellon, M., and S. Fondriest, Hogging it: Estimates of antimicrobial use in livestock. Union of Concerned Scientists, Cambridge, MA, 2001; available from www.ucsusa.org/assets/documents/food_and_environment/hog_front.pdf [cited September 3, 2007].

37. Ostfeld, R., and F. Keesing, The function of biodiversity in the ecology of vector-borne zoonotic diseases. *Canadian Journal of Zoology—Revue Canadienne de Zoologie*, 2000;78(12):2061–2078.

38. Combes, C., and H. Mone, Possible mechanisms of the decoy effect in Schistosoma-mansoni transmission. *International Journal for Parasitology*, 1987;17(4):971–975.

39. Yousif, F., M.E. Eman, and K.E. Sayed, Impact of two non-target snails on location and infection of Biomphalaria alexandrina with Schistosoma mansoni miracidia under simulated natural conditions. *Journal of Egyptian German Society of Zoology*, 1999;28(D):35–46.

40. Stauffer, J.R., et al., Controlling vectors and hosts of parasitic diseases using fishes—a case history of schistosomiasis in Lake Malawi. *Bioscience*, 1997;47(1):41–49.

41. Vilarinhos, P.T.R., and R. Monnerat, Larvicidal persistence of formulations of Bacillus thuringiensis var. israelensis to control larval Aedes aegypti. *Journal of the American Mosquito Control Association*, 2004;20(3):311–314.

42. Lacey, L.A., et al., Insect pathogens as biological control agents: Do they have a future? *Biological Control*, 2001;21(3):230–248.

43. Scholte, E.J., et al., Entomopathogenic fungi for mosquito control: A review. *Journal of Insect Science*, 2004;4(19):1–24.

44. Nisbet, D., Defined competitive exclusion cultures in the prevention of enteropathogen colonisation in poultry and swine. *Antonie Van Leeuwenhoek International Journal of General and Molecular Microbiology*, 2002;81(1–4):481–486.

45. Nam, V.S., et al., Elimination of dengue by community programs using Mesocyclops (Copepoda) against Aedes aegypti in central Vietnam. *American Journal of Tropical Medicine and Hygiene*, 2005;72(1):67–73.

46. Collins, L., and A. Blackwell, The biology of Toxorhynchites mosquitoes and their potential as biocontrol agents. *Biocontrol*, 2000;21(4):105N–116N.

47. Ghosh, S.K., et al., Larvivorous fish in wells target the malaria vector sibling species of the Anopheles culicifacies complex in villages in Karnataka, India. *Transactions of the Royal Society of Tropical Medicine and Hygiene*, 2005;99(2):101–105.

48. Kumar, A., et al., Field trials of biolarvicide Bacillus thuringiensis var. israelensis strain 164 and the larvivorous fish Aplocheilus blocki against Anopheles stephensi for malaria control in Goa, India. *Journal of the American Mosquito Control Association*, 1998;14(4):457–462.

49. Hahn, B., et al., AIDS as a zoonosis: Scientific and public health implications. *Science*, 2000;287:8.

50. Keele, B.F., et al., Chimpanzee reservoirs of pandemic and nonpandemic HIV-1. *Science*, 2006;313(5786):523–526.

51. Korber, B., et al., Timing the ancestor of the HIV-1 pandemic strains. *Science*, 2000;288(5472):1789–1796.

52. Lemey, P., et al., Tracing the origin and history of the HIV-2 epidemic. *Proceedings of the National Academy of Sciences of the USA*, 2003;100(11):6588–6592.

53. Hahn, B.H., et al., AIDS—AIDS as a zoonosis: Scientific and public health implications. *Science*, 2000;287(5453):607–614.

54. Peeters, M., et al., Risk to human health from a plethora of Simian immunodeficiency viruses in primate bushmeat. *Emerging Infectious Diseases*, 2002;8(5):451–457.

55. Wolfe, N.D., et al., Naturally acquired simian retrovirus infections in central African hunters. *Lancet*, 2004;363(9413):932–937.

56. Wolfe, N.D., et al., Emergence of unique primate T-lymphotropic viruses among central African bushmeat hunters. *Proceedings of the National Academy of Sciences of the USA*, 2005;102(22):7994–7999.

57. Vrielink, H., and H.W. Reesink, HTLV-I/II prevalence in different geographic locations. *Transfusion Medicine Reviews*, 2004;18(1):46–57.

58. Matsuoka, M., and K.T. Jeang, Human T-cell leukemia virus type I at age 25: A progress report. *Cancer Research*, 2005;65(11):4467–4470.

59. Lobitz, B., et al., Climate and infectious disease: Use of remote sensing for detection of Vibrio cholerae by indirect measurement. *Proceedings of the National Academy of Sciences of the USA*, 2000;97(4):1438–1443.

60. Patz, J.A., et al., Impact of regional climate change on human health. *Nature*, 2005;438(7066):310–317.

61. Rodo, X., et al., ENSO and cholera: A nonstationary link related to climate change? *Proceedings of the National Academy of Sciences of the USA*, 2002;99(20):12901–12906.

62. Ko, A.I., et al., Urban epidemic of severe leptospirosis in Brazil. *Lancet*, 1999;354(9181):820–825.

63. Mackenzie, W.R., et al., A massive outbreak in Milwaukee of Cryptosporidium infection transmitted through the public water-supply. *New England Journal of Medicine*, 1994;331(3):161–167.

64. Bennet, L., A. Halling, and J. Berglund, Increased incidence of Lyme borreliosis in southern Sweden following mild winters and during warm, humid summers. *European Journal of Clinical Microbiology and Infectious Diseases*, 2006;25(7):426–432.

65. Lindgren, E., and R. Gustafson, Tick-borne encephalitis in Sweden and climate change. *Lancet*, 2001;358(9275):16–18.

66. Hjelle, B., and G.E. Glass, Outbreak of hantavirus infection in the four corners region of the United States in the wake of the 1997–1998 El Nino-southern oscillation. *Journal of Infectious Diseases*, 2000;181(5):1569–1573.

Box 7.1

a. Taylor, L.H., S.M. Latham, and M.E.J. Woolhouse, Risk factors for human disease emergence. *Philosophical Transactions of the Royal Society of London Series B—Biological Sciences*, 2001;356(1411):983–989.

Box 7.2

Picquet, M., et al., The epidemiology of human schistosomiasis in the Senegal River Basin. *Transactions of the Royal Society of Tropical Medicine and Hygiene*, 1996;90(4):340–346.

Southgate, V.R., Schistosomiasis in the Senegal River Basin: Before and after the construction of the dams at Diama, Senegal and Manantali, Mali and future prospects. *Journal of Helminthology*, 1997;71(2):125–132.

Talla, I., et al., Outbreak of intestinal schistosomiasis in the Senegal River Basin. *Annales de la Societe Belge de Medecine Tropicale*, 1990;70(3):173–180.

Box 7.3

Amerasinghe, F.P., et al., Anopheline ecology and malaria infection during the irrigation development of an area of the Mahaweli project, Sri-Lanka. *American Journal of Tropical Medicine and Hygiene*, 1991;45(2):226–235.

Amerasinghe, F.P., and N.G. Indrajith, Postirrigation breeding patterns of surface-water mosquitos in the Mahaweli Project, Sri-Lanka, and comparisons with preceding developmental phases. *Journal of Medical Entomology*, 1994;31(4):516–523.

Additional References

Akiba, T., et al., Analysis of Japanese encephalitis epidemic in western Nepal in 1997. *Epidemiology and Infection*, 2001;126(1):81–88.

Ashford, R.W., The leishmaniases as emerging and reemerging zoonoses. *International Journal for Parasitology*, 2000;30(12–13):1269–1281.

Chua, K.B., et al., Nipah virus: A recently emergent deadly paramyxovirus. *Science*, 2000;288(5470):1432–1435.

Daszak, P., A.A. Cunningham, and A.D. Hyatt, Anthropogenic environmental change and the emergence of infectious diseases in wildlife. *Acta Tropica*, 2001;78(2):103–116.

Daszak, P., A.A. Cunningham, and A.D. Hyatt, Wildlife ecology—emerging infectious diseases of wildlife—threats to biodiversity and human health. *Science*, 2000;287(5452):443–449.

Endy, T.P., and A. Nisalak, Japanese encephalitis virus: Ecology and epidemiology, in *Japanese Encephalitis and West Nile Viruses*. New York, Springer Verlag, 2002, 11–48.

Enria, D.A., A.M. Briggiler, and M.R. Feuillade, An overview of the epidemiological, ecological and preventive hallmarks of Argentine haemorrhagic fever (Junin virus). *Bulletin de l'Institut Pasteur*, 1998;96(2):103–114.

Gratz, N., The impact of rice production on vector-borne disease problems in developing countries, in *Vector Borne Disease Control in Humans Through Rice Agroecosystem Management*. International Rice Research Institute, Los Baños, Philippines, 1988, 7–12.

Harvell, C.D., et al., Ecology—climate warming and disease risks for terrestrial and marine biota. *Science*, 2002;296(5576):2158–2162.

LoGiudice, K., et al., The ecology of infectious disease: Effects of host diversity and community composition on Lyme disease risk. *Proceedings of the National Academy of Sciences of the USA*, 2003;100(2):567–571.

Mackenzie, J.S., D.J. Gubler, and L.R. Petersen, Emerging flaviviruses: The spread and resurgence of Japanese encephalitis, West Nile and dengue viruses. *Nature Medicine*, 2004;10(12):S98–S109.

McMichael, A.J., The urban environment and health in a world of increasing globalization: Issues for developing countries. *Bulletin of the World Health Organization*, 2000;78(9):1117–1126.

Mead, P.S., et al., Food-related illness and death in the United States. *Emerging Infectious Diseases*, 1999;5(5):607–625.

Molyneux, D.H., Common themes in changing vector-borne disease scenarios. *Transactions of the Royal Society of Tropical Medicine and Hygiene*, 2003;97(2):129–132.

Molyneux, D.H., Patterns of change in vector-borne diseases. *Annals of Tropical Medicine and Parasitology*, 1997;91(7):827–839.

Molyneux, D.H., Vector-borne infections in the tropics and health policy issues in the twenty-first century. *Transactions of the Royal Society of Tropical Medicine and Hygiene*, 2001;95(3):233–238.

Murcia, C., Edge effects in fragmented forests—implications for conservation. *Trends in Ecology and Evolution*, 1995;10(2):58–62.

Murua, R., et al., Hantavirus pulmonary syndrome: Current situation among rodent reservoirs and human population in the Xth Region, Chile. *Revista Medica de Chile*, 2003;131(2):169–176.

Ngonseu, E., G.J. Greer, and R. Mimpfoundi, Population-dynamics and infestation of Bulinus-truncatus and Bulinus-forskalii by schistosome larvae in the Sudan-Sahelian zone of Cameroon. *Annales de la Societe Belge de Medecine Tropicale*, 1992;72(4):311–320.

Nijera, J., Malaria and rice: Strategies for control, in *Vector-Borne Disease Control in Humans Through Rice Agroecosystem Management*. International Rice Research Institute, Los Banos, Philippines, 1988, 122–132.

Nupp, T.E., and R.K. Swihart, Effects of forest fragmentation on population attributes of white-footed mice and eastern chipmunks. *Journal of Mammalogy*, 1998;79(4):1234–1243.

Ostfeld, R.S., and F. Keesing, Biodiversity and disease risk: The case of lyme disease. *Conservation Biology*, 2000;14(3):722–728.

Patz, J.A., et al., Effects of environmental change on emerging parasitic diseases. *International Journal for Parasitology*, 2000;30(12–13):1395–1405.

Poon, L.L.M., et al., The aetiology, origins, and diagnosis of severe acute respiratory syndrome. *Lancet Infectious Diseases*, 2004;4(11):663–671.

Ramasamy, R., et al., Malaria transmission at a new irrigation project in Sri-Lanka—the emergence of Anopheles-annularis as a major vector. *American Journal of Tropical Medicine and Hygiene*, 1992;47(5):547–553.

Schmid, K.A., and R.S. Ostfeld, Biodiversity and the dilution effect in disease ecology. *Ecology*, 2001;82(3):609–619.

Southgate, V.R., et al., Observations on the compatibility between Bulinus spp. and Schistosoma haematobium in the Senegal River basin. *Annals of Tropical Medicine and Parasitology*, 2000;94(2):157–164.

Southgate, V.R., et al., Studies on the biology of schistosomiasis with emphasis on the Senegal river basin. *Memorias do Instituto Oswaldo Cruz*, 2001;96:75–78.

Vercruysse, J., V.R. Southgate, and D. Rollinson, The epidemiology of human and animal schistosomiasis in the Senegal River basin. *Acta Tropica*, 1985;42(3):249–259.

Wager, R., Elizabeth Springs goby and Edgbaston goby: Distribution and status, in *Endangered Species Unit Project Number 417*. Australian Nature Conservation Agency, Canberra, 1995.

Wager, R., Final Report Part B: The distribution of two endangered fish in Queensland. The distribution and status of the red-finned blue eye, in *Endangered Species Project Number 276*. Australian Nature Conservation Agency, Canberra, 1994, 32, 65A.

WHO, Hurricane Mitch—Update 5. 1998; available from www.who.int/csr/don/1998_11_24/en/index.html [cited August 17, 2006].

Wolfe, N.D., et al., Exposure to nonhuman primates in rural Cameron. *Emerging Infectious Diseases*, 2004;10(12):2094–2099.

Chapter 8

1. Cohen, J.E., Human population: The next half century. *Science*, 2003;302(5648):1172–1175.

2. World Resources Institute, EarthTrends. 2006; available from earthtrends.wri.org/ [cited September 26, 2006].

3. Arnold, M.L., Natural hybridization and the evolution of domesticated, pest and disease organisms. *Molecular Ecology*, 2004;13(5):997–1007.

4. Pollan, M., *The Omnivore's Dilemma: A Natural History of Four Meals*. Penguin Press, New York, 2006.

5. Smith, B., *The Emergence of Agriculture*. Scientific American Library, New York, 1995.

6. Diamond, J., *Guns, Germs, and Steel: The Fates of Human Societies*. W.W. Norton & Company, New York, 1999.

7. Larsen, C.S., Animal source foods and human health during evolution. *Journal of Nutrition*, 2003;133(11):3893S–3897S.

8. Leach, H.M., Human domestication reconsidered. *Current Anthropology*, 2003;44(3):349–368.

9. Foley, J.A., et al., Global consequences of land use. *Science*, 2005;309(5734):570–574.

10. U.N. Food and Agriculture Organization, Biological Diversity in Food and Agriculture: Crops. FAO, 2004; available from www.fao.org/biodiversity/crops_en.asp [cited October 12, 2006].

11. U.N. Food and Agriculture Organization, Biological Diversity in Food and Agriculture: Domestic Animal Genetic Diversity. FAO, 2004; available from www.fao.org/biodiversity/Domestic_en.asp [cited October 12, 2006].

12. Vandermeer, J., Biodiversity loss in and around agroecosystems, in *Biodiversity and Human Health*, F. Grifo and J. Rosenthal (editors). Island Press, Washington, DC, 1997.

13. Pollard, K.A., and J.M. Holland, Arthropods within the woody element of hedgerows and their distribution pattern. *Agricultural and Forest Entomology*, 2006;8(3):203–211.

14. Lande, R., Genetics and demography in biological conservation. *Science*, 1988;241(4872):1455–1460.

15. Di Falco, S., and J.P. Chavas, Crop genetic diversity, farm productivity and the management of environmental risk in rainfed agriculture. *European Review of Agricultural Economics*, 2006;33(3):289–314.

16. Mokyr, J., Irish Potato Famine. Encyclopedia Britannica Online, 2006; available from www.search.eb.com/eb/article-9003032 [cited October 12, 2006].

17. McMullen, M., R. Jones, and D. Gallenberg, Scab of wheat and barley: A re-emerging disease of devastating impact. *Plant Disease*, 1997;81(12):1340–1348.

18. Wanyera, R., et al., The spread of stem rust caused by Puccinia graminis f. sp tritici, with virulence on Sr31 in wheat in Eastern Africa. *Plant Disease*, 2006;90(1):113.

19. Expert Panel on the Stem Rust Outbreak in Eastern Africa, *Sounding the Alarm on Global Stem Rust: An Assessment of Race Ug99 in Kenya and Ethiopia and the potential for Impact in Neighboring Regions and Beyond*. International Maize and Wheat Improvement Center (CIMMYT), El Batan, Mexico, 2005.

20. Wolfe, M.S., Crop strength through diversity. *Nature*, 2000;406(6797):681–682.

21. Wolfe, M., Barley diseases: Maintaining the value of our varieties, in *Barley Genetics*, L. Munck (editor). Munksgaard International, Copenhagen, 1992, 1055–1067.

22. Jackson, L. (editor), *Ecology in Agriculture*. Elsevier Academic Press, San Diego, 1997.

23. Tonhasca, A., and D.N. Byrne, The effects of crop diversification on herbivorous insects—a metaanalysis approach. *Ecological Entomology*, 1994;19(3):239–244.

24. Ogol, C., J.R. Spence, and A. Keddie, Maize stem borer colonization, establishment and crop damage levels in a maize-leucaena agroforestry system in Kenya. *Agriculture Ecosystems and Environment*, 1999;76(1):1–15.

25. Ehler, L.E., Integrated pest management (IPM): Definition, historical development and implementation, and the other IPM. *Pest Management Science*, 2006;62(9):787–789.

26. Bianchi, F., C.J.H. Booij, and T. Tscharntke, Sustainable pest regulation in agricultural landscapes: A review on landscape composition, biodiversity and natural pest control. *Proceedings of the Royal Society of London Series B—Biological Sciences*, 2006;273(1595):1715–1727.

27. Nyffeler, M., and K.D. Sunderland, Composition, abundance and pest control potential of spider communities in agroecosystems: A comparison of European and US studies. *Agriculture Ecosystems and Environment*, 2003;95(2–3):579–612.

28. Donald, P.F., R.E. Green, and M.F. Heath, Agricultural intensification and the collapse of Europe's farmland bird populations. *Proceedings of the Royal Society of London Series B—Biological Sciences*, 2001;268(1462):25–29.

29. Sinclair, A.R.E., S.A.R. Mduma, and P. Arcese, Protected areas as biodiversity benchmarks for human impact: Agriculture and the Serengeti avifauna. *Proceedings of the Royal Society of London Series B—Biological Sciences*, 2002;269(1508):2401–2405.

30. Vibe-Petersen, S., H. Leirs, and L. De Bruyn, Effects of predation and dispersal on *Mastomys natalensis* population dynamics in Tanzanian maize fields. *Journal of Animal Ecology*, 2006;75(1):213–220.

31. Kearns, C.A., D.W. Inouye, and N.M. Waser, Endangered mutualisms: The conservation of plant-pollinator interactions. *Annual Review of Ecology and Systematics*, 1998;29:83–112.

32. Reddi, E., Under pollination a major constraint on cashewnut production. *Proceedings of the Indian Academy of Sciences*, 1987;B53:249–252.

33. Richards, A.J., Does low biodiversity resulting from modern agricultural practice affect crop pollination and yield? *Annals of Botany*, 2001;88(2):165–172.

34. Williams, I.H., Aspects of bee diversity and crop pollination in the European Union, in *The Conservation of Bees*, S. Buchmann, et al. (editors). Academic Press, London, 1996.

35. Corbet, S.A., I.H. Williams, and J.L. Osborne, Bees and the pollination of crops and wild flowers in the European Community. *Bee World*, 1991;72(2):47–59.

36. Barrioneuvo, A., Honeybees, gone with the wind, leave crops and keepers in peril. *New York Times*, February 27, 2007, A1.

37. vanEngelsdorp, D., et al., "Fall-Dwindle Disease": Investigations into the Causes of Sudden and Alarming Colony Losses Experienced by Beekeepers in the Fall of 2006. 2006; available from www.ento.psu.edu/MAAREC/pressReleases/ColonyCollapseDisorderWG.html [cited March 11, 2007].

38. Halm, M.P., et al., New risk assessment approach for systemic insecticides: The case of honey bees and imidacloprid (Gaucho). *Environmental Science and Technology*, 2006;40(7):2448–2454.

39. Cox-Foster, D.L., et al., A metagenomic survey of microbes in honey bee colony collapse disorder. *Sciencexpress* (epublication ahead of print), 2007; available from www.sciencemag.org [cited September 13, 2007].

40. Wilcock, C., and R. Neiland, Pollination failure in plants: Why it happens and when it matters. *Trends in Plant Science*, 2002;7(6):270–277.

41. Biesmeijer, J.C., et al., Parallel declines in pollinators and insect-pollinated plants in Britain and the Netherlands. *Science*, 2006;313(5785):351–354.

42. Steffan-Dewenter, I., and T. Tscharntke, Effects of habitat isolation on pollinator communities and seed set. *Oecologia*, 1999;121(3):432–440.

43. Mänd, M., R. Mänd, and I.H. Williams, Bumblebees in the agricultural landscape of Estonia. *Agriculture Ecosystems and Environment*, 2002;89(1–2):69–76.

44. Ellstrand, N., *Dangerous Liaisons? When Cultivated Plants Mate with Their Wild Relatives*. Johns Hopkins University Press, Baltimore, MD, 2003.

45. Ladizinsky, G., Founder effect in crop-plant evolution. *Economic Botany*, 1985;39(2):191–199.

46. Ellstrand, N.C., and K.A. Schierenbeck, Hybridization as a stimulus for the evolution of invasiveness in plants? *Proceedings of the National Academy of Sciences of the USA*, 2000;97(13):7043–7050.

47. Oka, H., Ecology of Wild-Rice Planted in Taiwan. 2. Comparison of 2 Populations with Different Genotypes. *Botanical Bulletin of Academia Sinica*, 1992;33(1):75–84.

48. Chang, T., Rice, in *Evolution of Crop Plants*, J. Smartt and N. Simmonds (editors). Longman, Harlow, UK, 1995, 147–155.

49. Small, E., Hybridization in the domesticated-weed-wild complex, in *Plant Biosystematics*, W. Grant (editor). Academic Press, Toronto, 1984, 195–210.

50. Saeglitz, C., M. Pohl, and D. Bartsch, Monitoring gene flow from transgenic sugar beet using cytoplasmic male-sterile bait plants. *Molecular Ecology*, 2000;9(12):20352040.

51. U.S. Department of Agriculture, Plant Breeding Genetics and Genomics: National Plant Germplasm System. 2006; available from www.csrees.usda.gov/nea/plants/in_focus/pbgg_if_npgs.html [cited October 12, 2006].

52. Malawi Government, *Malawi: Country Report to the FAO International Technical Conference on Plant Genetic Resource*. Food and Agriculture Organization of the United Nations, Li Longwe, Malawi, 1996.

53. International Plant Genetic Resources Institute, Home page. 2006; available from www.bioversityinternational.org/ [cited October 12, 2006].

54. Pearce, F., Doomsday vault to avert world famine. *New Scientist*, 2006;189(2534):12.

55. Brussaard, L., et al., Biodiversity and ecosystem functioning in soil. *Ambio*, 1997;26(8):563–570.

56. Wall, D.H., and R.A. Virginia, The world beneath our feet: Soil biodiversity and ecosystem functioning, in *Nature and Human Society: The Quest for a Sustainable World*, P. Raven and T. Williams (editors). National Academy of Sciences and National Research Council, Washington, DC, 2000, 225–241.

57. Groombridge, B. (editor), *Global Biodiversity: Status of the Earth's Living Resources*. World Conservation Monitoring Center. 1992, Chapman and Hall, London.

58. Wall, D.H. (editor), *Sustaining Biodiversity and Ecosystem Services in Soils and Sediments*. Island Press, Washington, DC, 2004.

59. Coleman, D.C., D.A. Crossley, and P.F. Hendrix (editors), *Fundamentals of Soil Ecology*, 2nd ed. Elsevier Press, San Diego, 2004.

60. Wall-Freckman, D.W., et al., Linking biodiversity and ecosystem functioning of soils and sediments. *Ambio*, 1997;26(8):556–562.

61. Overgaard-Nielsen, C., Studies on Enchytraeidae 2: Field studies *Natura Jutlandica* 1955;4:5–58.

62. Bardgett, R.D., et al., The influence of soil biodiversity on hydrological pathways and the transfer of materials between terrestrial and aquatic ecosystems. *Ecosystems*, 2001;4(5):421–429.

63. Wardle, D.A., *Communities and Ecosystems: Linking the Aboveground and Belowground Components*. Princeton University Press, Princeton, NJ, 2002.

64. Wall, D.H., P. Snelgrove, and A. Covich, Conservation priorities for soil and sediment invertebrates, in *Conservation Biology: Research Priorities for the next Decade*, M. Soulé and G. Orians (editors). Island Press, Washington, DC, 2001.

65. Eggleton, P., et al., The species richness and composition of termites (Isoptera) in primary and regenerating lowland dipterocarp forest in Sabah, East Malaysia. *Ecotropica*, 1997;3:119–128.

66. Freckman, D.W., and C.H. Ettema, Assessing nematode communities in agro-ecosystems of varying human intervention. *Agriculture Ecosystems and Environment*, 1993;45(3–4):239–261.

67. Nestel, D., F. Dickschen, and M.A. Altieri, Diversity patterns of soil macro-coleoptera in Mexican shaded and unshaded coffee agroecosystems—an indication of habitat perturbation. *Biodiversity and Conservation*, 1993;2(1):70–78.

68. Perfecto, I., and R. Snelling, Biodiversity and the transformation of a tropical agroecosystem—ants in coffee plantations. *Ecological Applications*, 1995;5(4):1084–1097.

69. Thompson, J., Decline of vesicular-arbuscular mycorrhizas in long fallow disorder of field crops and its expression in phosphorus deficiency of sunflower. *Australian Journal of Agricultural Research* 1987;38:847–867.

70. Wasilewska, L., The relationship between the diversity of soil nematode communities and the plant species richness of meadows. *Ekologia Polska*, 1997;45:719–732.

71. Chauvel, A., et al., Pasture damage by an Amazonian earthworm. *Nature*, 1999;398(6722):32–33.

72. Anderson, D., Below-ground herbivory in natural communities: A review emphasizing fossorial animals. *Quarterly Review of Biology*, 1987;62:261–286.

73. Fragoso, C., et al., Agricultural intensification, soil biodiversity and agroecosystem function in the tropics: The role of earthworms. *Applied Soil Ecology*, 1997;6(1):17–35.

74. Hendrix, P.F., et al., Detritus food webs in conventional and no-tillage agroecosystems. *Bioscience*, 1986;36(6):374–380.

75. Edwards, C.A., and J.R. Lofty, Effects of earthworm inoculation upon the root-growth of direct drilled cereals. *Journal of Applied Ecology*, 1980;17(3):533–543.

76. Rovira, A., The effect of farming practices on the soil biota, in *Soil Biota Management in Sustainable Farming Systems*, C.E. Pankhurst, et al. (editors). CSIRO, East Melbourne, Victoria, Australia, 1994, 81–87.

77. Pankhurst, C., and J. Lynch, The role of the soil biota in sustainable agriculture, in *Soil Biota Management in Sustainable Farming Systems*, C.E. Pankhurst, et al. (editors). CSIRO, East Melbourne, Victoria, Australia, 1994, 3–9.

78. Blair, J., P. Bohlen, and D. Freckman, Soil invertebrates as indicators of soil quality, in *Methods for Assessing Soil Quality*. Soil Science Society of America, Madison, WI, 1996, 273–291.

79. Lal, R., Soil carbon sequestration impacts on global climate change and food security. *Science*, 2004;304(5677):1623–1627.

80. Delgado, C., et al., Livestock to 2020: The Next Food Revolution (Food, Agriculture and the Environment Paper 28). International Food Policy Research Institute, Washington, DC, 1999.

81. Moss, A.R., J.P. Jouany, and J. Newbold, Methane production by ruminants: Its contribution to global warming. *Annales De Zootechnie*, 2000;49(3):231–253.

82. UNDP, *Globalization with a Human Face. Human Development Report 1999*. U.N. Development Programme, New York, 1999.

83. Pinstrup-Andersen, P., R. Pandya-Lorch, and M.W. Rosengrant, The world food situation: Recent developments, emerging issues, and long-term prospects, in *Food Policy Report*. International Food Policy Research Institute, Washington, DC, 1997.

84. U.N. Population Fund, *State of the World Population 2002: People, Poverty, and Possibilities*. U.N. Population Fund, New York, 2002.

85. Horrigan, L., R.S. Lawrence, and P. Walker, How sustainable agriculture can address the environmental and human health harms of industrial agriculture. *Environmental Health Perspectives*, 2002;110(5):445–456.

86. Ponting, C., *A Green History of the World. The Environment and the Collapse of Great Civilizations*. Penguin Books, New York, 1991.

87. Simon, M.F., and F.L. Garagorry, The expansion of agriculture in the Brazilian Amazon. *Environmental Conservation*, 2005;32(3):203–212.

88. Henderson, J.P., Anaerobic Digestion in Rural China. 2006; available from www.epa.gov/agstar/resources/biocycle2.html [cited September 30, 2006].

89. Mukiibi, J. Opening remarks, in *Modernizing Agriculture: Visions and Technologies for Animal Traction and Conservation Agriculture*. Uganda Network for Animal Traction and Conservation Agriculture, Jinja, Uganda, 2002.

90. U.N. Food and Agriculture Organization, *Protecting Animal Genetic Diversity for Food and Agriculture*. FAO, Rome, 2006.

91. U.S. EPA, Ag 101: Dairy Production Systems. 2006; available from www.epa.gov/agriculture/ag101/dairysystems.html [cited September 24, 2006].

92. Easterling, W., and M. Apps, Assessing the consequences of climate change for food and forest resources: A view from the IPCC. *Climatic Change*, 2005;70(1–2):165–189.

93. Tews, J., et al., Linking a population model with an ecosystem model: Assessing the impact of land use and climate change on savanna shrub cover dynamics. *Ecological Modelling*, 2006;195(3–4):219–228.

94. Rosensweig, C., et al., *Climate Change and U.S. Agriculture: The Impacts of Warming and Extreme Weather Events on Productivity, Plant Diseases, and Pests*. Center for Health and the Global Environment, Harvard Medical School, Boston, MA, 2000.

95. Aitken, I., Environmental change and animal disease, in *The Advancement of Veterinary Science: The Bicentenary Symposium Series. Veterinary Medicine Beyond 2000*, A. Michell (editor). CAB International, Wallingford, UK, 1993, 179–193.

96. Jenkins, E.J., et al., Climate change and the epidemiology of protostrongylid nematodes 86 northern ecosystems: Parelaphostrongylus adocoilei and Protostrongylus stilesi in Dall's sheep (Ovis d. dalli). *Parasitology*, 2006;132:387–401.

97. Purse, B.V., et al., Climate change and the recent emergence of bluetongue in Europe. *Nature Reviews Microbiology*, 2005;3(2):171–181.

98. de Haan, C.S.H, and H. Blackburn, *Livestock and the Environment: Finding the Balance*. WRENmedia, Suffolk, UK, 1997.

99. Hughes, T.P., et al., Climate change, human impacts, and the resilience of coral reefs. *Science*, 2003;301(5635):929–933.

100. Hutchings, P., M. Peyrot-Clausade, and A. Osnorno, Influence of land run-off on rates and agents of bioerosion of coral substrates. *Marine Pollution Bulletin*, 2005;51(1–4):438–447.

101. Mallin, M.A., et al., Comparative effects of poultry and swine waste lagoon spills on the quality of receiving streamwaters. *Journal of Environmental Quality*, 1997;26(6):1622–1631.

102. Pinckney, J.L., et al., Responses of phytoplankton and Pfiesteria-like dinoflagellate zoospores to nutrient enrichment in the Neuse River Estuary, North Carolina, USA. *Marine Ecology Progress Series*, 2000;192:65–78.

103. Batoreu, M.C.C., et al., Risk of human exposure to paralytic toxins of algal origin. *Environmental Toxicology and Pharmacology*, 2005;19(3):401–406.

104. Friedman, M.A., and B.E. Levin, Neurobehavioral effects of harmful algal bloom (HAB) toxins: A critical review. *Journal of the International Neuropsychological Society*, 2005;11(3):331–338.

105. National Oceanic and Atmospheric Administration, Harmful Algal Blooms. 2006; available from www.cop.noaa.gov/stressors/extremeevents/hab/ [cited September 30, 2006].

106. Phoofolo, P., Face to face with famine: The BaSotho and the rinderpest, 1897–1899. *Journal of Southern African Studies*, 2003;29(2):503–527.

107. Plowright, W., Rinderpest in the world today—control and possible eradication by vaccination. *Annales de Medecine Veterinaire*, 1985;129(1):9–32.

108. Plowright, W., Effects of rinderpest and rinderpest control on wildlife in Africa. *Symposium of the Zoological Society of London*, 1982;50:1–27.

109. Roeder, P., Infectious diseases: Preparing for the future. A case study of Rinderpest in Africa, in *Foresight*. Office of Science and Innovation, London, 2006.

110. Roelke-Parker, M.E., et al., A canine distemper virus epidemic in Serengeti lions (Panthera leo). *Nature*, 1996;379(6564):441–445.

111. Schaftenaar, W., Use of vaccination against foot and mouth disease in zoo animals, endangered species and exceptionally valuable animals. *Revue Scientifique et Technique de l'Office International des Epizooties*, 2002;21(3):613–623.

112. Horton, M., The human-settlement of the Red Sea, in *Key Environments: Red Sea*, A. Edwards and A. Head (editors). Pergamon Press, Oxford, 1987, 339–362.

113. Jackson, J.B.C., Reefs since Columbus. *Coral Reefs*, 1997;16:S23–S32.

114. U.N. Food and Agriculture Organization, The State of World Fisheries and Aquaculture (SOFIA) 2002. FAO, 2002; available from www.fao.org/docrep/005/y7300e/y7300e00.htm [cited October 12, 2006].

115. U.N. Food and Agriculture Organization, *World Watch List for Domestic Animal Diversity*, 3rd ed. FAO, Rome, 2000.

116. Trites, A., et al., Ecosystem change and the decline of marine mammals in the Eastern Bering Sea: Testing the ecosystem shift and commercial whaling hypotheses. *Fisheries Centre Research Reports*, 1999;7(1):1–107.

117. New, M., Aquaculture and the capture fisheries—Balancing the scales. *World Aquaculture*, 1997;28(2):11–30.

118. Westlund, L. Apparent historical consumption and future demand for fish and fishery products—exploratory calculations, in *International Conference on the Sustainable Contribution of Fisheries to Food Security*. Kyoto Declaration and Plan of Action, Kyoto, Japan, 2000.

119. U.N. Food and Agriculture Organization, Strategies for increasing the sustainable contribution of small-scale fisheries to food security and poverty alleviation, in *Committee on Fisheries*. FAO, Rome, 2003.

120. Thorpe, A., Mainstreaming Fisheries into National Development and Poverty Reduction Strategies: Current Situation and Opportunities (FAO Fisheries Circular No. 997). FAO, Rome, 2005.

121. Price, A., et al., *Coasts: Environment and Development Briefs*. UNESCO, Paris, 1993, 16.

122. Morris, A.V., C.M. Roberts, and J.P. Hawkins, The threatened status of groupers (Epinephelinae). *Biodiversity and Conservation*, 2000;9(7):919–942.

123. Pauly, D., et al., Fishing down marine food webs. *Science*, 1998;279(5352):860–863.

124. Barnes, R., and R. Hughes, *An Introduction to Marine Ecology*. Oxford, Blackwell Scientific Publications, 1982.

125. de Roos, A.M., D.S. Boukal, and L. Persson, Evolutionary regime shifts in age and size at maturation of exploited fish stocks. *Proceedings of the Royal Society of London Series B—Biological Sciences*, 2006;273(1596):1873–1880.

126. Brashares, J.S., et al., Bushmeat hunting, wildlife declines, and fish supply in West Africa. *Science*, 2004;306(5699):1180–1183.

127. Peeters, M., et al., Risk to human health from a plethora of Simian immunodeficiency viruses in primate bushmeat. *Emerging Infectious Diseases*, 2002;8(5):451–457.

128. Wolfe, N.D., et al., Emergence of unique primate T-lymphotropic viruses among central African bushmeat hunters. *Proceedings of the National Academy of Sciences of the USA*, 2005;102(22):7994–7999.

129. Leidy, R., and P. Moyle, Conservation of the world's freshwater fish fauna: An overview, in *Conservation Biology: For the Coming Decade*, P. Fiedler and P. Karieva (editors). Chapman & Hall, New York, 1997.

130. International Union for Conservation of Nature and Natural Resources, Red List of Threatened Species. 2006; available from www.iucnredlist.org/ [cited October 20, 2006].

131. Meybeck, M., Global analysis of river systems: From Earth system controls to Anthropocene syndromes. *Philosophical Transactions of the Royal Society of London Series B—Biological Sciences*, 2003;358(1440):1935–1955.

132. Ormerod, S.J., Current issues with fish and fisheries: Editor's overview and introduction. *Journal of Applied Ecology*, 2003;40(2):204–213.

133. U.N. Food and Agriculture Organization, The State of World Fisheries and Aquaculture (SOFIA) 2004. FAO, 2004; available from www.fao.org/DOCREP/007/y5600e/y5600e00.htm [cited October 12, 2006].

134. Cabello, F.C., Heavy use of prophylactic antibiotics in aquaculture: A growing problem for human and animal health and for the environment. *Environmental Microbiology*, 2006;8(7):1137–1144.

135. Christensen, A.M., F. Ingerslev, and A. Baun, Ecotoxicity of mixtures of antibiotics used in aquacultures. *Environmental Toxicology and Chemistry*, 2006;25(8):2208–2215.

136. Akinbowale, O.L., H. Peng, and M.D. Barton, Antimicrobial resistance in bacteria isolated from aquaculture sources in Australia. *Journal of Applied Microbiology*, 2006;100(5):1103–1113.

137. Le, T.X., Y. Munekage, and S. Kato, Antibiotic resistance in bacteria from shrimp farming in mangrove areas. *Science of the Total Environment*, 2005;349(1–3):95–105.

138. Costanzo, S.D., M.J. O'Donohue, and W.C. Dennison, Assessing the influence and distribution of shrimp pond effluent in a tidal mangrove creek in north-east Australia. *Marine Pollution Bulletin*, 2004;48(5–6):514–525.

139. Sorokin, Y.I., P.Y. Sorokin, and G. Ravagnan, Hypereutrophication events in the Ca'Pisani lagoons associated with intensive aquaculture. *Hydrobiologia*, 2006;571:1–15.

140. Lotz, J.M., and M.A. Soto, Model of white spot syndrome virus (WSSV) epidemics in Litopenaeus vannamei. *Diseases of Aquatic Organisms*, 2002;50(3):199–209.

141. Nadala, E.C.B., and P.C. Loh, A comparative study of three different isolates of white spot virus. *Diseases of Aquatic Organisms*, 1998;33(3):231–234.

142. Hilborn, R., Salmon-farming impacts on wild salmon. *Proceedings of the National Academy of Sciences of the USA*, 2006;103(42):15277.

143. Krkosek, M., et al., Epizootics of wild fish induced by farm fish. *Proceedings of the National Academy of Sciences of the USA*, 2006;103(42):15506–15510.

144. Porter, G., *Protecting Wild Atlantic Salmon from Impacts of Salmon Aquaculture.* World Wildlife Fund and Atlantic Salmon Federation, Washington, DC, 2005

145. Gross, M.R., One species with two biologies: Atlantic salmon (Salmo salar) in the wild and in aquaculture. *Canadian Journal of Fisheries and Aquatic Sciences*, 1998;55:131–144.

146. Hansen, P., J. Jacobsen, and R. Und, High numbers of farmed Atlantic salmon, Salmo salar, observed in oceanic waters north of the Faroe Islands. *Aquaculture Fisheries Management*, 1993;24:777–781.

147. Naylor, R.L., J. Eagle, and W.L. Smith, Salmon aquaculture in the Pacific Northwest—a global industry. *Environment*, 2003;45(8):18–39.

148. Naylor, R.L., et al., Effect of aquaculture on world fish supplies. *Nature*, 2000;405(6790):1017–1024.

149. Mukerjee, M., Pink gold—the trials and tribulations of shrimp farming. *Scientific American*, 1996;275(1):24–26.

150. Naylor, R.L., et al., Ecology—nature's subsidies to shrimp and salmon farming. *Science*, 1998;282(5390):883–884.

151. Barbier, E.B., Natural barriers to natural disasters: Replanting mangroves after the tsunami. *Frontiers in Ecology and the Environment*, 2006;4(3):124–131.

152. Thampanya, U., et al., Coastal erosion and mangrove progradation of Southern Thailand. *Estuarine Coastal and Shelf Science*, 2006;68(1–2):75–85.

153. Tacon, A., Feeding tomorrow's fish. *Aquaculture*, 1996;27:20–32.

154. Martins, D.A., et al., Growth, digestibility and nutrient utilization of rainbow trout (Oncorhynchus mykiss) and European sea bass (Dicentrarchus labrax) juveniles fed different dietary soybean oil levels. *Aquaculture International*, 2006;14(3):285–295.

155. Bell, J.G., et al., Dioxin and dioxin-like polychlorinated biphenyls (PCBS) in Scottish farmed salmon (Salmo salar): Effects of replacement of dietary marine fish oil vegetable oils. *Aquaculture*, 2005;243(1–4):305–314.

156. Neori, A., et al., Integrated aquaculture: Rationale, evolution and state of the art emphasizing seaweed biofiltration in modem mariculture. *Aquaculture*, 2004;231(1–4):361–391.

157. Liu, F., and W.Y. Han, Reuse strategy of wastewater in prawn nursery by microbial remediation. *Aquaculture*, 2004;230(1–4):281–296.

158. Tacon, A.G.J., and S.S. DeSilva, Feed preparation and feed management strategies within semi-intensive fish farming systems in the tropics. *Aquaculture*, 1997;151(1–4):379–404.

159. Hagiwara, H., and W.J. Mitsch, Ecosystem modeling of a multispecies integrated aquaculture pond in South China. *Ecological Modelling*, 1994;72(1–2):41–73.

160. Troell, M., et al., Integrated marine cultivation of Gracilaria chilensis (Gracilariales, Rhodophyta) and salmon cages for reduced environmental impact and increased economic output. *Aquaculture*, 1997;156(1–2):45–61.

161. Castro, R.S., C. Azevedo, and F. Bezerra-Neto, Increasing cherry tomato yield using fish effluent as irrigation water in Northeast Brazil. *Scientia Horticulturae*, 2006;110(1):44–50.

Box 8.1

a. Wolfe, M.S., Crop strength through diversity. *Nature*, 2000;406(6797):681–682.

b. Zhu, Y.Y., et al., Genetic diversity and disease control in rice. *Nature*, 2000;406(6797):718–722.

Box 8.2

a. Buchori, D., and S. Manuwoto, The role of crop protection in agricultural development in Indonesia, in *AP31 GIFAP (International Group of National Associations of Manufacturers of Agrochemical Products) Asian Working Group*. Jakarta, Indonesia, 1995.

b. Rubia, E., et al., Stemborer damage and grain yield of flooded rice. *Journal of Plant Protection in the Tropics*, 1989;6:205–211.

c. Mochida, O., Brown planthopper "hama wereng" problems on rice in Indonesia, in *Cooperative CRIA-IRRI Program, Sukamandi, West Java. Indonesia*. International Rice Research Institute, Los Banos, Phillipines, 1978.

d. Settle, W.H., et al., Managing tropical rice pests through conservation of generalist natural enemies and alternative prey. *Ecology*, 1996;77(7):1975–1988.

e. Gallagher, K., *Effects of Host Plant Resistance on the Microevolution of Rice Brown Planthopper, Nilaparvata lugens (Stahl)*. University of California, Berkeley, 1988.

f. Shepard, B., A. Barrion, and J. Litsinger, *Helpful Insects, Spiders and Pathogens*. International Rice Research Institute, Los Banos, Phillipines, 1994.

g. Buchori, D., and H. Triwidodo, Conserving diversity and sustainability in tropical agriculture: On farm experiences with Karawang rice farmers, in *The South East Asia Regional Workshop on Sustainable Agriculture: Toward a Sustainable Food Supply for All*. Third World Network, Konphalindo (National Consortium for Forest and Nature Conservation in Indonesia), Mojokerto, East Java, Indonesia, 1997.

h. Way, M.J., and K.L. Heong, The role of biodiversity in the dynamics and management of insect pests of tropical irrigated rice—a review. *Bulletin of Entomological Research*, 1994;84(4):567–587.

Box 8.3

a. DeFoliart, G.R., An overview of the role of edible insects in preserving biodiversity. *Ecology of Food and Nutrition*, 1997;36(2–4):109–132.

b. Pemberton, R.W., The revival of rice-field grasshoppers as human food in South-Korea. *Pan-Pacific Entomologist*, 1994;70(4):323–327.

c. Gorton, P., Villagers turn "foe" into food. *Agricultural Information Development Bulletin*, May 1988, 19–20.

d. Litton, E., Letters. *The food insects newsletter*, 1993;6(1):3.

Box 8.5

a. Taylor, T.N., and M. Krings, Fossil microorganisms and land plants: Associations and interactions. *Symbiosis*, 2005;40(3):119–135.

b. Pirozynski, K.A., Interactions between fungi and plants through the ages. *Canadian Journal of Botany—Revue Canadienne De Botanique*, 1981;59(10):1824–1827.

c. Hawksworth, D.L., The magnitude of fungal diversity: The 1.5 million species estimate revisited. *Mycological Research*, 2001;105:1422–1432.

d. Pennisi, E., The secret life of fungi. *Science*, 2004;304(5677):1620–1622.

e. Jones, M.D., D.M. Durall, and P.B. Tinker, Comparison of arbuscular and ectomycorrhizal Eucalyptus coccifera: Growth response, phosphorus uptake efficiency and external hyphal production. *New Phytologist*, 1998;140(1):125–134.

f. Heinrich, B., *The Trees in My Forest*. Harper Collins, New York, 1998.

g. Smith, S., and D. Read, *Mycorrhizal Symbiosis*, 2nd ed. Academic Press, London, 1997.

h. Mohammad, M.J., W.L. Pan, and A.C. Kennedy, Chemical alteration of the rhizosphere of the mycorrhizal-colonized wheat root. *Mycorrhiza*, 2005;15(4):259–266.

i. Al Agely, A., D.M. Sylvia, and L.Q. Ma, Mycorrhizae increase arsenic uptake by the hyperaccumulator Chinese brake fern (Pteris vittata L.). *Journal of Environmental Quality*, 2005;34(6):2181–2186.

j. Borowicz, V.A., Do arbuscular mycorrhizal fungi alter plant-pathogen relations? *Ecology*, 2001;82(11):3057–3068.

k. Selosse, M.A., E. Baudoin, and P. Vandenkoornhuyse, Symbiotic microorganisms, a key for ecological success and protection of plants. *Comptes Rendus Biologies*, 2004;327(7):639–648.

l. van der Heijden, M.G.A., et al., Mycorrhizal fungal diversity determines plant biodiversity, ecosystem variability and productivity. *Nature*, 1998;396(6706):69–72.

m. Kernaghan, G., Mycorrhizal diversity: Cause and effect? *Pedobiologia*, 2005;49(6):511–520.

n. Wardle, D.A., et al., Ecological linkages between aboveground and belowground biota. *Science*, 2004;304(5677):1629–1633.

o. Egerton-Warburton, L.M., et al., Reconstruction of the historical changes in mycorrhizal fungal communities under anthropogenic nitrogen deposition. *Proceedings of the Royal Society of London Series B—Biological Sciences*, 2001;268(1484):2479–2484.

p. Lilleskov, E.A., et al., Belowground ectomycorrhizal fungal community change over a nitrogen deposition gradient in Alaska. *Ecology*, 2002;83(1):104–115.

q. Johnson, N.C., et al., Nitrogen enrichment alters mycorrhizal allocation at five mesic to semiarid grasslands. *Ecology*, 2003;84(7):1895–1908.

r. Chung, H.G., D.R. Zak, and E.A. Lilleskov, Fungal community composition and metabolism under elevated CO_2 and O_3. *Oecologia*, 2006;147(1):143–154.

Box 8.6

a. Mulder, L., et al., Integration of signalling pathways in the establishment of the legume-rhizobia symbiosis. *Physiologia Plantarum*, 2005;123(2):207–218.

b. Karr, D.B., N.W. Oehrle, and D.W. Emerich, Recovery of nitrogenase from aerobically isolated soybean nodule bacteroids. *Plant and Soil*, 2003;257(1):27–33.

c. Angelini, J., S. Castro, and A. Fabra, Alterations in root colonization and nodC gene induction in the peanut-rhizobia interaction under acidic conditions. *Plant Physiology and Biochemistry*, 2003;41(3):289–294.

d. Ponsone, L., A. Fabra, and S. Castro, Interactive effects of acidity and aluminium on the growth, lipopolysaccharide and glutathione contents in two nodulating peanut rhizobia. *Symbiosis*, 2004;36(2):193–204.

e. Abd-Alla, M.H., S.A. Omar, and S. Karanxha, The impact of pesticides on arbuscular mycorrhizal and nitrogen-fixing symbioses in legumes. *Applied Soil Ecology*, 2000;14(3):191–200.

f. McInnes, A., and R.A. Date, Improving the survival of rhizobia on Desmanthus and Stylosanthes seed at high temperature. *Australian Journal of Experimental Agriculture*, 2005;45(2–3):171–182.

Box 8.7

a. National CBD and Biosafety Office, *Biodiversity Clearing-House Mechanism of China*. 2006; available from english.biodiv.gov.cn/ [cited September 26, 2006].

Box 8.8

a. Bren, L., Antibiotic resistance from down on the chicken farm. *FDA Consumer Magazine*, 2001;35(1).

b. Ferber, D., Antibiotic resistance: WHO advises kicking the livestock antibiotic habit. *Science*, 2003;301(5636):1027.

Additional References

Alpine, A.E., and J.E. Cloern, Trophic interactions and direct physical effects control phytoplankton biomass and production in an estuary. *Limnology and Oceanography*, 1992;37(5):946–955.

Ampofo, J.K.O., Maize stalk borer (Lepidoptera, Pyralidae) damage and plant-resistance. *Environmental Entomology*, 1986;15(6):1124–1129.

Anderson, J., The soil system, in *Global Biodiversity Assessment*, V. Haywood (editor). Cambridge University Press, Cambridge, UK, 1995.

Baker, B., et al., Signaling in plant-microbe interactions. *Science*, 1997;276(5313):726–733.

Benfey, T., Environmental impacts of genetically modified animals, in *FAO/WHO Expert Consultation of Safety Assessment of Foods Derived from Genetically Modified Animals Including Fish*. Food and Agriculture Organization of the United Nations, Rome, 2003.

Brown, L., *Who Will Feed China: Wake Up Call for a Small Planet*. W.W. Norton & Company, New York, 1995.

Brown, L., M. Renner, and B. Halweil, *Vital Signs 2000*. W.W. Norton & Company, New York, 2000.

Brown, P., et al., Bovine spongiform encephalopathy and variant Creutzfeldt-Jakob disease: Background, evolution, and current concerns. *Emerging Infectious Diseases*, 2001;7(1):6–16.

Buchmann, S., and G. Nabhan, *The Forgotten Pollinators*. Island Press, Washington, DC, 1996.

Caddy, J.F., Fisheries management in the twenty-first century: Will new paradigms apply? *Reviews in Fish Biology and Fisheries*, 1999;9(1):1–43.

Caddy, J.F., and L. Garibaldi, Apparent changes in the trophic composition of world marine harvests: The perspective from the FAO capture database. *Ocean and Coastal Management*, 2000;43(8–9):615–655.

Caldararo, N., Human ecological intervention and the role of forest fires in human ecology. *Science of the Total Environment*, 2002;292(3):141–165.

Clarke, K.R., and R.M. Warwick, Quantifying structural redundancy in ecological communities. *Oecologia*, 1998;113(2):278–289.

Collins, W.W., and C.O. Quaslet (editors), *Biodiversity in Agroecosystems*. Lewis Publishers, New York, 1998.

Daily, G., P. Matson, and P. Vitousek, Ecosystem services supplied by soil, in *Natures Services: Societal Dependence on Natural Ecosystems*, G. Daily (editor). Island Press, Washington, DC, 1997, 113–132.

Des Clers, S., Sustainability of the Falklands Islands Loligo squid fishery, in *Conservation of Biological Resources*, E. Milner-Gulland and R. Mace (editors). Blackwell Science, Oxford, UK, 1998, 225–241.

Dunbar, R., Scapegoat for a thousand deserts. *New Scientist*, 1984;104:30–33.

Dyson, T., World food trends and prospects to 2025. *Proceedings of the National Academy of Sciences of the USA*, 1999;96(11):5929–5936.

Edwards, C.A., et al., The role of agroecology and integrated farming systems in agricultural sustainability. *Agriculture Ecosystems and Environment*, 1993;46(1–4):99–121.

Ellstrand, N.C., and K.A. Schierenbeck, Hybridization as a stimulus for the evolution of invasiveness in plants? *Proceedings of the National Academy of Sciences of the USA*, 2000;97(13):7043–7050.

Ferguson, N.M., C.A. Donnelly, and R.M. Anderson, The foot-and-mouth epidemic in Great Britain: Pattern of spread and impact of interventions. *Science*, 2001;292(5519):1155–1160.

Frey, S.D., E.T. Elliott, and K. Paustian, Bacterial and fungal abundance and biomass in conventional and no-tillage agroecosystems along two climatic gradients. *Soil Biology and Biochemistry*, 1999;31(4):573–585.

Giller, K.E., et al., Agricultural intensification, soil biodiversity and agroecosystem function. *Applied Soil Ecology*, 1997;6(1):3–16.

Gjedrem, T., Selective breeding to improve aquaculture production. *World Aquaculture*, 1997;28(1):33–45.

Goni, R., Fisheries effects on ecosystems, in *Seas at the Millennium*, C. Sheppard (editor). Elsevier Science Ltd., Oxford, 2000, 117–133.

Gupta, A., et al., Antimicrobial resistance among Campylobacter strains, United States, 1997–2001. *Emerging Infectious Diseases*, 2004;10(6):1102–1109.

Hallerman, E., Hazards associated with transgenic methods, in *FAO/WHO Expert Consultation of Safety Assessment of Foods Derived From Genetically Modified Animals Including Fish*. Food and Agriculture Organization of the United Nations, Rome, 2003.

Hammond, K., Animal genetic resources for the twenty-first century. *Acta Agriculturae Scandinavica Section A—Animal Science*, 1998;48:11–18.

Hammond, K. Development of the global strategy for the management of farm animal genetic resources, in *Proceedings of the 6th World Congress on Genetics Applied to Livestock Production*. Animal Genetics and Breeding Unit, Armidale, NSW, Australia, 1998.

Haughton, A.J., et al., Invertebrate responses to the management of genetically modified herbicide-tolerant and conventional spring crops. II. Within-field epigeal and aerial arthropods. *Philosophical Transactions of the Royal Society of London Series B—Biological Sciences*, 2003;358(1439):1863–1877.

Hawes, C., et al., Responses of plants and invertebrate trophic groups to contrasting herbicide regimes in the Farm Scale Evaluations of genetically modified herbicide-tolerant crops. *Philosophical Transactions of the Royal Society of London Series B—Biological Sciences*, 2003;358(1439):1899–1913.

Hellmich, R.L., et al., Monarch larvae sensitivity to Bacillus thuringiensis-purified proteins and pollen. *Proceedings of the National Academy of Sciences of the USA*, 2001;98(21):11925–11930.

Hendrix, P.F., D.A.J. Crossley, J.M. Blair, and D.C. Coleman, Soil biota as components of sustainable agroecosystems, in *Sustainable Agricultural Systems*. Soil and Water Conservation Society, Ankeny, IA, 1990, 637–654.

Hillel, D., *Environmental Soil Physics*. Academic Press, San Diego, 1998.

Hooper, D.U., et al., Interactions between aboveground and belowground biodiversity in terrestrial ecosystems: Patterns, mechanisms, and feedbacks. *Bioscience*, 2000;50(12):1049–1061.

Hunt, H.W., et al., The detrital food web in a shortgrass prairie. *Biology and Fertility of Soils*, 1987;3(1–2):57–68.

Jenkins, M., Prospects for biodiversity. *Science*, 2003;302(5648):1175–1177.

Jennings, S., and M.J. Kaiser, The effects of fishing on marine ecosystems. *Advances in Marine Biology*, 1998;34:201–352.

Johnson, D.G., Sustainable agriculture and resistance: Transforming food production in Cuba. *Economic Development and Cultural Change*, 2003;51(4):1023–1025.

Kang, J.X., et al., Transgenic mice—Fat-1 mice convert n-6 to n-3 fatty acids. *Nature*, 2004;427(6974):504–504.

Kislev, M.E., E. Weiss, and A. Hartmann, Impetus for sowing and the beginning of agriculture: Ground collecting of wild cereals. *Proceedings of the National Academy of Sciences of the USA*, 2004;101(9):2692–2695.

Knibb, W., G. Gorshkova, and S. Gorshkov, Selection for growth in the gilthead seabream, Sparus aurata L. *Israeli Journal of Aquaculture-Bamidgeh*, 1997;49(2):57–66.

Ladizinsky, G., Ecological and genetic considerations in collecting and using wild relatives, in *The Use of Plant Genetic Resources*, A. Brown, et al. (editors). Cambridge University Press, Cambridge, UK, 1989, 297–305.

Lalli, C., and T. Parsons, *Biological Oceanography: An Introduction*. Pergamon Press, Oxford, 1993.

Lauenroth, W.K., and D.G. Milchunas, Short-grass steppe, in *Ecosystems of the World 8A*, R. Coupland (editor). New York, Elsevier, 1993, 183–226.

Losey, J.E., L.S. Rayor, and M.E. Carter, Transgenic pollen harms monarch larvae. *Nature*, 1999;399(6733):214.

Losordo, T., M. Masser, and J. Rakocy, Recirculating aquaculture tank production systems. *World Aquaculture*, 2001;32(1):18–22.

Markowitz, T.M., et al., Dusky dolphin foraging habitat: Overlap with aquaculture in New Zealand. *Aquatic Conservation-Marine and Freshwater Ecosystems*, 2004;14(2):133–149.

McGlade, J., Integrated fisheries management models: Understanding the limits to marine resource exploitation. *American Fisheries Society Symposium*, 1989;6:139–165.

McGlade, J., et al., *Rediscovery Plans for the North Sea Ecosystem, with Special Reference to Cod, Haddock and Plaice*. UK World Wildlife Fund, Godalming, UK, 1997, 33.

Miller, G.T.J., *Living in the Environment*. Wadsworth Publishing Company, Belmont, CA, 1996.

National Oceanic and Atmospheric Administration, Economic Statistics for NOAA. 2006; available from www.publicaffairs.noaa.gov/pdf/economic-statistics-may2006.pdf [cited September 26, 2006].

Nixon, S.W., Replacing the Nile: Are anthropogenic nutrients providing the fertility once brought to the Mediterranean by a great river? *Ambio*, 2003;32(1):30–39.

Orskov, E., *Reality in Rural Development Aid with Emphasis on Livestock*. Rowett Research Services Ltd., Aberdeen, UK, 1993.

Ou, S.H., Pathogen variability and host-resistance in rice blast disease. *Annual Review of Phytopathology*, 1980;18:167–187.

Pauly, D., et al., Towards sustainability in world fisheries. *Nature*, 2002;418(6898):689–695.

Petersen, H., and M. Luxton, A comparative-analysis of soil fauna populations and their role in decomposition processes. *Oikos*, 1982;39(3):287–388.

Phelps, H.L., The Asiatic clam (Corbicula-fluminea) invasion and system-level ecological change in the Potomac River estuary near Washington, DC. *Estuaries*, 1994;17(3):614–621.

Pimentel, D., *Techniques for Reducing Pesticide Use. Economic and Environmental Benefits*. John Wiley & Sons, New York, 1997.

Pimentel, D., and N. Kounang, Ecology of soil erosion in ecosystems. *Ecosystems*, 1998;1(5):416–426.

Pimentel, D., et al., Economic and environmental benefits of biodiversity. *Bioscience*, 1997;47(11):747–757.

Postel, S., Securing water for people, crops, and ecosystems: New mindset and new priorities. *Natural Resources Forum—a United Nations Journal*, 2003;27(2):89–98.

Price, A.R.G., Distribution of penaeid shrimp larvae along the Arabian Gulf coast of Saudi-Arabia. *Journal of Natural History*, 1982;16(5):745–757.

Price, L.B., et al., Fluoroquinolone-resistant Campylobacter isolates from conventional and antibiotic-free chicken products. *Environmental Health Perspectives*, 2005;113(5):557–560.

Qi, B.X., et al., Production of very long chain polyunsaturated omega-3 and omega-6 fatty acids in plants. *Nature Biotechnology*, 2004;22(6):739–745.

Rana, K., and A. Immink, Farming of aquatic organisms, particularly the Chinese and Thai experience, in *Seas at the Millenium*, C. Sheppard (editor). Elsevier Science Ltd., Oxford, UK, 2000, 165–167.

Risch, S.J., D. Andow, and M.A. Altieri, Agroecosystem diversity and pest-control—data, tentative conclusions, and new research directions. *Environmental Entomology*, 1983;12(3):625–629.

Robertson, G.P., and D.W. Freckman, The spatial-distribution of nematode trophic groups across a cultivated ecosystem. *Ecology*, 1995;76(5):1425–1432.

Root, R.B., Organization of a plant-arthropod association in simple and diverse habitats—fauna of collards (Brassica-oleracea). *Ecological Monographs*, 1973;43(1):95–120.

Rossiter, P.B., et al., Re-emergence of rinderpest as a threat in East-Africa since 1979. *Veterinary Record*, 1983;113(20):459–461.

Royal Society of London, *Genetically Modified Plants for Food Use and Human Health—an Update*. Royal Society, London, 2002.

Royal Society of London, et al., *Transgenic Plants and World Agriculture*. National Academy Press, Washington, DC, 2000.

Sasser, J. Managing nematodes by plant breeding, in *Annual Tall Timbers Conference on Ecological Animal Control by Habitat Management*. Tall Timbers Research Association, Lubbock, TX, 1972.

Sears, M.K., et al., Impact of Bt corn pollen on monarch butterfly populations: A risk assessment. *Proceedings of the National Academy of Sciences of the USA*, 2001;98(21):11937–11942.

Sherman, B., Marine ecosystem health as an expression of morbidity, mortality and disease events, in *Seas at the Millenium*, C. Sheppard (editor). Elsevier Science Ltd., Oxford, UK, 2000, 211–234.

Smith, N.J.H., et al., Agroforestry developments and potential in the Brazilian Amazon. *Land Degradation and Rehabilitation*, 1995;6(4):251–263.

Snelgrove, P.V.R., The biodiversity of macrofaunal organisms in marine sediments. *Biodiversity and Conservation*, 1998;7(9):1123–1132.

Snelgrove, P., et al., The importance of marine sediment biodiversity in ecosystem precesses. *Ambio*, 1997;26(8):578–583.

Squire, G.R., et al., On the rationale and interpretation of the farm scale evaluations of genetically modified herbicide-tolerant crops. *Philosophical Transactions of the Royal Society of London Series B—Biological Sciences*, 2003;358(1439):1779–1799.

Staskawicz, B.J., et al., Molecular-genetics of plant-disease resistance. *Science*, 1995;268(5211):661–667.

Steadman, D.W., Prehistoric extinctions of pacific island birds—biodiversity meets zoo-archaeology. *Science*, 1995;267(5201):1123–1131.

Swift, M.J., O.W. Heal, and J.M. Anderson, *Decomposition in Terrestrial Ecosystems*. Blackwell, Oxford, UK, 1979.

Thompson, P., *Agricultural Ethics: Research, Teaching, and Public Policy*. Iowa State University Press, Ames, IA, 1998.

Todd, E., The cost of marine diseases, in *Global Changes and Emergence of Infectious Diseases*, M. Wilson (editor). New York Academy of Sciences, New York, 1994, 423–435.

U.N. Development Programme, *Human Development Report 2005*. UNDP, New York, 2005.

U.N. Food and Agriculture Organization, *FAO yearbook. Fishery statistics: Capture Production*. FAO, Rome, 1998.

U.N. Food and Agriculture Organization, *Livestock Breeds of China* (FAO Animal Production and Health Papers No. 46). FAO, Rome, 1984, 217.

U.N. Food and Agriculture Organization, *Review of the State of the World Fishery Resources: Marine Fisheries* (FAO Fisheries Circular). FAO, Rome, 1995.

U.N. Food and Agriculture Organization, Soil biodiversity and sustainable agriculture, in *Convention on Biological Diversity*. FAO, Montreal, 2001.

U.N. Food and Agriculture Organization, World review of fisheries and aquaculture: Fisheries resources: Trends in production, utilization, and trade, in *The State of World Fisheries and Aquaculture*. FAO Fisheries Department, Rome, 2000.

U.N. Food and Agriculture Organization and World Health Organization, Safety aspects of genetically modified foods of plant origin, in *Report of a Joint FAO/WHO Expert Consultation on Foods Derived from Biotechnology*. WHO, Geneva, 2000.

Verdegem, M., et al., Comparison of effluents from pond and recirculating production systems receiving formulated diets. *World Aquaculture*, 1999;30(4):28–33.

Wagener, S.M., M.W. Oswood, and J.P. Schimel, Rivers and soils: Parallels in carbon and nutrient processing. *Bioscience*, 1998;48(2):104–108.

Wardle, D.A., K.E. Giller, and G.M. Barker, The regulation and functional significance of soil biodiversity in agroecosystems, in *Agrobiodiversity*, D. Wood and J. Lenne (editors). CAB International, Wallingford, UK, 1999, 87–121.

Wardle, D.A., and P. Lavelle, Linkages between soil biota, plant litter quality and decomposition, in *Driven by Nature—Plant Litter Quality and Decomposition*, G. Cadisch and K.E. Giller (editors). CAB International, Wallingford, UK, 1997, 107–124.

Wardle, D.A., H.A. Verhoef, and M. Clarholm, Trophic relationships in the soil microfood-web: Predicting the responses to a changing global environment. *Global Change Biology*, 1998;4(7):713–727.

Wolters, V., et al., Effects of global changes on above- and belowground biodiversity in terrestrial ecosystems: Implications for ecosystem functioning. *Bioscience*, 2000;50(12):1089–1098.

CHAPTER 9

1. Serageldin, I., Biotechnology and food security in the 21st century. *Science*, 1999;285:387–389.

2. International Service for the Acquisition of Agri-biotech Applications, Global Status of Commercialized Biotech/GM Crops: 2005 (ISAAA Brief No. 34). ISAAA, 2005; available from www.isaaa.org/resources/publications/briefs/34/default.html [cited September 10, 2007].

3. Pew Charitable Trust, Fact Sheet: Genetically Modified Crops in the United States. Pew Charitable Trust, 2005; available from pewagbiotech.org/resources/factsheets/display.php3?FactsheetID=1 [cited September 10, 2007].

4. Dill, G., Glyphosate-resistant crops: History, status and future. *Pest Management Science*, 2005;61(3):219–224.

5. International Service for the Acquisition of Agri-biotech Applications, Global Status of Commercialized Biotech/GM Crops: 2006 (ISAAA Brief No. 35). ISAAA, 2006; available from www.isaaa.org/resources/publications/briefs/35/default.html [cited September 10, 2007].

6. Bates, S., et al., Insect resistance management in GM crops: Past, present and future. *Nature Biotechnology*, 2005;23(1):57–62.

7. Owen, M., and I. Zelaya, Herbicide-resistant crops and weed resistance to herbicides. *Pest Management Science*, 2005;61(3):301–311.

8. Potrykus, I., Nutritionally enhanced rice to combat nutrition disorders of the poor. *Nutrition Reviews*, 2003;61(6 pt 2, suppl S):S101–S104.

9. Maruta, Y., et al., Transgenic rice with reduced glutelin content by transformation with glutelin A antisense gene. *Breeding*, 2002;8(4):273–284.

10. Panos, A., et al., Dramatic post-cardiotomy outcome, due to severe anaphylactic reaction to prolamine. *European Journal of Cardio-Thoracic Surgery*, 2003;24(2):325–327.

11. Pinstrup-Andersen, P., and R. Pandya-Lorch, Food security and sustainable use of natural resources: A 2020 Vision. *Ecological Economics*, 1998;26(1):1–10.

12. Waara, S., and K. Glimelius, The potential of somatic hybridization in crop breeding. *Euphytica*, 1995;85(1–3):217–233.

13. Dantas, A., J. Miranda, and M. Alleoni, Diallel cross analysis for young plants of brachytic maize (Zea mays L.) varieties. *Brazilian Journal of Genetics*, 1997;20(3):453–458.

14. Gepts, P., A comparison between crop domestication, classical plant breeding, and genetic engineering: Review and interpretation. *Crop Science*, 2002;42:1780–1790.

15. Dyson, T., World food trends and prospects to 2025, *Proceedings of the National Academy of Sciences of the USA*, 1999;96(11)5929–5936.

16. World Resources Institute, Earth Trends. World Resources Institute, 2006; available from earthtrends.wri.org/ [cited 2006 September 26].

17. Wolfenbarger, L., and P. Phifer, The ecological risks and benefits of genetically engineered plants. *Science*, 2000;290:2088–2093.

18. Cattaneo, M., et al., Farm-scale evaluation of the impacts of transgenic cotton on biodiversity, pesticide use, and yield, *Proceedings of the National Academy of Sciences of the USA*, 2006;103(20)7571–7576.

19. Huang, J., et al., Plant biotechnology in China. *Science*, 2002;295:674–677.

20. Huang, J., et al., Insect-resistant GM rice in farmers' fields: Assessing productivity and health effects in China. *Science*, 2005;308:688–690.

21. Bennett, R., S. Morse, and Y. Ismael, The economic impact of genetically modified cotton on South African smallholders: Yield, profit and health effects. *Journal of Development Studies*, 2006;42(4):662–677.

22. Mamy, L., E. Barriuso, and B. Gabrielle, Environmental fate of herbicides trifluralin, metazachlor, metamitron and sulcotrione compared with that of glyphosate, a substitute broad spectrum herbicide for different glyphosate-resistant crops. *Pest Management Science*, 2005;61(9):905–916.

23. Smith, E., and F. Oehme, The biological activity of glyphosate to plants and animals—a literature review. *Veterinary and Human Toxicology*, 1992;34(6):531–543.

24. Cox, C., Glyphosate (Roundup). *Journal of Pest Reform*, 1998;18:3–17.

25. Relyea, R., The lethal impact of Roundup on aquatic and terrestrial amphibians. *Ecological Applications*, 2005;15(4):1118–1124.

26. Howe, C., et al., Toxicity of glyphosate-based pesticides to four North American frog species, *Environmental Toxicology and Chemistry*, 2004;23(8):1928–1938.

27. Wang, S., D.R. Just, and P. Pinstrup-Andersen, Tarnishing silver bullets: Bt technology adoption, bounded rationality and the outbreak of secondary pest infestations in China, in *American Agricultural Economics Association Annual Meeting*. Long Beach, CA, 2006.

28. Coghlan, A., China's GM cotton battles a new bug. *New Scientist*, 2006; available from www.newscientist.com/article/dn9614.html [cited September 10, 2007].

29. Tabashnik, B., et al., Insect resistance to transgenic Bt crops: Lessons from the laboratory and field. *Journal of Economic Entomology*, 2003;96:1031–1038.

30. Snow, A., et al., Genetically engineered organisms and the environment: Current status and recommendations. *Ecological Applications*, 2005;15(2):377–404.

31. Song, J., et al., Gene RB cloned from Solanum bulbocastanum confers broad spectrum resistance to potato late blight. *Proceedings of the National Academy of Sciences of the USA*, 2003;100(16)9128–9133.

32. Garg, A., et al., Trehalose accumulation in rice plants confers high tolerance levels to different abiotic stresses. *Proceedings of the National Academy of Sciences of the USA*, 2002;99(25)15898–15903.

33. Zhang, H.-X., and E. Blumwald, Transgenic salt-tolerant tomato plants accumulate salt in foliage but not in fruit. *Nature Biotechnology*, 2001;19:765–768.

34. Fernandez-Cornejo, J., W.D. McBride, Genetically engineered crops for pest management in U.S. agriculture. Agricultural Economic Report No. 786, U.S. Department of Agriculture, 2000.

35. Grancova, K., et al., Transgenic plants—a potential tool for decontamination of environmental pollutants. *Chemicke Listy*, 2001;95(10):630–637.

36. Giddings, G., et al., Transgenic plants as factories for biopharmaceuticals. *Nature Biotechnology*, 2000;18(11):1151–1155.

37. Hudert, C., et al., Transgenic mice rich in endogenous omega-3 fatty acids are protected from colitis. *Proceedings of the National Academy of Sciences of the USA*, 2006;103(30):11276–11281.

38. Ponnampalam, E., N.J. Mann, and A.J. Sinclair, Effect of feedings systems on omega-3 fatty acids, conjugated linoleic acid and trans fatty acids in Australian beef cuts: Potential impact on human health. *Asia Pacific Journal of Clinical Nutrition*, 2006;15(1):21–29.

39. Robert, S., et al., Metabolic engineering of Arabidopsis to produce nutritionally important DHA in seed oil. *Functional Plant Biology*, 2005;32(6):473–479.

40. Niemann, H., and W.A Kues, Application of transgenesis in livestock for agriculture and biomedicine. *Animal Reproduction Science*, 2003;79(3–4):291–317.

41. Ellstrand, N., H.C. Prentice, and J.F. Hancock, Gene flow and introgression from domestic plants into their wild relatives. *Annual Review of Ecology and Systematics*, 1999;30:539–563.

42. Pilson, D., and H. Prendeville, Ecological effects of transgenic crops and the escape of transgenes into wild populations. *Annual Review of Ecology, Evolution, and Systematics*, 2004;35:149–174.

43. Snow, A., et al., A Bt transgene reduces herbivory and enhances fecundity in wild sunflower. *Ecological Applications*, 2003;13:279–286.

44. Crawley, M., et al., Transgenic crops in natural habitats. *Nature*, 2001;409:682–683.

45. Marvier, M., Ecology of transgenic crops. *American Scientist*, 2001;89:160–167.

46. Hall, L., et al., Pollen flow between herbicide-resistant Brassica napus is the cause of multiple-resistant B. napus volunteers. *Weed Science*, 2000;48:688 694.

47. Comission for Environmental Cooperation, *Maize and Biodiversity: The Effects of Transgenic Maize in Mexico*. Commission for Environmental Cooperation, 2004; available from www.cec.org/files/PDF//Maize-and-Biodiversity_en.pdf [cited September 10, 2007].

48. Benz, B., Archaeological evidence of teosinte domestication from Guilá Naquitz, Oaxaca. *Proceedings of the National Academy of Sciences of the USA*, 2001;98(4):2104–2106.

49. Dutton, A., et al., Prey-mediated effects of Bacillus thuringiensis spray on the predator Chrysoperla carnea in maize. *Biological Control*, 2003;26:209–215.

50. Stotzky, G., Persistence and biological activity in soil of the insecticidal proteins from Bacillus thuringiensis, especially from transgenic plants. *Plant and Soil*, 2004;266(1 2):77 89.

51. Tapp, H., and S. G, Persistence of the insecticidal toxin from Bacillus thuringiensis subsp. Kurstaki in soil. *Soil Biology and Biochemistry*, 1998;30(4):471–476.

52. Saxena, D., and G. Stotzky, Bt corn has a higher lignin content than non-Bt corn. *American Journal of Botany*, 2001;88(9):1704–1706.

53. Firbank, L., Introduction. The farm scale evaluations of genetically modified herbicide-tolerant crops. *Philosophical Transactions of the Royal Society of London Series B—Biological Sciences*, 2003;358(1439):1777–1779.

54. Andow, D., UK farm-scale evaluations of transgenic herbicide-tolerant crops. *Nature Biotechnology*, 2003;21(12):1453–1454.

55. Tabashnik, B., Evolution of resistance to Bacillus thuringiensis. *Annual Review of Entomology*, 1994;39:47–79.

56. Powles, S., and C. Preston, Evolved glyphosate resistance in plants: Biochemical and genetic basis of resistance. *Weed Technology*, 2006;20(2):282–289.

57. Owen, M., and I. Zelaya, Herbicide-resistant crops and weed resistance to herbicides. *Pest Management Science*, 2005;61(3):301–311.

58. Landrigan, P., and A. Garg, Chronic effects of toxic environmental exposures on children's health. *Journal of Toxicology—Clinical Toxicology*, 2002;40(4):449–456.

59. DellaPenna, D., Nutritional genomics: Manipulating plant micronutrients to improve human health. *Science*, 1999;285:375–379.

60. Marvier, M., and R. Van Acker, Can crop transgenes be kept on a leash? *Frontiers in Ecology and the Environment*, 2005;3(2):93–100.

61. Gay, P., and S. Gillespie, Antibiotic resistance markers in genetically modified plants: A risk to human health? *Lancet Infectious Diseases*, 2005;5(10):637–646.

62. Royal Society of London, *Genetically Modified Plants for Food Use*. Royal Society, London, 1998.

63. Medical Research Council of the United Kingdom, Research into the Potential Health Effects of Genetically Modified (GM) Foods. 2001; available from www.biotech-info.net/GM_research_med.html [cited September 10, 2007].

64. Bernstein, J., et al., Clinical and laboratory investigation of allergy to genetically modified foods. *Environmental Health Perspectives*, 2003;111(8):1114–1121.

65. Matsuda, T., T. Matsubara, and H.N.O. Shingo, Immunogenic and allergenic potentials of natural and recombinant innocuous proteins. *Journal of Bioscience and Bioengineering*, 2006;101(3):203–211.

66. Richard, S., et al., Differential effects of glyphosate and roundup on human placental cells and aromatase. *Environmental Health Perspectives*, 2005;113(6):716–720.

67. Kiely, T., D. Donaldson, and A. Grube, *Pesticides Industry Sales and Usage: 2000 and 2001 Market Estimates*. U.S. Environmental Protection Agency, Washington, DC, 2004.

68. U.S. Geological Survey, *Glyphosate Herbicide Found in Many Midwestern Streams, Antibiotics Not Common*. Toxic Substances Hydrology Program, USGS, 2006; available from toxics.usgs.gov/highlights/glyphosate02.html [cited September 10, 2007].

69. Lotter, D., Organic agriculture. *Journal of Sustainable Agriculture*, 2003;21(4):59–128.

70. International Federation of Organic Agriculture Movements, Swiss Research Institute of Organic Agriculture, and Foundation Ecology and Farming, *The World of Organic Agriculture 2006—Statistics and Emerging Trends*, 8th rev. ed. IFOAM, Swiss Research Institute of Organic Agriculture, and Foundation Ecology and Farming, Frick, Switzerland, 2006.

71. Zörb, C., et al., metabolite profiling of wheat grains (Triticum aestivum l.) from organic and conventional agriculture. *Journal of Agricultural and Food Chemistry*, 2006;54(21):8301–8306.

72. Worthington, V., Nutritional quality of organic versus conventional fruits, vegetables, and grains. *Journal of Alternative and Complementary Medicine*, 2001;7(2):161–173.

73. Brandt, K., and J. Mølgaard, Organic agriculture: Does it enhance or reduce the nutritional value of plant foods? *Journal of the Science of Food and Agriculture*, 2001;81:924–931.

74. Asami, D., et al., Comparison of the total phenolic and ascorbic acid content of freeze-dried and air-dried marionberry, strawberry, and corn grown using conventional, organic, and sustainable agricultural practices. *Journal of Agriculture and Food Chemistry*, 2003;51:1237–1241.

75. Peluso, M., Flavonoids attenuate cardiovascular disease, inhibit phosphodiesterase, and modulate lipid homeostasis in adipose tissue and liver. *Experimental Biology and Medicine*, 2006;231(8):1287–1299.

76. Delmas, D., et al., Resveratrol as a chemopreventive agent: A promising molecule for fighting cancer. *Current Drug Targets*, 2006;7(4):423–442.

77. Baker, D., and P. Landrigan, Occupational exposures and human health, in *Critical Condition: Human Health and the Environment*, E. Chivian, M. McCally, H. Hu, and A. Haines (editors). MIT Press, Cambridge, MA, 1993.

78. Curl, C., R.A. Fenske, and K. Elgethun, Organophosphorus pesticide exposure of urban and suburban preschool children with organic and conventional diets. *Environmental Health Perspectives*, 2003;111(3):377–382.

79. National Research Council, *Pesticides in the Diets of Infants and Children*. National Academy Press, Washington, DC, 1993.

80. Slotkin, T., et al., Organophosphate insecticides target the serotonergic system in developing rat brain regions: Disparate effects of diazinon and parathion at doses spanning the threshold for cholinesterase inhibition. *Environmental Health Perspectives*, 2006;114(10):1542–1546.

81. Sanborn, M., et al., Identifying and managing adverse environmental health effects: 4. Pesticides. *Canadian Medical Association Journal*, 2002;166(11):1431–1436.

82. Damgaard, I., et al., Persistent pesticides in human breast milk and cryptorchidism. *Environmental Health Perspectives*, 2006;114(7):1133–1138.

83. Colborn, T., A case for revisiting the safety of pesticides: A closer look at neurodevelopment. *Environmental Health Perspectives*, 2006;1114(1):10–17.

84. Hole, D., et al., Does organic farming benefit biodiversity? *Biological Conservation*, 2005;122:113–130.

85. Fuller, R., et al., Benefits of organic farming to biodiversity vary among taxa. *Biology Letters*, 2005;1(4).

86. Mäder P, et al., Soil fertility and biodiversity in organic farming. *Science*, 2002;296:1694–1697.

87. Purin, S., O. Klauberg Filho, and S.L. Sturmer, Mycorrhizae activity and diversity in conventional and organic apple orchards in Brazil. *Soil Biology and Biochemistry*, 2006;38(7):1831–1839.

88. Gosling, P., et al., Arbuscular mycorrhizal fungi and organic farming. *Agriculture Ecosystems and Environment*, 2006;113(1–4):17–35.

89. Nelson L, et al., Organic FAQs. *Nature*, 2004;428:796–798.

90. Reganold, J., et al., Sustainability of three apple production systems. *Nature*, 2001;410(6831):926–930.

91. Green, R.E., et al., Farming and the fate of wild nature. *Science*, 2005;307(5709):550–555.

92. Smil, V., *Enriching the Earth: Fritz Haber, Carl Bosch, and the Transformation of World Food Production*. MIT Press, Cambridge, MA, 2004.

93. Rothamsted's Classical Experiments, The Rothamsted Archive. 2007; available from www.rothamsted.ac.uk/resources/ClassicalExperiments.html [cited September 10, 2007].

94. Lotter, D., R. Seidel, and W. Liebhardt, The performance of organic and conventional cropping systems in an extreme climate year. *American Journal of Alternative Agriculture*, 2003;18(3):146–154.

95. Pretty, J., and R. Hine, *Reducing Food Poverty with Sustainable Agriculture: A Summary of New Evidence*. 2006; available from www.essex.ac.uk/ces/esu/occasionalpapers/ SAFErepSUBHEADS.shtm [cited September 10, 2007].

96. Badgley, C., et al., Organic agriculture and the global food supply. *Renewable Agriculture and Food Systems*, 2007;22(2):86–108.

97. Ehler, L., Integrated pest management (IPM): Definition, historical development and implementation, and the other IPM. *Science*, 2006;62:787–789.

98. Dasgupta, S., C. Meisner, and D. Wheeler, Is environmentally friendly agriculture less profitable for farmers? Evidence on integrated pest management in Bangladesh. *Review of Agricultural Economics*, 2007;29(1):103–118.

99. Kogan, M., Integrated pest management: Historical perspectives and contemporary developments. *Annual Review of Entomology*, 1998;43:243–270.

100. Lu, J., and X. Li, Review of rice-fish-farming systems in China—one of the Globally Important Ingenious Agricultural Heritage Systems (GIASH). *Aquaculture*, 2006;260(1–4):106–113.

101. Badgley C., Perfecto I. Can organic agriculture feed the world? *Renewable Agriculture and Food Systems*, 2007;22(2):80–85.

Box 9.1

a. Reichhardt, T., Will souped up salmon sink or swim? *Nature*, 2000;406:10–12.

b. Bessey, C., et al., Reproductive performance of growth-enhanced transgenic coho salmon. *Transactions of the American Fisheries Society*, 2004;133(5):1205–1220.

Box 9.2

a. Losey, J., Rayor, L.S., and M.E. Carter, Transgenic pollen harms monarch larvae. *Nature*, 1999;399:214.

b. Snow, A., et al., Genetically engineered organisms and the environment: Current status and recommendations. *Ecological Applications*, 2005;15(2):377–404.

Box 9.4

F. Runes, et al. (editors), *Sustainable Agriculture and Resistance: Transforming Food Production in Cuba*. Food First Books, Milford, CT, 2002.

Box 9.5

a. Clancy, S., et al., *Farming Practices for a Sustainable Agriculture in North Dakota*. North Dakota State University, Carrington Research Extension Center, Carrington, ND, 1993.

Box 9.6

a. Furuno, T., *The Power of Duck: Integrated Rice and Duck Farming*. Tagari Publications, Tasmania, Australia, 2001.

Additional References

Andow, D., and C. Zwahlen, Assessing environmental risks of transgenic plants. *Ecology Letters*, 9:196–214.

Barton, J., and M. Dracup, Genetically modified crops and the environment. *Agronomy Journal*, 2000;92:797–803.

Williams, I., Cultivation of GM crops in the EU, farmland biodiversity and bees. *Bee World*, 2002;83(3):119–133.

CHAPTER 10

1. Millenium Ecosystem Assessment: Synthesis 2005; available from www.millenniumassessment.org/ [cited August 21, 2006].

2. Rees, W., M. Wackernagel, and P. Testernale, *Our Ecological Footprint: Reducing Human Impact on the Earth*. New Society Publishers, Galbriola Island, BC, 1995.

3. Myers, N., and J. Kent, *The New Consumers: The Influence of Affluence on the Environment*. Island Press, Washington, DC, 2004.

4. Brown, L.R., *Plan B: Rescuing a Planet under Stress and a Civilisation in Trouble*. W.W. Norton & Company, New York, 2003.

5. Halwell, B., et al., *State of the World: Special Focus: The Consumer Society*, L. Starke (editor). W.W. Norton & Company, New York, 2004.

6. Gorman, J., Plastic surgery gets a new look. *New York Times*, April 27, 2004, F1.

7. Melnick, D., et al., *Environment and Human Well-being: A Practical Strategy*. U.N. Development Programme, New York, 2005.

8. Diamond, J., *Collapse: How Societies Choose to Fail or Succeed*. Viking Penguin, New York, 2005.

9. Brower, M., and W. Leon, *The Consumer's Guide to Effective Environmental Choices: Practical Advice from the Union of Concerned Scientists.* W.W. Norton & Company, New York, 1999.

10. World Bank, *Environmental Fiscal Reform: What Should Be Done and How to Achieve It.* International Bank for Reconstruction and Development/World Bank, Washington, DC, 2005.

11. Kinga, S., et al. (editors), *Gross National Happiness: A Set of Discussion Papers.* Center for Bhutan Studies, Thimphu, Bhutan, 1999.

12. Goldsmith, E., *The Way: An Ecological World View.* Green Books, Foxhole, UK, 1992.

13. Wolfe, N.D., et al., Bushmeat hunting, deforestation, and prediction of zoonotic disease. *Emerging Infectious Diseases,* 2005;11(12):1822–1827.

14. Alverson, D.L., et al., A global assessment of fisheries bycatch and discards. Food and Agriculture Organization of the United Nations, Rome, 1996.

15. Danielsen, F., et al., The Asian tsunami: A protective role for coastal vegetation. *Science,* 2005;310(5748):643–643.

16. Ankarberg, E., et al., Study of dioxin and dioxin-like PCB levels in fatty fish from Sweden 2000–2002. *Organohalogen Compounds,* 2004;66:2035–2039.

17. Hole, D.G., et al., Does organic farming benefit biodiversity? *Biological Conservation,* 2005;122(1):113–130.

18. Roseland, M., *Toward Sustainable Communities: Resources for Citizens and Their Governments.* New Society Publishers, Sony Creek, CT, 1998.

19. Thomas, C.D., et al., Extinction risk from climate change. *Nature,* 2004;427(6970):145.

20. McMichael, A., et al., *Climate Change and Human Health: Risks and Responses.* World Health Organization, Geneva, 2003.

21. Energy Information Administration, Residential Energy Consumption Survey. 2001; available from www.eia.doe.gov/emeu/recs/ [cited August 23, 2006].

22. Bernasek, A., Real energy savers don't wear cardigans. Or do they? *New York Times,* November 13, 2005, 5.

23. American Water Works Association, Stats on Tap. 2006; available from www.awwa.org/Advocacy/pressroom/STATS.cfm [cited August 24, 2006].

24. Center for a New American Dream, Just the Facts: Junk Mail Facts and Figures. 2003; available from www.newdream.org/junkmail/facts.php [cited August 24, 2006].

25. Speth, J.G., *Red Sky at Morning: America and the Crisis of the Global Environment.* Yale University Press, New Haven, CT, 2004.

26. Rosenberg, D.K., B.R. Noon, and E.C. Meslow, Biological corridors: Form, function, and efficacy. *Bioscience,* 1997;47(10):677–687.

27. McNeely, J., and S. Scherr, *Ecoagriculture: Strategies for Feeding the World and Conserving Wild Biodiversity.* Island Press, Washington, DC, 2003.

Additional References

Buchmann, S., and B. Nabhan, *The Forgotten Pollinators.* Island Press, Washington, DC, 1997.

Energy Information Administration, Annual Energy Review: Energy Overview, 1949–2005. 2006; available from www.eia.doe.gov/emeu/aer/overview.html [cited August 23, 2006].

International Energy Agency, *Things That Go Blip in the Night: Standby Power and How to Limit It.* IEA, Paris, 2001.

International Union for the Conservation of Nature and Natural Resources, *Guidelines for Protected Area Management Categories*. IUCN, Gland, Switzerland, 1994.

International Union for the Conservation of Nature and Natural Resources, *Vision for Water and Nature: A World Strategy for Conservation and Sustainable Management of Water Resources in the 21st Century*. IUCN, Gland, Switzerland, 2000.

Murphy, D., Challenges to biological diversity in urban areas, in *Biodiversity*, E. Wilson (editor). National Academy of Sciences, Washington, DC, 1988, 71–76.

Prescott-Allen, R., *The Well-being of Nations*. Island Press, Washington, DC, 2001.

U.N. Environment Programme, *Global Environmental Outlook—3*. Earthscan, London, 2002.

CHAPTER AUTHORS

AARON BERNSTEIN, MD, has been affiliated with the Center for Health and the Global Environment since 2001 and is currently a resident in the Boston Combined Residency in Pediatrics at Harvard Medical School and the Boston University School of Medicine. He received his undergraduate degree from Stanford University and medical degree from the University of Chicago Pritzker School of Medicine.

ERIC CHIVIAN, MD, is the founder and Director of the Center for Health and the Global Environment at Harvard Medical School. In 1980, he co-founded, with three other Harvard faculty members, International Physicians for the Prevention of Nuclear War, which won the 1985 Nobel Peace Prize. He is the senior editor and author of *Last Aid: The Medical Dimensions of Nuclear War* (W.H. Freeman, 1982) and of *Critical Condition: Human Health and the Environment* (MIT Press, 1993), which appeared in German, Spanish, Chinese, Japanese, and Persian editions. For the past 14 years, Dr. Chivian has run an almost fully organic orchard of heirloom apples, peaches, pears, Asian pears, apricots, plums, cherries, and grapes.

MARIA ALICE S. ALVES, PhD, is a professor of ecology at the University of Rio de Janeiro State (UERJ). She is also a researcher of the National Research Council–Conselho Nacional de Desenvolvimento Científico e Tecnológico (CNPq). Her research interests include behavioral ecology, bird species and bird–plant interactions, distribution patterns of endemic and threatened vertebrate species, and their conservation.

DANIEL HILLEL, PhD, is Professor Emeritus of Crop, Soil, and Environmental Sciences at the University of Massachusetts and is currently a senior research scientist at the Center for Climate Systems Research at Columbia University. He has served as advisor to the Environmental Department of the World Bank and to the Land and Water Department of the U.N. Food and Agricultural Organization. He is a Fellow of the American Association for the Advancement of Science, the Soil Science Society of America, the American Society of Agronomy, and the American Geophysical Union.

JOHN KILAMA, PhD, was born in Uganda and trained in the United States, with BS degrees in both pharmacy and chemistry. Dr. Kilama received his PhD in medicinal chemistry from the University of Arizona in Tucson. He was with the DuPont Company for many years as a senior medicinal research chemist, where he worked on new classes of chemicals for crop protection and helped establish collaborations with developing countries. He has recently been appointed as Consultant Managing Director of African Biotherapeutics Institute, a consortium of universities from the United States and Africa. He is the founder and President of the Global Bioscience Development Institute.

JEFFREY A. MCNEELY is chief scientist at the International Union for the Conservation of Nature and Natural Resources (IUCN), where he has worked since 1980 in addressing a wide range of conservation issues. Prior to joining IUCN, he worked in Asia for twelve years, conducting research on the mammals of Thailand, studying the relationship between people and Nature in the eastern Himalayas, designing a system of protected areas in the Lower Mekong Basin and establishing the World

Wildlife Fund–IUCN Programme in Indonesia. He is the author or editor of some forty books and serves on the editorial board of seven international journals.

JERRY MELILLO, PhD, is Co-director of the Ecosystems Center of the Marine Biological Laboratory in Woods Hole, Massachusetts. He just completed terms as President of the Ecological Society of America and President of the Scientific Committee on Problems of the Environment, which is made up of the national science academies from 38 countries and 22 international scientific unions. He conducts field studies and simulation modeling analyses of human impacts on the environment. Much of his current work focuses on the ecological impacts of climate change and human disruption of the nitrogen cycle.

DAVID H. MOLYNEUX, PhD, is a professor of tropical medicine, Director of the Lymphatic Filariasis Support Centre, and former Dean of the Liverpool School of Tropical Medicine. His major research interests are leishmaniasis, lymphatic filariasis, malaria control, onchocerciasis, parasitic and vector-borne disease control, and trypanosomiasis.

KALEMANI JO MULONGOY, PhD, is the head of the Division of Scientific, Technical, and Technological Matters of the Secretariat of the U.N. Convention on Biological Diversity (CBD). Before this position at the CBD, he was an associate professor at the National University of Kinshasa in the Democratic Republic of Congo, head of the Department of Microbiology at the International Institute of Tropical Agriculture, head of the Plant Biotechnology Department at the International Institute for Research for Development in Africa, and Director of the Biodiversity and Biotechnology Programme at the International Academy of the Environment. He is author and co-author of many scientific publications and has edited a number of books.

DAVID J. NEWMAN, DPhil, received his degree in microbial chemistry from the University of Sussex in the United Kingdom in 1968. He worked for small and large pharmaceutical companies on antibiotic and antitumor drug discovery before moving in 1991 to the Natural Products Branch, where he is currently chief, at the National Cancer Institute of the U.S. National Institutes of Health. He is also an adjunct professor at the Center of Marine Biotechnology of the University of Maryland.

RICHARD S. OSTFELD, PhD, is senior scientist and animal ecologist at the Institute of Ecosystem Studies in Millbrook, New York, and is adjunct professor at Rutgers University and the University of Connecticut. His research focuses on the ways in which complex interactions in ecological communities influence risk of exposure to zoonotic diseases.

STUART L. PIMM, PhD, is the Doris Duke Chair of Conservation Ecology at the Nicholas School of the Environment at Duke University. His research interests involve understanding the patterns of species extinction in order to be able to prevent them.

JOSHUA P. ROSENTHAL, PhD, Deputy Director of the Division of International Training and Research at the Fogarty International Center of the National Institutes of Health, manages two interagency research and capacity-building programs. The first is the International Cooperative Biodiversity Groups, which are cooperative agreements for research and development in drug discovery and biological inventory in fifteen countries; the second is the Ecology of Infectious Diseases program, which supports research to develop integrated methods, in relation to environmental change, for the prediction of infectious disease dynamics. Dr. Rosenthal has authored a wide variety of publications and serves on a number of advisory panels, on the subjects of biodiversity conservation, bioinformatics, genetic resources, and biomedicine.

CYNTHIA ROSENZWEIG, PhD, is head of the Climate Impacts Group at NASA Goddard Institute for Space Studies and a senior research scientist at the Earth Institute of Columbia University. She is a Fellow of the American Society of Agronomy.

OSVALDO SALA, PhD, is a professor of ecology in the Department of Ecology and Evolutionary Biology at Brown University, where he also directs both the Environmental Change Initiative and the Center for Environmental Studies. He is currently President of the Scientific Committee on Problems of the Environment and a member of the American Academy of Arts and Sciences. His interests in ecology include the arid ecosystems of Patagonia and global change issues.

ELEANOR STERLING, PhD, directs the Center for Biodiversity Conservation at the American Museum of Natural History in New York City, where she oversees biodiversity research projects in the Americas, Africa, Asia, and the Pacific. She is also an adjunct professor at Columbia University and is the Director of Graduate Studies for the Department of Ecology, Evolution, and Environmental Biology.

Contributing Authors

Chapter 1

Callum M. Roberts, PhD, professor of biology at the University of York, York, England, wrote the "Marine Species" section.

Chapter 2

Stuart L. Pimm, PhD, Doris Duke Chair of Conservation Ecology at the Nicholas School of the Environment at Duke University, contributed to the sections on habitat loss and overexploitation on land, and that on invasive species.

Maria Alice S. Alves, PhD, professor of ecology at the University of Rio de Janeiro State and a researcher of the National Research Council—Conselho Nacional de Desenvolvimento Científico e Tecnológico, contributed to the section on habitat loss and overexploitation on land, and to the section on invasive species.

Callum M. Roberts, PhD, professor of biology at the University of York, York, England, wrote the section on habitat loss in the oceans.

Jerry M. Melillo, PhD, Co-director of the Ecosystem Center at the Marine Biological Laboratories, Woods Hole, Massachusetts, wrote the overview of climate change science in the section on climate change and species loss.

Judy Oglethorpe, MSc, Director of the Ecoregion Support Unit, World Wildlife Fund, contributed to and reviewed the section on war and conflict.

Melanie L.J. Stiassny, PhD, Axelrod Research Curator in the Department of Ichthyology at the American Museum of Natural History and an adjunct professor at Columbia University, wrote the section on freshwater habitat loss.

Chapter 4

Gordon M. Cragg, DPhil, retired chief of the Natural Products Branch of the National Cancer Institute, wrote the section on the history of plant drug use and the stories of paclitaxel and the calanolides.

Elaine Elisabetsky, PhD, professor at the Laboratório de Ethnofarmacologia, Universidade Federal do Rio Grande do Sul in Brazil, wrote the section on South American medicines.

William Fenical, PhD, Director of the Scripps Institute of Marine Biotechnology and Biomedicine, University of California—San Diego, wrote the section on marine microbes.

CHAPTER 5

KENNETH PAIGEN, PhD, senior staff scientist and former director of the Jackson Laboratory, contributed to the section on mouse genetics.

GARY RUVKUN, PhD, professor of genetics at Harvard Medical School, contributed to the introduction and the section on *C. elegans*.

CHAPTER 6

MARC R.L. CATTET, DVM, PhD, professional research associate at the Department of Veterinary Pathology, University of Saskatchewan Western College of Veterinary Medicine, contributed to the section on medical research in denning bears.

JOHN W. DALY, PhD, scientist emeritus at the Laboratory of Bioorganic Chemistry, National Institute of Diabetes and Digestive and Kidney Diseases of the U.S. National Institutes of Health, contributed to the section on the use of amphibians in medical research, and contributed to and reviewed the section on medicines from amphibians.

ANDREW G. HENDRICKX, PhD, professor at the Center for Health and the Environment and former Director of the California Regional Primate Research Center, University of California Davis, contributed to the section on primates.

JOHN J. MARCHALONIS, PhD, professor of microbiology and immunology at the University of Arizona College of Medicine, contributed to the section on shark immune systems.

RALPH A. NELSON, MD, PhD, Emeritus Professor of Medicine at the University of Illinois at Urbana Champaign, contributed to the section on denning bears.

CHAPTER 7

JONATHAN H. EPSTEIN, DVM, MPH, senior research scientist at the Consortium for Conservation Medicine, wrote the section on Nipah virus.

PAUL R. EPSTEIN, MD, MPH, Associate Director of the Center for Health and the Global Environment, Harvard Medical School, contributed to the section on climate change and infectious disease.

THOMAS K. KRISTENSEN, PhD, professor and head of the Mandahl-Barth Research Center for Biodiversity and Health, Denmark, wrote the sections on schistosomiasis.

CHAPTER 8

AARON BERNSTEIN, MD, wrote box 8.8 on the use of antibiotics in livestock.

DAMAYANTI BUCHORI, PhD, professor and Chair of the Department of Plant Protection and Director of the Center for Conservation and Insect Studies at the Bogor

Agricultural University in Bogor, Indonesia, wrote box 8.2 on beneficial insects and rice production in Indonesia.

Eric Chivian, MD, wrote box 8.4 on how a fallen leaf decomposes, box 8.5 on mycorrhizae, and box 8.6 on nitrogen-fixing bacteria, and contributed to the section on aquaculture.

Andrew R. Price, PhD, reader in the Ecology and Epidemiology Group, Department of Biological Sciences, University of Warwick, Coventry, United Kingdom, wrote the section on food from aquatic ecosystems.

David M. Sherman, DVM, MS, Country Program Director for the Dutch Committee for Afghanistan, former Director of the Division of Animal Health, Biosecurity, and Dairy Services of the Commonwealth of Massachusetts, and former head and associate professor in the Section on International Veterinary Medicine at Tufts University School of Veterinary Medicine, wrote the section on livestock production.

Amos Tandler, PhD, senior scientist and head of the National Center for Mariculture, Eilat, Israel, wrote the section on aquaculture.

Diana H. Wall, PhD, professor and Director of the Natural Resource Ecology Laboratory at Colorado State University, wrote the section on soil biodiversity.

Chapter 9

Daniel Hillel, PhD, Professor Emeritus of Crop, Soil, and Environmental Sciences at the University of Massachusetts and senior research scientist at the Center for Climate Systems Research at Columbia University, contributed to the section on genetically modified crops.

Frederick L. Kirschenmann, PhD, Distinguished Fellow and former Director of the Aldo Leopold Institute, Iowa State University, wrote the section on large-scale organic farming in the United States and reviewed the section on organic farming.

Richard Levins, PhD, John Rock Professor of Population Sciences, Harvard School of Public Health, contributed to the section on Cuban agriculture.

John P. Reganold, PhD, Regents Professor of Soil Science, Washington State University, contributed to and reviewed the section on organic farming.

REVIEWERS

CHAPTER 1

NORMAN R. PACE, PHD, professor of Molecular, Cellular, and Developmental Biology, University of Colorado at Boulder, reviewed the section on the three-domain map and that on the microbial world.

CHAPTER 2

ROBERT J. DIAZ, PHD, professor of marine science, University of William and Mary at the Virginia Institute of Marine Science, reviewed the section on dead zones.

GIOVANNI DI GUARDO, DVM, Diplomate of the European College of Veterinary Pathologists and professor of general pathology and veterinary pathophysiology at the University of Teramo, Italy, reviewed the section on marine mammal die-offs.

STEVEN H. FERGUSON, PHD, research scientist at Fisheries and Oceans Canada, reviewed the section on Ringed Seals.

THOMAS E. LOVEJOY, PHD, president of the H. John Heinz III Center for Science, Economics, and the Environment, former chief biodiversity advisor at the World Bank, former science advisor to the Secretary of the Interior, and former assistant secretary and counselor to the Secretary at the Smithsonian Institution, reviewed the section on global climate change and species loss.

CALLUM M. ROBERTS, PHD, professor of biology at the University of York, York, England, reviewed the section on overexploitation in the oceans.

DAVID M. SHERMAN, DVM, MS, Country Program Director of the Dutch Committee for Afghanistan and former associate professor of veterinary medicine at Tufts University, reviewed the section on pharmaceuticals.

BRIAN R. SILLIMAN, PHD, assistant professor in the Department of Zoology, University of Florida, reviewed the paragraph on drought and wetlands in the section on climate change and species loss.

CHAPTER 3

VIRGINIA R. BURKETT, PHD, Coordinator of Global Change Research, U.S. Geological Survey, reviewed the sections on freshwater wetlands.

DONALD A. KLEIN, PHD, professor of microbiology, immunology and pathology at Colorado State University, reviewed box 3.1 on microbial ecosystems.

PAUL B. ECKBURG, MD, postdoctoral research fellow in the Department of Microbiology and Immunology at Stanford University, reviewed box 3.1 on microbial ecosystems.

BRUCE J. PASTER, PHD, professor of oral and developmental biology at Harvard School of Dental Medicine, reviewed box 3.1 on microbial ecosystems.

DAVID A. RELMAN, MD, associate professor of microbiology and immunology at Stanford University, reviewed box 3.1 on microbial ecosytems.

MICHAEL A. ZASLOFF, MD, PHD, professor in the Departments of Surgery and Pediatrics, Georgetown University School of Medicine, reviewed box 3.1 on microbial ecosystems.

CHAPTER 4

DAVID O. CARPENTER, MD, professor of environmental health and toxicology at State University of New York at Albany, reviewed the section on omega-3 fatty acids.

ALEXANDER LEAF, MD, Jackson Professor of Clinical Medicine, Emeritus, Harvard Medical School and Massachusetts General Hospital, reviewed the section on omega-3 fatty acids.

CHAPTER 5

ALEJANDRO SANCHEZ ALVARADO, PHD, professor of neurobiology and anatomy at the University of Utah School of Medicine, reviewed the section on planarium regeneration.

ADAM AMSTERDAM, PHD, research scientist at the Center for Cancer Research, Massachusetts Institute of Technology, reviewed the section on Zebrafish genetics.

TOSHITAKA FUJISAWA, PHD, associate professor in the Department of Developmental Genetics, National Institute of Genetics, Japan, reviewed the section on regeneration in hydras.

PHILIP C. HANAWALT, PHD, Howard H. and Jessie T. Watkins University Professor, Department of Biological Sciences, Stanford University, reviewed the section on E. coli.

NANCY HOPKINS, PHD, Amgen, Inc. Professor of Biology at the Massachusetts Institute of Technology, reviewed the section on Zebrafish genetics.

CARL A. HUFFMAN, PHD, Robert Stockwell Professor of Greek Language and Literature, Depauw University, reviewed the paragraphs on the history of medical research in ancient Greece.

DOUGLAS A. MELTON, PHD, Thomas Dudley Cabot Professor of Natural Sciences, Harvard University and investigator at Howard Hughes Medical Institute, reviewed the section on stem cell research.

FERNANDO NOTTEBOHM, PhD, professor and Director of the Rockefeller University Field Research Center, reviewed the section on neurogenesis.

KENNETH D. POSS, PhD, assistant professor of cell biology at Duke University Medical Center, reviewed the section on Zebrafish regeneration.

GARY RUVKUN, PhD, professor of genetics at Harvard Medical School, reviewed the section on genetics.

MINORU SAITOE, PhD, investigator at the Tokyo Metropolitan Institute of Neuroscience, reviewed the sections on the Fruit Fly.

ANJA O. SAURA, PhD, docent with the Department of Biological and Environmental Sciences/Genetics, University of Helsinki, Finland, reviewed the section on Fruit Fly genetics.

ANN E. SHINNAR, PhD, associate professor at Lander College for Men of Touro College, reviewed the section on hagfish.

MICHAEL A. ZASLOFF, MD, PhD, professor in the Departments of Surgery and Pediatrics, Georgetown University School of Medicine, reviewed the sections on innate immunity and on the agnathans.

CHAPTER 6

ROBERT B. BARLOW, PhD, Director of the Center for Vision Research and professor of ophthalmology at the State University of New York Upstate Medical University, reviewed the section on horseshoe crabs.

JULIA K. BAUM, PhD candidate at Dalhousie University, Halifax, Canada, reviewed the section on sharks.

ANDREW E. DEROCHER, PhD, Chair of the International Union for the Conservation of Nature and Natural Resources Species Survival Commission Polar Bear Specialist Group and professor of biological sciences at the University of Alberta, reviewed the section on Polar Bear endangerment.

FRANKLIN H. EPSTEIN, MD, William Applebaum Professor of Medicine at Harvard Medical School, reviewed the section on the Dogfish Shark.

BEATRICE H. HAHN, PhD, professor in the Departments of Medicine and Microbiology at the University of Alabama at Birmingham, reviewed the section on HIV research with primates.

JAMES HANKEN, PhD, Alexander Agassiz Professor of Zoology and Curator of Herpetology in the Museum of Comparative Zoology, Harvard University, reviewed the section on amphibian loss.

MICHAEL J. LANNOO, PhD, professor of anatomy at Indiana University School of Medicine and U.S. National Coordinator of the Declining Amphibian

Populations Task Force of the International Union for the Conservation of Nature and Natural Resources Species Survival Commission, reviewed the section on amphibian loss.

RICHARD LEVINS, PhD, John Rock Professor of Population Sciences, Department of Population and International Health, Harvard School of Public Health, reviewed the section on amphibian loss.

RICHARD LEWIS, PhD, associate professor at the Institute for Molecular Bioscience, University of Queensland, and head of pharmacology at Xenome Ltd., reviewed the section on cone snail medicines.

BALDOMERO OLIVERA, PhD, Distinguished Professor of Biology, University of Utah, reviewed the sections on threats to cone snails and cone snail medicines.

PASQUALE J. PALUMBO, MD, Emeritus Professor of Medicine, Mayo Clinic College of Medicine, reviewed the section on denning bears.

DALE PETERSEN, PhD, author of *Eating Apes*, reviewed the section on endangered primates.

CARL SAFINA, PhD, co-founder and President of Blue Ocean Institute, reviewed the section on endangered sharks.

WILLIAM SARGENT, author of *Crab Wars: A Tale of Horseshoe Crabs, Bioterrorism, and Human Health*, consultant for the *NOVA* science series, and former Director of the Baltimore Aquarium and research assistant at the Woods Hole Oceanographic Institution, reviewed the section on horseshoe crabs.

SCOTT L. SCHLIEBE, PhD, project leader of Polar Bear Project, U.S. Fish and Wildlife Service, Alaska, reviewed the section on Polar Bear endangerment.

CHRIS SHAW, PhD, professor of drug discovery at the School of Pharmacy, Queen's University, Belfast, North Ireland, reviewed the section on medicines from amphibians.

LOUIS R. SIBAL, PhD, special consultant to the U.S. National Institutes of Health and former Director of the Office of Laboratory Animal Research at the National Institutes of Health, reviewed the section on research in nonhuman primates.

BURT E. VAUGHAN, PhD, professor of biological sciences at Washington State University–Tricities, reviewed the section on turrid and terebra snails.

DAVID B. WAKE, PhD, professor of integrative biology and Curator of Herpetology, Museum of Vertebrate Zoology, University of California–Berkeley, reviewed the section on amphibians.

MARK S. WALLACE, MD, associate professor, and Director of the Center for Pain and Palliative Medicine, University of California–San Diego, reviewed the section on cone snail medicines.

RICHARD W. WRANGHAM, PhD, professor of biological anthropology at Harvard University, reviewed the section on endangered primates.

MICHAEL A. ZASLOFF, MD, PhD, professor in the Departments of Surgery and Pediatrics at Georgetown University School of Medicine, reviewed sections both on medicines from amphibians and medicines from sharks.

CHAPTER 7

BEATRICE H. HAHN, PhD, professor in the Departments of Medicine and Microbiology at the University of Alabama at Birmingham, reviewed the section on bushmeat and HIV.

WALID HENEINE, PhD, chief of the HIV Drug Resistance and Retroviral Zoonoses Laboratory at the Centers for Disease Control and Prevention, reviewed the section on HTLVs.

WILLIAM B. KARESH, DVM, Director of the Field Veterinary Program, Wildlife Conservation Society, and Co-chair of the International Union for the Conservation of Nature and Natural Resources Species Survival Commission Wildlife Health Specialist Group, reviewed the section on avian flu.

JUNKO YASUOKA, DSc, MPH, scientist in charge of Malaria Prevention and Control at the World Health Organization Office in Cambodia, reviewed sections on mosquito ecology and malaria.

CHAPTER 8

ROSAMUND L. NAYLOR, PhD, the Julie Wrigley Senior Fellow at the Center for Environmental Science and Policy, Director of the Program on Food Security and the Environment, and associate professor of economics at Stanford University, reviewed the section on aquaculture.

DONALD H. PFISTER, PhD, Asa Gray Professor of Systematic Botany and Curator of the Harvard University Herbaria, reviewed the section on mycorrhizae.

ANNE PRINGLE, PhD, assistant professor in the Department of Organismic and Evolutionary Biology at Harvard University, reviewed the section on mycorrhizae.

JANICE E, THIES, PhD, associate professor of soil biology at Cornell University, reviewed the section on nitrogen-fixing bacteria.

ELSE C. VELLINGA, PhD, lecturer in the Department of Plant and Microbial Biology at the University of California–Berkeley, reviewed the section on mycorrhizae.

CHAPTER 9

L. LaREESA WOLFENBARGER, PhD, associate professor at the University of Omaha, Omaha, Nebraska, reviewed the section on genetically modified crops.

MICHELLE MARVIER, PHD, associate professor and Executive Director of the Environmental Studies Institute, Santa Clara University, reviewed the section on genetically modified crops.

HANS R. HERREN, PHD, Director of the Millennium Institute and former Director of the International Centre for Insect Physiology and Ecology (ICIPE), reviewed the section on genetically modified crops.

Index

Note: page numbers followed by *f*, *t*, and *b* indicate figures, tables, and boxes.